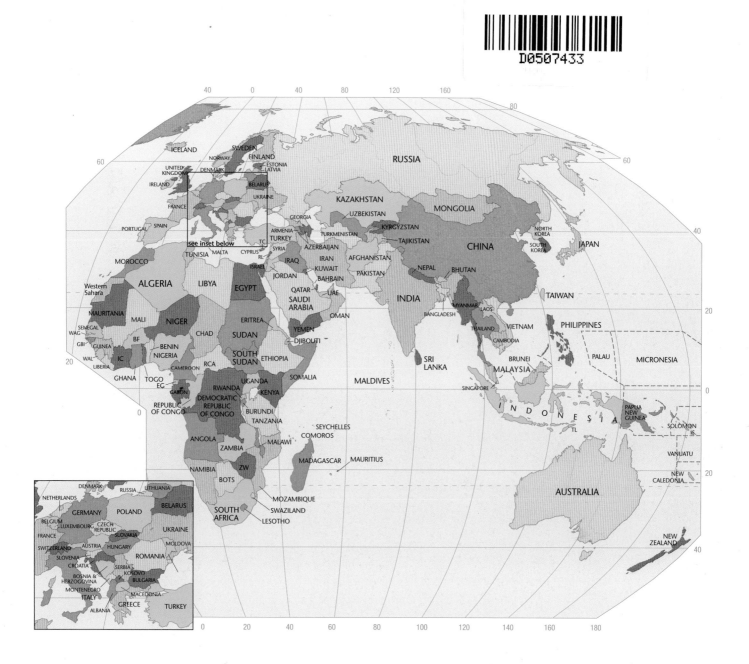

CONTEMPORARY
HUMAN GEOGRAPHY

CONTEMPORARY HUMAN GEOGRAPHY

Culture, Globalization, Landscape

Mona Domosh
Dartmouth College

Roderick P. Neumann
Florida International University

Patricia L. Price
Florida International University

W. H. Freeman and Company

A Macmillan Education Company

Vice President, Editoral: Charles Linsmeier
Developmental Editor: Janice Wiggins
Marketing Manager: Taryn Burns
Marketing Assistant: Bailey James
Media Editor: Lindsay Neff
Editorial Assistant: Carlos Marin
Project Editor: Liz Geller
Text and Cover Designer: Blake Logan
Illustration Coordinator: Janice Donnola
Illustrations: maps.com
Photo Editor: Bianca Moscatelli
Production Manager: Susan Wein
Composition and Layout: Sheridan Sellers
Printing and Binding: C.O.S. Printers Pte Ltd - Singapore

Cover photo: John P. Kelly/Getty Images

Library of Congress Control Number: 2014951268
ISBN-13: 978-1-4641-3344-2
ISBN-10: 1-4641-3344-1

W. H. Freeman and Company
41 Madison Ave.
New York, NY 10010
www.whfreeman.com/geography

CONTENTS IN BRIEF

CONTENTS

Chapter 6

**Political Geography
A Divided World 223**

Chapter 7

**The Geography of Religion
Spaces and Places of Sacredness 271**

Chapter 8

Agriculture
The Geography of the Global Food System 317

Chapter 11

One World or Many?
The Cultural Geography of the Future   457

PREFACE

This first edition of *Contemporary Human Geography* is a result of the evolution of the human geography field as a whole. As we worked on the most recent revision of *The Human Mosaic*, it became clear that over the past few editions, the book had become something new and distinct. The textbook's evolution in no small measure reflects wider changes in human geography and, more specifically, in cultural geography over the past two decades. Most significantly, the metaphor of culture that provided the textbook's title—humanity as a mosaic of discrete, homogeneous cultures—is no longer salient. Cultural geography and the cultural theory that informs it are guided by new ideas of heterogeneity, contingency, fluidity, encounter, and hybridity. Areas of inquiry that were widely pursued in cultural geography 20 or 30 years ago are no longer prominent. Human geography has moved forward and we believe this textbook should reflect the field's vibrancy and dynamism.

We are therefore launching the 1st edition of *Contemporary Human Geography*. Two goals guide our vision for the new textbook. First, we want to present the latest, cutting-edge work in human geography in language that is accessible to undergraduates. Each of the authors is actively engaged in conducting new research, publishing in geography journals, and serving on editorial boards. We believe our active scholarship provides *Contemporary Human Geography* with a distinct identity and best positions us to ensure that the textbook keeps pace with advances in the field. Moreover, we place great emphasis on writing that is clear, lively, and peppered with fresh examples to help students relate to abstract concepts. Our choice of the word "contemporary" for the book title reflects the importance we place on this goal. Second, we want the textbook to actively facilitate student learning. Collectively, we have six decades of experience teaching and mentoring students from lower-division undergraduate studies through the doctoral level. This experience informs our understanding of what is important for our students of human geography to learn and how to best ensure that they learn it.

Adopters of *The Human Mosaic* will still see much of what drew them to use the book in the past. Extending the evolution metaphor, *Contemporary Human Geography* shares much of *The Human Mosaic*'s DNA. For example, the five themes, which we had revised for the 11th edition of *The Human Mosaic* to reflect advances in the discipline, will provide the new textbook's underlying organizational structure. Above all, *Contemporary Human Geography* retains our enduring view of human geography as seen through the lens of culture. We identify as cultural geographers and have endeavored to produce a textbook suitable for both introductory human geography and cultural geography courses.

We hope that you enjoy teaching with *Contemporary Human Geography* as much as we enjoyed writing it.

Organization: The Five Themes

Contemporary Human Geography is organized around five themes. These themes are introduced and explained in the first chapter and serve as the framework for the 10 topical chapters that follow. Each theme is applied to a variety of human geographical topics, such as language, ethnicity, politics, religion, and agriculture. This thematic organization allows students to relate to the most important aspects of human geography at every point in the text. As instructors, we have found that beginning students learn best when provided with a precise and useful framework, and the five-themes approach provides such a framework for understanding human geography. The themes are identified by colored banners that serve as a visual reminder to students when they appear throughout the book. The five themes are:

REGION

MOBILITY

GLOBALIZATION

NATURE-CULTURE

CULTURAL LANDSCAPE

Our *region* theme appeals to students' curiosity about the differences among places. *Mobility* conveys the dynamic aspect of human geographical phenomena, a theme particularly relevant in a world of incessant and rapid change. Students acquire an appreciation of how people, things, and ideas move (or do not move). The topics employed to illustrate the concepts of mobility include many popular cultural references familiar in students' daily lives. *Globalization* permits students to understand the complex processes that link the various economies, cultures, and societies around the world. An understanding of globalizing processes is necessary for explaining how those linkages can create economic and cultural similarities as well as disparities. *Nature–culture* addresses the complicated relationship between culture and the physical environment. With today's complex and often controversial relationship between the natural environment and our globalizing world, both the tensions and the alliances that arise with regard to this relationship are now at the forefront of this theme. Finally, the theme of *cultural landscape* heightens students' awareness of the visible character of places and regions.

Special Features of This First Edition

As we made the transition to *Contemporary Human Geography*, we paid close attention to the instructor, student, and reviewer feedback on *The Human Mosaic* that we have received over the years. Past adopters of *The Human Mosaic* will find in *Contemporary Human Geography* content that has been significantly revised as well as entirely new material.

- Changes in geographic thought and teaching on culture are reflected in a thoroughly rewritten **Chapter 2**.

- We removed the "Geography of Economies" chapter, added a totally new chapter on development (**Chapter 9**), and consolidated the previous two urban geography chapters into one (**Chapter 10**). We thus reduced the total number of chapters from 12 to 11 while adding new material.

- A new UNESCO **World Heritage Site** feature has been included in each chapter. Given the enormous cultural significance of these places to humankind, this feature will enhance student appreciation of the importance of place and landscape. Each one illustrates a particular constellation of themes relevant to the chapter in which it is situated.

- **The Video Connection** is a new video-based learning feature added to each chapter. It is focused around brief, timely videos housed on the *New York Times* web site. Each video speaks to an important issue in the chapter through lived, human experiences. This feature provides opportunities to apply human geographic concepts and engage in critical thinking.

- Each chapter now features **Learning Objectives** to aid in student learning and assessment. Learning objectives are presented at the beginning of each chapter, and then revisited at the chapter's end, with specific questions and prompts that student can use to reinforce and demonstrate learning.

- The new feature **Geography at Work** helps answer the question "How can a degree in geography lead to a satisfying career?" It highlights the career paths of recent geography graduates who in their own words describe the work they do and how they found meaningful employment in a wide variety of occupations.

- **Every map and photo** in the textbook has been individually assessed to ensure that it is up to date with the latest data and visually appealing. Many new maps and illustrations have also been added throughout.

- **A new appendix on how to read maps** offers students guidance and strategies for reading maps and thinking critically about them.

We have retained those features of *The Human Mosaic* that now make *Contemporary Human Geography* relevant and accessible to today's student of human geography:

- Students begin their exploration of geographical concepts at the start of each chapter with a critical thinking question based on the chapter-opening photo. Students then contemplate this question throughout the chapter and explore it in greater depth at the end of the chapter with the **Seeing Geography** feature. Most Seeing Geography features have been enhanced with updated photographs; many now include new exploration and analysis by the authors.

- **Subject to Debate** boxes present key chapter concepts as questions with an examination of different sides of the surrounding debate. These boxes conclude with a set of critical thinking questions that ask students to explore the debated topic in greater depth.

- The **Author's Notebook** features explore the personal travels and experiences of the authors as they discover and analyze various concepts related to human geography.

- The **Doing Geography** boxes give students the opportunity to apply the concepts presented in a chapter.

Media and Supplements

Contemporary Human Geography is accompanied by a media and supplements package that facilitates student learning and enhances the teaching experience.

LaunchPad: Resources for Students and Instructors

Macmillan Education **LaunchPad**

www.macmillanhighered.com/launchpad/DomoshCHG1e

Our new coursespace, LaunchPad, combines an interactive e-Book with high-quality multimedia content and ready-made assessment options, including LearningCurve adaptive quizzing. Pre-built, curated units are easy to assign or adapt with your own material, such as readings, videos, quizzes, discussion groups, and more. LaunchPad also provides access to a gradebook that provides a clear window on performance for your whole class, for individual students, and for individual assignments. The following resources are available on LaunchPad.

For Students

Macmillan Education **LearningCurve**

LearningCurve is an adaptive quizzing engine that automatically adjusts questions to the student's mastery level. With LearningCurve activities, each student follows a unique path to understanding the material. The more questions a student answers correctly, the more difficult the questions become. Each question is written specifically for the text and is linked to the relevant e-Book section. LearningCurve also provides a personal study plan for students as well as complete metrics for instructors. Proven to raise student performance, LearningCurve serves as an ideal formative assessment and learning tool. For detailed information, visit http://learningcurveworks.com

Chapter Quizzes provide students with a multiple-choice review to help them assess their mastery of each chapter.

Map Activities, using Map Builder, allow students to create layered thematic maps on their own and complete related activities specific to the chapter.

Videos and Video Activities. More than 200 videos and video activities are available, relating to primarily human and cultural geography with a selection of physical geography clips as well.

Focus on Geography Articles and Quizzes. Selected articles from *Focus on Geography* magazine and accompanying quizzes bring contemporary geographical issues to students' attention.

For Instructors

Test Bank

The Test Bank, written by Keith Harrington, provides a wide range of questions appropriate for assessing your students' comprehension, interpretation, analysis, and synthesis skills. The Test Bank offers approximately 200 multiple-choice and true/false questions per chapter designed for comprehensive coverage of the text concepts. Each question is tagged for difficulty; level on Bloom's Taxonomy; book sections and page number; as well as learning objectives. The computerized Test Bank, available on CD or as a download, also allows for exporting into a variety of formats compatible with many Internet-based products.

Presentation Slides

- Illustration slides feature all of the text art and illustrations in PowerPoint format.
- Lecture slides focus on key concepts and themes from the text, and feature tables, graphs, and figures from the text.
- Chapter photos, figures, and tables give you access to all of the images from the text, organized by chapter.

Clicker Questions, by Keith Harrington, allow instructors to jump-start discussions, illuminate important points, and promote better conceptual understanding during lectures.

Acknowledgements

An introductory text covering a wide range of topics must draw heavily on the research and help of others. Many geographers contributed advice, comments, ideas, and assistance as this book moved from outline through draft to

publication. We would like to thank those colleagues who contributed their helpful feedback and opinions:

Michael Cote, University of Southern Maine
Seth Giertz, University of Nebraska, Lincoln
Shawn Knabb, Western Washington University
Marc Law, University of Vermont
Michael Makowsky, Johns Hopkins University
Robert McComb, Texas Tech University
Paul Menchik, Michigan State University
Erin Moody, University of Maryland
Matthew Rutledge, Boston College
Daniel Sacks, Indiana University, Bloomington

We would also like to thank those colleagues who offered helpful comments during the preparation of earlier editions of *The Human Mosaic,* whose foundation supports *Contemporary Human Geography* in many ways:

Jennifer Adams, Pennsylvania State University
Paul C. Adams, University of Texas, Austin
W. Frank Ainsley, University of North Carolina, Wilmington
Christopher Airriess, Ball State University
Nigel Allan, University of California, Davis
Thomas D. Anderson, Bowling Green State University
Timothy G. Anderson, Ohio Wesleyan University
Patrick Ashwood, Hawkeye Community College
Nancy Bain, Ohio University
Timothy Bawden, University of Wisconsin, Eau Claire
Brad A. Bays, Oklahoma State University
A. Steele Becker, University of Nebraska, Kearney
Sarah Bednarz, Texas A&M University
Gigi Berardi, Western Washington University
Daniel Borough, California State University, Los Angeles
Mathias Le Bossé, Kutztown University of Pennsylvania
Patricia Boudinot, George Mason University
Wayne Brew, Montgomery Community College
Michael J. Broadway, Northern Michigan University
Sarah Osgood Brooks, Central Michigan University
Larry E. Brown, University of Missouri, Columbia
Scott S. Brown, Francis Marion University
Craig S. Campbell, Youngstown State University
Kimberlee J. Chambers, Sonoma State University
Wing H. Cheung, Palomar Community College
Carolyn A. Coulter, Atlantic Cape Community College
Merel J. Cox, Pennsylvania State University, Altoona
Marcelo Cruz, University of Wisconsin, Green Bay
Christina Dando, University of Nebraska, Omaha
Robin E. Datel, California State University, Sacramento
James A. Davis, Brigham Young University
Richard Deal, Western Kentucky University
Jeff R. DeGrave, University of Wisconsin, Eau Claire
Lorraine Dowler, Pennsylvania State University
Christine Drake, Old Dominion University
Owen Dwyer, Indiana University–Purdue University Indianapolis
Anthony J. Dzik, Shawnee State University
Matthew Ebiner, El Camino College
James D. Ewing, Florida Community College, Jackson
Maria Grace Fadiman, Florida Atlantic University
David Albert Farmer, Wilmington College
Kim Feigenbaum, Santa Fe Community College
D. J. P. Forth, West Hills Community College
Carolyn Gallaher, American University
Charles R. Gildersleeve, University of Nebraska, Omaha
M. A. Goodman, Grossmont College
Jeffrey J. Gordon, Bowling Green State University

Richard J. Grant, University of Miami
Charles F. Gritzner, South Dakota State University
Sally Gros, University of Oklahoma, Norman
Qian Guo, San Francisco State University
Joshua Hagen, Marshall University
Daniel J. Hammel, University of Toledo
Ellen R. Hansen, Emporia State University
Jennifer Helzer, California State University, Stanislaus
Andy Herod, University of Georgia
Elliot P. Hertzenberg, Wilmington College
Deryck Holdsworth, Pennsylvania State University
Cecelia Hudleson, Foothill College
Ronald Isaac, Ohio University
Gregory Jean, Samford University
Kari B. Jensen, Hofstra University
Brad Jokisch, Ohio University
Edward L. Kinman, Longwood University
James R. Keese, California Polytechnic State University
Artimus Keiffer, Wittenberg University
Edward L. Kinman, Longwood University
Marti L. Klein, Saddleback College
Vandara Kohli, California State University, Bakersfield
Jennifer Kopf, West Texas A&M
John C. Kostelnick, Illinois State University
Debra Kreitzer, Western Kentucky University
Olaf Kuhlke, University of Minnesota, Duluth
Michael Kukral, Ohio Wesleyan University
Hsiang-te Kung, Memphis University
William G. Laatsch, University of Wisconsin, Green Bay
Paul R. Larson, Southern Utah University
Ann Legreid, Central Missouri State University
Peter Li, Tennessee Technological University
Ronald Lockmann, California State University, Dominiguez Hills
Jose Javier Lopez, Minnesota State University
Michael Madsen, Brigham Young University, Idaho
Debra J. Taylor Matthews, Boise State University
Blake L. Mayberry, University of Kansas
Vincent Mazeika, Ohio University

Jesse O. McKee, University of Southern Mississippi
Wayne McKim, Towson State University
Douglas Meyer, Eastern Illinois University
Klaus Meyer-Arendt, Mississippi State University
John Milbauer, Northeastern State University
Cynthia A. Miller, Syracuse University
Glenn R. Miller, Bridgewater State University
Don Mitchell, Syracuse University
Norman Moline, Augustana College
Edris Montalvo, Texas State University, San Marcos
Karen M. Morin, Bucknell University
James Mulvihill, California State University, San Bernardino
Douglas Munski, University of North Dakota
Gareth A. Myers, University of Kansas
David J. Nemeth, University of Toledo
James W. Newton, University of North Carolina, Chapel Hill
Michael G. Noll, Valdosta State University
Ann M. Oberhauser, West Virginia University
Stephen M. O'Connell, Oklahoma State University
Thomas Orf, Las Positas College
Brian Osborne, Queen's University
Kenji Oshiro, Wright State University
Bimal K. Paul, Kansas State University
Paul E. Phillips, Ft. Hays (KS) State University
Eric Prout, Texas A&M University
Darren Purcell, University of Oklahoma
Virginia M. Ragan, Maple Woods Community College
Jeffrey P. Richetto, University of Alabama
Henry O. Robertson, Louisiana State University, Alexandria
Robert Rundstrom, University of Oklahoma
Norman H. Runge, University of Delaware

Stephen Sandlin, California State University, Pomona
Lydia Savage, University of Southern Maine
Steven Schnell, Kutztown University of Pennsylvania
Andrew Schoolmaster III, University of North Texas
Cynthia S. Simmons, Michigan State University
Emily Skop, University of Colorado, Colorado Springs
Christa Smith, Clemson University
Anne K. Soper, Indiana University
Roger W. Stump, State University of New York, Albany
Ray Sumner, Long Beach City College
Mary Kimsey Tacy, James Madison University
David A. Tait, Rogers State University
Jonathan Taylor, California State University, Fullerton
Thomas A. Terich, Western Washington University
Thomas M. Tharp, Purdue University
Benjamin F. Timms, California Polytechnic State University
Ralph Triplette, Western Carolina University
Daniel E. Turbeville III, Eastern Washington University
Elisabeth S. Vidon, Indiana University
Ingolf Vogeler, University of Wisconsin
Timothy M. Vowles, Colorado State University
Philip Wagner, Simon Fraser State University
Barney Warf, Florida State University
Henry Way, University of Kansas
Barbara Weightman, California State University, Fullerton
John Western, Syracuse University
W. Michael Wheeler, Southwestern Oklahoma State University
David Wilkins, University of Utah
Douglas Wilms, East Carolina State University
Donald Zeigler, Old Dominion University

In addition, special thanks to Diana Ter-Ghazaryan, Charles Heck, and Jackal Tanelorn for their invaluable research and assistance throughout the developmental process.

Our thanks also go to various staff members of W. H. Freeman and Company, whose encouragement, skills, and suggestions have created a special working environment and to whom we express our deepest gratitude. In particular, we thank Steven Rigolosi; Janice Wiggins, developmental editor par excellence; Taryn Burns, marketing manager; Lindsay Neff, media editor; Liz Geller, project editor; Blake Logan, interior and cover designer; Sheridan Sellers, page makeup designer; Janice Donnola, illustration coordinator; Bianca Moscatelli, photo editor; Susan Wein, production manager; Lisa Kinne, managing editor; and Tracey Kuehn, director of editing, design, and media production. In addition, we'd like to thank several others whose work helped ensure the overall quality of the book: Patti Brecht, copyeditor; Liz Marraffino, proofreader; and Ellen Brennan, indexer. The fine work of all these people can be detected throughout the book.

ABOUT THE AUTHORS

Courtesy of Mona Domosh

Mona Domosh is professor of geography at Dartmouth College. She earned her PhD at Clark University. Her research has examined the links between gender ideologies and the cultural formation of large American cities in the nineteenth century, particularly with regard to such critical but vexing distinctions as consumption/production, public/private, masculine/feminine. She is currently engaged in research that takes the ideological association of women, femininity, and space in a more postcolonial direction by asking what roles nineteenth-century ideas of femininity, masculinity, consumption, and "whiteness" played in the crucial shift from American nation-building to empire-building. Domosh is the author of *American Commodities in an Age of Empire* (2006) and *Invented Cities: The Creation of Landscape in 19th-Century New York and Boston* (1996); the coauthor, with Joni Seager, of *Putting Women in Place: Feminist Geographers Make Sense of the World* (2001); and the coeditor of the *Handbook of Cultural Geography* (2002). At present, she is the president of the Association of American Geographers.

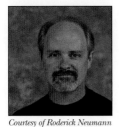

Courtesy of Roderick Neumann

Roderick P. Neumann is professor of geography and chair of the Department of Global and Sociocultural Studies in the School of International and Public Affairs at Florida International University. He earned his PhD at the University of California, Berkeley. He studies the complex interactions of culture and nature through a specific focus on protected areas such as national parks. In his research, he combines the analytical tools of cultural and political ecology with landscape studies. He has pursued these investigations through historical and ethnographic research mostly in East Africa, with comparative work in North America and England. He has published in many leading geography journals including the *Annals of the Association of American Geographers and Progress in Human Geography*. His recent research explores interwoven narratives of nature, landscape, and racial and national identity in the European Union and the American West. His scholarly books include *Imposing Wilderness: Struggles over Livelihoods and Nature Preservation in Africa* (1998); *Making Political Ecology* (2005); and *The Commercialization of Non-Timber Forest Products* (2000), coauthored with Eric Hirsch.

Courtesy of American Council on Education

Patricia L. Price is professor of geography at Florida International University. She earned her PhD at the University of Washington. Her field research interests encompass critical geographies of race and ethnicity and Latinos/as in the United States. She is currently researching Cuban exile landscapes in Miami and how nostalgia shapes and connects places. She has authored many journal articles and three books: *Dry Place* (Minnesota, 2004), *The Cultural Geography Reader* (Routledge, 2008), and *The Human Mosaic*, 12th edition (W. H. Freeman, 2012). Landscape, place, and belonging are the longstanding themes in Price's research, which has been funded by the National Science Foundation and the National Endowment for Humanities. The American Council on Education selected Price as a Fellow from 2013 to 2014. Her fellowship-year project focused on diversity in higher education.

MALOELAP ATOLL

WOTJE ATOLL

MEJIT ISLAND

AILUK ATOLL

UTRIK ATOLL

LIKIEP ATOLL

KWAJALEIN ATOLL

NAMU ATOLL

LIB ISLAND

UJAE ATOLL

Why is it difficult for most of us to interpret this image as a map?

Navigational chart representing a portion of the Marshall Islands in Micronesia.
(Courtesy of Roderick Neumann)

Go to "Seeing Geography" on page 38 to learn more about this image.

HUMAN GEOGRAPHY
A Cultural Approach

Most of us are born geographers. We are curious about the distinctive character of places and peoples. We think in terms of territory and space. Take a look outside your window right now. The houses and commercial buildings, streets and highways, gardens and lawns all tell us something interesting and profound about who we are as a culture. If you travel down the road, or on a jet to another region or country, that view outside your window will change, sometimes subtly, sometimes drastically. Our geographical imaginations will push us to look and think and begin to make sense of what is going on in these different places, environments, and landscapes. It is this curiosity about the world—about how and why it is structured the way it is, what it means, and how we have changed it and continue to change it—that is at the heart of human geography. You are already geographers; we hope that our book will make you better ones.

If all places on Earth were identical, we would not need geography, but each is unique. Every place, however, does share characteristics with other places. Geographers define the concept of **region** as a grouping of similar places or of places with similar characteristics. The existence of different regions endows the Earth's surface with a mosaiclike quality. Geography as an academic discipline is an outgrowth of both our curiosity about lands and peoples other than our own and our need to come to grips with the place-centered element within ourselves. When professional,

region A grouping of like places or the functional union of places to form a spatial unit.

academic geographers consider the differences and similarities among places, they want to understand what they see. They first find out exactly what variations exist among regions and places by describing them as precisely as possible. Then they try to decide what forces made these areas different or alike. Geographers ask *what? where? why?* and *how?*

Our natural geographical curiosity and intrinsic need for identity were long ago reinforced by pragmatism, the practical motives of traders and empire builders who wanted information about the world for the purposes of commerce and conquest. This concern for the practical aspects of geography first arose thousands of years ago among the ancient Greeks, Romans, Mesopotamians, and Phoenicians, the greatest traders and empire builders of their times. They cataloged factual information about locations, places, and products. Indeed, **geography** is a Greek word meaning literally "to describe the Earth." Not content merely to chart and describe the world, these ancient geographers soon began to ask questions about why cultures and environments differ from place to place, initiating the study of what today we call geography.

> **geography** The study of spatial patterns and of differences and similarities from one place to another in environment and culture.

WHAT IS A CULTURAL APPROACH TO HUMAN GEOGRAPHY?

> **human geography** The study of the relationships between people and the places and spaces in which they live.

Human geography forms one part of the discipline of geography, complementing physical geography (which deals with the natural environment). Human geography examines the relationships between people and the places and spaces in which they live using a variety of scales ranging from the local to the global. Human geographers explore how these relationships create the diverse spatial arrangements that we see around us, arrangements that include homes, neighborhoods, cities, nations, and regions. A cultural approach to the study of human geography implies an emphasis on the meanings, values, attitudes, and beliefs that different groups of people around the world lend to and derive from places and spaces. To understand the scope of a cultural approach to human geography, we must first discuss the various meanings of **culture.**

Culture is a very complex term and there are many different definitions. In this book, we think of culture in two ways. In one sense, we use culture to refer to a variety of human behavioral traits and beliefs (e.g., marriage rites), which we define as learned, as opposed to innate, or inborn.

Defined in this way, culture involves a means of communicating these learned beliefs, practices, memories, perceptions, and values that serves to shape individual and collective behaviors. As geographers, we tend to be interested in how these various aspects of culture shape, and take shape in, particular places, environments, and landscapes.

Culture is not a static, fixed phenomenon, and it does not always govern its members. Rather, as

> **culture** A total way of life held in common by a group of people, including such learned features as speech, ideology, behavior, livelihood, technology, and government; or the local, customary way of doing things—a way of life; an ever-changing process in which a group is actively engaged; a dynamic mix of symbols, beliefs, speech, and practices.

cultural practices The social activities and interactions—ranging from religious rituals to food preferences to clothing—that collectively distinguish group identity.

geographers Kay Anderson and Fay Gale put it, "Culture is a *process* in which people are actively engaged." Individuals can and do change **cultural practices,** those social activities and interactions—ranging from religious rituals to food preferences to clothing—that collectively distinguish group identity. This understanding of culture recognizes that ways of life constantly change and that tensions between opposing views are usually present.

Throughout the book, we will also use culture in the second sense to refer to a distinctive group identity (e.g., Navaho culture). By this we mean that collectively shared beliefs, practices, language, technologies, and so forth produce characteristic ways of life for groups of people. Thus, we speak of cultures in the plural. Geographers have historically thought in terms of both categories of cultures (e.g., tribal, urban) and also specific cultures associated with specific places and regions (e.g., Tuareg camel pastoralists). There are an almost infinite number of ways of defining human groups in cultural terms or of creating categories of cultures. Today we are accustomed to hearing about not only Tuareg culture or Navaho culture, but also hip-hop culture or gay culture. The now-familiar appeals to multiculturalism

and cultural diversity reflect a growing appreciation of the complex, ongoing processes of cultural differentiation that are characteristic of human society.

Cultures, however we may categorize and label them, are never internally homogeneous because individual humans never think or behave in exactly the same manner. Sometimes internal differences can be so great that groups experience a fission that results in the formation of new and distinct cultures. Historically, such fissions were often associated with mass human migrations to new lands and environments. Neither are cultural groups consistent across time and space. Cultures are defined as much by dynamism and change as by fixity and stability. In short, we follow Anderson and Gale in taking the view of culture as a process rather than as a thing.

A cultural approach to human geography, then, studies the relationships among space, place, environment, and culture. It examines the ways in which culture is expressed and symbolized in the landscapes we see around us, including homes, commercial buildings, roads, agricultural patterns, gardens, and parks. It analyzes the ways in which language, religion, the economy, government, and other cultural phenomena vary or remain constant from one place to another, and provides a perspective for understanding how people function spatially and identify with place and region (**Figure 1.1**).

FIGURE 1.1   Two traditional houses of worship. Geographers seek to learn how and why cultures differ, or are similar, from one place to another. Often the differences and similarities have a visual expression. *In what ways are these two structures—one a Catholic church in Honduras and the other a Buddhist temple in Laos—alike and different? (Left: Jon Arnold Images Ltd/Alamy; Right: Bruno Morandi/The Image Bank/Getty Images.)*

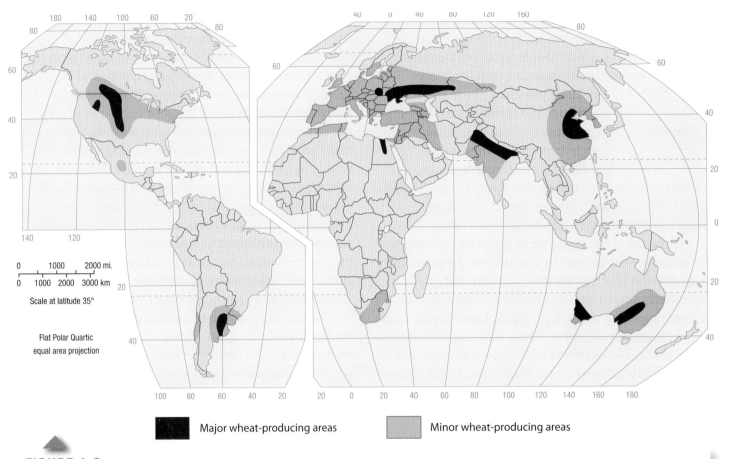

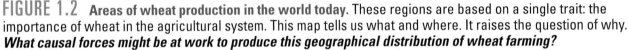

Major wheat-producing areas   Minor wheat-producing areas

FIGURE 1.2   **Areas of wheat production in the world today.** These regions are based on a single trait: the importance of wheat in the agricultural system. This map tells us what and where. It raises the question of why. ***What causal forces might be at work to produce this geographical distribution of wheat farming?***

physical environment
All aspects of the natural physical surroundings, such as climate, terrain, soils, vegetation, and wildlife.

In seeking explanations for cultural diversity and place identity, geographers consider a wide array of factors that cause this diversity. Some of these involve the **physical environment:** terrain, climate, natural vegetation, wildlife, variations in soil, and the pattern of land and water. Because we cannot understand a culture removed from its physical setting, human geography offers not only a spatial perspective but also an ecological one.

Many complex forces are at work on cultural phenomena, and all of them are interconnected in very complicated ways. The complexity of the forces that affect culture can be illustrated by an example drawn from agricultural geography: the distribution of wheat cultivation in the world. If you look at

Figure 1.2, you can see important areas of wheat cultivation in Australia but not in Africa, in the United States but not in Chile, in China but not in Southeast Asia. Why does this spatial pattern exist? Partly, it results from environmental factors such as climate, terrain, and soils. Some regions have always been too dry for wheat cultivation. The land in others is too steep or infertile. Indeed, strong correlation exists between wheat cultivation and midlatitude climates, level terrain, and good soil.

Still, we should not place exclusive importance on such physical factors. People can modify the effects of climate through irrigation; the use of hothouses; or the development of new, specialized strains of wheat. They can conquer slopes by terracing, and they can make poor soils productive with fertilization. For example, farmers in mountainous parts of Greece traditionally wrested an annual harvest of wheat from tiny terraced plots where soil had been trapped behind hand-built stone retaining walls. Even in the United States, environmental factors alone cannot explain the curious fact that major wheat cultivation is concentrated in the semiarid Great Plains, some distance from states such as Ohio and Illinois, where the climate for growing wheat is better. Human geographers know that wheat has to survive in a cultural environment as well as a physical one.

Ultimately, agricultural patterns cannot be explained by the characteristics of the land and climate alone. Many factors complicate the distribution of wheat, including people's tastes and traditions. Food preferences and taboos, often backed by religious beliefs, strongly influence the choice of crops to plant. Where wheat bread is preferred, people are willing to put great efforts into overcoming physical surroundings hostile to growing wheat. They have even created new strains of wheat, thereby decreasing the environment's influence on the distribution of wheat cultivation. Other factors, such as public policies, can also encourage or discourage wheat cultivation. For example, tariffs protect the wheat farmers of France and other European countries from competition with more efficient American and Canadian producers.

This is by no means a complete list of the forces that affect the geographical distribution of wheat cultivation. The distribution of all cultural elements is a result of the constant interplay of diverse factors. Human geography is the discipline that seeks such explanations and understandings.

HOW TO UNDERSTAND HUMAN GEOGRAPHY

Generally speaking, geographers have adopted three different perspectives in studying and understanding the complexity of the human mosaic. Each of these perspectives brings a different emphasis to studying the diversity of human patterns on the Earth.

Spatial Models Some geographers seek patterns and regularities amid the complexity and apply the scientific method to the study of people. Emulating physicists

and chemists, they devise theories and seek regularities or universal spatial principles that apply across cultural lines, explaining all of humankind. These principles ideally become the basis for laws of human spatial behavior. **Space**—a term that refers to an abstract location on a map—is the word that perhaps best connotes this approach to cultural geography (see Doing Geography, on page 37).

Social scientists face a difficult problem because, unlike physical scientists, they cannot limit the effects of diverse factors by running experiments in controlled laboratories. One solution to this problem is the technique known as **model** building. Aware that many causal forces are involved in the real world, they set up artificial situations to focus on one or more potential factors. That is, they abstract a limited set of factors from the complexities of social life in an effort to uncover fundamental spatial patterns and forms.

Torsten Hägerstrand's diagrams of different ways in which ideas and people move from one place to another are examples of spatial models. Some model-building geographers devise culture-specific models to describe and explain certain facets of spatial behavior within specific cultures. They still seek regularities and spatial principles but within the bounds of individual cultures. For example, several geographers proposed a model for Latin American cities in an effort to stress similarities among them and to understand why cities are formed the way they are (Figure 1.3). Obviously, no actual city in Latin America conforms precisely to their uncomplicated geometric plan. Instead, as in all model building, they deliberately generalized and simplified so that an urban type could be recognized and studied. The model will look strange to a person living in a city in the United States or Canada, for it describes a very different kind of urban environment, based in another culture.

Sense of Place Other geographers seek to understand the uniqueness of each region and place. Just as *space* identifies the perspective of the model-building

> **space** Connotes the objective, quantitative, theoretical, model-based, economics-oriented type of geography that seeks to understand spatial systems and networks through application of the principles of social science.

> **model** An abstraction, an imaginary situation, proposed by geographers to simulate laboratory conditions so they can isolate certain causal forces for detailed study.

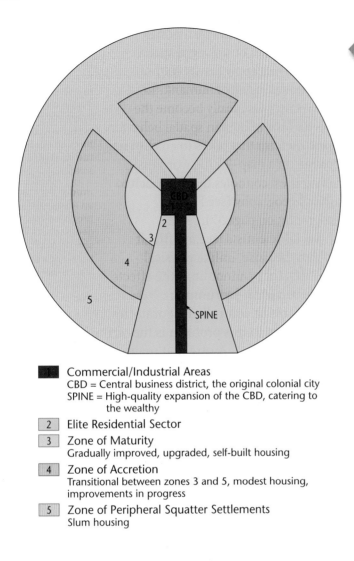

CBD
2
3
4
5
SPINE

1 Commercial/Industrial Areas
CBD = Central business district, the original colonial city
SPINE = High-quality expansion of the CBD, catering to
the wealthy

2 Elite Residential Sector

3 Zone of Maturity
Gradually improved, upgraded, self-built housing

4 Zone of Accretion
Transitional between zones 3 and 5, modest housing,
improvements in progress

5 Zone of Peripheral Squatter Settlements
Slum housing

FIGURE 1.3 **A generalized model of the Latin American city.** Urban structure differs from one culture to another, and in many ways the cities of Latin America are distinctive, sharing much in common with one another. Geographers Ernst Griffin and Larry Ford developed the model diagrammed here to help describe and explain the processes at work shaping the cities of Latin America. *In what ways would this model not be applicable to cities in the United States and Canada?* *(After Griffin and Ford, 1980, p. 406.)*

in people, even though those people may have little direct connection with those places.

Denis Cosgrove, for example, studied why Venice, Italy, continues to stir people's imaginations—people as diverse as tourists from Japan and farmers from Iowa—despite the fact that the city hasn't held any political or economic power in hundreds of years and that the cultures out of which it was formed have long since ceased to exist. Such is the symbolic importance of Venice that in the 1960s millions of dollars came from all around the world in a massive international campaign to save the city from decay and flooding. Similar successful campaigns in other places led the United Nations in 1975 to establish a program to coordinate efforts to identify, catalog, and protect sites of worldwide significance. Known as **World Heritage Sites,** these are places (e.g., buildings, cities, forests, lakes, deserts, archeological ruins) that the UN's International Heritage Programme judges to possess outstanding cultural or natural importance to the common heritage of humanity. Venice was among the earliest of the UN's additions to its list, which numbered 981 sites in 2014. The designation has served to draw even more international tourists to Venice, which now receives roughly 50 times as many visitors per year as the number of permanent residents in the historic city.

Some cultural geographers, however, are interested in the opposite kind of places—ordinary places. They ask how and why people become attached to and derive meaning from their local

place Connotes the subjective, idiographic, humanistic, culturally oriented type of geography that seeks to understand the unique character of individual regions and places, rejecting the principles of science as flawed and unknowingly biased.

World Heritage Sites Places (e.g., buildings, cities, forests, lakes, deserts, archeological ruins) that the UN's International Heritage Programme judges to possess outstanding cultural or natural importance to the common heritage of humanity.

geographer, **place** is the key word connoting this more humanistic view of geography. The geographer Yi-Fu Tuan coined the word *topophilia*, literally "love of place," to describe the characteristic of people who exhibit a strong sense of place and the geographers who are attracted to the study of such places and peoples. Geographer Edward Relph tells us that "to be human is to have and know your place" in the geographical sense. This perspective on cultural geography values subjective experience over objective scientific observation. It focuses on understanding the complexity of different cultures and how those cultures give meaning to and derive meaning from particular places. Many geographers are interested in understanding how and why certain places continue to evoke strong emotions neighborhoods or communities, and explore how those meanings can often come into conflict with each other. Many of the debates that you learn about in newspapers or online—debates on the construction of a new high-rise building or the location of a highway, for example—can only be understood by examining the meanings and values different groups of people give to and derive from particular places.

Power and Ideology Cultures are rarely, if ever, homogeneous. Often certain groups of people have more power in society, and their beliefs and ways of life dominate and are considered the norm, whereas other groups of people with less power may participate in alternative cultures. These divisions are often based on gender, economic class, racial categories, ethnicity, or sexual orientation. The social hierarchies that result are maintained, reinforced, and challenged through many means. Those means can include such things as physical violence, but often social hierarchies are maintained in ways far more subtle.

For example, some geographers study ideology—a set of dominant ideas and beliefs—in relationship to place, environment, and landscape in order to understand how power works culturally. For instance, most nations maintain a set of powerful beliefs about their relationships to the land, some holding to the idea that a deep and natural connection exists between a particular territory and the people who have inhabited it. These ideas often form part of a national identity and are expressed so routinely in poems, music, laws, and rituals that people accept these ideas as truths. Many American patriotic songs, for example, express the idea that the country naturally spreads from "sea to sea." Yet, immigrants to that culture and country, and people who have been marginalized by that culture, may hold very different ideas of identity with the land. Native Americans have claims to land that far predate those of the U.S. government, and they would argue against an American national identity that includes a so-called natural connection to all the land between the Atlantic and the Pacific. Uncovering and analyzing the connections between ideology and power, then, are often integral to the geographer's task of understanding the diversity within a culture.

These different approaches to thinking about human geography are both necessary and healthy.

These groups ask different questions about place and space; not surprisingly, they often obtain different answers. The model-builders tend to minimize diversity through their search for universal causal forces; the humanists examine diversity *among* cultures and strive to understand the unique; those who look to power and ideology focus on diversity and contestation *within* cultures. All lines of inquiry yield valuable findings. We present all these perspectives throughout *Contemporary Human Geography.*

THEMES IN HUMAN GEOGRAPHY

Our study of the human mosaic is organized around five geographical concepts or themes: region, mobility, globalization, nature–culture, and cultural landscape. We use these themes to organize the diversity of issues that confront human geography and have selected them because they represent the major concepts that human geographers discuss. Each of them stresses one particular aspect of the discipline, and even though we have separated them for purposes of clarity, it is important to remember that the concepts are related to each other. When discussing the theme of mobility, for example, we will inevitably bring up issues related to globalization, and vice versa. These themes give a common structure to each chapter and are stressed throughout the book.

REGION

1.1 **LEARNING OBJECTIVE**
Identify and define different types of regions.

Phrased as a question, the theme of region could be "How are people and their traits grouped or arranged geographically?" Places and regions provide the essence of geography. How and why are places alike or different? How do they mesh together into functioning spatial networks? How do their inhabitants perceive them and identify with them? These are central geographical questions. A region, then, is a geographical unit based on characteristics and functions of culture. Geographers recognize three types of regions: formal, functional, and vernacular.

FORMAL REGIONS

> **formal region** A cultural region inhabited by people who have one or more cultural traits in common.

A **formal region** is an area inhabited by people who have one or more traits in common, such as language, religion, or a system of livelihood. It is an area, therefore, that is relatively homogeneous with regard to one or more cultural traits. Geographers use this concept to map spatial differences throughout the world. For example, an Arabic-language formal region can be drawn on a map of languages and would include the areas where Arabic is the native language of the majority of the population. Similarly, a wheat-farming formal region would include the parts of the world where wheat is a predominant crop (look again at Figure 1.2).

The examples of Arabic speech and of wheat cultivation represent the concept of formal region at its simplest level. Each is based on a single cultural trait. More commonly, formal regions depend on multiple related traits (**Figure 1.4**). Thus, an Inuit (Eskimo) culture region might be based on language, religion, economy, social organization, and type of dwellings. The region would reflect the spatial distribution of these five Inuit cultural traits. Districts in which all five of these traits are present would be part of the culture region.

Formal regions are the geographer's somewhat subjective creations. No two cultural traits have the

FIGURE 1.4 **An Inuit hunter with his dogsled team.** Various facets of a multitrait formal region can be seen here, including the clothing, the use of a dogsled as transportation, and hunting as a livelihood system. **What other traits can you see in this photo?** (Louise Murray/Alamy.)

same distribution, and the territorial extent of a culture region depends on what and how many defining traits are used. Why *five* Inuit traits, not four or six? Why not *foods* instead of (or in addition to) dwelling types? Consider, for example, Greeks and Turks, who differ in language and religion. Formal regions defined on the basis of speech and religious faith would separate these two groups. However, Greeks and Turks hold many other cultural traits in common. Both groups are monotheistic, worshipping a single god. In both groups, male supremacy and patriarchal families are the rule. Both enjoy certain traditional foods, such as shish kebab. Whether or not Greeks and Turks are placed in the same formal region or in different ones depends entirely on how the geographer chooses to define the region. That choice, in turn, depends on the specific purpose of research or teaching that the region is designed to serve. Thus, an infinite number of formal regions can be created. It is unlikely that any two geographers would use exactly the same distinguishing criteria or place cultural boundaries in precisely the same location.

The geographer who identifies a formal region must locate borders. Because cultures overlap and mix, such boundaries are rarely sharp, even if only a single cultural trait is mapped. For this reason, we find **border zones** rather than lines. These zones broaden with each additional trait considered because no two traits have the same spatial distribution. As a result, instead of having clear borders, formal regions reveal a center or core where the defining traits are all present. Moving away from the central core, the characteristics weaken and disappear. Thus, many formal regions display a **core-periphery** pattern. This refers to a situation where a region can be divided into two sections: one near the center where the particular attributes that define the region (in this case, language and religion) are strong, and other portions of the region farther away from the core, called the periphery, where those attributes are weaker.

border zones The areas where different regions meet and sometimes overlap.

core-periphery The tendency of both formal and functional culture regions to consist of a core or node, in which defining traits are purest or functions are headquartered, and a periphery that is tributary and displays fewer of the defining traits.

In a real sense, then, the human world is chaotic and undergoing constant change. No matter how closely related two elements of culture seem to be, careful investigation always shows that they do not cover exactly the same area. This is true regardless of the degree of detail involved. What does chaos and change mean to the human geographer? First, it tells us that every cultural trait is spatially unique and that the explanation for each spatial variation differs in some degree from all others. Second, it means that culture changes continually throughout an area and that every inhabited place on Earth has a unique combination of cultural features. No place is exactly like another.

Does this cultural uniqueness of each place prevent geographers from seeking explanatory theories? Does it doom them to explaining each locale separately? The answer must be no. The fact that no two hills or rocks, no two planets or stars, no two trees or flowers are identical has not prevented geologists, astronomers, and botanists from formulating theories and explanations based on generalizations. So it is with human geography.

FUNCTIONAL REGIONS

The hallmark of a formal region is cultural homogeneity. By contrast, a **functional region** need not be culturally homogeneous; instead, it is an area that has been organized to function politically, socially, or economically as one unit. A city, an independent state, a precinct, a church diocese or parish, a trade area, a farm, and a Federal Reserve Bank district are all examples of functional regions. Functional regions have **nodes,** or central points where the functions are coordinated and directed. Examples of such nodes are city halls, national capitals, precinct voting places, parish churches, factories, and banks. In this sense, functional regions also possess a core-periphery configuration, in common with formal regions.

functional region A cultural area that functions as a unit politically, socially, or economically.

node A central point in a functional culture region where functions are coordinated and directed.

Many functional regions have clearly defined borders. A metropolitan area is a functional region that includes all the land under the jurisdiction of a particular urban government (**Figure 1.5**). The borders of this functional region may not be so apparent from a car window, but they will be clearly delineated on a regional map by a line distinguishing one jurisdiction from another. Similarly, each state in the United States and each Canadian province is a functional region, coordinated and directed from a capital, with government control extended over a fixed area with clearly defined borders.

Not all functional regions have fixed, precise borders, however. A good example is a daily newspaper's circulation area. The node for the paper would be the plant where it is produced. Every morning, trucks move out of the plant to distribute the paper throughout the city and surrounding suburbs and countryside. The sales territory of one newspaper may thus overlap with the sales territories of other city newspapers, making boundaries difficult to define. Take, for example, the *Kansas City Star,* which is distributed in 33 counties in Kansas and Missouri. To the east, the paper's distribution overlaps with that of

the *St. Louis Post-Dispatch,* which has a much larger distribution area including Kansas City itself.

Functional regions generally do not coincide spatially with formal regions, and this disjuncture often creates problems for the functional region. Germany provides an example (**Figure 1.6**). As an independent state, Germany forms a functional region. Language provides a substantial basis for political unity. However, the formal region of the German language extends beyond the political borders of Germany. It includes part or all of eight other independent states. More important, numerous formal regions have borders cutting through German territory. Some of these have endured for millennia, causing differences among northern, southern, eastern, and western Germans. These contrasts make the functioning of the German state more difficult and help explain why Germany has been politically fragmented more often than unified.

VERNACULAR REGIONS

A **vernacular region** is one that is *perceived* to exist by its inhabitants, as evidenced by the widespread acceptance and use of a special regional name. We might think of vernacular regions as the most democratic type of all. Vernacular regions are named and their boundaries imagined through popular consensus. (Indeed, the term "popular region" is sometimes used in a similar way.) Like old gospel or folk songs passed down over the years from one musician to another, it is often difficult to trace the precise origins of a vernacular regional designation. Typically, the name of a vernacular region first becomes widely known through a local newspaper column, popular song, blog site, or some other form of popular mass communication. The vernacular name then "sticks." That is, it becomes widely adopted in everyday speech, if not in official documents.

Some vernacular regions are based on physical environmental features. For example, there are many regions called simply "the valley." Wikipedia lists more than 30 different regions in the United States and Canada referred to as such, places as varied as the

> ■ **vernacular region**
> A culture region perceived to exist by its inhabitants, based in the collective spatial perception of the population at large and bearing a generally accepted name or nickname (such as "Dixie").

FIGURE 1.5 Aerial view of Denver. This image clearly illustrates the node of a functional region—here, the dense cluster of commercial buildings—that coordinates activities throughout the area surrounding it. ***Can you identify the border of this functional region? Why or why not?*** *(Jim Wark/Airphoto.)*

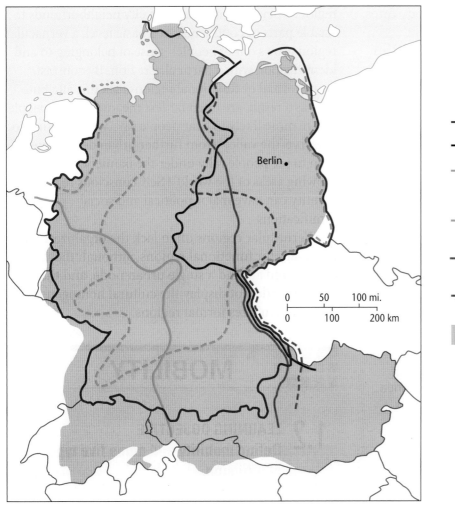

Present borders of Germany

"Iron Curtain," 1945–1990

Northern limit of divided inheritance (derived from Romans)

Northern limit of Catholic majority

Western limit of surviving rural feudal estates, 1800

German-Slav, Christian-pagan border, A.D. 800

German-speaking area

Berlin

0   50   100 mi.

0   100   200 km

FIGURE 1.6 East versus west and north versus south in Germany. As a political unit and functional culture region, Germany must overcome the disruptions caused by numerous formal regions that tend to make the sections of Germany culturally different. Formal and functional regions rarely coincide spatially. *How might these sectional contrasts cause problems for modern Germany?*

Sudbury Basin in Ontario and the Lehigh Valley in Pennsylvania. In the 1980s, "the valley" became synonymous with the San Fernando Valley in Southern California and became associated with a type of landscape (suburban), person (a white, teenage girl, called the "valley girl"), and way of speaking ("valspeak").

Other vernacular regions find their basis in economic, political, or historical characteristics. "Redneck Riviera" is a popular designation for the coastal area of the Florida panhandle, running from Panama City in the east to Pensacola and beyond in the west. "Redneck" is a term referring to politically conservative, working-class, rural culture in the United States. "Riviera" refers, ironically, to the affluent resort area of the French Mediterranean coast. Historically, the numerous military bases strung up and down the panhandle coastline propelled the region's economy. The bases provided jobs, while the gentle climate, inexpensive real estate, and white sand beaches encouraged retired military personnel to put down roots. The politically conservative culture of military retirees combined with the remnants of rural southern vernacular speech and cuisine creates a distinctive beach vacation destination (Figure 1.7). In 1996 country-and-western singer Tom T. Hall released a song entitled "Redneck Riviera" that included the lyrics:

"They got beaches of the whitest sand / Nobody cares if gramma's got a tattoo." The region's local governments and chambers of commerce created an alternative designation for it, "Emerald Coast," and have worked hard to eliminate all references to the Redneck Riviera. In 2010 the Gulf Shores, Alabama, city government even passed a prohibitive ordinance when a television production company scouted the town as a location for a proposed reality TV show to be called *Redneck Riviera.* City ordinances aside, we would be willing to wager a week's salary that far more Floridians can identify the vernacular Redneck Rivera than can identify the officially sanctioned Emerald Coast.

Vernacular regions, like most regions, generally lack sharp borders, and the inhabitants of any given area may claim residence in more than one such region. They vary in scale from city neighborhoods to sizable parts of continents. At a basic level, a vernacular region grows out of people's sense of belonging to and identification with a particular region. By contrast, many formal or functional regions lack this attribute and, as a result, are often far less meaningful for people. You're more likely to hear people say "we're fighting to preserve 'the valley' from further urban development" than to see people rally under the banner of "wheat-growing areas of the world"! Self-conscious regional identity can have major political and social ramifications.

Vernacular regions often lack the organization necessary for functional regions, although they may be centered around a single urban node, and they frequently do not display the cultural homogeneity that characterizes formal regions.

FIGURE 1.7 **The Redneck Riviera.** The coastline of Florida's panhandle region is popularly known in Florida and beyond as the Redneck Riviera, though local city governments in the region prefer to call it the Emerald Coast. *Can you think of other regions in the United States, or elsewhere, that have such popular designations?* (Nik Wheeler/Corbis.)

MOBILITY

1.2 LEARNING OBJECTIVE
Define mobility and name five types of diffusion.

The concept of regions helps us see that similar or related sets of elements are often grouped together in space. Equally important in geography is understanding how and why these different cultural elements move through space and locate in particular settings. Regions themselves, as we have seen, are not stable but are constantly changing as people, ideas, practices, and technologies move around in space. Are there some patterns to these movements? These questions define our second theme, **mobility.**

One important way to study mobility is through the concept of diffusion. **Diffusion** can be defined as the movement of people, ideas, or things from one location outward toward other locations where these items are not initially found. Through the study of diffusion, the human

> **mobility** The relative ability of people, ideas, or things to move freely through space.
>
> **diffusion** The movement of people, ideas, or things from one location outward toward other locations.

geographer can trace how spatial patterns in culture emerged and evolved. After all, any culture is the product of almost countless innovations that spread from their points of origin to cover a wider area. Some innovations occur only once, and geographers can sometimes trace a cultural element back to a single place of origin. In other cases, **independent invention** occurs: the same or a very similar innovation is separately developed at different places by different peoples.

> ■ **independent invention** A cultural innovation developed in two or more locations by individuals or groups working independently.

TYPES OF DIFFUSION

Examining spatial models developed by Hägerstrand, geographers recognize several different kinds of diffusion (**Figure 1.8**).

Relocation diffusion occurs when individuals or groups with a particular idea or practice migrate from one location to another, thereby bringing the idea or practice to their new homeland. Religions frequently spread this way. An example is the migration of Christianity with European settlers who came to America. In **expansion diffusion,** ideas or practices spread throughout a population, from area to area, in a snowballing process, so that the total number of knowers or users and the areas of occurrence increase.

> ■ **relocation diffusion** The spread of an innovation or other element of culture that occurs with the bodily relocation (migration) of the individual or group responsible for the innovation.
>
> **expansion diffusion** The spread of innovations within an area in a snowballing process, so that the total number of knowers or users becomes greater and the area of occurrence grows.

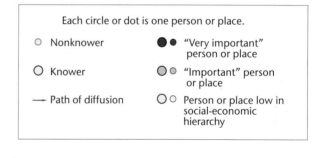

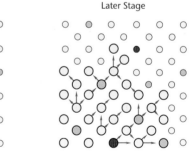

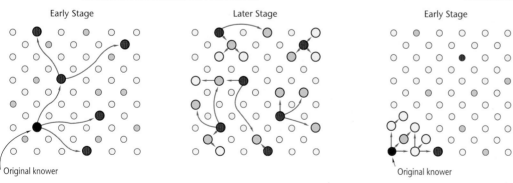

FIGURE 1.8 **Types of cultural diffusion.** These diagrams are merely suggestive; in reality, spatial diffusion is far more complex. In hierarchical diffusion, different scales can be used, so that, for example, the category "very important person" could be replaced by "large city." *In terms of the spread of disease, which type of diffusion do you think is most common?*

hierarchical diffusion
A type of expansion diffusion in which innovations spread from one important person to another or from one urban center to another, temporarily bypassing other persons or rural areas.

contagious diffusion
A type of expansion diffusion in which cultural innovation spreads by person-to-person contact, moving wavelike through an area and population without regard to social status.

Expansion diffusion can be further divided into three subtypes. In **hierarchical diffusion,** ideas leapfrog from one important person to another or from one urban center to another, temporarily bypassing other persons or rural territories. We can see hierarchical diffusion at work in everyday life by observing the acceptance of new modes of dress or foods. For example, sushi restaurants originally diffused from Japan in the 1970s very slowly because many people were reluctant to eat raw fish. In the United States, the first sushi restaurants appeared in the major cities of Los Angeles and New York. Only gradually throughout the 1980s and 1990s did eating sushi become more common in the less urbanized parts of the country. By contrast, **contagious diffusion** involves the wavelike spread of ideas in the manner of a contagious disease, moving throughout space without regard to hierarchies.

Hierarchical and contagious diffusion often work together. The 2009 influenza pandemic provides a sobering example of how these two types of diffusion can reinforce each other. A pandemic is a worldwide disease epidemic. In March 2009 Mexico announced that a new influenza virus strain, known as H1N1, had reached epidemic proportions in Mexico City. Many countries cut off airline traffic to Mexico in an attempt to inhibit the spread of H1N1. However, the virus had already been carried out of the country by air travelers. Through hierarchical diffusion it began to appear in major cities worldwide, and from there it spread through contagious diffusion to the countryside. China offers a clear example of this process. Chinese authorities first detected the virus in May 2009 in cities with international airports, where it remained confined for the next four months. Four to six months after the initial cases, H1N1 experienced its greatest geographic expansion as it spread across all of China (**Figure 1.9**).

Sometimes a specific trait is rejected but the underlying idea is accepted, resulting in **stimulus diffusion.** For example, early Siberian peoples domesticated reindeer only after exposure to the domesticated cattle raised by cultures to their south. The Siberians had no use for cattle, but the idea of domesticated herds appealed to them, and they began domesticating reindeer, an animal they had long hunted.

If you throw a rock into a pond and watch the spreading ripples, you can see them become gradually weaker as they move away from the point of impact. In the same way, diffusion becomes weaker as a cultural innovation moves away from its point of origin. That is, diffusion decreases with distance. An innovation will usually be accepted most thoroughly in the areas closest to where it originates. Because innovations take increasing time to spread outward, time is also a factor. Acceptance generally decreases with distance and time, producing what geographers call **time–distance decay.** Modern mass media, however, along with relatively new technologies like the Internet and cell phones, have greatly accelerated diffusion, diminishing the impact of time–distance decay.

In addition to the gradual weakening or decay of an innovation through time and distance, barriers can retard its spread. **Absorbing barriers** completely halt diffusion, allowing no further progress. For example, many of us have become accustomed to using Google's "Street View" mapping function that provides panoramic and zoom-in views from positions along many streets in the world. This technology allows users to plan everything from dining out to purchasing real estate. Street View, however, has raised privacy questions worldwide. Austria and Greece have chosen to prohibit the use of Street View technology within their borders because of privacy concerns. Among other effects on these countries, the ban will likely slow or halt the

stimulus diffusion
A type of expansion diffusion in which a specific trait fails to spread but the underlying idea or concept is accepted.

time–distance decay The decrease in acceptance of a cultural innovation with increasing time and distance from its origin.

absorbing barrier
A barrier that completely halts diffusion of innovations and blocks the spread of cultural elements.

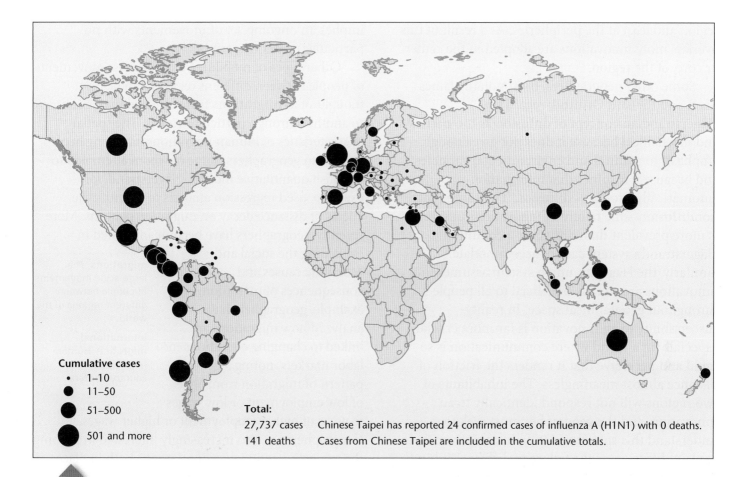

FIGURE 1.9 **Diffusion of influenza infections during the H1N1 pandemic of 2009.** The influenza outbreak started in Mexico City. Global air travel facilitated hierarchical diffusion to international airports around the world. From those points, the virus spread within countries through contagious diffusion. ***How might a country contain the spread of influenza?***

innovative uses of Street View now being explored in the visual arts.

Such examples aside, few absorbing barriers are effective or persist for long. More commonly, barriers are permeable, allowing part of the innovation wave to diffuse through them but acting to weaken or retard its continued spread. When a school board objects to students with tattoos or body piercings, the principal of a high school may set limits by mandating that these markings be covered by clothing. However, over time, those mandates may change as people get used to the idea of body markings. More likely than not, though, some mandates will remain in place. In this way, the principal and school board act as a **permeable barrier** to cultural innovations.

Although all places and communities hypothetically have equal potential to adopt a new idea or practice, diffusion typically produces a core-periphery spatial arrangement. Hägerstrand offered an explanation of how diffusion produces such a regional configuration. The distribution of innovations can be random, but the overlap of new ideas and traits as they diffuse through space and time is greatest toward the center of the

permeable barrier
A barrier that permits some aspects of an innovation to diffuse through it but weakens and retards continued spread; an innovation can be modified in passing through a permeable barrier.

region and least at the peripheries. As a result of this overlap, more innovations are adopted in the center, or core, of the region.

Some other geographers, most notably James Blaut and Richard Ormrod, regard the Hägerstrandian concept of diffusion as too narrow and mechanical because it does not give enough emphasis to cultural and environmental variables and because it assumes that information automatically produces diffusion. They argue that nondiffusion—the failure of innovations to spread—is more prevalent than diffusion, a condition Hägerstrand's system cannot accommodate. Similarly, the Hägerstrandian system assumes that innovations are equally beneficial to all people throughout geographical space. In reality, susceptibility to an innovation is far more crucial, especially in a world where communication is so rapid and pervasive that it renders the friction of distance almost meaningless. The inhabitants of two regions will not respond identically to an innovation, and the geographer must seek to understand this spatial variation in receptiveness to explain diffusion or the failure to diffuse. Within the context of their culture, people must perceive some advantage before they will adopt an innovation. As with most human-geographic models, diffusion helps us recognize persistent or recurring spatial patterns, but does not offer universally valid explanations.

MIGRATION

Increasingly, we see many other examples of mobility, such as the almost instant global communication about new ideas and technologies through computers and other digital media, the rapid movement of goods from the place of production to that of consumption, and the seemingly nonstop movement of money worldwide through digital financial networks. These types of movements through space do not necessarily follow the pattern of core-periphery but instead create new and different types of patterns. The term **circulation** might better fit many of these forms of mobility because this term

> **circulation** An ongoing set of movements of people, ideas, or things that have no particular center or periphery.

implies an ongoing set of movements with no particular center or periphery.

Other types of mobilities, such as mass movements of people between different regions, can be best thought of as **migrations** from one region or country to another through particular routes. The spatial characteristics of human migrations have long been of interest to geographers, some of whom attempted to establish quantitative models. Hägerstrand, for example, used regression analysis to quantify the effects of distance decay on migration patterns. More recently, geographers have become interested in explaining the social and economic causes and consequences of migration. For example, geographers have analyzed how migrations are linked to changing conditions in labor markets, noting a general pattern of migration from areas of low employment or low wages to areas of higher employment or higher wages. Geographers are also increasingly interested in probing the subjective experience of migrants, both in places of origin and places of destination. Jeremy Foster, for example, studied the experiences of European migrants in colonial South Africa to reveal how painting, photography, poetry, and other forms of artistic expression are being used to transform a foreign land into a homeland. That is, Foster is interested in how migrants come to feel "at home" (or not) in a foreign land through engagement with certain kinds of cultural practices.

> **migrations** The large-scale movements of people between different regions of the world.
>
> **international migration** Human migration across country borders.

Migration implies the crossing of boundaries of some sort. Migration that crosses country borders is considered **international migration.** All residents of the United States, with the exception of Native Americans, can trace their presence in the country to international migration, often multiple migrations from different points of origin, via different routes, and at different times. One's "American" ancestry might easily involve early-eighteenth-century migration from western Europe, nineteenth-century migration from North Africa, and twentieth-century migration from East Asia. This example illustrates some of the primary characteristics geographers use to categorize migration, including spatial scale (regional/global), time (temporary/permanent), and distance (long/short).

internal migration Human migration that occurs within the borders of a country.

Great Migration The twentieth-century movement of 6 million African Americans from the rural southern states to the cities of the midwestern and northeastern states.

Migration that occurs *within* the borders of a country is classified as **internal migration.** A classic example of internal migration from the United States is the **Great Migration.** It refers to the twentieth-century movement of 6 million African Americans from the rural southern states to the cities of the midwestern and northeastern states. Such mass migrations have deep, long-lasting impacts on social, economic, and cultural conditions in both the places of origin and of destination. We can view the Great Migration as a significant, but relatively small, part of a global-scale pattern of mass migration from rural to urban areas sparked by the Industrial Revolution. In 2007 the United Nations determined that for the first time in human history, the majority of the world's population lived in cities. This historic urbanization is primarily the result of rural to urban migration now experienced by nearly every country in the world.

Whether international or internal, geographers have a system for categorizing types of migration. **Stepwise migration** is migration conducted in a series of stages. Migration might begin with a move from the countryside to a provincial capital. From there the migration continues to a country's capital city and then might proceed across international borders to another country. Such stepwise migrations can take place over the course of one or more generations. **Return migration** refers to the phenomenon of migrants going back to their place of origin after long-term residency elsewhere. For example, it is not uncommon for migrants to return to their home countries upon reaching retirement age, even after spending most of their adult lives working abroad. Finally, **seasonal migration** refers to annual cycles of movement based on the time of year. Agricultural workers are often seasonal migrants because labor demand is seasonally variable and highly concentrated

stepwise migration Human migration conducted in a series of stages.

return migration The phenomenon of migrants returning to their place of origin after long-term residency elsewhere.

seasonal migration Usually associated with crop harvest periods, migrants move according seasonal changes in weather.

FIGURE 1.10 **Demonstrations in Hamburg, Germany, against the deportation of immigrants.** Many migrant-worker groups are fighting to maintain the rights of transnational migrants who would like to be able to move freely between their home countries and those that currently provide work for them. *Do you think transnational migrants should be able to move without restriction between their home countries and where they work? Why or why not?* (Vario Images GmbH & Co. KG/Alamy.)

during the harvest periods of particular crops in particular places.

In today's globalizing world, with better and faster communication and transportation technologies, many migrants more easily maintain ties to their homelands even after they have migrated, and some may move back and forth between their home countries and those to which they have migrated. Scholars refer to this kind of migration as **transnational migration** (Figure 1.10). We will discuss these different patterns of mobility in greater detail throughout the rest of the book.

transnational migration The movement of groups of people who maintain ties to their homelands after they have migrated.

West Virginia, Still Home
http://tinyurl.com/kerscuv

This video focuses on McDowell County, West Virginia, describing the recent decline of the region's coal industry and resulting outmigration of most of its population. Despite the economic hardship and social upheaval, many remaining residents cannot imagine living elsewhere and remain deeply attached to the landscape. One resident tells the camera, "We are very connected to these mountains," a sentiment that reveals a strong sense of place, belonging and regional identity. As we have learned in this chapter, the relationship between people and the places and landscapes in which they live is a main topic of study for human geographers. Quite a few other basic ideas in human geography are expressed by the video as well.

Thinking Geographically:

1. The video shows how social and economic forces operate across geographic scales—the global slump in the demand for coal and mechanization in the coal industry nationwide have led to the decline of once prosperous and well-populated towns in a southern corner of West Virginia. Can you think of similar examples in the area where you live or with which you are familiar?

2. One of the residents in the video spoke of a "brain drain" and another of her hopes that some of those people will "have the desire to come back to this community." How do these and other statements about the town's population relate to the discussion of migration in this textbook? How many types of migration are illustrated in the video?

3. Recall the various ways that the concept of region was discussed in the chapter. How many of those concepts are illustrated by the case of McDowell County?

4. How does this video illustrate the concepts of place and topophilia? Think about the personal histories of some of the residents interviewed and their onscreen statements about the town and surrounding landscape. What sorts of emotions and sentiments do people express when discussing their home?

GLOBALIZATION

1.3 LEARNING OBJECTIVE
Describe globalization and explain its importance to cultural geography.

globalization
Processes of economic, political, and social integration that operate on a global scale and have collectively created ties that make a difference to lives around the planet.

The modern technological age, in which improved worldwide transport and communications allow the instantaneous diffusion of ideas and innovations, has accelerated the phenomenon called **globalization.** Geographer Matthew Sparke states that the term refers to "processes of economic, political, and social integration that have collectively created ties that make a difference to lives around the planet." Simply put, actions and conditions in one place are increasingly linked to actions and conditions in other places around the globe.

This interconnected and interdependent world has been created from a set of factors: faster and more reliable transportation, particularly the jet plane; the almost-instantaneous communication that computers, phones, faxes, and so on have allowed; and the creation of digital sources of information and media, such as the Internet. Thus, globalization in this sense is a rather recent phenomenon, dating from the late twentieth century. Yet, we know that long before that time, different countries and different parts of the world were linked. For example, in early medieval times overland trade routes connected China with other parts of Asia. The British East India Company maintained maritime trading routes between England and large portions of South Asia as early as the seventeenth century, thereby creating interdependencies between the regions that persisted for centuries.

Interdependence in the context of colonial empire refers to the reciprocal ties established between regions that over time collectively create a global economic system. Interdependence does not mean equality. It simply means that there is a two-way flow of goods, money, people, and ideas that create dependencies at both ends of the connection, but not dependencies of equal consequence. Spain, Portugal, and England all established interdependent global empires rooted in Europe's Age of Exploration and Discovery. Some geographers refer to such moments as early global encounters and suggest that they set the stage for contemporary globalization. That is, they trace today's interdependencies under globalization to the interdependencies established under yesterday's colonial empires.

> **interdependence**
> Relations between regions or countries of mutual, but not necessarily equal, dependence.

While some may locate its historical roots in European empires, most agree that globalization is a fundamentally different phenomenon. Beginning around the 1970s, new and advanced communication technologies allowed encounters between different cultures to take place not face-to-face but rather through technologies such as film, computers, the Internet, and, most recently, smart phones. These new media forms allowed such encounters to happen simultaneously in many different places, at any time. Moreover, multinational corporations play a leading role in globalization, sometimes exceeding the wealth and power of national governments. Collectively, these trends have significantly diminished the relevance of national borders to the global circulation of people, goods, and ideas. According to geographers R. J. Johnston, Peter Taylor, and Michael Watts, with globalization, activities and outcomes "do not merely cross borders, these processes operate as if borders were not there." Most importantly from a human geography perspective, globalization means that an ever-greater proportion of social life is organized on a global scale.

These increasingly linked and interdependent economic, political, and cultural networks around the world might lead many to believe that different groups of people around the globe are becoming more and more alike. In some ways this is true, but what these new global encounters have enabled is an increasing recognition of the differences between groups of people. And some of those differences have been caused by globalization itself.

Some groups of people have access to advanced technologies, better health care, and education, whereas others do not. Even within our own neighborhoods and cities, we know that there are people who are less able to afford these things. If we mapped certain indicators of human well-being on a global scale, such as life expectancy, literacy, and standard of living, we would find quite an uneven distribution. **Figure 1.11** shows us that different cultures around the world have different access to these types of resources. These differences are what scholars mean when they refer to stages of development. In Figure 1.11, you can see regions of the world that have a fairly high Human Development Index (HDI) and those that have a relatively low HDI. Scholars often refer to these two types of regions as developed (relatively high HDI) and developing (relatively low HDI). This inequitable distribution of resources is referred to as **uneven development.** We will discuss development in depth in Chapter 9.

> **uneven development**
> The tendency for industry to develop a core-periphery pattern, enriching the industrialized countries of the core and impoverishing the less industrialized periphery. Also used to describe urban patterns in which suburban areas are enriched, while the inner city is impoverished.

Globalization helps make us aware of these uneven developments and contributes to some of them. How does this happen? Globalization may be thought of as both a set of processes that are economic, political, and cultural in nature, and as the effects of those processes. For example, economic globalization refers to the interlinked networks of money, production, transportation, labor, and consumption that allow, say, the parts of an automobile to be manufactured in several countries, assembled in yet another, and then sold throughout large portions of the world. These economic networks and processes, in turn, have significant and often uneven effects on the economies of different countries and regions. Some regions gain employment, whereas others lose jobs; some consumers are able to afford these cars because they are less expensive, whereas other potential consumers who have become unemployed cannot afford to purchase them. These global economic processes and effects are, in turn, linked to politics and culture. In other words, globalization entails not only certain processes and effects but also the relationships among these things. For example, those countries chosen by the automobile manufacturer as sites of production might see their standards of living improve, leading to larger consumer markets, better communications and media, and often changing political sensibilities. And these changing political ideas, in turn, will shape economic decisions and other factors. Globalization, therefore, involves looking at complex interconnections between a set of related processes and their effects.

Culture, of course, is a key variable in these interactions and interconnections. In fact, as we have just suggested, globalization is occurring through cultural media, for example, in films, on television, and on the Internet (see Mona's Notebook, page 22). In addition, if we consider culture as a way of life, then globalization is a key shaper of culture and is shaped by it. Some scholars have suggested that globalizing processes and an increase in mobility will work to homogenize different peoples, breaking down culture regions and eventually producing a single global culture. Other scholars see a different picture, one where new forms of media and communication will allow local cultures to maintain

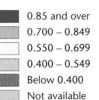

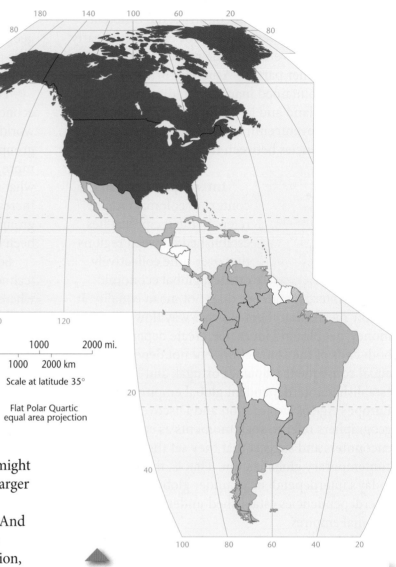

FIGURE 1.11 **World map of the Human Development Index, 2012.** The Human Development Index (HDI) includes life expectancy, adult literacy, educational participation, and gross domestic product (GDP). These statistics are brought together to create a measure of development that is more balanced than one based only on economic growth. The United Nations Development Programme (UNDP), in its Human Development Reports, calculates the HDI yearly. ***In what countries or regions does the HDI measure surprise you? Why?*** *(Source: Human Development Report, 2012. Accessed at http://hdr.undp.org.)*

their distinct identities, reinforcing the diversity of cultures around the world. Throughout *Contemporary Human Geography*, we will return to these issues, asking and considering the complex role of culture and cultures in an increasingly global world.

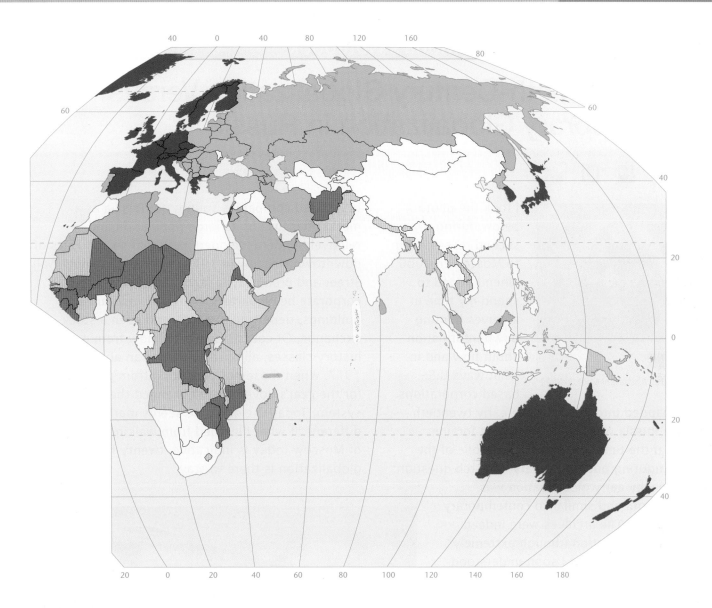

NATURE–CULTURE

1.4 LEARNING OBJECTIVE
Define the different models of nature–culture relations.

The themes of region, mobility, and globalization help us understand patterns, movements, and interconnections that characterize the ways in which people create spatial patterns on the Earth. Our fourth theme, **nature–culture,** adds a different dimension to this analysis; it focuses our attention literally on how people inhabit the Earth, their relationships to the physical environment. This theme helps us investigate how groups of people interact with the Earth's biophysical environment and examine how the culture, politics, and economies of those groups affect their ecological situation and resource use. Human geographers view the relationship between people and nature as a two-way interaction. People's cultural values, beliefs, perceptions, and practices have ecological impacts, and ecological conditions, in turn, influence cultural perceptions and practices. The human geographer must study the interaction

nature–culture
Refers to the complex relationships between people and the physical environment, including how culture, politics, and economies affect people's ecological situation and resource use.

Mona's Notebook

Early-Twentieth-Century Globalization Versus Contemporary Globalization in Russia

Mona Domosh. *(Courtesy of Mona Domosh.)*

I took the photo below during my second research trip to Moscow in 2006, when I hoped to spend my time in archives hunting down information about how and in what ways U.S.-based corporations had expanded into Russia in the early twentieth century. Russia was America's largest foreign market at the time, so I knew it held one of the keys to figuring out my primary research question: how and why early globalization was different from and similar to contemporary globalization. The archives were indeed filled with information (though extremely difficult to access), but I soon understood that I could learn almost as much from walking the streets as I could from sitting, squinting, and typing in closed-off archival rooms. What I saw was a city in the midst of a huge consumer revolution, funded with money from Russian natural gas and oil, and capitalized by global corporations. Construction cranes were everywhere because large, globally based development companies saw profitable opportunities in the skyrocketing demand for housing, entertainment, and shopping spaces.

One hundred years earlier, as I found out in the archives, Moscow was in a similar situation, though the scale and pace of globalization was much smaller and slower. In addition to German, French, and British companies, U.S.-based companies sought out Russia as a market because of its large population and relatively underdeveloped industrial sector. Parts of downtown Moscow in 1915 began to resemble Wall Street and Broadway: a landscape filled with corporate headquarters, bank and insurance buildings, department stores, and even a stock exchange. Of course, as we all know from our history classes, all of this came to an abrupt halt in 1917, when those not benefiting from globalization (or the czar's government!) changed the entire system. Today the outcome seems markedly different. If what I learned from walking the streets of Moscow today is indicative, twenty-first-century globalization is there to stay.

Nineteenth-century commercial buildings being renovated in downtown Moscow. What other signs of globalization can you see in this image? *(Courtesy of Mona Domosh.)*

between culture and environment to understand spatial variations in culture.

The term *ecology* was coined in the nineteenth century to refer to a new biological science concerned with studying the complex relationships among living organisms and their physical environments.

> **cultural ecology**
> Broadly defined, the study of the relationships between the physical environment and culture; narrowly (and more commonly) defined, the study of culture as an adaptive system that facilitates human adaptation to nature and environmental change.

Geographers borrowed this term in the mid-twentieth century and joined it with the term *culture* to identify a field of study—**cultural ecology**— that dealt with the interaction between culture and physical environments. Later, another concept, the ecosystem, was introduced to describe a territorially bounded system consisting of interacting organic and inorganic components. Plant and animal species were said to be adapted to specific conditions in the ecosystem and to function so as to keep the system stable over time.

It soon became clear, however, that human cultural interactions with the environment were far too complex to be analyzed with concepts borrowed from biology. Furthermore, the idea that groups of people interacted with their ecosystems in isolation from larger-scale political, economic, and social forces was difficult to defend. We can readily observe, for example, that trade goods come into communities, agricultural commodities flow out, money circulates, taxes are collected, and people migrate in and out for work. So, geographers now use the term *nature–culture* to refer to the complex interactions among all these variables and to reflect the fact that studies of local human–environment relations need to include political, economic, and social forces operating on national, and even global, scales.

The theme of nature–culture, the meeting ground of cultural and physical geographers, has traditionally provided a focal point for the academic discipline of geography. In fact, some geographers have proposed that *the* theme of geography is nature–culture: that the study of the intricate relationships between people and their physical environments unites cultural and physical geography to form the entire academic discipline. Although few accept this narrow definition of geography, most will agree that an appreciation of the complex people–environment relationship is necessary for concerned citizens of the twenty-first century.

Through the years, human geographers have developed various perspectives on the interaction between humans and the land. Four schools of thought have developed: environmental determinism, possibilism, environmental perception, and humans as modifiers of the Earth.

ENVIRONMENTAL DETERMINISM

During the late nineteenth and early twentieth centuries, many geographers accepted **environmental determinism:** the belief that the physical environment is the dominant force shaping cultures and that humankind is essentially a passive product of its physical surroundings. Humans are clay to be molded by nature. Similar physical environments produce similar cultures.

> **environmental determinism** The belief that cultures are directly or indirectly shaped by the physical environment.

For example, environmental determinists believed that peoples of the mountains were predestined by the rugged terrain to be simple, backward, conservative, unimaginative, and freedom loving. Desert dwellers were likely to believe in one God but to live under the rule of tyrants. Temperate climates produced inventiveness, industriousness, and democracy. Coastlands pitted with fjords produced great navigators and fishers. Environmental determinism had serious consequences, particularly during the time of European colonization in the late nineteenth century. For example, many Europeans saw Latin American native inhabitants as lazy, childlike, and prone to vices such as alcoholism because of the tropical climates in much of this region. Living in a tropical climate supposedly ensured that people didn't have to work very hard for their food. Europeans were able to rationalize their colonization of large portions of the world in part along climatic lines. Because the natives were "naturally" lazy and slow, the European reasoning went, they would benefit from the presence of the "naturally" stronger, smarter, and more industrious Europeans who came from more temperate lands.

Determinists overemphasize the role of environment in human affairs. This does not mean that environmental influences are inconsequential or that geographers should not study such influences. Rather, the physical environment is only one of many forces affecting human culture and is never the sole determinant of behavior and beliefs.

POSSIBILISM

possibilism A school of thought based on the belief that humans, rather than the physical environment, are the primary active force; that any environment offers a number of different possible ways for a culture to develop; and that the choices among these possibilities are guided by cultural heritage.

After the 1920s, **possibilism** became a prominent view among geographers. Possibilists claim that any physical environment offers a number of possible ways for a culture to develop. In this way, the local environment helps shape its resident culture. However, a culture's way of life ultimately depends on the choices people make among the possibilities offered by the environment. These choices are guided by cultural heritage and are shaped by a particular political and economic system. Possibilists see the physical environment as offering opportunities and limitations; people make choices among these to satisfy their needs. Figure 1.12 provides an interesting example: the cities of San Francisco and Chongqing both were built on similar physical terrains that dictated an overall form, but differing cultures lead to very different street patterns, architecture, and land use. In short, local traits of culture and economy are the products of culturally based decisions made within the limits of possibilities offered by the environment.

Most possibilists believe that the higher the technological level of a culture, the greater the number of possibilities and the weaker the influences of the physical environment. Technologically advanced cultures, in this view, have achieved some mastery over their physical surroundings. Geographers Jim Norwine and Thomas Anderson, however, warn that even in these advanced societies "the quantity and quality of human life are still strongly influenced by the natural environment," especially climate. They argue that humankind's

control of nature is anything but supreme and perhaps even illusory. One only has to think of the devastation caused by the January 2010 earthquake in Haiti or the March 2011 earthquake and tsunami in Japan to underscore the often-illusory character of the control humans think they have over their physical surroundings.

ENVIRONMENTAL PERCEPTION

Another approach to the theme of nature–culture focuses on how humans perceive nature. Each person and cultural group has mental images of the physical environment, shaped by knowledge, ignorance, experience, values, and emotions. To describe such mental images, human geographers use the term **environmental perception.** Whereas the possibilist sees humankind as having a choice of different possibilities in a given physical setting, the environmental perceptionist declares that the choices people make will depend more on what they perceive the environment to be than on the actual character of the land. Perception, in turn, is colored by the teachings of culture.

environmental perception The belief that culture depends more on what people perceive the environment to be than on the actual character of the environment; perception, in turn, is colored by the teachings of culture.

natural hazard An inherent danger present in a given habitat, such as floods, hurricanes, volcanic eruptions, or earthquakes; often perceived differently by different peoples.

Some of the most productive research done by geographers who are environmental perceptionists has been on the topic of **natural hazards,** such as floods, hurricanes, volcanic eruptions, earthquakes, insect infestations, and droughts. Natural hazards are sometimes called "acts of God," to highlight that such disasters are beyond human control. A closer look, however, reveals that there is usually some human input into the making of disasters. Hazards are often well known, but economic interests, poverty, or political problems inhibit proper planning. For example, towns in the upper Mississippi River drainage basin were built close to the riverbanks to facilitate commerce, but were then vulnerable to the flooding common in the region. Sometimes hazards are known, but collectively forgotten because the time between disastrous occurrences is great and people become complacent. And sometimes human-made hazards

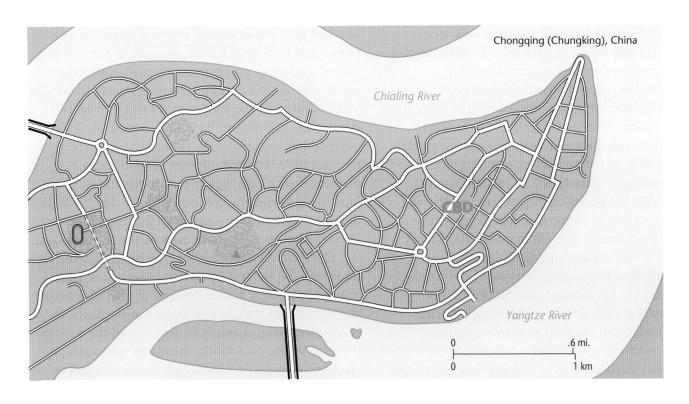

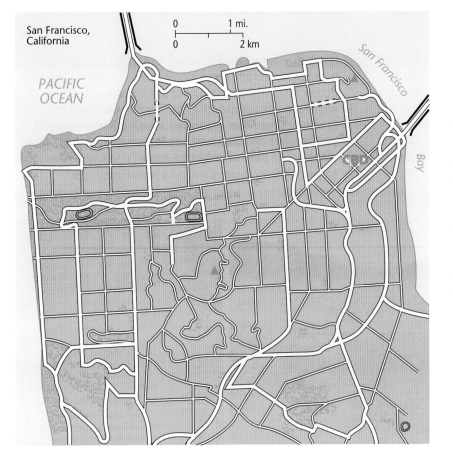

FIGURE 1.12 Chongqing (Chungking) and San Francisco. Both these cities are among the largest in their respective countries. Both developed on elongated, hilly sites flanked on all but one side by water, and both were connected in the twentieth century by bridges leading to adjacent land across the water. In certain other respects, too—such as the use of tunnels for arterial roads—the cities are similar. Note, however, the contrast in street patterns. In Chongqing, the streets were laid out to accommodate the rugged terrain, but in San Francisco, relatively little deviation from a gridiron pattern was permitted. Note, too, that although San Francisco has a much smaller population than Chongqing, it covers a far larger area. *What do these contrasts suggest about the relative merits of environmental determinism and possibilism? About the role of culture?*

combine with natural hazards to create even worse disasters. Human-made hazards include such things as pollution and contamination from industrial production. Thus, it is often difficult and even misguided to attempt to separate the natural from the human in understanding vulnerability to hazards.

Japan's so-called triple disaster illustrates the extent and complexity of nature–culture coupling inherent in natural hazards (**Figure 1.13**). On the afternoon of March 11, 2011, a strong earthquake occurred in the northwestern Pacific Ocean near Japan's coast. The resulting shock waves were felt in major cities across Japan, but government planning minimized the impacts. Because the country is prone to earthquakes, the government has invested in new construction technologies, enforced some of the world's strictest building codes, and installed an early earthquake warning system. These initiatives prevented property damage and saved many lives in Tokyo and other cities.

The earthquake, however, produced an 8-meter displacement in the seafloor, resulting in a tsunami (a wave produced by the displacement of a large volume of water). Within an hour of the initial earthquake, the tsunami began flooding coastal areas at heights ranging from 3 to 7 meters. Though the government also had installed a tsunami warning system, it was inadequate to deal with the magnitude of the flooding. Indeed, dozens of government evacuation shelters, which were supposedly out of harm's way, were flooded and destroyed. It was later revealed that previous generations of residents had marked the coastline with stone tablets warning of the tsunami hazard. Some markers were hundreds of years old and advised residents to not build below certain elevations. The Japanese government and most present-day residents did not heed these warnings, with the result that whole towns and ports were swept away by the 2011 tsunami.

The third disaster followed the tsunami, which destroyed the electrical power infrastructure and inundated nuclear power facilities along the coast. At the Fukushima Daiichi and Daini power plants, the waves swept over seawalls and destroyed the backup power systems. As a result, the plants were without electrical power and unable to maintain control of radioactive materials on site. Several explosions and radioactivity leakage followed, leading to the evacuation of over 200,000 people. As the nuclear reactors melted down, it became clear that the company in charge had failed to adequately plan for the tsunami hazard. The releases of radioactivity left

FIGURE 1.13   **Japan's 2011 "triple disaster."** On March 11, a very strong 9.0 magnitude earthquake occurred undersea off the Pacific Coast of Japan. This initiated the first part of the disaster. Shock waves traveled through the Earth's crust and caused widespread structural damage (left). The accompanying seafloor displacement generated a tsunami that swept away entire communities. All that remained of the coastal town of Minamisnriku was rubble (center). The tsunami also triggered the third part of the disaster: a meltdown and radiation release at the Fukushima Nuclear Power Plant (right). The power plant was not designed to withstand a tsunami or the flooding that damaged the reactors. ***Given that Japan has a well-known history of earthquakes and tsunamis, would you classify this "triple disaster" as natural or human-made?*** *(The Asahi Shimbury/Getty Images; AFP/Getty Images; Gamma-Rapho via Getty Images.)*

many areas uninhabitable, caused the closure of local fisheries, and spurred the government to abandon nuclear power as an energy source for the nation. Japan's triple disaster demonstrates how labeling a hazard "natural" or an "act of God" masks the human component present in all disasters.

In virtually all cultures, people knowingly inhabit hazard zones, especially floodplains, exposed coastal sites, drought-prone regions, and the environs of active volcanoes. In the United States, for example, every year more people live in areas subject to frequent hurricanes along the coast of the Gulf of Mexico and atop earthquake faults in California. How accurately do they perceive the hazard involved? Why have they chosen to live in such places? How might we minimize the eventual disasters? Human geographers seek the answer to these questions and aspires, with other experts, to mitigate the inevitable disasters through such strategies as land-use planning.

Perhaps the most fundamental expression of environmental perception lies in the way different cultures see nature itself. We must understand at the outset that nature is a culturally derived concept that has different meanings to different peoples. In the **organic view of nature,** people are part of nature. In many cultures around the globe, elements of the natural environment—trees, animals, landscape features, and so forth—are believed to be animated by various types of spirits. Thus, human lives are seen as intimately intertwined with natural phenomena. Many indigenous peoples around the world hold this view, but increasingly many environmental activists in Westernized cultures have adopted similar positions. By contrast, the dominant Western philosophical tradition offers a **mechanistic view of nature.** Humans are separate from and hold dominion over nature. This philosophy sees nature as an integrated system of mechanisms governed by external forces that can be rendered into natural laws and understood and controlled by people.

organic view of nature The view that humans are part of, not separate from, nature and that elements of the natural environment are animated by various types of spirits.

mechanistic view of nature The view that humans are separate from nature and hold dominion over it, and that nature is an integrated system of mechanisms governed by external forces which humans can render into natural laws and control.

HUMANS AS MODIFIERS OF THE EARTH

Many human geographers, observing the environmental changes people have wrought, emphasize humans as modifiers of the habitat. This presents yet another facet of the theme of nature–culture. In a sense, the human-as-modifier school of thought is the opposite of environmental determinism. Whereas the determinists proclaim that nature molds humankind, and possibilists believe that nature presents possibilities for people, those geographers who emphasize the human impact on the land assert that humans mold nature.

In addition to deliberate modifications of the Earth through such activities as mining, logging, and irrigation, we now know that even seemingly innocuous behavior, repeated for millennia, for centuries, or in some cases for mere decades, can have catastrophic effects on the environment. Plowing fields and grazing livestock can eventually denude regions (**Figure 1.14**).

FIGURE 1.14 **Human modification of the Earth includes severe soil erosion.** This erosion could have been caused by road building or poor farming methods. The scene is in the Amazon Basin of Brazil. ***How can we adopt less destructive ways of modifying the land?*** (Michael Nichols/National Geographic.)

FIGURE 1.15 **Terraced cultural landscape of an irrigated rice district in Indonesia.** In such areas, the artificial landscape made by people overwhelms nature and forms a human mosaic on the land. *Why is rice cultivated in such hilly areas in Asia, whereas in the United States rice farming is confined to flat plains?* (Nacivet/Photographer's Choice/Getty Images.)

The use of certain types of air conditioners or spray cans apparently has the potential to destroy the planet's very ability to support life. And the increasing release of fossil-fuel emissions from vehicles and factories—what are known as greenhouse gases—is leading to global warming, with potentially devastating effects on the Earth's environment (see Subject to Debate, page 29). Clearly, access to energy and technology is the key variable that controls the magnitude and speed of environmental alteration. Geographers seek to understand and explain the processes of environmental alteration as they vary from one culture to another and, through applied geography, to propose alternative, less destructive modes of behavior.

Gender differences can also play a role in the human modification of the Earth. **Ecofeminism,** a term derived from a book by Karen Warren, maintains that because of socialization, women have been better

> **ecofeminism** A doctrine proposing that women are inherently better environmental preservationists than men because of their traditional roles of creating and nurturing life, whereas the traditional roles of men too often necessitated death and destruction.

ecologists and environmentalists than men. Traditionally, women—as childbearers, gardeners, and nurturers of the family and home—dealt with the daily chores of gathering food from the Earth, whereas men—as hunters, fishers, warriors, and forest clearers—were involved with activities more associated with destruction. Regardless of whether we agree with this rather deterministic and essentializing viewpoint (i.e., understanding gender differences as biologically determined rather than culturally constructed), we can see that in many situations through time, and around the globe, women and men have had different relationships to the natural world.

CULTURAL LANDSCAPE

1.5 **LEARNING OBJECTIVE**
Explain the concept of cultural landscape.

Our fifth and final theme is **cultural landscape.** The human or cultural landscape is made up of all the built forms that cultural groups create in inhabiting the Earth— roads, agricultural fields, cities, houses, parks, gardens, commercial buildings, and so on.

> **cultural landscape** The visible human imprint on the land.

Every inhabited area has a cultural landscape, fashioned from the natural landscape, and each uniquely reflects the culture or cultures that created it (**Figure 1.15**). Landscape mirrors a culture's needs, values, and attitudes toward the Earth, and the human geographer can learn much about a group of people by carefully observing and studying the landscape. Indeed, so important is this visual record of cultures that some geographers regard landscape study as geography's central interest.

When the membership of the United Nations initiated the system of World Heritage Sites (see page 30), it was a clear signal that landscape is universally valued as an expression of culture and identity. World Heritage Sites simultaneously celebrate both the vast diversity of cultural achievements worldwide and also the singular impulse to creatively manipulate

HUMAN ACTIVITIES AND GLOBAL CLIMATE CHANGE

One of the most important and vexing scientific and political issues of the early twenty-first century is understanding the causes of recent global climate change and deciding what policies to pursue to lessen its effects. It's also an issue that human geographers, with their emphasis on understanding nature–culture relationships, are well prepared to discuss.

According to a report by the National Academies of the United States (a joint body made up of the National Academy of Science, the National Academy of Engineering, the Institute of Medicine, and the National Research Council), the Earth's temperatures are rising. Since the early twentieth century, the surface temperature of the Earth has risen 1.4°F, with the greatest increase occurring since 1978. Global climate, of course, is always changing. What is critical today, however, is the degree to which scientists have been able to correlate global warming trends with the rise in the levels of carbon dioxide in the atmosphere. Carbon dioxide is one of several greenhouse gases that keep radiative energy (and therefore warmth) trapped in the Earth's atmosphere. It occurs naturally in the atmosphere but is also released when fossil fuels such as coal, oil, and natural gas are burned. Changes in the levels of carbon dioxide in the atmosphere, therefore, lead to changes in the Earth's temperature. Changes in temperature, in turn, lead to other changes in the environment, such as the melting of glacial caps and the rising of sea levels.

To what degree are our activities—particularly our energy demands that lead to the burning of fossil fuels— responsible for this climate change? This is where the real debate starts. Some scientists are wary of pointing the finger at carbon dioxide emissions as the primary culprit, arguing along several lines that it is far from certain whether human activities have had such impacts on the Earth's climate. Some are critical of the data that show increases in surface temperature. Some believe that the recent fluctuation in climate is a far more natural occurrence than one induced by humans. And some believe that the Earth's atmospheric and climatic systems are simply so complex that it is premature to isolate one factor. Yet, there is growing worldwide consensus that human activities—particularly our reliance on the burning of fossil fuels—are the primary factors responsible for recent global warming. The Intergovernmental Panel on Climate Change (IPCC), a group of scientists from many different countries, concluded that because of the increase in greenhouse gases in the atmosphere, by 2100 average temperatures on the surface of the Earth are likely to rise between 2.5°F and 10.4°F above 1990 levels. The question for this group of scientists is not what is causing these changes but what to do about it. The first step, these scientists argue, is to find ways to decrease levels of carbon dioxide released by looking to new technologies and alternative energy sources. But because the changes in climate are already occurring, the second step is finding ways to deal with the effects of global warming.

Geography is a discipline well suited to addressing this debate because, as we have seen, one of its primary activities is understanding nature–culture relationships. Figuring out to what degree, how, and why human activities interact with our physical environment is clearly the heart of what is being disputed here.

Continuing the Debate

As geographers, we know that various cultures interact with the environment differently and have varied beliefs and ideas about the role of science in explaining physical phenomena. Keeping this all in mind, consider these questions:

- How might such cultural differences affect people's conclusions about the causes and effects of global climate change?

- How are your ideas about global climate change impacted by your position in the world?

Polar bear on breaking ice floe, North Pole. One of the many consequences of global climate change is the disappearance of habitats for many species, including the polar bear. *Are there any endangered habitats in your local region?* (© Yi Lu/Corbis.)

World Heritage Site

The Great Zimbabwe National Monument was constructed between 1100 and 1450 C.E. and encompasses over 700 hectares in a region populated by the Shona people of Zimbabwe. Placed on the list of World Heritage Sites in 1986, Great Zimbabwe contains the best example of African architecture south of the Sahara.

In the section known as the Great Enclosure, imposing dry masonry stone walls rise 11 meters, enclosing numerous dwellings and other stone structures. In addition to the Great Enclosure, two other distinct zones comprise the complex: the Hill Ruins and the Valley Ruins.

For nearly four centuries, the site served as the principle city of the Kingdom of Zimbabwe, housing 10,000 to 20,000 residents and functioning as the center of an international trading network that stretched across the Indian Ocean and, ultimately, to China. The kingdom ruled over a gold-rich plateau covering a portion of present-day Zimbabwe and Mozambique. Locally mined gold was smelted at the site into ingots for trade with coastal cities, which imported porcelain, glass beads, and other luxury goods.

Racist nineteenth-century theories of history prevented Europeans from recognizing the ruins as the grand achievement of an African civilization. They wildly speculated that ancient Phoenicians, Egyptians, and even the Queen of Sheba had built it. Even in the 1970s, the white-minority government of Zimbabwe (then Southern Rhodesia) officially denied the site's black African origins. Twentieth-century archaeological studies have confirmed that the Shona's ancestors built and continuously occupied the site for nearly four centuries.

In the context of the twentieth-century pan-African struggle against European colonialism and white-minority rule, the site came to serve as a key source of cultural pride among black Africans across the continent. It is symbolically important in the formation of both Shona ethnic identity and postcolonial Zimbabwean national identity. Indeed, the present-day country of Zimbabwe took its name from the site and adopted the Zimbabwean bird as its national symbol after gaining independence from white-minority rule in 1980.

Built between 1100 and 1450 C.E. by the Shona people of Zimbabwe, the site is a unique example of African architecture.

There are three main areas to the site: the Great Enclosure, the Hill Ruins, and the Valley Ruins.

Placed on the list of World Heritage Sites in 1986, the site has been visited by thousands of tourists, but tourism has slowed due to political and economic instability in the country.

www.whc.unesco.org/en/list/364

(Richard I'Anson/Lonely Planet Images/Getty Images.)

The Great Zimbabwe National Monument

THE GREAT ENCLOSURE: The most remarkable feature of the site is the Great Enclosure, an area dominated by a monumental, oval-shaped outer wall spanning 250 meters in length and 3 meters in width. The craftsmanship of this area is stunning, beginning with the skillful splitting of granite blocks excavated from a nearby quarry. The granite split easily along fracture planes, resulting in cubelike blocks that were stacked and fitted together after craftsmen carefully smoothed the surfaces. This technique resulted in a finished quality that rivals modern brick walls and allowed the stone city to be built without mortar, yet remain intact for centuries.

(Danita Delimont/Gallo Images/Getty Images.)

■ Cut stones set in herringbone and chevron patterns accent the top of the outer wall. Within the wall are a smaller stone enclosure and a narrow stone passageway leading to the beehive-shaped conical tower. The tower is constructed of solid stone blocks and assumed to have had a primarily symbolic or ceremonial purpose.

■ The enclosure once contained houses built of clay and gravel, of which only traces remain. These were organized into walled-off community areas composed of a kitchen, dwelling huts, and a court. One key theory suggests that the enclosure contained the royal residence of the kingdom's ruler. Thus, the Great Enclosure represents the seat of political and economic power for the expansive Zimbabwe kingdom.

TOURISM: The site is readily accessible and also relatively close to other attractions such as Victoria Falls (another World Heritage Site) and many of southern Africa's wildlife parks.

■ International tour companies include it on their packaged travel itineraries. Visitors freely roam the site and explore the very different perspectives and features offered by the Great Enclosure, the Hill Ruins, and Valley Ruins.

■ Political and economic instability in the country, however, has resulted in a significant drop-off in tourist activity in Zimbabwe. Although ruinous to the local economy, the lull in tourist traffic may be giving this archeological wonder a needed respite from overuse.

(Colin Hoskins/Alamy.)

► **FIGURE 1.16** **Chinatown, San Francisco.** Gates are symbolically important in denoting the entry into sacred space in traditional Chinese culture; they are often used in American cities like San Francisco to mark a similar transition into a neighborhood dominated by immigrant Chinese. Most of our landscapes are dotted with such cultural symbols. ***Can you identify some in your neighborhood?*** *(© Ron Niebrugge/Alamy.)*

pledge precious government resources to forever protect landscapes within their borders and to respond with emergency assistance when threats to sites arise in any other member state.

Why is such importance attached to the human landscape? Perhaps part of the answer is that it visually reflects the most basic strivings of humankind: shelter, food, and clothing. In addition, the cultural landscape reveals people's different attitudes toward the modification of the Earth. It also contains valuable evidence about the origin, spread, and development of cultures because it usually preserves various types of archaic forms. Dominant and alternative cultures use, alter, and manipulate landscapes to express their diverse identities (**Figure 1.16**).

the material environment that unites all humanity. Since the program's establishment, countries have lobbied hard to receive World Heritage Site status for their landscapes and take pride in the global recognition of local cultural achievement. Members

FIGURE 1.17 **Landscape triptych panel by Fra Angelico, fifteenth century.** Notice the depiction of the beautiful and orderly agricultural landscape outside the city walls but the absence of people actually doing the work to maintain that order. ***Why aren't the laborers depicted?*** *(Archivo Iconografico, S.A./Corbis.)*

FIGURE 1.18 **Yokohama at dusk.** This skyline is a powerful symbol of the economic importance of the world's largest city, Tokyo-Yokohama. ***What landscape form best symbolizes your town or city?*** *(Jose Fuste Raga/Corbis.)*

Aside from containing archaic forms, landscapes also convey revealing messages about their present-day inhabitants and cultures. According to geographer Pierce Lewis, "The cultural landscape is our collective and revealing autobiography, reflecting our tastes, values, aspirations, and fears in tangible forms." Cultural landscapes offer "texts" that geographers read to discover dominant ideas and prevailing practices within a culture as well as less dominant and alternative forms within it. This "reading," however, is often a very difficult task, given the complexity of cultures, cultural change, and recent globalizing trends that can obscure local histories.

Geographers have pushed the idea of "reading" the landscape further to focus on the symbolic and ideological qualities of landscape. In fact, as geographer Denis Cosgrove has suggested, the very idea of landscape itself was ideological, in that its development in the Renaissance served the interests of the new elite class for whom agricultural land was valued not for its productivity but for its use as a visual subject. Land, in other words, was important to look at as a scene, and the actual activities necessary for agriculture were thus hidden from these views. If you go to an art museum, for example, it will be difficult, if not impossible, to find in the Italian Renaissance room any landscape paintings that depict agricultural laborers (**Figure 1.17**).

Closer to home, we need only to look outside our windows to see other symbolic and ideological landscapes. One of the most familiar and obvious **symbolic landscapes** is the urban skyline. Composed of tall buildings that normally house financial service industries, it represents the power and dominance of finance and economics within that culture (**Figure 1.18**). However, other cities are dominated by tall structures that have little to do with economics but more with religion. In medieval Europe, for example, cathedrals and churches rose high above other buildings, symbolizing the centrality and dominance of Catholicism in this culture (**Figure 1.19**).

Even the most mundane landscape element can be interpreted as symbolic and ideological. The typical, middle-class, suburban, American or Canadian home, for example, expresses a dominant set of ideas about culture and family structure (**Figure 1.20**). These homes often include a living room, dining room, kitchen, and bedrooms, all separated by walls. The cultural assumptions built into this division of space

> **symbolic landscapes** Landscapes that express the values, beliefs, and meanings of a particular culture

FIGURE 1.19 **Prague's skyline is dominated by church spires.** Here, St. Vitus Cathedral sits majestically overlooking the Vltava River and the Small Town. *What other skylines of the world might include a large number of church spires?* (Courtesy of Mona Domosh.)

FIGURE 1.20 **American ranch house.** The horizontally expansive ranch is a common house form in the United States and Canada. *Can you think of other countries where ranch houses are also common?* (iStockphoto/Thinkstock/Getty Images.)

are the assumed value of individual privacy (everyone has his or her own bedroom); the idea that certain functions should be spatially separate from others (cooking, eating, socializing, sleeping); and the notion that a family is composed of a mother, father, and children (indicated by the "master" bedroom and smaller "children's" bedrooms). Thus, even the most common of landscapes can be seen as symbolic of a particular culture and built from ideological assumptions.

As we have seen, the physical content of the cultural landscape is both varied and complex. To better study and understand these complexities, geographical studies focus on three principal aspects of landscape: settlement forms, land-division patterns, and architectural styles. In the study of **settlement forms,** human geographers describe and explain the spatial arrangement of buildings, roads, and other features that people construct while inhabiting an area. One of the most basic ways in which geographers categorize settlement forms is to examine their degree of **nucleation,** a term that refers to the relative density of landscape elements. Urban centers are of course very nucleated, whereas rural farming areas tend to be much less nucleated, what geographers call **dispersed.** Another common way to think about settlement forms is the degree to which they appear standardized and planned, such as the grid form of much of the American West (**Figure 1.21**), versus the degree to which the forms appear to be organic, that is, built without any apparent geometric plan, such as the central areas of most European cities. Thinking about settlement forms in terms of these two basic categories helps geographers begin their analysis of the relationships between cultures and the landscapes they produce.

Land-division patterns indicate the uses of particular parcels of land and, as such, reveal the way people have divided the land for economic, social, and political uses. Within a particular nucleated settlement form—a city, for example—you can see different patterns of land use. Some areas are devoted

> **settlement forms** The spatial arrangement of buildings, roads, towns, and other features that people construct while inhabiting an area.
>
> **nucleation** A relatively dense settlement form.
>
> **dispersed** A type of settlement form in which people live relatively distant from each other.

> **land-division patterns** Refers to the spatial patterns of different land uses.

to economic uses, others to residential, political (city hall, for example), social, and cultural uses. Each of these areas can be further divided. Economic uses may include offices for financial services, retail stores, warehouses, and factories. Residential areas are often divided into middle-class, upper-class, and working-class districts and/or are grouped by ethnicity and race (see Chapter 10). Such patterns, of course, vary a great deal from place to place and culture to culture, as we will see throughout this book. One of the best ways to glimpse settlement and land-division patterns is through an airplane window. Looking down, you can see the multicolored abstract patterns of planted fields, as vivid as any modern painting, and the regular checkerboard or chaotic tangle of urban streets.

Perhaps no other aspect of the human landscape is as readily visible from ground level as the architectural style of a culture. Geographers look at the exterior materials and decoration, as well as the layout and design of the interiors. Styles tend to vary both through time, as cultures change, and across space, in the sense that different cultures adopt and invent their own stylistic detailing according to their own particular needs, aesthetics, and desires. Thus, examining architectural style is often useful when trying to date a particular landscape element or to understand the particular values and beliefs that cultures may hold. In North American culture,

FIGURE 1.21 **The town of Westmoreland in the Imperial Valley of California.** It's difficult to find a more regularized, geometric land pattern than this. ***Why do you think much of the American West was divided into these rectangles?*** *(Jim Wark/Airphoto.)*

different building styles catch the eye: modest white New England churches and giant urban cathedrals, hand-hewn barns and geodesic domes, wooden one-room schoolhouses and the new windowless school buildings of urban areas, shopping malls and glass office buildings. Each tells us something about the people who designed, built, or inhabit these spaces. Thus, architecture provides a vivid record of the resident culture (**Figure 1.22**). For this reason, cultural geographers have traditionally devoted considerable attention to examining architecture and style in the cultural landscape.

CONCLUSION

As we have seen and will continue to see, the interests of human geographers are diverse. It might seem to you, confronted by the various themes, subject matter, viewpoints, and methodologies described in this chapter, that geographers run off in all directions, lacking unity of purpose. What does a geographer who studies architecture have in common with a colleague who studies the political and cultural causes of environmental degradation? What interests do an environmental perceptionist and a student of diffusion share? Why do scholars with such

apparently different interests belong in the same academic discipline? Why are they all geographers?

The answer is that regardless of the particular topic the human geographer studies, he or she necessarily touches on several or all of the five themes we have discussed. The themes are closely related segments of a whole. Spatial patterns in culture, as revealed by maps of regions, are reflected in and expressed through the cultural landscape, require an ecological interpretation, are the result of mobility, and are inextricably linked with globalization.

As an example of how the various themes of human geography overlap and intertwine, let's look at one element of architecture that most North Americans will be familiar with: the ranch-style, single-family house (Figure 1.20). This house type is defined by its one-story height and its linear form. Ranch houses may be found throughout much of the United States and to a lesser extent in Canada, though they are rare in other countries. They are obviously part of the cultural landscape, and their spatial distribution constitutes a formal region that can be mapped.

Geographers who study such houses also need to employ the other themes of human geography to gain a complete understanding. They can use the concept of diffusion to learn when and by what routes this building style emerged and diffused, and what

FIGURE 1.22 **Architecture as a reflection of culture.** This log house (left), near Ottawa in Canada, is a folk dwelling and stands in sharp contrast to the professional architecture of the Toronto skyline (right). ***What conclusion might a perceptive person from another culture reach (considering*** the "virtues" of height, durability, and centrality) about the ideology of the culture that produced the Toronto landscape?* (Left: Courtesy of Terry G. Jordan-Bychkov; Right: Michael Mahovitch/Radius Images/Getty Images.)

GEOGRAPHY @ WORK

Matt Rosenberg. *(Courtesy of Matt Rosenberg.)*

Matt Rosenberg
Geography Expert for About.com

Education:
BA Geography—University of California, Davis
MA Geography—California State University, Northridge

Q. *Why did you major in geography and decide to pursue a career in the field?*

A. I "stumbled" into the field after taking a course on urban and economic geography my freshman year in college and loving it. I like that geography is a wide-ranging discipline that requires students to be well versed in a variety of topics; ultimately the study produces well-rounded individuals (which I think employers especially appreciate).

Q. *Please describe your job.*

A. I think of myself as a "geographical research library." I distill research, offer guidance, and answer questions related to all aspects of the geography field. I also serve as editor-in-chief for the About.com Geography site and write a regular blog featured on the site.

Q. *What types of people contact you via the web site?*

A. I get a lot of high school and college students from around the world that pose questions based on their studies. I also hear from plenty of "armchair geographers" who ask a variety of questions, such as "Which countries are not competing in the Olympic games?" "What is the newest nation in the world?" and "Why are there climate extremes in the world?"

Q. *What advice do you have for college students considering a career in geography?*

A. I think it is very important to do one (or more) internships while in college to help you figure out what area of geography you want to work within. When I was in college, I interned at a city planning department and a GIS department, which helped me decide what I wanted to do in the field.

---

barriers hindered its diffusion. In this particular case, geographers would be led back to the early years of the twentieth century, when the first suburban houses were built outside of urban centers. In this case, land was relatively inexpensive, allowing for a house type that occupied a wide expanse of space. What's more, they would learn that a dominant design motif in the United States in the early twentieth century was based on the notion that buildings should fit in with their natural surroundings instead of dominate them. As a result, low-slung housing styles like the ranch house were particularly popular. Further, the geographer would need an ecological interpretation of the ranch house. What materials were required to build such a house? Did the style vary across the different climatic and ecological regions in which such houses were built? Finally, the human geographer would want to know how the use of ranch houses was related to globalization. Did economic changes in the world raise the standard of living, leading people to accept ranch houses? Did changes in technology lead to more elaborate houses? Why did it become the quintessential house type in post–World War II America, featured in many of the popular TV shows

DOING GEOGRAPHY

Space, Place, and Knowing Your Way Around

We started this chapter by saying that most of us are born geographers, with a sense of curiosity about the places and spaces around us. Think, for example, of the place you call home. You are probably familiar enough with its streets and buildings and greenspaces, and with the people who inhabit these spaces, to make connections among them—you know how to "read" the place. There are many other places, however, where this is not the case. Most of you have had the experience of going somewhere new, where it is difficult to find your way around, literally and metaphorically. Many of you, for example, are attending a college or university far from home, whereas others have experienced the disorienting feeling of moving from one home to another, whether across town or across continents. How did you find your way? How did the "strange" space become a familiar place?

This activity requires you to draw on your own experiences to understand two fundamental concepts in human geography: space and place. Geographers tend to use the term *space* in a much more abstract way than *place*—as a term that describes a two-dimensional location on a map. *Place,* however, is a less dry term, one used to describe a location that has meaning. Your college campus, for example, may have been simply an abstract space located on a map when you applied to the school, yet now it is a place because you have filled it with your own meanings. Here are the four steps in your activity.

Steps to Understanding Space, Place, and Knowing Your Way Around

Step 1: Draw on your own experiences by picking one particular space that has become a place for you.

Step 2: Identify what you knew about the space beforehand and how you knew this (e.g., perhaps you looked up the place in an atlas or saw it in a movie).

Step 3: In narrative form, describe the process whereby that space became a place for you.

Step 4: Identify the ways in which you learned how to "read" this place. In other words, how did you learn to see this particular location as a three-dimensional, meaningful place that is different from what you knew of it as a "space"?

Use your experiences to answer the following questions:

- Do you have to live in a place to really "know" it?
- Do you have to experience a space for it to become a place?

Father and son in front of college dorm. For many young people, the transition to college is the first time they have had to find their own way and create homes for themselves. *(Fuse/Getty Images.)*

of the time? Do these humble structures possess a symbolism related to traditional American values and virtues? Thus, the geographer interested in housing is firmly bound by the total fabric of human geography, unable to segregate a particular topic such as ranch houses from the geographical whole. Region, cultural landscape, nature–culture relationships, mobility, and globalization are interwoven.

In this manner, the human geographer passes from one theme to another, demonstrating the holistic nature of the discipline. In no small measure, it is this holism—this broad, multithematic approach—that distinguishes the human geographer from other students of culture. We believe that, by the end of the course, you will have gained a new perspective on the Earth as the home of humankind.

SEEING GEOGRAPHY
Navigational stick chart of the Marshall Islands
Why is it difficult for most of us to interpret this image as a map?

A stick chart used by Micronesian mariners to navigate the Marshall Islands in Micronesia. *(Courtesy of Roderick Neumann.)*

As the title indicates, this is a stick chart traditionally used by Micronesian mariners to navigate the Marshall Islands in the North Pacific Ocean. Like all maps, it is a two-dimensional rendering of three-dimensional space. In other words, it is an attempt to represent surface features and location on a flat plane. All cultures devise certain symbols that allow for these representations to be understood. On United States Geological Survey (USGS) maps, for example, a standardized set of symbols represent such things as roads, rivers, and cities. Similarly, this map is filled with symbols that represent such features as wave patterns (bent sticks), ocean currents (straight sticks), and islands (the cowry shells).

As we've learned throughout this chapter, different cultures create and experience landscapes differently. Here, we can see that different cultures also represent their landscapes differently. This image is part of a long tradition of Micronesian peoples' representations that reflect their particular culture and their views of the seascapes and landscapes they occupy. In general, these representations demonstrate that there is no universal system for finding one's way across the Earth's surface. Unlike Western sailors who navigated using the relative positions of celestial bodies, Micronesian sailors navigated by sensing the movement of the ocean's surface. Hence, these charts do not represent standardized measurements such as degrees of longitude and latitude or miles and kilometers. Instead, they use patterns of ocean swells and currents to help navigators position themselves on the Earth's surface.

To most Western eyes, then, this image does not look like a map because we neither recognize the symbols that Micronesian mariners used to translate three-dimensional spaces into two dimensions, nor think of maps as meaningful in and of themselves. Most likely, Western maps would have looked very odd to Micronesian eyes. One of the goals of studying human geography, as we will see throughout this book, is to appreciate this diversity of relationships between peoples and the places they inhabit, shape, and represent.

Chapter 1
LEARNING OBJECTIVES REEXAMINED

1.1 Identify and define different types of regions.
See pages 7–12. What are the three types of regions recognized by geographers?

1.2 Define mobility and name five types of diffusion.
See pages 12–18. Name five types of diffusion. How are these types of diffusion different from and related to each other?

1.3 Describe globalization and explain its importance to cultural geography.
See pages 18–21. What role does globalization play in the shaping of cultures, economies, and societies?

1.4 Define the different models of nature–culture relations.
See pages 21–28. Name three models of nature–culture relations. In your opinion, which of these models provides the clearest explanation of the relationships between people and the physical environment? Why?

1.5 Explain the concept of cultural landscape.
See pages 28–35. How are peoples' interactions with nature expressed in the landscape?

KEY TERMS

absorbing barrier	p. 14	environmental	
border zones	p. 9	perception	p. 24
circulation	p. 16	expansion diffusion	p. 13
contagious diffusion	p. 14	formal region	p. 8
core-periphery	p. 9	functional region	p. 9
cultural ecology	p. 23	geography	p. 2
cultural landscape	p. 28	globalization	p. 18
cultural practices	p. 3	Great Migration	p. 17
culture	p. 2	hierarchical diffusion	p. 14
diffusion	p. 12	human geography	p. 2
dispersed	p. 34	independent invention	p. 13
ecofeminism	p. 28	interdependence	p. 19
environmental		internal migration	p. 17
determinism	p. 23	international migration	p. 16
land-division patterns	p. 34	region	p. 1
mechanistic view of		relocation diffusion	p. 13
nature	p. 27	return migration	p. 17
migrations	p. 16	seasonal migration	p. 17
mobility	p. 12	settlement forms	p. 34
model	p. 5	space	p. 5
natural hazard	p. 24	stepwise migration	p. 17
nature–culture	p. 21	stimulus diffusion	p. 14
node	p. 9	symbolic landscapes	p. 33
nucleation	p. 34	time–distance decay	p. 14
organic view of nature	p. 27	transnational	
permeable barrier	p. 15	migration	p. 17
physical environment	p. 4	uneven development	p. 19
place	p. 6	vernacular region	p. 10
possibilism	p. 24	World Heritage Sites	p. 6

Geography on the Internet

You can learn more about the discipline of geography and the subdiscipline of human geography on the Internet at the following web sites:

American Geographical Society
http:// www.amergeog.org/
America's oldest geographical organization, with a long and distinguished record; publisher of the *Geographical Review*.

Association of American Geographers
http:// www.aag.org/
The leading organization of professional geographers in the United States. This site contains information about the discipline, the association, and its activities, including annual and regional meetings.

National Geographic Society
http:// www.nationalgeographic.com/
An organization that has, for more than a century, served to popularize geography with active programs of publishing and television presentations prepared for the public.

Royal Geographic Society/Institute of British Geographers
http:// www.rgs.org/
Explore the activities of these allied British organizations, whose collective history goes back to the Age of Exploration and Discovery in the 1800s . . . and don't forget to visit the *Contemporary Human Geography* LaunchPad at: http://www.macmillanhighered.com/launchpad/DomoshCHG1e.

Sources

Anderson, Kay, and Fay Gale. 1999. *Cultural Geographies.* Melbourne: Pearson Education.

Blaut, James M. 1977. "Two Views of Diffusion." *Annals of the Association of American Geographers* 67: 343–349.

Foster, Jeremy. 2008. *Washed with Sun: Landscape and the Making of White South Africa.* Pittsburgh, Pa.: University of Pittsburgh Press.

Gould, Peter. 1993. *The Slow Plague: A Geography of the AIDS Pandemic.* Cambridge, U.K.: Blackwell.

Griffin, Ernst, and Larry Ford. 1980. "A Model of Latin American City Structure." *Geographical Review* 70: 397–422.

Hägerstrand, Torsten. 1967. *Innovation Diffusion as a Spatial Process.* Allan Pred (trans.). Chicago: University of Chicago Press.

Johnston, R. J., Peter Taylor, and Michael Watts (eds.). 2002. *Geographies of Global Change: Remapping the World.* 2nd ed. Oxford: Blackwell.

Lewis, Peirce. 1983. "Learning from Looking: Geographic and Other Writing About the American Cultural Landscape." *American Quarterly* 35: 242–261.

Norwine, Jim, and Thomas D. Anderson. 1980. *Geography as Human Ecology?* Lanham, Md.: University Press of America.

Ormrod, Richard K. 1990. "Local Context and Innovation Diffusion in a Well-Connected World." *Economic Geography* 66: 109–122.

Relph, Edward. 1981. *Rational Landscapes and Humanistic Geography.* New York: Barnes & Noble.

Sparke, Matthew. 2013. *Introducing Globalization: Ties, Tensions, and Uneven Integration.* Oxford: John Wiley & Sons.

Staudt, Amanda, Nancy Huddleston, and Sandi Rudenstein. 2006. *Understanding and Responding to Climate Change, a Report Prepared by the National Research Council based on National Academies Reports.* Washington, D.C.: National Academy of Science.

Tuan, Yi-Fu. 1974. *Topophilia: A Study of Environmental Perception, Attitudes, and Values.* Englewood Cliffs, N.J.: Prentice-Hall.

Warren, Karen J. (ed.). 1997. *Ecofeminism: Women, Culture, Nature.* Bloomington: Indiana University Press.

Ten Recommended Books on a Cultural Approach to Human Geography

(For additional suggested readings, see the *Contemporary Human Geography* LaunchPad at: http://www.macmillanhighered.com/launchpad/DomoshCHG1e.)

Anderson, Kay, Mona Domosh, Steve Pile, and Nigel Thrift (eds.). 2003. *Handbook of Cultural Geography.* London: SAGE. An edited collection of essays that push the boundaries of cultural geography into such sub-disciplines as economic, social, and political geography.

Blunt, Alison, Pyrs Gruffudd, Jon May, Miles Ogborn, and David Pinder (eds.). 2003. *Cultural Geography in Practice.* London: Arnold. An edited collection of essays that take a very practical view of what it means to actually conduct research in the field of cultural geography.

Cosgrove, Denis. 1998. *Social Formation and Symbolic Landscape.* Madison: University of Wisconsin Press. The landmark study that outlines the relationships between the idea of landscape and social and class formation in such places as Italy, England, and the United States.

Cosgrove, Denis, and Stephen Daniels (eds.). 1990. *The Iconography of Landscape: Essays on the Symbolic Representation, Design and Use of Past Environments.* Cambridge, U.K.: Cambridge University Press. An important collection of essays that lay the foundation for an ideological reading of landscape.

Foote, Kenneth E., Peter J. Hugill, Kent Mathewson, and Jonathan M. Smith (eds.). 1994. *Re-Reading Cultural Geography.* Austin: University of Texas Press. A beautifully compiled representative collection of some of the best works in American cultural geography at the end of the twentieth century, and a useful companion to the book edited by Wagner and Mikesell.

Mitchell, Don. 1999. *Cultural Geography: A Critical Introduction.* New York: Blackwell. An introductory text on cultural geography that emphasizes the material and political elements of the discipline.

Oakes, Timothy, and Patricia Price (eds.). 2008. *The Cultural Geography Reader.* New York: Routledge. A set of landmark statements on cultural geography introduced by the editors' explanatory and contextualizing essays.

Radcliffe, Sarah (ed.). 2006. *Culture and Development in a Globalizing World: Geographies, Actors, and Paradigms.* London: Routledge. A series of essays that provide case studies from around the world showing the various ways in which culture and economic development are integrally related.

Tuan, Yi-Fu. 1974. *Topophilia: A Study of Environmental Perception, Attitudes, and Values.* Englewood Cliffs, N.J.: Prentice-Hall. A Chinese-born geographer's innovative and imaginative look at people's attachment to place, a central concern of the cultural approach to human geography.

Wagner, Philip L., and Marvin W. Mikesell (eds.). 1962. *Readings in Cultural Geography.* Chicago: University of Chicago Press. A classic collection, edited by two distinguished Berkeley-trained cultural geographers, presenting the subdiscipline as it existed in the mid-twentieth century and developing the device of five themes.

Journals in Human Geography

Annals of the Association of American Geographers. Volume 1 was published in 1911. The leading scholarly journal of American geographers.

Cultural Geographies (formerly known as *Ecumene*). Volume I was published in 1994.

Geographical Review. Published by the American Geographical Society. Volume 1 was published in 1916.

Journal of Cultural Geography. Published semiannually by the Department of Geography, Oklahoma State University, Stillwater, Oklahoma. Volume 1 was published in 1980.

Progress in Human Geography. A quarterly journal providing critical appraisal of developments and trends in the discipline. Volume 1 was published in 1977.

Social and Cultural Geography. Volume 1 was published in 2000 by Routledge, Taylor, & Francis in Great Britain.

What can this scene tell us about nature–culture relations in North American popular culture?

Enjoying nature in a national park campground.
(Courtesy of Roderick Neumann.)

Go to "Seeing Geography" on page 89 to learn more about this image.

MANY WORLDS
Geographies of Cultural Difference

No matter where we live, if you look carefully, you will see how important cultural identity is to our daily lives. Cultural difference is evident everywhere—not only in the geographic distribution of different cultures but also in the way that difference is created or reinforced by geography. For example, in the United States, the historical legacy of segregating "Whites" from "Blacks" has been important in establishing and maintaining cultural differences between these groups (**Figure 2.1**).

In Chapter 1, we noted that human geographers are interested in studying the geographic differences both among and within cultures. For example, using the concept of formal region (a region inhabited by people who have one or more cultural traits in common), we can identify and map differences among cultures. This sort of analysis is usually done on a very large geographic scale, such as a continent or even the entire world. But, geographers are also interested in analyses at smaller scales. When we look more closely at a formal culture region, we begin to see that differences appear along racial, religious, gender-related, and other lines of distinction. Sometimes groups within a dominant culture become distinctive enough that we label them **subcultures**. These can be the result of resistance to the dominant culture or of a separate religious, ethnic, or national group forming its own distinct community within a larger culture (**Figure 2.2**).

subcultures Groups of people with norms, values, and material practices that differentiate them from the dominant culture to which they belong.

We also noted in Chapter 1 that there are many ways of defining and categorizing cultures. In this chapter, we will begin to explore the vast range of geographies of culture difference. The term *difference* implies a relationship and a set of criteria for

LEARNING OBJECTIVES

2.1 Explain the role of cultural difference in shaping regions.

2.2 Describe the ways that mobility interacts with cultural difference.

2.3 Analyze the potential for globalizing processes to shape and be altered by cultural difference.

2.4 Identify the influence of culture on physical geography and physical geography on culture.

2.5 Recognize how landscapes reflect cultural differences.

▲ FIGURE 2.1 **Greyhound bus station in 1943 Memphis, Tennessee, with separate facilities for Blacks and Whites.** It exemplifies the racial segregation found historically throughout much of the United States. Though such practices have long been outlawed, historic patterns of segregation may be found in the landscape today. *(Buyenlarge/Getty Images.)*

▲ FIGURE 2.2 **Kuala Lumpur's "Chinatown" at dusk.** Such ethnic enclaves may be found in cities around the world. *(Jeremy Horner/LightRocket via Getty Images.)*

comparison and assessment. That is, cultures are defined as they relate to each other using a predetermined set of characteristics. But, what does it mean to speak of *geographies* in the plural? Isn't there only one geography? The plural form emphasizes that there is no single way of seeing and experiencing place and landscape. Recall from Chapter 1 our discussion on the concept of personal experience in the sense of place, which emphasizes the multiplicity of meanings versus a single, universally shared meaning. Cultural geography studies have shown, for example, that women and men often experience the same places in different ways. A certain street corner or tavern might be a comfortable and familiar hangout for men but a threatening or uncomfortable zone that women avoid (see Seeing Geography, Chapter 3). To speak of geographies, then, is to go beyond the idea of a single way of looking at the world and raise new questions about the different meanings that people give to places and landscapes: how these relate to their sense of self and belonging; and how, in multicultural societies, we deal with these different meanings politically and socially.

MANY CULTURES

We classify cultures using many different criteria, both material and nonmaterial. **Material culture** includes the physical objects made and/or used by members of a cultural group: buildings, furniture, clothing, artwork, musical instruments, and so forth. The elements of material culture are visible. **Nonmaterial culture** includes the wide range of beliefs,

■ **material culture**
All physical, tangible objects made and used by members of a cultural group, such as clothing, buildings, tools and utensils, instruments, furniture, and artwork; the visible aspect of culture.

nonmaterial culture The wide range of tales, songs, lore, beliefs, values, and customs that pass from generation to generation as part of an oral or written tradition.

values, myths, and symbolic meanings passed from generation to generation of a given society. Often it is difficult to draw clear lines between material and nonmaterial features of culture. Food, for example, is material, but it is also charged with symbolic meaning and valued differently in different cultures (**Figure 2.3**). Many Westerners would gag at the mere thought of eating the worms and insects that are the daily food staples and even much valued delicacies in non-Western, rural cultures around the world. Therefore, cultures can be thought of in terms of both material and nonmaterial characteristics.

Let's look briefly at how these material and nonmaterial characteristics were first used to identify, categorize, and map cultures. In eighteenth-century Europe and North America, people first began to speak of "cultures" in the plural form. Specifically, following Europe's Age of Exploration and Discovery (when Europeans began to explore the world by sea), they started thinking about "European culture" in relation to other cultures they encountered around the world. Although now discredited as ethnocentric and racist, hierarchical models—in which European culture was ranked above other cultures—were invented to categorize and rank cultures. Religion and technology were key criteria in constructing these hierarchies, with Christian and technologically developed cultures ranked at the top and non-Christian, nonindustrial cultures ranked at the bottom.

Folk Culture As Europe and North America industrialized and urbanized in the nineteenth century, people began thinking in terms of internal cultural differences between groups. Urban city dwellers started to view rural country spaces as inhabited by distinct cultures. Thus, a new term, **folk culture,** was invented to distinguish traditional ways of life in rural and agricultural spaces from ways of life in the new urban and industrial spaces. When Europe's kingdoms and empires fragmented into new countries throughout the nineteenth century and into the twentieth, the

folk culture Rural, unified, largely self-sufficient groups that share similar customs and ethnicity. In terms of material culture, many items of daily use such as clothing, furniture, and housing are handmade, often from raw materials found locally. Most food is grown and consumed locally.

FIGURE 2.3 **Would you eat these?** Food preferences are strong markers of culture difference. These water beetles are a common food source in Southeast Asia. *(Graham Day/Getty Images.)*

urbanized, cosmopolitan elites looked to the rural folk as the source of distinctive national identities (**Figure 2.4**). Similarly, as the United States urbanized and expanded westward, rural folk were seen as the

FIGURE 2.4 **Young men in Bavaria dance the "Schuhplattler,"** one of the traditional folk dances of Germany. The revival of such traditional folk dances by urban, educated classes has been important to the formation of modern national identities. *(Getty Images.)*

national culture The controversial idea that citizens possess a set of recognizable values, behaviors, and beliefs—often including the same ethnic and linguistic traits—that express the core culture of each modern nation.

source of an American national character defined by self-reliance, perseverance, and a pioneering spirit. Hence, the cowboy—an economically minor and short-lived rural occupation in the nineteenth-century western United States—came to symbolically embody the imagined American national character. Folk culture and **national culture** thus have been closely associated. The term *national culture*, sometimes regarded as a controversial idea, suggests that a country's population possesses a set of recognizable characteristics or traits—often including the same ethnic and linguistic traits—that express the core culture of each modern nation. Hence, many countries have museums and institutes dedicated to the celebration of folk culture and its relevance to contemporary national identity (**Figure 2.5**).

Folk culture is defined as rural, unified, largely self-sufficient groups that share similar customs and ethnicity. In terms of material culture, many items of daily use such as clothing, furniture, and housing are handmade, often from raw materials found locally. Most food is grown and consumed locally, which

FIGURE 2.5 **National Folk Museum of South Korea, Seoul.** While folk cultures have largely disappeared in industrialized and urbanized countries, they remain preserved in museums as a source of national pride. *(Jane Sweeney/AWL Images/Getty Images.)*

often has resulted in the formation of distinctive regional cuisines, such as the Japanese diet of rice and raw fish or the Mesoamerican diet of maize, beans, and spicy peppers. In terms of nonmaterial culture, folk cultures typically have strong family or clan structures and highly localized rituals, such as an annual blessing of water wells and springs or harvest dances. Generations of gifted individuals within folk communities create songs, musical styles, and storytelling that are unique to their region. As with food, folk cultures often produce distinctive, highly regionalized musical genres, Appalachian bluegrass music being one well-known example. In Europe and North America, folk cultures, strictly defined, no longer exist, though many material and nonmaterial traces can be found in the landscape. Moreover, the reasons for the creation of folk cultures —human inventiveness, curiosity, and need, uniquely expressed in specific places—endure, as we will illustrate throughout this textbook.

Indigenous Culture Today, geographers frequently use terms such as *local, place-based,* or *indigenous* when referring to some of the distinctive cultural traits typically associated with folk culture. Indeed, the concept of **indigenous culture** has emerged as a focus of study in cultural geography. A simple definition of *indigenous* is "native" or "of native origin." In the modern world of independent countries, the word has acquired much greater cultural and political meanings. The International Labour Organization's (ILO) Indigenous and Tribal Peoples Convention 169 (Article 1.1) presents a legal definition that recognizes indigenous peoples as comprising a distinct culture. According to the ILO, indigenous peoples are self-identified tribal peoples whose social, cultural, and economic conditions distinguish them from the national society of the country in which they live (**Figure 2.6**). As with folk cultures, indigenous cultures are generally associated with nonurban, rural spaces, particularly the isolated forests and mountain ranges of the world.

Indigenous peoples are regarded as descending from peoples present in a territory at the time of conquest or colonization by an external empire or

indigenous culture A culture group that constitutes the original inhabitants of a territory, distinct from the dominant national culture, which is often derived from colonial occupation.

FIGURE 2.6 **A Karen of Northern Thailand works on roof thatching.** The Karen and other indigenous, tribal peoples around the world have maintained cultures distinct from the dominant national cultures of their home countries. *(Nicholas J. Reid/The Image Bank/Getty Images.)*

are, in effect, those peoples who were colonized—mostly, but not exclusively—by European cultures and are now minorities in their homelands. Whereas folk culture was seen as providing the foundation for a national culture, indigenous culture is usually positioned in opposition to national culture.

FROM FOLK CULTURE TO POPULAR CULTURE

The notion of folk cultures emerged historically when urban elites began to fear that the ways of traditional, rural life were being erased and replaced by the rapid spread of **popular culture.** Popular culture refers to modern ways of life associated with the rise of mass-produced, machine-made goods and the invention of long-distance communication technologies (**Figure 2.7**). People sometimes use the terms *mass culture* and *consumer culture* to mean the same thing as popular culture. "Mass" emphasizes that popular culture is widespread, undifferentiated, and placeless. Mass media such as the Internet, film, print, television, and radio, and are influential in

> **■ popular culture**
> Broadly defined, the modern ways of life associated with the rise of mass-produced, machine-made goods and the invention of long-distance communication technologies that collectively shape cultural preferences and define cultural identity.

civilization. Although indigenous peoples may share some of the material and nonmaterial characteristics that define folk cultures, their histories (and geographies) are quite different. Indigenous cultures

FIGURE 2.7 **Popular culture is reflected in every aspect of life,** from the clothes we wear to the recreational activities that occupy our leisure time. Musician Kanye West performs with Jay-Z at American Airlines Arena in Miami. *(Left: Scott Olson/Getty Images; Right: Kevin Mazur/WireImage/Getty Images.)*

shaping popular culture. "Consumer" emphasizes that popular cultural identity is shaped through the purchase and consumption of mass-produced commodities. Finally, geographer Edward Relph suggests that popular culture is characterized by a profound **placelessness,** a feeling resulting from the standardization of the built environment that diminishes regional variation and eliminates the unique meanings associated with specific locations (Figure 2.8). Shopping malls, airports, and global hotel and fast-food chains are examples of placeless locations that could be anywhere. One could travel, for instance, to Beijing, China, stay overnight at the airport Hilton much like any other Hilton across the world, eat steak and mashed potatoes for dinner much like one would order in any major city, shop for shoes at the global chain stores in the airport mall while waiting for one's flight, and never experience anything Chinese or unique to Beijing.

In popular culture, people are more mobile and less likely to live their lives in a single place. Secular institutions of authority—such as schools and courts—become increasingly important for maintaining social order. Individual choice is highly valued. Social relationships multiply, but fewer of them are conducted face to face. Rather, relationships become "stretched" across time and space. Think of online credit card purchases in contrast to the bartering that took place in the local village markets of historic folk cultures. Whereas the latter exchange was conducted in person between two people, think of how many individuals and corporations are involved in an online credit card purchase and home delivery of a simple pair of shoes, and how geographically distant from one another they are.

Just as the line between the material and the nonmaterial elements of culture is blurry, so too is the line between elements of folk and popular culture. For example, blues music had its origins among African American folk culture in the U.S. South, particularly the Mississippi Delta region (Figure 2.9). Later, in the 1950s, blues evolved into rock and roll, a musical genre that has come to epitomize global popular culture. There are many more examples of folk or indigenous cultural characteristics that have been incorporated into popular culture, including foods, textiles, and architecture. Conversely, items of popular culture have been incorporated into rural, folk-oriented cultures in surprising new ways. For example, the introduction of cheap, mass-produced glass beads from Europe provided the basis for vibrant new art forms in many Native American

placelessness
A spatial standardization that diminishes regional variety; may result from the spread of popular culture, which can diminish or destroy the uniqueness of place through cultural standardization on a national or even worldwide scale.

FIGURE 2.8 **Placelessness in urban landscapes: scenes almost anywhere.** One of these photos was taken in the United States and the other two in the United Kingdom and Japan, respectively. It is difficult to detect many local distinctions in these popular urban landscapes. *(Left to right: Bloomberg via Getty Images; Adina Tovy/Lonely Planet Images/Getty Images; Photo Japan/Alamy.)*

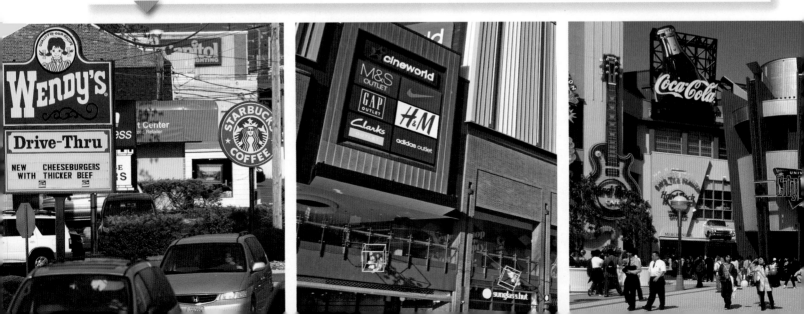

FIGURE 2.9 **McKinley Morganfield, better known as Muddy Waters, performing on stage.** Muddy Waters was born in the Mississippi Delta and moved as an adult to Chicago where he became a recording star after electrifying his blues guitar. He is widely viewed as a key artist in the transformation of the folk music traditions of the delta region into rock and roll. *(Hulton Archive/Getty Images.)*

to faraway regions and cities where they form recognizable enclaves and contribute greatly to cultural diversity. Cities around the world are dotted with neighborhoods called "Little Italy," "Chinatown," "Little Haiti," "Germantown," and so on (**Figure 2.11**). Geographers increasingly refer to such transplanted ethnic, racial, and national population concentrations as **diaspora cultures.** Historically, diaspora referred to a people (especially ancient Jewish culture) that had been displaced and geographically scattered. With the rise of globalization, however, international mobility has increased the size and number of diaspora cultures around the globe. Often diaspora cultures maintain strong social and economic ties to their homelands and therefore display hybrid or multiple cultural identities.

> ■ **diaspora culture** Ethnic, racial, and national population concentrations of people displaced and geographically scattered from their homelands. Such displaced groups often maintain strong social and economic ties to their homelands.

In addition to the diversity produced through increased global mobility, popular culture is subject to endless diversification into subcultures through the reuse and reinterpretation of mass-produced goods. Think of the importance of safety pins—a common household item originally associated with motherhood and diapers—to the material expression of punk culture

cultures (**Figure 2.10**). Thus, our interest in *Contemporary Human Geography* is not to draw distinct boundaries between different cultural types, but rather to explore how the materials, ideas, and symbols of one type of culture are adopted and reinterpreted by other cultures.

Mobility and Diversity To say that the elements of popular culture are defined by the rise of mass production, consumption, transportation, and communication is not to say that the growth of popular culture has followed a single, set pattern. Far from it. For example, greater mobility means that local, ethnically distinct cultures are frequently transplanted

FIGURE 2.10 **A Sioux amulet in the form of a turtle, made of hide decorated with beads imported from Europe.** Indigenous and folk cultures have often used such mass-produced objects to create new local art forms. *(SSPL via Getty Images.)*

FIGURE 2.11 **New York City's "Little Italy."** Many major cities in the United States have such neighborhoods named for the predominant (or once-predominant) immigrant population. *(Picture Partners/Alamy.)*

FIGURE 2.12 **Two punks in Osaka, Japan.** The use of safety pins—designed originally for cloth diapers—in body piercings became an iconic symbol of punk culture. *(Roger Ressmeyer/Corbis.)*

(Figure 2.12). The list of such subcultures emerging under globalization is nearly infinite. Gay, hip-hop, NASCAR, and grunge are just a few of the adjectives that we place before "culture" to talk about popular cultural identities, each with its particular geographic expression. Culture is as dynamic and limitless as the human imagination. We can use our five themes— region, mobility, globalization, nature–culture, and cultural landscape—to study the ongoing generation of geographies of cultural difference.

REGION

2.1 **LEARNING OBJECTIVE**
Explain the role of cultural difference in shaping regions.

COLLEGE SPORTS AS AN EXPRESSION OF CULTURE REGIONS

Attending or watching sporting events on television is a major part of popular culture. The biggest sports

spectacle in the United States is unquestionably the Super Bowl, the annual championship game of the National Football League, the country's premier professional (American) football association. The Super Bowl is so broadly popular across the country that it functions as an unofficial national holiday (Figure 2.13). Only on Thanksgiving is more food consumed on a single day in the United States. Television broadcasts of the game commonly record the largest viewership of the year for any single event or program. Many observers have come to view the sport as a symbol of American national culture.

When we think of college sports, chances are football comes to mind first. Successful college football programs tend to define universities and draw national publicity unmatched by that earned from other sports or academic achievements. Universities, however, compete with one another in dozens of other sports. Most of these sports do not have the broad national appeal of football. Rather, the vast majority of sports sponsored by the National Collegiate Athletic Association (NCAA) have distinctive regional associations. Frequently, these regional associations can be traced back in time to folk traditions emerging in specific places.

Let's look at just three of the dozens of sports that the NCAA sponsors for men in its Division I category: ice hockey, lacrosse, and volleyball. For each of these sports, a geographic pattern soon emerges: a clustering of universities that offer them.

Ice Hockey Beginning with ice hockey, there is a clear concentration of teams in the northern states east of the Rocky Mountains (**Figure 2.14**). Michigan counts 7 of the 59 college teams, while only one team, the University of Alabama-Huntsville, is located in the South. Only three teams may be found in the western United States, all of which are clustered around Denver, Colorado.

A clearly defined collegiate ice-hockey region thus exists in the northern United States. Furthermore, the number of ice hockey teams in the region—and the absence of the sport in other regions—are closely associated with the geographic origins and development of the sport. Modern ice hockey

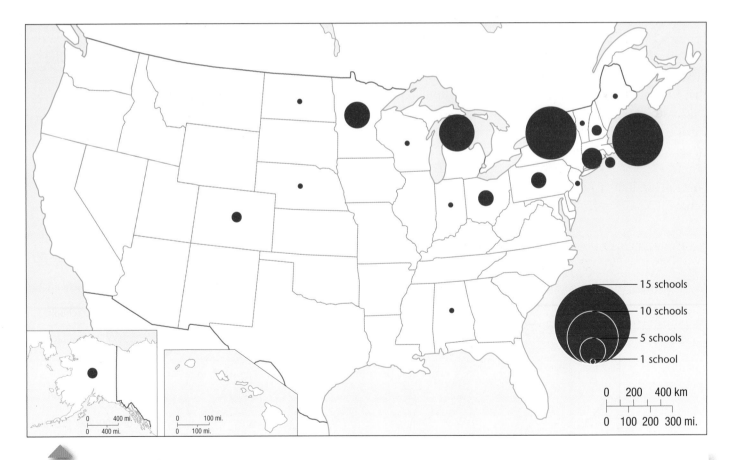

- 15 schools
- 10 schools
- 5 schools
- 1 school

0 200 400 km

0 100 200 300 mi.

FIGURE 2.14 **The distribution of NCAA Division I men's ice hockey teams.** There is a discernible concentration of teams in the northern midwestern and northeastern states, suggesting a core ice hockey region.

emerged in eastern Canada in the nineteenth century. Its invention appears to be rooted in the stick and ball games of the indigenous peoples of the region as well as winter folk traditions imported from Europe. Student enthusiasts formed the first ice hockey club with formal rules at Canada's McGill University in 1877. The sport spread from there, restricted almost exclusively to areas that experienced winters long and cold enough to form outdoor ice hockey rinks. Neighborhood kids organized games whenever and wherever it was cold enough for a local pond or creek to freeze over (**Figure 2.15**). By the time technology made indoor ice rinks feasible anywhere, the regional cultural preferences had been already established.

Lacrosse The NCAA's 63 college lacrosse teams also exhibit a distinctive regional clustering, with the heaviest concentration found in the Northeast (**Figure 2.16**). As with ice hockey, lacrosse has flourished close to its point of geographic origin. It is

FIGURE 2.15 **A backyard game of ice hockey.** *(Radius Images/Alamy.)*

derived from a game played by the indigenous peoples of eastern Canada. European Canadians created the formal rules for modern lacrosse in 1867 and intercollegiate competition followed soon

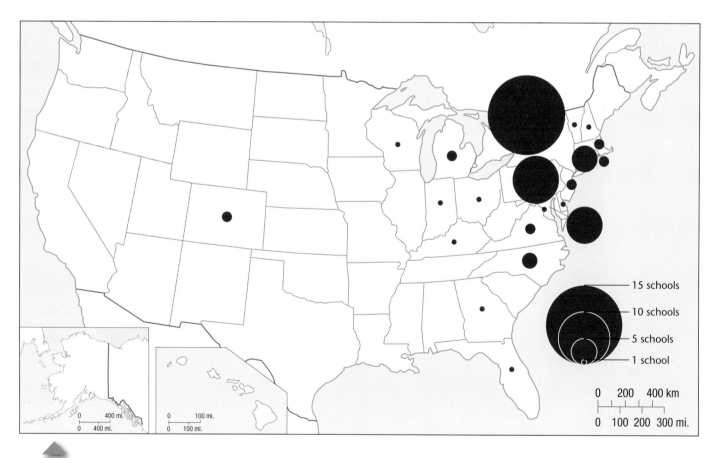

FIGURE 2.16 **The distribution of NCAA Division I men's lacrosse teams.** There is a discernible concentration of teams in the northeastern states, suggesting a core lacrosse region.

thereafter. The Ivy League universities and private liberal arts colleges of the northeast United States established the first teams, followed by a few large public universities, such as the University of Michigan (Figure 2.17). At the college level, lacrosse is virtually unknown west of the Rocky Mountains.

Volleyball Our final example, volleyball, has a history and regional pattern different from those of lacrosse and ice hockey. There are far fewer Division I college volleyball teams—only 33—and they are concentrated in three separate regions: the West Coast, the Midwest, and the Northeast (Figure 2.18). The game itself was invented by an educator/physician working for the Young Men's Christian Association (YMCA) in Massachusetts in 1896 as part of an adult health and fitness program.

Although the game has since spread throughout the United States and beyond, college-level competition developed a separate and unanticipated

FIGURE 2.17 **NCAA Division I men's lacrosse match between Syracuse and Duke universities. Syracuse,** which has won ten NCAA championships, has a history of recruiting Native Americans from the reservations throughout upstate New York. The game originated among Native American tribes in the region. *(Getty Images.)*

FIGURE 2.18 **The distribution of NCAA Division I men's volleyball teams.** There is a discernible concentration of teams on the Pacific coast, suggesting a core volleyball region.

geographic pattern. Volleyball was increasingly associated with a particular environment and culture far from its Massachusetts origin and with a particular set of regional cultural values very different from those of the nineteenth-century YMCA. By the 1960s, volleyball, like surfing, had become symbolically associated with Southern California's beach culture (Figure 2.19). As with hockey on winter days in the North, kids in Southern California spent their summer days playing pickup volleyball on the beach or at neighborhood outdoor sand courts. Today, there is a strong concentration of college teams in California, where 9 of the 33 Division I teams are located. While three separate regional concentrations

FIGURE 2.19 **Pick-up volleyball game on a southern California beach.** Though the game was invented in nineteenth-century Massachusetts as an indoor sport, it became a dominant symbol of California beach culture after the 1960s. *(Arthur Tilley/The Image Bank/Getty Images.)*

exist, California universities dominate the sport. In 43 years of Division I championships, California teams have won all but eight championships and all but one of those teams is located in the southern half of the state. College sports are clearly useful for thinking about and identifying regional cultural patterns.

INDIGENOUS CULTURE REGIONS

Indigenous peoples generally live in areas with few roads or modern communications systems, such as mountainous areas, vast arid and semiarid regions, or large expanses of forest or wetlands. These concentrations of people constitute indigenous culture regions. Worldwide, large concentrations of indigenous populations exist outside of the strong influence of national cultures and the control of governments located in faraway capital cities. Control of indigenous regions by a central government is often weakened by the minimal infrastructure, rough topography, or harsh environmental conditions of the region. In many cases, indigenous peoples either fled or were forcibly removed by central governments to environmentally marginal regions, such as arid lands.

In the United States and Canada, for example, this geographic pattern of marginalization is particularly evident. In the United States, the central government has had a complex and often contradictory relationship with indigenous peoples, sometimes treating them as independent nations and at other times as second-class citizens. The way in which indigenous peoples are distributed in the United States reflects both the history of the east-to-west movement of European settlers and nineteenth-century government policies (Figure 2.20). For example, the Indian Removal Act of 1830 was intended to make way for European settlers by relocating eastern Native American tribes west of the Mississippi River, many to Oklahoma. Western Native American tribes were also eventually forced onto government-created reservations, generally on the most unproductive and arid lands. Some of the larger reservation complexes in the arid West constitute an indigenous culture region (Figure 2.21).

The so-called Hill Tribes of South Asia are another good example of the regional marginalization

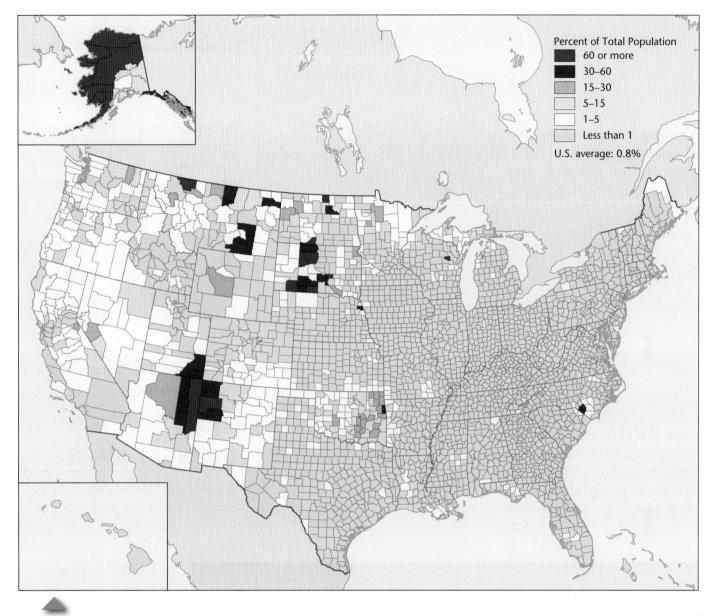

Percent of Total Population
- 60 or more
- 30–60
- 15–30
- 5–15
- 1–5
- Less than 1

U.S. average: 0.8%

FIGURE 2.20 **Indigenous Native American population distribution in the United States.** *(U.S. Census Bureau.)*

of indigenous peoples. Mountain ranges, including the Chittagong Hills, the Assam Hills, and the Himalayas, surround the fertile valleys and deltas around which the ancient South Asian Hindu and Islamic civilizations were centered. Various indigenous peoples occupy these highland regions, which are remote from the lowland centers of authority and culturally distinct from them (Figure 2.22). A series of indigenous culture regions ringing the valleys of South Asia thus exists, occupied by what the British colonial authorities referred to as

Hill Tribes. Most of these peoples practice some version of swidden agriculture, which involves multiyear cycles of forest clearing, planting, and fallowing (see Chapter 8). Most hold Christian or animist beliefs (see Chapter 7) and speak languages distinct from those spoken in the lowlands. A similar pattern of highland indigenous culture regions can also be identified in the countries of Southeast Asia, such as Myanmar and Thailand, where indigenous peoples such as the Shan and Karen have populations in the millions.

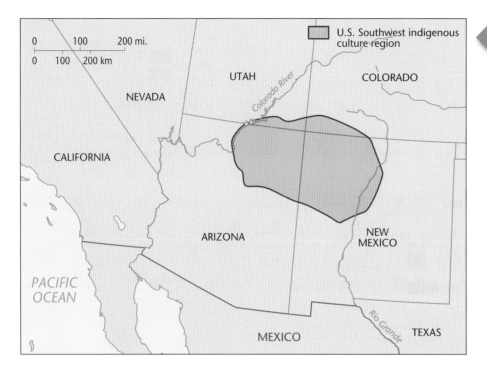

U.S. Southwest indigenous
culture region

UTAH

COLORADO

NEVADA

Colorado River

CALIFORNIA

ARIZONA

NEW
MEXICO

PACIFIC
OCEAN

Rio Grande

TEXAS

MEXICO

FIGURE 2.21 **An indigenous culture region in the United States.** *(U.S. Geological Survey.)*

peoples exists in sections of the Andes Mountains. This area constituted the geographic core of the Inca civilization, which thrived between 1300 and 1533 C.E. and incorporated several major linguistic groups under its rule. Today, up to 55 percent of the national populations of Andean countries, such as Bolivia, Peru, and Ecuador, are indigenous. On the slopes and in the high valleys of the Andes, Quechua and Aymara speakers constitute an overwhelming majority, signifying an indigenous culture region (**Figure 2.24**).

Indigenous culture regions also exist in Central and South America. There is a distinct Mayan culture region that encompasses parts of Mexico, Belize, Guatemala, and Honduras (**Figure 2.23**). Concentrations of Mayan speakers are especially common in rugged highlands and tropical forests. In South America, another concentration of indigenous

THE RISE OF THE LBGT DISTRICT

Sexuality, as with all aspects of human culture, is expressed geographically. One of the most dynamic

FIGURE 2.22 **A Murong grandmother with her grandchildren in the indigenous culture region of Bangladesh's Chittagong Hill Tracts.** *(Rashed Hasan/ agefotostock.)*

FIGURE 2.23 The **Mayan culture region in Middle America.** The ancient Mayan Empire collapsed centuries ago, but its Mayan-speaking descendants continue to occupy the region. In many cases Mayan communities, after centuries of political and economic marginalization, are today actively struggling to have their land rights recognized by their respective governments.

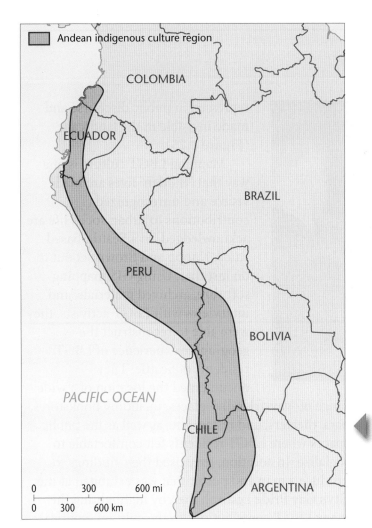

developments in cultural geography over the past two decades has been the emergence of a new field of study on sexuality and space, referred to as *queer geography* (or sometimes *gay geography*). The emergence of queer geography as a field of study reflects and is reflected in the increasing acceptance of lesbian, bisexual, gay, and transgendered (LBGT) identities in popular culture.

As geographers Larry Knopp and Michael Brown have noted, from the 1990s onward there has been an expanding interest in documenting the history of LBGT experience, most notably through a series of case studies of North American cities. These studies have coincided with a dramatic turnaround in popular attitudes toward sexualities that historically had been kept "in the closet." That is, human sexuality that was different from a narrowly prescribed heterosexual norm was until recently

FIGURE 2.24 **The indigenous culture region of the Andes, including Quechua- and Aymara-speaking peoples.** This is the core of the ancient Inca Empire, where today many of the indigenous people speak their mother language rather than Spanish. *(Adapted from de Blij & Muller, 2004.)*

THE VIDEO CONNECTION

An Attack on Equality
http://www.tinyurl.com/mfj3e45

This video centers on an interview with two young victims of a hate crime against same-sex partners. The attack took place in daylight on a busy public street near Madison Square Garden. Such hate crimes, the video reports, are on the rise in New York City, despite the city's reputation for tolerating diversity. The victims explain that the circumstances of the attack have caused them to reevaluate their ideas about New York City and its residents' readiness to accommodate and defend cultural diversity. They conclude that the State of New York's legalization of same-sex marriage has not meant the end of discrimination and violence against openly gay citizens.

Thinking Geographically:

1. Besides the hate crimes, what are some examples of unequal treatment of same-sex couples mentioned in the video? In particular, think about how same-sex couples may experience public space differently than heterosexual couples as a result of unequal treatment.

2. Sexuality, as one mode of cultural difference, is expressed geographically. How do the violent attacks discussed in the video help explain why so many cities have LBGT districts?

3. This chapter discusses the idea of place images. In the video, one person describes his childhood image of New York as "this dream metropolis" and "basically the Emerald City that I've always wanted to go to." What do you think he meant by that? How do you think the attacks have altered the place image that fueled his move to New York?

FIGURE 2.25 Stonewall Riots. Protestors attempt to impede police arrests outside the Stonewall Inn in Greenwich Village in 1969. These protests are viewed as the catalyst of the gay rights movement that continues to work to end all forms of repression and discrimination based on sexual preferences. *(NY Daily News via Getty Images.)*

repressed in public discourse and made invisible in the landscape (**Figure 2.25**).

Mapping LBGT culture is one way that "hidden" lives are made visible and unrecognized contributions to urban social life are acknowledged. In a Seattle-based study, Knopp and Brown set out to do just that. Using GIS mapping software, archived materials, and interviews with LBGT activists, they were able to reconstruct the geographic experience of LBGT residents of Seattle. They documented the location of a wide range of "gay-friendly" places, including bookstores, bars, theaters, and restaurants as well as the public spaces where LBGT residents felt comfortable to socialize. In addition, they used their findings to produce a map and poster that was exhibited at the city's Gay Pride Festival, thereby highlighting the many LBGT contributions to the cultural, political

◀ FIGURE 2.26 Seattle's Gay Pride Parade. Gay pride parades in the United States trace their origins to the 1970 marches in several cities to commemorate the first anniversary of the Stonewall riots. To understand the huge cultural shift that has occurred over four decades, compare the police in Figure 2.25 with the police in this figure. (AP Photo/ Elaine Thompson.)

and economic development of the city (Figure 2.26). Among Knopp and Brown's other findings was that LBGT influence was strong throughout the city and not just confined to the widely recognized gay-friendly neighborhood of Capitol City.

Neighborhoods such as Capitol City are so common across North America that a new terminology has grown to describe them. Known as a "gayborhood," "gay village," or simply, "LBGT district," such urban spaces may be found in many major cities. As with numerous urban subcultures, LBGT residents have tended to cluster in neighborhoods where they feel comfortable among accepting or like-minded people and are able to live without hiding their sexual identity. Today, rather than try to suppress or hide such districts, local governments are celebrating them as part of the cultural diversity of their cities. Chicago's "Boystown" is a good example. The city's official marketing organization, Choose Chicago, features Boystown at its promotional web site under the theme of "Chicago Pride" in its neighborhoods. Boystown is described as the first officially recognized LBGT district in the United States, signified by the city's installation of rainbow-colored Art Deco pillars to mark its borders (Figure 2.27). Local pride notwithstanding, Choose Chicago's mission is to "sell" the city to tourists, business investors, and prospective homeowners. It is interesting to note how an LBGT district has suddenly gone from being hidden to being featured in one city's official sales pitch.

FIGURE 2.27 Sidewalk towers with rainbow color bands mark Chicago's Boystown neighborhood. Rather than suppress and hide the presence of LBGT groups, as was done in the recent past, Chicago has today chosen to celebrate this neighborhood. (Charles Cook/Lonely Planet Images/Getty Images.)

VERNACULAR CULTURE REGIONS

> **vernacular culture region** A culture region perceived to exist by its inhabitants, based in the collective spatial perception of the population at large, and bearing a generally accepted name or nickname (such as "Dixie").

A **vernacular culture region** is the product of the spatial perception of the population at large—a composite of the mental maps of the people. Such regions vary greatly in size, from small districts covering only part of a city or town to huge, multistate areas. Like most other geographical regions, they often overlap and usually have poorly defined borders.

Even though cadastral surveys (surveys that define the boundaries and subdivisions of public lands) and official atlases with official place names are the basis for modern governments and businesses, people all over the world still use vernacular references to navigate everyday life. In conversation, people will often say, "She lives over in the Valley" or "He opened a business near the Old West End" rather than reciting a city name or street address (**Figure 2.28**). Such terms are embedded in the thick cultural and historical knowledge of local daily life, which we take for granted until we move to a new city or country and have to learn our way around.

Official maps and GPS instruments are not much help, because vernacular regional boundaries are fuzzy and indefinite and often their vernacular names are not even included. Great Britain's Ordnance Survey acknowledges that people navigate on a daily basis using a geography related to but distinct from the official maps. The Ordnance Survey is the government agency charged with surveying and mapping every aspect of the country. In the late 2000s it created the Vernacular Geography initiative. One project is to create an alternative name gazetteer whereby one could learn that "Spaghetti Junction" refers to the M6 Motorway Junction 6 in Birmingham (**Figure 2.29**). Another project links with geographers at the universities of Cardiff and Sheffield to gather geotagged data from the web, both by automatic means and online public surveys. The ultimate goal is to make vernacular regions and places more widely accessible so that, for instance, dispatchers for emergency responders can understand local descriptions of locations.

Not only are the boundaries of vernacular regions fuzzy, they are dynamic, changing as everyday

FIGURE 2.30 The use of vernacular geography in business. "Dixie" is commonly used in naming businesses in the southeastern states. We can map business names as one way to define the boundaries of a vernacular region. *(Courtesy of Roderick Neumann.)*

FIGURE 2.28 The Utah Hotel, an iconic building in the South of Market (SoMa) neighborhood of San Francisco. While the Utah serves to symbolize the heart of SoMa, the boundaries of vernacular neighborhoods such as this one are more indeterminate and typically not identified on official maps. *(Robin Allen/PhotoLibrary/Getty Images.)*

FIGURE 2.29 The United Kingdom's "Spaghetti Junction," the vernacular place name for the M6 motorway's Junction 6 in Birmingham. *(Duncan Wherrett/The Image Bank/Getty Images.)*

perceptions of geography and identity transform. Geographer Shrinidhi Ambinakudige's 2009 study revisited vernacular understandings of the U.S. "South" and "Dixie" first mapped in the 1970s and 1980s. Dixie is distinguished from the South because the terms convey different cultural meanings. Dixie is associated with the antebellum rural, plantation culture and the South with the more recent urbanizing and industrializing "new" South. Ambinakudige mapped the frequency of occurrence of business names that use one of these terms in the Yellow Pages. He found that over the past few decades, the boundaries of both regions are eroding, though Dixie is doing so at a much faster rate (Figure 2.30). Moreover, the strongest associations with Dixie and the South occur in rural areas, with metropolitan areas showing weakening associations. One explanation for the decline in Dixie and Southern identities, particularly in cities, is the large numbers of new residents migrating into the region from elsewhere. From this study we might generalize that vernacular regions are in flux, with established regions shrinking or expanding and new regions constantly coming into being.

MOBILITY

2.2 LEARNING OBJECTIVE
Describe the ways that mobility interacts with cultural difference.

The movement of things, ideas, and people shapes and is shaped by geography. For example, things and people in motion can minimize the relevance of nation-state borders or nation-state borders can greatly restrict mobility. The interactions of geography and mobility are complex and not altogether predictable.

VAMPIRE TOURISTS

No other practice better exemplifies mobility in popular culture than tourist travel. And no form of tourist travel provides a more surprising blend of reality and fiction, folk and popular culture, than vampire tourism. According to geographer Duncan Light, vampire tourism is a form of "media tourism,"

FIGURE 2.31 Tawhai Falls in New Zealand, also known as Gollum's Waterfall. Location sites such as this from the *Lord of the Rings* films have become destinations for tourists in New Zealand. *(Simon Browitt/Alamy.)*

a practice whereby fans of a fictional story or character—whether portrayed in a novel, film, or television program—travel to key places and landscapes in the plot. The New Zealand locations of the *Lord of the Rings* film series, for example, have become a key destination for the country's foreign tourists (Figure 2.31).

Vampires are popping up everywhere in popular culture. They appear in novels and cinema (the *Twilight* series), animated feature films (*Hotel Transylvania*), graphic novels (*Interview with the Vampire: Claudia's Story*), and cable television (*Vampire Diaries*). So saturated is popular culture with vampire stories that one would be hard pressed to find a mass entertainment medium in which they do not materialize. Interestingly, the undead lives of these ubiquitous blood suckers are anchored in real geography, resulting in the rapid growth in vampire tourism nationally and globally. The Lonely Planet, a well-known publisher of travel guides, features a list of the "world's best vampire-spotting locations" on its web site.

Fictional vampire characters appear in the most unexpected places, including Forks, Washington; New Orleans, Louisiana; Volterra, Italy; and Mystic Falls, Virginia. Actually, Mystic Falls, the setting for the *Vampire Diaries,* is fictional, too, but the television show is shot in real-life Covington, Georgia. Entrepreneurs in Covington have wasted no time creating "Mystic Falls" tour packages. As one web site advertises, "The Mystic Falls Tour is the perfect destination for travelers looking for a memorable Vampire Diaries adventure." Forks, Washington, where the *Twilight* series is based, has been overrun by hundreds of tourists daily—many of them young women around the age of Bella's character—a phenomenon that has significantly improved the fortunes of this economically depressed logging town.

The Curse of Dracula, Revisited Forks and Covington, however, are mere way stations on the pilgrimage route to the mecca of vampire tourism, the Transylvania region of Romania (**Figure 2.32**). Transylvania is home to the world's most famous vampire, Count Dracula, subject of Bram Stoker's 1897 novel *Dracula* and of the numerous subsequent film versions. Dracula already had an international reputation—the novel has been translated into numerous languages—when Francis Ford Coppola's 1992 film *Bram Stoker's Dracula* made the count a global personality. Stoker loosely based Dracula on vampire folktales that circulated throughout southeastern Europe in the eighteenth century. In his novel, Stoker located the count's home in Transylvania's Dran Castle, which makes it a global destination for vampire tourists (**Figure 2.33**). Ironically, no folk tradition of vampirism exists in the region's culture. This inconvenient fact did not prevent Stoker from vaguely modeling Count Dracula on Vald III (or Vlad Țepeș), a real-life Transylvanian ruler from the 1450s whom the Romanians revere as a nationalist hero for his role in repelling an Ottoman Turkish invasion.

Light shows in his study *The Dracula Dilemma* how vampire tourism raises difficult questions for the

FIGURE 2.32 **Count Dracula's imagined home.** Bran Castle on the edge of Romania's Transylvania region has become the mecca of vampire tourism.

Romanian government and its citizens on issues of national and cultural identity. The Dracula story, Light points out, positions Romania as a backward, superstitious region on the eastern margins of modern, civilized western Europe. Moreover, Stoker links the evil beast Dracula, who is threatening the English homeland, to Vlad III, a heroic figure in the creation of Romanian cultural and national identity. Imagine the world outside the United States believing George Washington was a vampire. In short, Stoker's character casts Romania in a negative light and challenges its citizens' national identity at its deepest level. Dracula is an awkward figure around which to build a Romanian tourist industry.

Cashing In on the Count Until 1989 Romania was a communist country where government censorship of popular media, including the Dracula stories, was widespread. When the country abandoned communism, it lifted censorship and unleashed private entrepreneurship. Stoker's novel and Coppola's film began to circulate widely in Romania. The borders were opened up to the global tourist industry. Businesses started trading on the Dracula name, which is used to market guesthouses, restaurants, and even beer and cigarettes (**Figure 2.34**). The main marketing target is foreign tourists arriving to visit the "home" of Dracula.

proud cultural heritage. The ironies and frustrations of being the nation and culture most identified with Dracula are manifold, not least being the fact that no tradition of vampires exists in Romanian folklore. According to Light, Romanians do not and never did believe in vampires, but the whole world thinks of Transylvania as the home of the most infamous vampire of all.

Vampire tourism illustrates a number of features of mobility in popular culture. There is the mobility of the vampire myth; the ways it is transported and transplanted from folk to popular culture; the ways it circulates globally, linking unlikely places as distant and dissimilar as Forks and Transylvania; and the ways it has set people in motion in search of direct experiences with these places. Fictional characters move through factual landscapes and consequently affect real-life political, economic, and cultural relations through media tourism. The mobility of vampire tourists following fictional plot lines, as Light emphasizes for Romania, "illustrates global inequalities and asymmetries in cultural power." The Romanian government and citizenry have fought a losing battle for control over representation of their cultural heritage. In popular culture, Transylvania

FIGURE 2.33 **Souvenir shops at Bran Castle in Romania's Transylvania region.** The building is popularly known as "Dracula's Castle" due to its resemblance to the description in Bram Stoker's novel *Dracula*. *(Robert B. Fishman/picture-alliance/dpa/AP Images.)*

The Romanian government and many citizens were not as quick as entrepreneurs to embrace Stoker's portrayal of Transylvania. As Light points out, the Romanian nation-state wishes "to project a sense of its own political and cultural identity to the wider world on its own terms." Those terms are defined by a desire to be politically and economically integrated into the **European Union** while maintaining a distinct and

European Union
The union of 28 European countries established through a set of political, cultural, and economic treaties and supranational institutions.

FIGURE 2.34 **Dracula mugs in a souvenir shop at Bran Castle.** Businesses in Romania are cashing in on the Dracula name and image. *(Sean Gallup/Getty Images.)*

FIGURE 2.35 **Refugees from the Darfur region of Sudan awaiting international aid in northeastern Chad.** Civil wars in Sudan spanning decades have displaced millions of people and produced a Sudanese diaspora. *(Scott Nelson/Getty Images.)*

appears doomed to an undying association with a superstition transported from folk traditions rooted in far-off places.

IDENTITY IN DIASPORA CULTURE

Diaspora cultures are, by definition, born of mobility. Civil war and international armed conflicts over the past two generations have set millions of people in motion around the world, creating many new diaspora cultures. The list is long and includes the Cuban diaspora (following the 1959 communist revolution in Cuba), the Hmong diaspora (following the U.S.-Vietnam War in 1975), and the Sudanese diaspora (following civil wars ongoing since the 1980s). The civil wars in Sudan displaced at least 4 million people, creating a new diaspora located on three different continents (**Figure 2.35**). Major armed conflict in Sudan developed around the economic, religious, cultural, and racial differences distinguishing the Arabic/Islamic northern region, where the military and government were centered, from the African/Christian

southern region, where rich oil deposits are found. Thus, a significant portion of the diaspora are southerners fleeing the fighting between the northern military and southern insurgents. The United States has been among the largest receiver countries, taking in approximately 150,000 southern Sudanese. Major Sudanese diaspora concentrations are located in Minnesota, New York, and Texas.

The Sudanese Diaspora and the Establishment of South Sudan Geographer Caroline Faria conducted research among the Sudanese diaspora communities in the United States to investigate how members forge cultural and national identities under conditions of displacement. Faria's study of the Sudanese diaspora took place during a critical historical juncture, ranging from the 2005 signing of a peace agreement to the establishment in 2011 of South Sudan, the world's newest nation-state (**Figure 2.36**). It was during this period that the diaspora community vigorously debated what a new nation should be and what it would mean, culturally, to be a South Sudanese citizen.

FIGURE 2.36 **Southern Sudanese celebrate their first independence day in the capital city of Juba on Saturday, July 9, 2011.** Members of the Sudanese diaspora watched the festivities on television and conducted their own celebrations in cities around the world. *(AP Photo/Pete Muller.)*

Faria analyzed the first and second "Miss South Sudan" contests held in Washington, D.C., in 2006 and Kansas City in 2007 to demonstrate the importance of diasporic beauty pageants for "promoting and imagining a distant home and nation." Beauty pageants, in general, are fruitful sites for studying debates around gender, race, and national identity because the contest winner is assumed to represent, even embody, the essential qualities of the nation. Thus in the Miss South Sudan contests, organizers, participants, and spectators struggled to identify an ideal type of woman who would signify the new nation.

As Faria discovered after analyzing interviews and public commentary and debate, the ideal South Sudanese woman must conform to certain racial, religious, and gender identities. First, differences in religious beliefs both supported and were accentuated by the north–south armed conflict in Sudan. Christianity has become an important marker of difference to distinguish southern Sudanese cultural identity from the north. Miss South Sudan pageant commentators and judges thus emphasized "a strong Christian Faith as a vital part of any perspective winner." Second, southern Sudan is imagined as "culturally and pheno-typically oriented toward Sub-Saharan Africa while the North is associated with an Arab influence and a lighter skin tone." Consequently, participants and judges in the Miss South Sudan contests linked the notion of feminine beauty to "blackness," with the 2006 winner opting for the "natural" look of a simple hair weave that projected a more "African" identity for the new nation (Figure 2.37). Finally, the pageant exposed an unresolved contradiction in the Sudanese diaspora's ideas of womanhood and women's ideal role in the new South Sudan nation. Contestants were expected on the one hand to delay marriage and aspire to higher education and professional careers, while on the other to maintain a traditional role as mothers who would reproduce the new nation.

The case of the Miss South Sudan pageants highlights the challenge facing many diaspora

FIGURE 2.37 **Contestants at the inaugural Miss South Sudan Beauty Contest:** Rita Magoh, left, and Grace Bok, both South Sudanese from Kansas City. Such cultural events are important to diaspora communities for defining national identities. *(Jahi Chikwendiu/The Washington Post/Getty Images.)*

cultures around the world: how to establish a coherent identity under conditions of displacement and uncertainty. In imagining national and cultural identities from a distance, diaspora cultures must confront basic questions about national ideals of masculinity and femininity and the relationship of race and religion to nationhood. Showcases of national culture, such as beauty pageants, are not merely sources of popular entertainment. Rather, they are cultural cauldrons in which national identities are cemented, reproduced, and reimagined.

BARRIERS TO INFORMATION MOBILITY

Although mass media create the potential for the almost instant diffusion of information over very large areas, this potential can be greatly impeded and slowed if access is limited or denied. Access to the Internet is a case in point. Though older forms of mass communication, such as television and newspapers, are unlikely to disappear, an increasing proportion of people worldwide turn to the Internet.

digital divide A pattern of unequal access to advanced information technologies produced by socioeconomic inequalities and measured at scales ranging from the individual to countries and world regions.

The shift in technology has created a so-called **digital divide** between those who have ready access to the Internet information stream and those who have limited or no access (**Figure 2.38**). Many factors contribute to the digital divide, but one of the most basic is lack of Internet infrastructure in many parts of the world. Thus, while information may diffuse globally on the Internet at near light speed, it is irrelevant to those without a computer and the infrastructure to connect. A historical analogy might be the invention of the printing press when most of the world's population was illiterate.

Recognition of the digital divide and potential solutions have arisen and spread quickly, precisely because information moves around the globe at great speed. New initiatives meant to address the gap spring up continually. For example, the Digital Divide

Institute operates as a policy think tank to advise companies and governments on solutions and Pew Research conducts surveys on who does and does not use the Internet and why. The Internet has both exacerbated socioeconomic inequalities and allowed for a rapid response to try to correct them.

Governmental Bans on Technology In addition to the existing inequalities in Internet access, governments around the world periodically implement technology bans that have the effect of limiting the diffusion of information and ideas (**Figure 2.39**). For years Cuba banned cell phones, fearing that citizens would have access to information unfavorable to the government. A huge black market in cell phones forced the government to lift the ban in 2008. The government of the Czech Republic in 2010 banned the use of Google's Street View technology over privacy concerns. It has since lifted the ban after negotiating very strict rules on how Google is allowed to use the technology. The United Arab Emirates

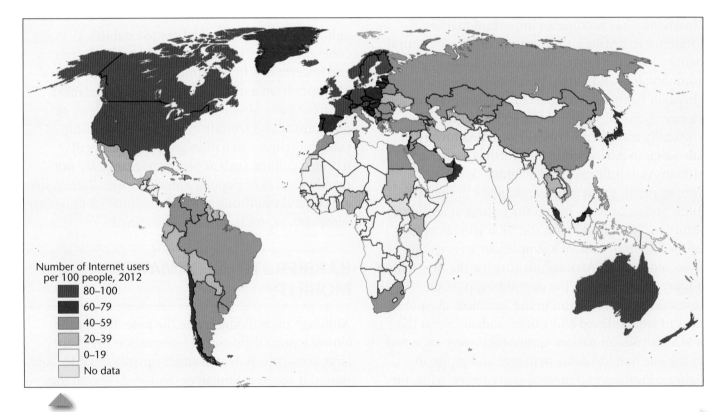

Number of Internet users per 100 people, 2012
80–100
60–79
40–59
20–39
0–19
No data

FIGURE 2.38 **One way to assess the digital divide on a global scale.** While the gaps among regions are closing, major hurdles to more equitable access remain.

FIGURE 2.39 Some countries have attempted to ban communication technologies in an effort to control citizens' access to information. For years, the Cuban government tried to ban cell phones. The ban ultimately failed because people were able turn to an illegal market in technology. *(AFP/Getty Images.)*

FROM DIFFERENCE TO CONVERGENCE

Globalization is most directly and visibly at work in popular culture. Increased leisure time, instant communications, greater affluence for many people, heightened mobility, and weakened attachment to family and place—all features of popular culture—have the potential, through interaction, to cause massive spatial restructuring. Most social scientists long assumed that the result of such globalizing forces and trends, especially mobility and the electronic media, would be the homogenization of culture, wherein the differences among places are reduced or eliminated. This assumption is called the **convergence hypothesis;** that is, cultures are converging, or becoming more alike. In the geographical sense, this would yield placelessness, a concept discussed earlier in the chapter on page 48.

> ■ **convergence hypothesis** A hypothesis holding that cultural differences among places are being reduced by improved transportation and communications systems, leading to a homogenization of popular culture.

briefly banned the BlackBerry in 2010 because the device's encryption makes it difficult for governments to monitor content for what they consider objectionable material. Though all these bans were ultimately lifted, they suggest that governments will continue to try to limit or control the diffusion of information across nation-state borders.

▦ GLOBALIZATION

**2.3 LEARNING OBJECTIVE
Analyze the potential for globalizing processes to shape and be altered by cultural difference.**

The effects of globalization on cultural difference and vice versa are a much discussed and debated issue for geographers and other social scientists. As we look for evidence in real-world examples, we find that the answer is less than straightforward.

DIFFERENCE REVITALIZED

Globalization, we should remember, is an ongoing process or, more accurately, set of processes. It is incomplete and its outcome far from predetermined. Geographer Peter Jackson is a strong proponent of the position that cultural differences are not simply obliterated under the wave of globalization. For Jackson, globalization is best understood as a "site of struggle." He means that cultural practices rooted in place shape the effects of globalization through resistance, transformation, and hybridization. In other words, globalization is not an all-powerful force. People in different places respond in different ways, rejecting outright some of what globalization brings while transforming and absorbing other aspects into local culture. Rather than one homogeneous globalized culture, Jackson sees multiple **local consumption cultures.**

Local consumption culture refers to the consumption practices and

> ■ **local consumption cultures** Distinct consumption practices and preferences in food, clothing, music, and so forth formed in specific places and historical moments.

preferences—in food, clothing, music, and so on—formed in specific places and historical moments. These local consumption cultures often shape globalization and its effects. In some ways, globalization revitalizes local difference. That is, people reject or incorporate into their cultural practices the ideas and artifacts of globalization and in the process reassert place-based identities.

For example, when England-based Cadbury decided to market its line of chocolates and confections in China, the company was forced to conform to local practices. Unlike much of Europe and North America, China had no culture of impulse buying and no tradition of self-service. Cadbury had to change the names of its products, eschew mass marketing, focus on a small group of high-end consumers, and even change product content. Jackson suggests that the introduction of Cadbury's chocolate to China is more than simply another sign of globalization. He argues that the case "demonstrates the resilience of local consumption cultures to which transnational corporations must adapt."

In cases where companies' products are negatively associated with their place of origin, such as exports from apartheid South Africa, the global ambitions of multinational companies can be thwarted. In such cases, labeling and advertising must mask the geographic origin of products lest global markets reject them. "Local" circumstances thus can make a difference to the outcomes of globalization.

Local resistance to globalization often takes the form of **consumer nationalism.** This occurs when local consumers avoid foreign companies or imported products and favor domestic businesses and products (**Figure 2.40**). India and China, in particular, have a long history of resisting outside domination through boycotts of imported goods. Jackson discusses a recent case in China in which Chinese entrepreneurs invented a local alternative to Kentucky Fried Chicken called Ronghua Fried Chicken Company. The company uses what it claims are traditional Chinese herbs in its recipe, delivering a product more suitable to Chinese cultural tastes.

> ■ **consumer nationalism** A situation in which local consumers favor nationally produced goods over imported goods as part of a nationalist political agenda.

FIGURE 2.40 **Residents protest against Pepsi in Ahmedabad, Gujarat, India.** Highly visible global brands, such as Pepsi, are often targeted outside the United States as symbols of foreign intrusion. *(Shailesh Raval/The India Today Group/Getty Images.)*

PLACE IMAGES

The same media that serve and reflect the rise of personal preference—movies, television, photography, music, advertising, art, and others—often produce place images, a subject studied by geographers Brian Godfrey and Leo Zonn, among others. Place, portrayer, and medium interact to produce the image, which, in turn, colors our perception of and beliefs about places and regions we have never visited. The focus on place images highlights the role of the collective imagination in the formation and dissolution of culture regions. It also explores the degree to which the image of a region fits the reality on the ground. That is, in imagining a region or place, often certain regional characteristics are stressed, whereas others are ignored (see Subject to Debate).

The images may be inaccurate or misleading, but they nevertheless create a world in our minds that has an array of unique places and place meanings. Our decisions about tourism and migration can be

SUBJECT TO *Debate*

MOBILE IDENTITIES: Questions of Culture and Citizenship

One of the most dynamic features of cultural difference in the world today is mobility. Diaspora communities have sprung up around the globe. Some of these communities have arisen quite recently and are historically unprecedented, such as the movement of Southeast Asians into U.S. cities in the late twentieth century. Some are artifacts of European colonialism, such as the large populations of South Asians in England or West and North Africans in France. Others are deeply historical and express a centuries-old interregional linkage, such as the contemporary movement of North Africans into southern Spain.

The movement and settlement of large populations of migrants have raised questions of belonging and exclusion. How do transplanted populations become English, French, or Spanish, not only in terms of citizenship but also in terms of belonging to that culture? Some observers have argued that diaspora populations find ways to blend symbols from their cultures of origin with those of their host cultures. Thus, people develop a sense of belonging through a process of cultural hybridization. Other observers point to long-standing situations of cultural exclusion. In 2005 in France, for example, riots broke out in more than a dozen cities in suburban enclaves of West African and North African populations. The reasons for the riots are complex. However, many observers pointed out that underprivileged youth of African descent feel excluded from mainstream French culture, even though many are second- and third-generation French citizens.

The debate over how to address questions of citizenship and cultural belonging is played out in many venues. In terms of policy, the French, for example, emphasize cultural integration; for example, the French government approved a law in 2010 banning full-face veils (burqas) in public, whereas Britain and the United States promote multiculturalism. But many questions remain about the effectiveness of state policies toward diaspora cultures.

Continuing the Debate

Based on the discussion presented above, consider these questions:

- Do the geographic enclaves of Asian and African diaspora populations in the former colonial capitals of Europe reflect an effort by migrants to retain a distinct cultural identity? Or do they reflect persisting racial and ethnic prejudices and efforts to segregate "foreigners"? Or is it a combination of factors?

- How long does it take an Asian immigrant community to become "English" or a West African immigrant community to become "French"? One generation? Two? Never?

- Does the presence of Asians and Africans make the landscape of England and France appear less "English" or less "French"? Why or why not?

Riots and protests, such as this memorial march for two dead teenagers in Clichy sous Bois, spread across several cities in France in 2005. Protesters' complaints centered on the discriminatory treatment of French citizens of African descent. *(Copyright Jean-Michel Turpin/Corbis.)*

influenced by these images. For example, through the media, Hawaii has become in the American mind a sort of earthly paradise peopled by scantily clad, eternally happy, invariably good-looking natives who live in a setting of unparalleled natural beauty and idyllic climate. People have always formed images of faraway places. Through the interworkings of popular culture, these images proliferate and become more vivid, if not more accurate.

LOCAL INDIGENOUS CULTURES GO GLOBAL

The world's indigenous peoples often interact with globalization in interesting ways. On the one hand, new global communications systems, institutions of global governance, and international nongovernmental organizations (NGOs) are providing indigenous peoples with extraordinary networking possibilities. Local indigenous peoples around the world are now linked in global networks that allow them to share strategies, rally international support for local causes, and create a united front to defend cultural survival. On the other hand, globalization brings the world to formerly isolated cultures. Global mass communications introduce new values, and multinational corporations' search for new markets and new sources of gas, oil, genetic, forest, and other resources can threaten local economies and environments.

Both aspects of indigenous peoples' interactions with globalization were evident at the World Trade Organization's (WTO) Ministerial Conference in Cancún, Mexico, in 2003. Indigenous peoples' organizations from around the world gathered for the conference, hosted by the Mayan community in nearby Quintana Roo. Though not officially part of the conference, they came together there to strategize ways to forward their collective cause of cultural survival and self-determination, gain worldwide publicity, and protest the WTO's vision of globalization (Figure 2.41). One outcome of this meeting was the International Cancún Declaration of Indigenous Peoples (ICDIP), a document highly critical of current trends in globalization.

According to the ICDIP, the situation of indigenous peoples globally "has turned from bad to

FIGURE 2.41 **A group of indigenous Filipinos participate in the opening of the Forum for Indigenous People** at the Casa de la Cultura in Cancún, Quintana Roo, Mexico, during the World Trade Organization ministerial meetings. Indigenous peoples' groups from around the world organized the forum as a counterpoint to the WTO talks. *(Jack Kurtz/The Image Works.)*

worse" since the establishment of the WTO. Indigenous rights organizations claim that "our territories and resources, our indigenous knowledge, cultures and identities are grossly violated" by international trade and investment rules. The document urges governments worldwide to make no further agreements under the WTO and to reconsider previous agreements. The control of plant genetic resources is a particularly important concern. Many indigenous peoples argue that generations of their labor and cumulative knowledge have gone into producing the genetic resources that transnational corporations are trying to privatize for their own profit. The ICDIP asks that future international agreements ensure "that we, Indigenous Peoples, retain our rights to have control over our seeds, medicinal plants and indigenous knowledge."

Globalization is clearly a critical issue for indigenous cultures. Some argue that globalization, because it facilitates the creation of global networks that provide strength in numbers, may ultimately improve indigenous peoples' efforts to control their own destinies. The future of indigenous cultural survival ultimately will depend on how globalization is structured and for whose benefit.

NATURE–CULTURE

2.4 LEARNING OBJECTIVE
Identify the influence of culture on physical geography and physical geography on culture.

People who depend directly on the land for their livelihoods—farmers, hunters, loggers, and ranchers, for example—often have a different view of nature from those who work in the offices, schools, factories, and shopping malls of the cities. Let's now look at the relationship between nature and cultural differences and how different cultures and subcultures differ in their interactions with the physical environment.

INDIGENOUS ECOLOGY

Many observers believe that indigenous peoples possess a close relationship with and a great deal of knowledge about their physical environment. In many cases, indigenous cultures have developed sustainable land-use practices over generations of experimentation in a particular environmental setting. As a consequence, academics, journalists, and even corporate advertisers frequently portray indigenous peoples as defenders of endangered environments, such as tropical rain forests. It was not always this way. Especially during the height of European **colonialism,** indigenous populations (then considered colonial subjects) were often accused of destroying the environment. The then-common belief that indigenous land-use practices were destructive helped Europeans justify colonialism by claiming they were saving colonial subjects from themselves. In hindsight, it is easy to see that this belief was related to now-discredited European ideas about the racial inferiority of colonized peoples.

colonialism The forceful appropriation of a territory by a distant state, often involving the displacement of indigenous populations to make way for colonial settlers.

Debate continues today, with some observing that, although indigenous cultures may once have lived sustainably, globalization is making their knowledge and practices less useful. That is, globalization introduces new markets, new types of crops, and new technologies that displace existing land-use practices. Others note that it is impossible to generalize about sustainability in indigenous cultures because the way indigenous peoples use their environments varies from place to place and the indigenous societies are internally heterogeneous. A key discussion centers on the role of indigenous peoples in conserving the environment. The discussion is important because indigenous peoples often occupy territories of high **biodiversity** (**Figure 2.42**). Biodiversity refers to the biological diversity of the entire living world, as measured at various scales including diversity among individuals, populations, species, communities, and ecosystems. Biodiversity thus includes genetic diversity, species diversity, and habitat diversity.

biodiversity Biological diversity of the entire living world, as measured at various scales including diversity among individuals, populations, species, communities, and ecosystems.

For example, many of the most biologically diverse areas in Latin America are under government protection in national parks and reserves. Eighty-five percent of these protected areas in Central America and 80 percent in South America contain resident indigenous populations. There is also a close geographic correspondence between indigenous territories and tropical rain forests in not only Latin America but also Africa and Southeast Asia. Tropical rain forests, although they cover only 6 percent of the Earth's surface, are estimated to contain 60 percent of the world's biodiversity. Thus, some see indigenous cultures as traditional caretakers of a precious resource. With the rise of genetic engineering, conservationists and corporations alike view tropical forests as in situ gene banks. That is, genetic diversity can be held in place in standing tropical forests as a reserve to be tapped in the future. As multinational biotechnology companies look to the tropics for genetic resources for use in developing new medicines or crop seeds, indigenous peoples are increasingly vocal about their proprietary rights over the biodiversity of their homelands.

Faced with these issues, cultural geographers emphasize the value of both the knowledge indigenous peoples bring to environmental management and of their land-use practices for sustainable development.

Concentrations of indigenous peoples

Concentrations of high biological diversity

FIGURE 2.42 **The global congruence of cultural and biological diversity.** Conservationists are aware that many of the world's most biologically diverse regions are occupied by indigenous cultures. Many suggest, therefore, that cultural preservation and biological preservation should go hand in hand. *(Adapted from IDRC, 2004.)*

Initially, geographers focused on how indigenous cultures adapted to ecological conditions. For example, some studied the social norms and land-use practices that helped certain cultures adapt to periodic drought. Later, they came to realize that external political and economic forces were just as important in shaping nature–culture relationships. A look at the work of a few key cultural geographers will illustrate the implications of this idea.

LOCAL KNOWLEDGE

Much of the interest in indigenous perceptions and practices falls under the category of **indigenous technical knowledge (ITK).** This is a concept that anthropologists and geographers developed to describe the detailed local knowledge about the environment and land use that is part of many indigenous cultures. Geographer Paul Richards, for example,

indigenous technical knowledge (ITK) Highly localized knowledge about environmental conditions and sustainable land-use practices.

suggested that ITK is, in many cases, superior to Western scientific knowledge and, therefore, should be considered in environmental management and agricultural planning. For example, in his study of West African cultures, *Indigenous Agricultural Revolution,* Richards documented the subtle and extensive knowledge about local soils, climate, and plant life. This local-scale knowledge provides the foundation for people to experiment with new crops and agricultural techniques while also allowing them to adjust successfully to changing social and environmental conditions.

GLOBAL ECONOMY

ITK is place based. It is produced in particular places and environments through a process of trial and error that often spans generations. Thus, these local systems of knowledge are highly adapted to local conditions. Increasingly, however, the power of ITK is weakened through exposure to the economic forces of globalization. Sometimes the global economy applies such pressure to local **subsistence economies** that they become ecologically unsustainable. Subsistence economies are those that are oriented primarily toward production for local consumption, rather than the production of commodities for sale on the market. When an indigenous society organized for subsistence production begins producing for an external market, social, ecological, and economic difficulties often ensue.

> **subsistence economies**
> Economies in which people seek to consume only what they produce and to produce only for local consumption rather than for exchange or export.

Geographer Bernard Nietschmann's classic study of the indigenous Miskito communities living along the Caribbean coast of Nicaragua showed how external markets can undermine local subsistence economies. Miskito communities had developed a subsistence economy founded on land-based gardening and the harvesting of marine resources, including green turtles. Marine resources were harvested in seasons in which agriculture was less demanding. The value of green turtles increased dramatically when companies moved in to process and export turtle products (meat, shells, leather). They paid cash and extended credit so that the Miskito could harvest turtles year-round instead of seasonally. Subsistence production in other areas suffered as labor was directed to harvesting turtles. Turtles became scarce, so more labor time was required to hunt them in a desperate effort to pay debts and buy food. Ultimately, the turtle population was decimated and the subsistence production system collapsed.

This study might sound like yet another tragic story of "disappearing peoples" or a "vanished way of life," but it didn't end there. Nietschmann continued his research with Miskito communities into the 1990s (he died in 2000), discovering, among other things, that the Miskito people did not disappear but continued to defend their cultural independence in the face of great external pressures from globalization. They responded to the collapse of their resource base in 1991 by creating, in cooperation with the Nicaraguan government, a protected area as part of a local environmental management plan. They were supported in this endeavor by academics, international conservation NGOs, and the Nicaraguan government. Known as the Cayos Miskitos and Franja Costera Marine Biological

Reserve, it encompasses 5019 square miles (13,000 square kilometers) of coastal area and offshore keys with 38 Miskito communities. Through this program, the Miskito were able to regulate and control their own exploitation of marine resources while reducing pressures from outsiders. This is an ongoing experiment. Miskito communities continue to struggle with outsiders for control over their land and resources in the reserve. There are hopeful signs, however, that the government is cooperating in this struggle and that both the natural resource base and the Miskito people will benefit from this project.

The Miskito case demonstrates the resiliency of indigenous cultures, the limits of ITK, and the strength of global economic forces. It reflects recent studies of the cultural and political ecology of indigenous peoples, such as those conducted by geographer Anthony Bebbington. Bebbington conducted research among the indigenous Quichua populations in the Ecuadorian Andes to assess how they interact with modernizing institutions and practices. He found that, although the Quichua people often possess extensive knowledge about local farming and resource management, ITK alone is not sufficient to allow them to prosper in a global economy. For example, there is little indigenous knowledge about the way international markets work and thus little understanding of how to price and market their own produce. As a consequence, they have sought the support and knowledge of government agencies, the Catholic Church, and NGOs. He further found that indigenous Quichua communities use outside ideas and technologies to promote their own cultural survival, attempting, in essence, to negotiate their interactions with globalization on their own terms.

FOLK ECOLOGY

As with indigenous cultures, ideas persist about the particular abilities of folk cultures to sustainably manage the environment. Although the attention to conservation varies from culture to culture, folk cultures' close ties to the land often produce detailed local ecological knowledge within folk groups. This becomes particularly evident when they migrate. Typically, they seek new lands similar to the one left behind. A good example can be seen in the migrations of Upland Southerners from the mountains of Appalachia between 1830 and 1930. As the Appalachians became increasingly populous, many Upland Southerners began looking elsewhere for similar areas to settle. Initially, they found an environmental twin of the Appalachians in the Ozark-Ouachita Mountains of Missouri and Arkansas. Somewhat later, others sought out the hollows, coves, and gaps of the central Texas Hill Country. The final migration of Appalachian hill people brought some 15,000 members of this folk culture to the Cascade and Coast mountain ranges of Washington State between 1880 and 1930 (**Figure 2.43**).

Although folk cultures have largely disappeared, the human inventiveness, curiosity, and need related to local ecologies remain, as we noted. Thus, there are many examples where local people derive what we might call folk solutions when confronted with a challenge of physical geography. The history of

mountain bikes provides an example. Mountain bikes were invented when young men in the mountain regions of the Western United States attempted to figure out how to adopt existing technology to a new enterprise. Bikes were cannibalized for parts and reassembled into something new and local. Enthusiasts rode these single-speed "clunkers" with big balloon tires, careening down mountain trails and dirt roads in a sport that did not yet have a name. Eventually, these locally imagined, hand-assembled clunkers became the rough prototypes of today's mass-produced mountain bikes. Another good example is the swamp buggy, a folk solution to the problem of navigating in and around the Everglades and associated wetlands of southern Florida (**Figure 2.44**). Local hunters in southwest Florida needed a way to access wildlife. In the 1940s they came up with a solution that involved installing huge balloon tires, oversized suspension, and other modifications on pickup trucks. Jacked up high, these vehicles are able to move through standing water and over rough terrain in a way that no mass-produced vehicle can. They remain the best solution to the challenge of local geography and are a common sight in the south Florida wetlands landscape.

NATURE AND GENDER

We noted in Chapter 1 and earlier in this chapter that cultures are heterogeneous and that gender is a key marker of difference within places and regions. Geographers and other social scientists have documented significant differences between

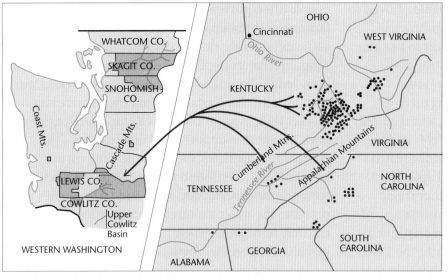

- Appalachian place of origin of families or individuals migrating to the Upper Cowlitz Basin
- Settlement areas of Appalachian hill folk in Washington State

FIGURE 2.43 **The relocation diffusion of Upland Southern hill folk from Appalachia to western Washington.** Each dot represents the former home of an individual or family that migrated to the Upper Cowlitz River basin in the Cascade Mountains of Washington State between 1884 and 1937. Some 3000 descendants of these migrants lived in the Cowlitz area by 1940. *What does the high degree of clustering of the sources of the migrants and subsequent clustering in Washington suggest about the processes of folk migrations? How should we interpret their choices of familiar terrain and vegetation for a new home? Why might members of a folk society who migrate choose a new land similar to the old one?* (After Clevinger, 1938, p. 120; Clevinger, 1942, p. 4.)

FIGURE 2.44 **A folk solution to navigating Florida's swamplands.** In the 1940s, local residents of the greater Everglades region began modifying truck chassis to better navigate the swampy landscape. The oversized tires and suspensions allowed hunters to drive through inundated land. The solution was widely adopted and is used today by hunters and tour operators. *(Time & Life Pictures/ Getty Images.)*

men's and women's relationships with the environment. This observation holds for many different types of cultures. **Ecofeminism** is one way of thinking about how gender influences our interactions with nature. This concept, however, might seem to suggest that there is something inherent or essential about men and women that makes them think about and behave toward the environment in particular and different ways. Although cultural geographers would argue against such essentialism, few would disagree that gender is an important variable in nature–culture relations.

Diane Rocheleau's work on women's roles in the management of **agroforestry** systems makes this point. Agroforestry systems are farming systems that combine the growing of trees with the cultivation of agricultural crops. Agroforestry is practiced by folk and indigenous cultures across the tropical world and has been shown to be a highly productive and ecologically sustainable practice. It is common in these production systems for men and women to have very distinct roles. Generally, for example, women are involved in seeding, weeding, and harvesting, whereas men take responsibility for clearing and cultivation. Gender differences also often exist in the types of crops men and women control and in the marketing of produce.

After conducting studies for many years—first in East Africa and later in the Dominican Republic—Rocheleau was able to see general themes regarding the way human–environment relations are different between men and women in not only agroforestry systems but also many rural and urban environments. Together with two colleagues, she identified three themes: gendered knowledge, gendered environmental rights, and gendered environmental politics. First, because women and men often have different tasks and move in different spaces, they possess different and even distinct sets of knowledge about the environment (gendered knowledge). Second, men and women have different rights, especially with regard to the ownership and control of land and resources (gendered environmental rights). Third, for reasons having to do with their responsibilities in their families and communities, women are often the main leaders and activists in political movements concerned with environmental issues (gendered environmental politics). Taken together, these themes suggest that environmental planning or resource management projects that do not address issues of gender are likely to have unintended consequences, some of them negative for both women and environmental quality.

NATURE IN POPULAR CULTURE

Popular culture is less directly tied to the physical environment than are folk and indigenous cultures, which is not to say that it does not have an enormous impact on the environment. City dwellers generally do not draw their livelihoods from the land. They have no direct experience with farming, mining, or

ecofeminism A doctrine proposing that women are inherently better environmental preservationists than men because the traditional roles of women involve creating and nurturing life, whereas the traditional roles of men too often necessitate death and destruction.

agroforestry A cultivation system that features the interplanting of trees with field crops.

FIGURE 2.45 **Traffic jam in Yosemite National Park at the height of the summer tourist season.** *(Tom Meyers Photography.)*

logging activities, though they could not live without the products and materials produced from those activities. Gone is the intimate association between people and land once known by past folk generations. Gone, too, is our direct vulnerability to many environmental forces, although this security is counter-balanced by new risks. Because popular culture is so tied to mass consumption, it can have enormous environmental impacts, such as the production of air and water pollution and massive amounts of solid waste. Also, because popular culture fosters limited contact with and knowledge of the physical world, usually through recreational activities, our environmental perceptions can become quite distorted. (See Rod's Notebook.)

Popular culture makes heavy demands on ecosystems. This is true even in the seemingly harmless activity of outdoor recreation. Recreational activities have increased greatly in the world's economically affluent regions. Many of these activities require machines, such as snowmobiles, off-road vehicles, and jet skis, that are powered by internal combustion engines and have numerous adverse ecological impacts ranging from air pollution to soil erosion. In national parks and protected areas worldwide, affluent tourists in search of nature have overtaxed protected environments and wildlife and produced levels of congestion approaching those of urban areas (**Figure 2.45**).

Such a massive presence of people in our recreational areas inevitably results in damage to the physical environment. A study by geographer Jeanne Kay and her students in Utah revealed substantial environmental damage done by off-road recreational vehicles, including "soil loss and long-term soil deterioration." One of the paradoxes of the modern age and popular culture seems to be that the more we cluster in cities and suburbs, the greater our impact on open areas because we carry our popular culture with us when we vacation in such regions.

CULTURAL LANDSCAPE

2.5 **LEARNING OBJECTIVE**
Recognize how landscapes reflect cultural differences.

The theme of cultural landscape reveals the important differences within and between cultures.

FOLK ARCHITECTURE IN NORTH AMERICA

Every folk culture produces a highly distinctive landscape. One of the most visible aspects of these landscapes is **folk architecture.** These traditional buildings illustrate the theme of folk cultural landscape. Folk

folk architecture Structures built by members of a folk society or culture in a traditional manner and style, without the assistance of professional architects or blueprints, using locally available raw materials.

Rod's Notebook

Encountering Nature

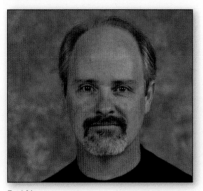

Rod Neumann. *(Courtesy of Roderick Neumann.)*

In popular culture, fewer and fewer of us have direct contact with what we might call wild nature. We don't make our livings by hunting, clearing fields, or pasturing livestock in the mountains. Rather, mass media, particularly television and film documentaries, shape our understanding of nature and wild places. Wild nature has been transformed into a spectacle in popular culture, something for our amusement and entertainment.

I was conducting research in the historical records of Tanzania's Serengeti National Park when these observations were driven home to me in a very powerful way. Serengeti, like the Amazon, is one of those iconic wild places in popular culture. Who hasn't seen a Serengeti cheetah blazing across the television screen in pursuit of a panicked wildebeest?

I spent my days looking through archived files at the park's research headquarters and my evenings at the Seronera Wildlife Lodge, a luxury hotel in the heart of the African "wilds." I would watch vanloads of American and European (but no African) tourists return to the lodge from their daily wildlife safaris. We would all perch on the balcony with beers or cocktails, watching the elephants, gazelles, and baboons gather at the artificial watering hole just meters away. Drawn to water in the arid landscape, these beasts daily provided us with an entertaining spectacle of wild African nature without our having to leave the bar!

After dark, people would move inside and gather around a television in the lounge and watch—wait for it!—television documentaries about wildlife in Serengeti. I was witnessing an almost surreal feedback loop in which urbanized tourists, drawn to East Africa by the television documentaries they'd viewed in their living rooms in the United States, now sat in front of a television set in the middle of Serengeti watching the wildlife they'd seen that day or hoped to see tomorrow. It seemed as if the television in the lounge was needed to reinforce the reality of the actual experience of viewing African wildlife firsthand. Such is the power of mass media in shaping human encounters with wild nature in popular culture.

The situation at Tarangire Safari Lodge in Tarangire National Park, Tanzania, where tourists are able to view wildlife from the comfort of a patio bar, is common in many African parks. *(Christina Micek.)*

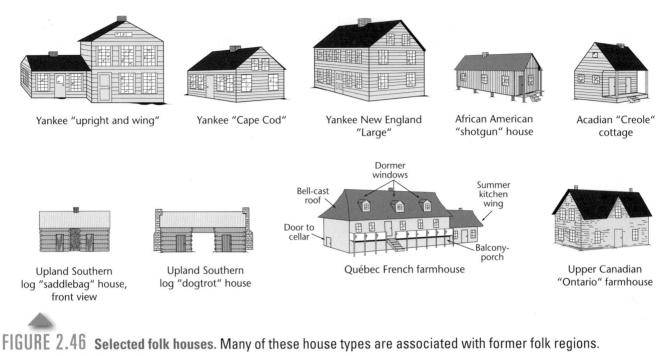

FIGURE 2.46 **Selected folk houses.** Many of these house types are associated with former folk regions. *(After Glassie, 1968; Kniffen, 1965.)*

architecture springs not from the drafting tables of professional architects but from the collective memory of groups of a people (**Figure 2.46**). Material composition, floor plan, and layout are important ingredients of folk architecture, but numerous other characteristics help classify farmsteads and dwellings. The form or shape of the roof, the placement of the chimney, and a building's location and orientation in the physical environment are important classifying criteria as well.

In the United States and Canada, folk architecture today is a relic. Traces can still be found here and there and provide both a sense of past cultural landscapes and enduring symbols of cultural identity. One needs to know what to be on the lookout for in folk houses, however, in order to properly "read" the landscape. Yankee folk houses, for example, are of wooden frame construction, and shingle siding often covers the exterior walls. They are built with a variety of floor plans, including the New England "large" house, a huge two-and-a-half-story house built around a central chimney and two rooms deep. As the Yankee folk migrated westward, they developed the upright-and-wing dwelling. These particular Yankee houses are frequently massive, in part because

the cold winters of the region forced most work to be done indoors.

By contrast, Upland Southern folk houses are smaller and built of notched logs. Many houses in this folk tradition consist of two log rooms, with either a double fireplace between, forming the saddlebag house, or an open, roofed breezeway separating the two rooms, a plan known as the dogtrot house (**Figure 2.47**). An example of an African American folk dwelling is the shotgun house, a narrow structure only one room in width but two, three, or even four rooms in depth. Acadiana, a French-derived folk region in Louisiana, is characterized by the half-timbered Creole cottage, which has a central chimney and built-in porch. Scores of other folk house types survive in the American landscape, although most such dwellings are now abandoned.

Canada also offers a variety of traditional folk houses (see Figure 2.46). In French-speaking Québec, one of the common types consists of a main story atop a cellar, with attic rooms beneath a curved, bell-shaped (or bell-cast) roof. A balcony-porch with railing extends across the front, sheltered by the overhanging eaves. Attached to one side of this type of French-Canadian folk house is a summer kitchen

FIGURE 2.47 **A dogtrot house, typical of the Upland Southern folk region.** The distinguishing feature is the open-air passageway, or dogtrot, between the two main rooms. This house is located in central Texas. *(Courtesy of Terry G. Jordan-Bychkov.)*

FOLK ARCHITECTURE AROUND THE WORLD

that is sealed off during the long, cold winter. Often the folk houses of Québec are built of stone. To the west, in the Upper Canadian folk region, one type of folk house occurs so frequently that it is known as the Ontario farmhouse. One-and-a-half stories in height, the Ontario farmhouse is usually built of brick and has a distinctive gabled front dormer window. Now that we have reviewed a few of the common styles of North American folk architecture, you can test your ability to identify them in the landscape (**Figure 2.48**).

Although in North America folk architecture tends to be found only as a cultural relic of the past, it continues to thrive in regions where agriculture is the main occupation. Most construction is done with local materials such as clay, stone, wood, bamboo, leaves, and grasses. Distinctive cultural landscapes emerge from the collective use of folk architecture, just two examples of which we present here.

FIGURE 2.48 **Four folk houses in North America.** Such distinctive house types often serve as visual clues to former folk regions. *(Courtesy of Terry G. Jordan-Bychkov.)*

World Heritage Site

Folk architecture has given way to professionally built and mass-produced housing. Recognizing that important cultural heritage is vanishing, in 2012 UNESCO designated seven large farmhouses in northeastern Sweden's Hälsingland region as the Decorated Farmhouses of Hälsingland World Heritage Site. Out of more than 1000 possibilities, the committee chose these seven as outstanding examples of a regional folk architecture rooted in the Middle Ages. The timber farmhouses were constructed mostly between 1800 and 1870 C.E. and are richly decorated, inside and out, following a local folk art tradition

■ Hälsingland, a small region in Sweden bordering the Gulf of Bothnia, contains a concentration of such farmhouses. The seven UNESCO-designated farmhouses are spread over an area stretching 100 by 50 kilometers. Both they and the surrounding landscapes are protected under Swedish cultural heritage and environmental laws, reflecting their importance to that nation's cultural identity.

■ The Decorated Farmhouses of Hälsingland represent the final flourishing of a centuries-old local tradition of timber construction. Beginning in the early nineteenth century, farmers in the region experienced a new period of prosperity based on forestry and flax production. Farmers invested their surplus earnings into building and decorating farmhouses. Most of these structures remain under the private ownership of farmers who continue to manage the surrounding lands for agriculture.

■ The cultural landscape of Hälsingland has evolved over centuries and bears the imprint of occupation by a population of independent farmers. The decorated farmhouses are located on active farmsteads and within a larger agrarian landscape. The mixed agrarian livelihood practices of cattle breeding, crop cultivation, forestry, and hunting have visibly shaped the land. Historically, farmers used the pastures and woodlands communally and shared cultivated fields. Changes in Swedish law in the nineteenth century led to individual privatization of lands, which brought prosperity to some farmers and a flowering of folk architecture.

(age fotostock/Alamy.)

Decorated Farmhouses of Hälsingland

IN THE FARMHOUSES:

The seven designated farmhouses are the Kristofers farm in Stene, Järvsö; the Gästgivars farm in Vallstabyn; the Pallars farm in Långhed; the Jon-Lars farm in Långhed; the Bortom åa farm in Gammelgården; the Bommars farm in Letsbo, Ljusdal; and the Erik-Anders farm in Askesta, Söderala. Each contains a number of decorated rooms for festivities and some very prosperous farmsteads have entire houses, a *herrstuga,* dedicated solely to festive celebrations. The farmsteads feature numerous intact buildings that were built and elaborately painted in a manner meant to display the prosperity and social status of the farmer.

■ The houses are made entirely of wood, most constructed in a building tradition that may be traced to the Middle Ages, which features jointed horizontal timbers of pine or spruce from nearby forests. Prosperous farmers of the nineteenth century enlarged existing houses from one to two or two and a half stories or built entirely new houses on the farmstead. The emphasis was on creating exceptionally large structures using the most talented local carpenters and craftsmen.

■ Each farmhouse is elaborately decorated. On the outside, local craftsmen carved decorative elements around the front

(Photolibrary/Getty Images.)

porch and main entrance. Inside, the houses are decorated with textile paintings hung on the walls or with paintings applied directly to the wooden walls and ceilings. The rooms dedicated solely for festivities and special occasions are typically the most highly decorated.

■ The paintings fuse local folk art traditions with more universal styles, such as Baroque and Rococo, favored by the landed gentry of nineteenth-century Sweden. Wealthy farmers commissioned itinerant artists to paint their farmhouse interiors with depictions of landscapes and biblical characters in contemporary fashion or with detailed floral patterns. Ten of the painters who worked on the World Heritage farmhouses are known, but most remain anonymous.

TOURISM:

The seven UNESCO-designated farmhouses are located in a scenic coastal region of northeastern Sweden, which has long been an attraction for both domestic and foreign tourists.

(Hans Strand/Corbis.)

■ Regional public and business groups, as well as individual farmers, actively collaborate to keep decorated farmhouses open to visitors. Museums in the region feature displays on the history of this folk tradition.

■ The maintenance and conservation of Swedish cultural heritage is a popular movement in the country. Many craftsmen and women in Hälsingland carry on folk handicrafts, such as textiles, furniture making, and basketry, which draw many visitors to the region.

■ Established between 1800 and 1870 C.E., these farmsteads in northeastern Sweden represent the final flowering of a centuries-old tradition of folk architecture.

■ In 2012 seven farmhouses were designated as World Heritage Sites, being noteworthy examples of the elaborately decorated interiors characteristic of the Hälsingland's region's folk culture.

■ Visitors are attracted to the region by its scenic beauty, agrarian landscapes, and celebration of local folk traditions and handicrafts.

http://whc.unesco.org/en/list/1282/

FIGURE 2.49 **Nipa hut in Baguio City, Philippines.** Folk architecture, such as the nipa hut, typically uses local materials and is designed to adapt to local environmental conditions. *(Getty Images/Flickr RF.)*

The characteristic folk housing of the Philippines is the nipa hut, still found in the countryside with many variations resulting from the mix of Malaysian, Spanish, Muslim, and American influences. Most commonly, it is constructed primarily of bamboo framing fastened and roofed with nipa leaves (Figure 2.49). Philippine folk architecture conveys a feeling of lightness (through, for example, the placement of large open-air windows) combined with a quality of warmth produced from the use of wood inside. The nipa hut is often raised on stilts, particularly in coastal areas and flood plains. The combination of large windows, wide roof eaves, and its elevated position results in the free flow of cooling breezes even during rainy weather. The nipa hut is widely considered as the Philippine national dwelling and thus is closely associated with cultural and national identity.

In southern Africa, one of the most distinctive house types is found in the Ndebele culture region, which stretches from South Africa north into southern Zimbabwe. In the rural parts of the Ndebele region, people are farmers and livestock keepers and live in traditional compounds. What makes these houses distinctive is the Ndebele custom of painting brightly colored designs on the exterior house walls and sometimes on the walls and gates surrounding the grounds (Figure 2.50). The precise origins of this custom are unclear, but it seems to date to the mid-nineteenth century. Some suggest that it was an

assertion of cultural identity in response to local residents' displacement and domination at the hands of white settlers. Others point to a religious or sacred role.

What is clear is that the custom has always been handled by women, a skill and practice passed down from mother to daughter. Many of the symbols and patterns are associated with particular families or clans. Initially, women used natural pigments from clay, charcoal, and local plants, which restricted their palette to earth tones of brown, red, and black. Today, many women use commercial paints—expensive, but longer lasting—to apply a range of bright colors, limited only by the imagination. Another new development is the types of designs and symbols used. People are incorporating modern machines such as automobiles, televisions, and airplanes into their designs. In many cases, traditional paints and symbols are blended with the modern to produce a synthetic design of old and new. Ndebele house painting is developing and evolving in new directions, all the while continuing to signal a persistent cultural identity to all who pass through the region.

LANDSCAPES OF CONSUMPTION

It is difficult to single out one type of popular landscape, just as it is difficult to distinguish one type of folk landscape. Popular culture has been, nonetheless, commonly associated with mass-produced suburban housing, commercial strips, and large indoor shopping

FIGURE 2.50 **Ndebele village in South Africa.** While the origins and meaning of Ndebele house painting remain matters of debate, there is no question that it creates a visually distinct cultural landscape. *(Ariadne Van Zandbergen/Alamy.)*

FIGURE 2.51 Indoor shopping mall at the foot of the Petronas Towers, Kuala Lumpur, Malaysia. Most of the world's largest malls are now located outside of North America. Malaysia boasts 4 of the top 15 largest in the world. *(Robert Harding World Imagery/Getty Images.)*

malls. These latter retail districts have come to be known by geographer Robert Sack's term, "landscapes of consumption" (refer back to Figure 2.8). Not only are such landscape features common in North America, they are now also seen in Brazil, China, India, and many other countries with rapidly growing economies increasingly linked through globalization (**Figure 2.51**). Indeed, the world's ten largest (in terms of leasable area) indoor shopping malls may all be found outside of North America.

Of these, North America's largest indoor mall is West Edmonton Mall in the Canadian province of Alberta (**Figure 2.52**). Enclosing some 5.3 million square feet (493,000 square meters) and opened in 1986, West Edmonton Mall employs 23,500 people in more than 800 stores and services, accounts for nearly one-fourth of the total retail space in greater Edmonton, earned 42 percent of the dollars spent in local shopping centers, and experienced 2800 crimes in its first nine months of operation. Beyond its sheer size, West Edmonton Mall also boasts a water park, a sea aquarium, an ice-skating rink, a miniature golf course, a roller coaster, 21 movie theaters, and a 360-room hotel. Its "streets" feature motifs from such distant places as New Orleans, represented by a Bourbon Street complete with fiberglass ladies of the evening. Jeffrey Hopkins, a geographer who studied this mall, refers to this as a "landscape of myth and elsewhereness," a "simulated landscape" that reveals the "growing intrusion of spectacle, fantasy, and escapism into the urban landscape."

LEISURE LANDSCAPES

Another common feature of popular culture is what geographer Karl Raitz labeled **leisure landscapes.** Leisure landscapes are designed to entertain people on weekends and vacations; often they are included as part of a larger tourist experience. Golf courses and theme parks such as Disney World are good examples of such landscapes. **Amenity landscapes** are a related landscape form. These are regions with attractive natural features such as forests, scenic mountains, or lakes and rivers that have become

> **leisure landscapes**
> Landscapes planned and designed primarily for entertainment purposes, such as ski and beach resorts.
>
> **amenity landscapes**
> Landscapes prized for their natural and cultural aesthetic qualities by the tourism and real estate industries and their customers.

FIGURE 2.52 The enormous West Edmonton Mall represents the growing infusion of spectacle and fantasy into the urban landscape. In addition to retail outlets, the mall includes a sea aquarium, an ice-skating rink, a miniature golf course, and the Submarine Ride shown here. *(James Marshall/Corbis.)*

desirable locations for retirement or vacation homes. In much of Europe, rural regions that can no longer economically support agricultural production are increasingly viewed as leisure landscapes. Geographer Arjen Buijs and colleagues have documented in France and the Netherlands a shift in public understanding of rural landscape from one of a requirement for livelihood to one of an amenity offering comfort and convenience. Such shifts have strong implications for the future of rural life in western Europe, not least of which is the growing use of the countryside for vacation and retirement homes (**Figure 2.53**).

The past, reflected in surviving buildings, has also been incorporated into leisure landscapes. Most often, folk or historic buildings that cannot be preserved in place are relocated to create a sort of outdoor museum. Many small towns across North America feature such historical villages. Examples include the Heritage Historical Village of New London, Wisconsin, which features five relocated and restored historic buildings, and the Historical Museum and Village of Sanibel, Florida, which includes seven building relocated from around the island (**Figure 2.54**). In cases where no historical

FIGURE 2.54 **Sanibel Island, Florida Historical Village and Museum.** Heritage villages like this one are created by relocating historic buildings to a single site to facilitate preservation and tourism. *(M. Timothy O'Keeffe/Alamy.)*

buildings exist, history parks have been entirely reconstructed—for example, at Jamestown, Virginia, or Louisbourg on Cape Breton Island, Nova Scotia. Some of the larger re-created historical villages feature *living history*, which frequently includes both actors in period costume and skilled craftspeople practicing lost arts such as blacksmithing or candle making. Such sites can play an important role in developing and maintaining connections to the historic roots in contemporary cultural identities.

ELITIST LANDSCAPES

A characteristic of popular culture is the development of social classes. A small elite group—consisting of persons of wealth, education, and expensive tastes—occupies the top economic position in popular cultures. The important geographical fact about such people is that because of their wealth, desire to be around similar people, and affluent lifestyles, they can and do create distinctive cultural landscapes, often over fairly large areas.

Daniel Gade, a cultural geographer, coined the term *elitist space* to describe such landscapes, using the French Riviera as an example (**Figure 2.55**). In that district of southern France, famous for its stunning natural beauty and idyllic climate, the

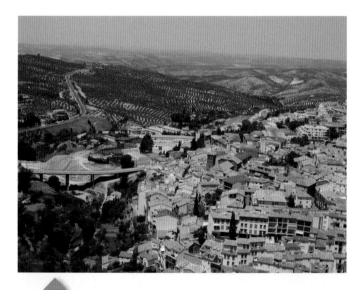

FIGURE 2.53 **New second-home sprawl in Cazorla, Spain.** Cazorla is in the heart of what has long been an important agricultural region. As agricultural industrialization reduced rural employment opportunities, the economy shifted toward tourism. Cazorla has experienced a mini real estate boom in retirement and vacation home construction as a result. *(Courtesy of Roderick Neumann.)*

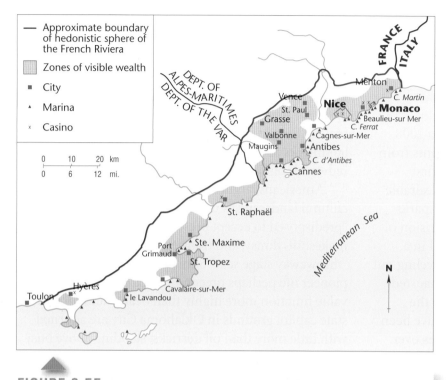

FIGURE 2.55 **The distribution of an elitist or hedonistic cultural landscape on the French Riviera.** *What forces in the popular culture generate such landscapes?* (Adapted from Gade, 1982, p. 22.)

laborers in these extractive industries. The New West refers to an amenity-based economy that features luxury second homes, outdoor recreational resorts, and international art and film festivals. The New Westerners are affluent professionals who earned their money in distant cities and are attracted to the natural beauty of the mountainous West. They are creating elite landscapes in spaces formerly dedicated to resource extraction.

Geographers John Harner and Bradley Benz conducted a study in southern Colorado that perfectly demonstrates how New West landscapes emerge. They focused on the area surrounding Durango, a former mining town that is now a tourist mecca featuring a steam-train, gourmet restaurants, ski shops, and boutique clothing stores. There, they measured the growth of *ranchettes*,

French elite applied "refined taste to create an aesthetically pleasing cultural landscape" characterized by the preservation of old buildings and town cores, a sense of proportion, and respect for scale. Building codes and height restrictions, for instance, are strictly enforced. Land values, in response, have risen, making the Riviera ever more elitist, far removed from the folk culture and poverty that prevailed there before 1850. Farmers and fishers have almost disappeared from the region, though one need drive but a short distance, to Toulon, to find a working seaport. It seems, then, that the different social classes generated within popular culture become geographically segregated, each producing a distinctive cultural landscape (**Figure 2.56**).

The United States, too, offers elitist landscapes. The economic and cultural changes unfolding in the western part of the country offer a vivid example. This phenomenon is popularly known as the "Old West" versus the "New West." The Old West is the historic regional economy and corresponding cultural landscapes based on cattle ranching, mining, and logging. Most Old Westerners worked as wage

FIGURE 2.56 **The cultural landscape of the affluent in Port of Fontvieille, Monaco.** The huge yachts and Mediterranean residences are out of reach for all but the world's wealthiest. *(Sergio Pitamitz/Age Fotostock.)*

35- to 70-acre private parcels of land too small to be economically productive or legally subdivided further. These new landholdings are created for wealthy New Westerners who vacation in or retire to their new luxury homes in the wide-open spaces.

Harner and Benz found that the number of ranchettes increased 67 percent from 1988 to 2008, which signals a huge influx of affluent migrants from the cities. Their study documented that the first ranchettes were constructed near the most desirable natural settings: streams and rivers, national parks and forests, and mountain vistas. The subdivision of large landholdings into ranchettes continued in a pattern of contiguous diffusion, gradually leveling out as the supply of amenity-rich spaces was exhausted.

Such fragmentation of landholdings and the corresponding creation of elite landscapes have been repeated across the western states many times over. Frequently, the process has led to clashes between newcomers and established residents. Geographers Peter Walker and Louise Fortmann studied the social conflicts emerging from the New West phenomenon in the foothills of California's Sierra Nevada range. They determined that the most contentious political debates focused not on economic issues, but on cultural values expressed in landscape preferences. Affluent newcomers and established residents made different aesthetic judgments of landscape, and this was what drove efforts to gain political control over land-use decisions. The New West phenomenon highlights the relevance of cultural landscape to regional politics and economies.

THE AMERICAN POPULAR LANDSCAPE

In an article entitled "The American Scene," geographer David Lowenthal attempted to analyze the visible impact of popular culture on the American countryside. Lowenthal identified the main characteristics of popular landscape in the United States, including the "cult of bigness"; the tolerance of present ugliness to achieve a supposedly glorious future; an emphasis on individual features at the expense of aggregates, producing a "casual chaos"; and the preeminence of function over form.

The American fondness for massiveness is reflected in structures such as the Empire State Building, the Pentagon, the San Francisco–Oakland Bay Bridge, and Salt Lake City's Mormon Temple. Americans have dotted their cultural landscape with the world's largest of this or that, perhaps in an effort to match the grand scale of the physical environment, which includes such landmarks as the Grand Canyon, the towering redwoods of California, and the Rocky Mountains.

Americans, argues Lowenthal, tend to regard their cultural landscape as unfinished. As a result, they are "predisposed to accept present structures that are makeshift, flimsy, and transient," resembling "throwaway stage sets." Similarly, the hardships of pioneer life perhaps preconditioned Americans to value function more highly than beauty and form. The state capitol grounds in Oklahoma City are adorned with little more than oil derricks, standing above busy pumps drawing oil and the wealth that comes with it from the Sooner soil—an extreme but revealing view of the American landscape (**Figure 2.57**).

In summary, American popular culture seems to have produced a built landscape that stresses bigness,

FIGURE 2.57 **Oil derrick on the Oklahoma state capitol grounds.** The landscape of American popular culture is characterized by such functionality. The public and private sectors of the economy are increasingly linked in the popular culture. ***Is criticism of such a landscape elitist and snobbish?*** *(See also Robertson, 1996. Photo courtesy of Terry G. Jordan-Bychkov.)*

FIGURE 2.58 **Walmart is one example among many of the "big box" stores now ubiquitous in suburban landscapes.** Such buildings incorporate the ideals of bigness, transience, and utilitarianism common in the architecture of popular culture. *(Courtesy of Roderick Neumann.)*

utilitarianism, and transience. Sometimes these are opposing trends, as in the case of many of the massive structures previously cited, which are clearly built to last. Sometimes the trends mesh, as in the recent explosion of "big box" retail chains (**Figure 2.58**). These giant retail buildings are no more than oversized metal sheds that one can easily imagine being razed overnight, only to be replaced by the next big thing.

CONCLUSION

In this chapter, we have just scratched the surface of the complex nature of the geographies of cultural difference. Using just a few of the multiple categories of cultures and subcultures, we learned that there are many ways of perceiving and being in the landscape. We have also seen how important geography is in the construction of national and cultural identities and vice versa. Religion, often a defining element of cultural difference and cultural identity, is also vitally important. Chapter 7 is devoted to this major cultural trait.

DOING GEOGRAPHY

Self-Representation of Indigenous Culture

Indigenous cultures are reasserting themselves after 500 years of marginalization. For centuries, the roar of dominant national cultures drowned out indigenous peoples' voices. Members of dominant national cultures—academics, missionaries, government officials, and journalists—have been largely responsible for writing about the histories and cultures of indigenous peoples. This is changing as some indigenous groups have prospered economically and as the indigenous rights movement gains increasing support worldwide. Today, more and more indigenous peoples are taking charge of representing themselves and their cultures to the outside world by building museums, producing films, hosting conferences, and creating web sites.

This exercise requires you to carefully study one of these platforms for cultural expression: *self-produced* indigenous peoples' web sites. To do so, follow the steps below.

Steps to Understanding Indigenous Culture

Step 1: Do some background research on the names and locations of major indigenous cultures. You can use the web sites listed at the end of this chapter to get started. It is important to verify that the web sites you decide to study are self-produced. Confirm that the site is produced by a tribal or indigenous organization, not by an external NGO, national government, corporation, or university.

Step 2: Think about how you are going to analyze the content of the web site in order to draw conclusions about the self-representation of indigenous cultures. Here are a few suggestions and possibilities: focus on questions of geography such as territorial claims, rights over natural resources, culturally significant relations with nature, and homeland self-rule.

Based on your research and analysis, systematically analyze how indigenous populations represent their cultural identities. Consider these questions:

- How do indigenous groups speak about their relationship to the land and the environment?

- What do they say about territorial claims and homelands? What roles do maps play on the web sites?

- What are their ideas on biodiversity conservation and bioprospecting (the search for genetic resources and other biological resources)?

In addition, look for discussions of conflict, cooperation, or dis-agreement with national governments or multinational corporations:

■ Is there a project or policy (e.g., disposal of radioactive material) under dispute?

■ What position is taken on the web site? How is the position framed in relation to indigenous rights and culture?

■ What major issues and challenges does the site highlight and how do these relate to globalization?

Finally, think about possibilities for comparison:

■ Are there regional (on either the U.S. national or global scale) differences in terms of the quantity and content of web sites?

■ Did you find common themes across or within regions?

■ Are there indigenous cultures that produce contrasting or competing representations?

■ Do some indigenous cultures have more than one self-generated web site and do those sites present different ideas?

Jim Enote, executive director of the A:shiwi A:wan Museum and Heritage center in New Mexico, discusses a Zuni map art painting at an exhibition. Many Native American tribes build museums to represent their culture and way of life. *(AFP/Getty Images.)*

Chapter 2
LEARNING OBJECTIVES REEXAMINED

2.1 Explain the role of cultural difference in shaping regions.
See pages 50–61. What is the difference between indigenous culture regions and vernacular culture regions?

2.2 Describe the ways that mobility interacts with cultural difference.
See pages 61–67. What is the *digital divide* and how has mobility created this divide?

2.3 Analyze the potential for globalizing processes to shape and be altered by cultural difference.
See pages 67–71. How are the forces of globalization reflected in popular culture?

2.4 Identify the influence of culture on physical geography and physical geography on culture.
See pages 71–76. How is the relationship between culture and nature changing over time?

2.5 Recognize how landscapes reflect cultural differences.
See pages 76–87. How is popular culture reflected in the landscape?

KEY TERMS

agroforestry	p. 75	indigenous technical	
amenity landscapes	p. 83	knowledge (ITK)	p. 72
biodiversity	p. 71	leisure landscapes	p. 83
colonialism	p. 71	local consumption	
consumer nationalism	p. 68	cultures	p. 67
convergence		material culture	p. 44
hypothesis	p. 67	national culture	p. 46
diaspora culture	p. 49	nonmaterial culture	p. 44
digital divide	p. 66	placelessness	p. 48
ecofeminism	p. 75	popular culture	p. 47
European Union	p. 63	subcultures	p. 43
folk architecture	p. 76	subsistence economies	p. 73
folk culture	p. 45	vernacular culture	
indigenous culture	p. 46	region	p. 60

Cultures on the Internet
You can learn more about some of the categories of culture difference discussed in this chapter on the Internet at the following web sites:

American Memory, Library of Congress
http://memory.loc.gov
A project of the Library of Congress that presents a history of American popular culture, complete with documentation and maps.

SEEING GEOGRAPHY Camping in the "Great Outdoors"

What can this scene tell us about nature–culture relations in North American popular culture?

Enjoying nature in a national park campground.

(Courtesy of Roderick Neumann.)

How we think about and interact with nature reveals a great deal about how we understand our place in the world. For example, within contemporary popular culture we often think of our relationship with nature as something outside the routine of daily life. We seek out nature on weekends, during vacations, or in retirement. Nature is equated with pretty scenery, which in popular culture typically means forests, mountains, and wide-open spaces. This scene from Everglades National Park in Florida reveals a great deal about one of the main ways within popular culture that we express this understanding of nature: camping in the "great outdoors."

Of course, there are many ways to camp. This particular form, recreational vehicle (RV) or motor home camping, is extremely energy intensive. Unseen in this photo is the supporting industrial manufacturing complex organized to produce a fleet of vehicles, trailers, and equipment, all of which are intended to be situated on an asphalt slab and connected to an electrical power grid. A lot of natural resources have to be consumed to experience nature in this way. Such an experience is shaped less by direct physical interaction with nature than by interaction with mass-produced commodities. Nature serves mainly as background scenery.

Thinking of nature as background scenery suggests a stage set for actors to interact on. Look carefully at the photo and you will see that this campground is a stage for both extremely private and highly public interactions. On one hand, every set of "campers" has its own private home completely sealed from the outside (e.g., note the satellite dish and rooftop air conditioners). Each RV is self-sufficient, requiring interactions neither with nature nor with human neighbors. On the other hand, the RVs are extremely closely spaced. Interactions among campers are almost forced, for merely stepping out of the RV puts one in public view. Conversations with strangers—even sharing drinks and meals—become a cultural norm in such a setting. Indeed, meeting new people is one of the reasons campers give when explaining why they enjoy camping. Could it be that getting in touch with nature really means getting in touch with one another in ways that would be difficult in the daily routines of popular culture?

Center for Folklife and Cultural Heritage

http://www.folklife.si.edu/

A research and educational unit of the Smithsonian Institution promoting the understanding and continuity of diverse, contemporary grassroots cultures in the United States and around the world. The center produces exhibitions, films and videos, and educational materials.

Cultural Survival

http://www.culturalsurvivalorg/

The interactive web site of Cultural Survival, an organization that promotes the human rights and goals of indigenous peoples. Many timely indigenous cultural issues and important links may be found here.

First Nations Seeker

http://www.firstnationsseeker.ca

This site is a directory of North American Indian portal web sites. Tribes of Canada and the United States are ordered linguistically.

First Peoples Worldwide

http://www.glbthistory.org/

This group promotes an indigenous-controlled international organization that advocates for indigenous self-governance and culturally appropriate economic development.

GLBT Historical Society

http://www.glbthistory.org/

The Gay, Lesbian, Bisexual Transgender Historical Society collects, preserves, and interprets the history of GLBT people and the communities that support them.

Popular Culture Association

http://pcaaca.org/

A multidisciplinary organization dedicated to the academic discussion of popular culture; the site enumerates the various activities of the association.

World Diaspora Organization

http://www.unaoc.org/ibis/

World Diaspora Organization is a transnational citizens movement, a social movement for the right to have transnational or multicultural identity and citizenship. It provides a forum for diaspora umbrella organizations with a World Congress format. The WDO forum is a networking place for national Diaspora World Congress presidents—also known as Statesmen Without States.

Sources

Alcorn, Janice. 1994. "Noble Savage or Noble State? Northern Myths and Southern Realities in Biodiversity Conservation." *Ethnoecologica* 2(3): 7–19.

Ambinakudige, Shrinidhi. 2009. "Revisiting 'the South' and 'Dixie': Delineating Vernacular Regions Using GIS." *Southeastern Geographer* 49(3): 240–250.

Bebbington, Anthony. 1996. "Movements, Modernizations, and Markets: Indigenous Organizations and Agrarian Strategies in Ecuador." In R. Peet and M. Watts (eds.), *Liberation Ecologies: Environment, Development, Social Movements,* pp. 86–109. London: Routledge.

Bebbington, Anthony. 2000. "Reencountering Development: Livelihood Transitions and Place Transformations in the Andes." *Annals of the Association of American Geographers* 90(3): 495–520.

Buijs, Arjen E., Bas Pedroli, and Yves Luginbu. 2006. "From Hiking through Farmland to Farming in a Leisure Landscape: Changing Social Perceptions of the European Landscape." *Landscape Ecology* 21: 375–389.

Clevinger, Woodrow R. 1938. "The Appalachian Mountaineers in the Upper Cowlitz Basin." *Pacific Northwest Quarterly* 29: 115–134.

Clevinger, Woodrow R. 1942. "Southern Appalachian Highlanders in Western Washington." *Pacific Northwest Quarterly* 33: 3–25.

Crang, Mike. 2011. "Tourist: Moving Places, Becoming Tourist, Becoming Ethnographer." In T. Cresswell and P. Merriman (eds.), *Geographies of Mobilities: Practices, Spaces, Subjects,* pp. 205–224. Farnham, U.K.: Ashgate.

de Blij, Harm, and Peter Muller. 2004. *Geography: Realms, Regions and Concepts.* 11th ed. New York: John Wiley & Sons.

Faria, Caroline. 2010. "Contesting Miss Sudan." *International Feminist Journal of Politics* 12(2): 222–243.

Faria, Caroline. 2011. "Staging a New South Sudan in the USA: Men, Masculinities, and Nationalist Performance at a Diasporic Beauty Pageant." *Gender, Place, and Culture* 20(1): 87–106.

Gade, Daniel W. 1982. "The French Riviera as Elitist Space." *Journal of Cultural Geography* 3: 19–28.

Glassie, Henry. 1968. *Pattern in the Material Folk Culture of the Eastern United States.* Philadelphia: University of Pennsylvania Press.

Godfrey, Brian J. 1993. "Regional Depiction in Contemporary Film." *Geographical Review* 83: 428–440.

Goss, Jon. 1999. "Once-upon-a-Time in the Commodity World: An Unofficial Guide to the Mall of America." *Annals of the Association of American Geographers* 89: 45–75.

Harner, John, and Bradley Benz. 2013. "The Growth of Ranchettes in La Platta County, Colorado, 1988–2008." *The Professional Geographer* 65(2): 329–344.

Hecock, Richard D. 1987. "Changes in the Amenity Landscape: The Case of Some Northern Minnesota Townships." *North American Culture* 3(1): 53–66.

Hopkins, Jeffrey. 1990. "West Edmonton Mall: Landscape of Myths and Elsewhereness." *Canadian Geographer* 34: 2–17.

International Cancún Declaration of Indigenous Peoples. 2003. Prepared during the Fifth WTO Ministerial Conference, Cancún, Mexico. http://www.ifg.org/programs/indig/CancunDec.html.

International Development Research Centre (IDRC). 2004. http://web.idrc.ca/en/ev-1248-201-1-DO_TOPIC.html.

International Labour Organization (ILO). 1989. Indigenous and Tribal Peoples Convention. http://members.tripod.com/PPLP/ILOC169.html.

Jackson, Peter. 2004. "Local Consumption Cultures in a Globalizing World." *Transactions of the Institute of British Geographers* NS 29: 165–178.

Kay, Jeanne, et al. 1981. "Evaluating Environmental Impacts of Off-Road Vehicles." *Journal of Geography* 80: 10–18.

Kniffen, Fred B. 1965. "Folk Housing: Key to Diffusion." *Annals of the Association of American Geographers* 55: 549–577.

Knopp, Lawrence, and Michael Brown. 2008. "Queering the Map: The Productive Tensions of Colliding Epistmologies." *Annals of the Association of American Geographers* 98(1): 40–58.

Kunstler, James H. 1993. *The Geography of Nowhere: The Rise and Decline of America's Man-Made Landscape.* New York: Simon & Schuster.

Light, Duncan. 2012. *The Dracula Dilemma: Tourism, Identity and the State in Romania.* Farnham, U.K.: Ashgate.

Lowenthal, David. 1968. "The American Scene." *Geographical Review* 58: 61–88.

McDowell, Linda. 1994. "The Transformation of Cultural Geography." In Derek Gregory, Ron Martin, and Graham Smith (eds.), *Human Geography: Society, Space, and Social Science*, pp. 146–173. Minneapolis: University of Minnesota Press.

Mings, Robert C., and Kevin E. McHugh. 1989. "The RV Resort Landscape." *Journal of Cultural Geography* 10: 35–49.

Nietschmann, Bernard. 1973. *Between Land and Water: The Subsistence Ecology of the Miskito Indians, Eastern Nicaragua*. New York: Seminar Press.

Nietschmann, Bernard. 1979. "Ecological Change, Inflation, and Migration in the Far West Caribbean." *Geographical Review* 69: 1–24.

Raitz, Karl B. 1987. "Place, Space and Environment in America's Leisure Landscapes." *Journal of Cultural Geography* 8(1): 49–62.

Relph, Edward. 1976. *Place and Placelessness*. London: Pion.

Richards, Paul. 1975. "'Alternative' Strategies for the African Environment. 'Folk Ecology' as a Basis for Community Oriented Agricultural Development." In Paul Richards (ed.), *African Environment, Problems and Perspectives*, pp. 102–117. London: International African Institute.

Richards, Paul. 1985. *Indigenous Agricultural Revolution*. London: Hutchinson.

Robertson, David S. 1996. "Oil Derricks and Corinthian Columns: The Industrial Transformation of the Oklahoma State Capitol Grounds." *Journal of Cultural Geography* 16: 17–44.

Rocheleau, Diane, Barbara Thomas-Slayter, and Esther Wangari (eds.). 1996. *Feminist Political Ecology: Global Issues and Local Experiences*. New York: Routledge.

Sack, Robert D. 1992. *Place, Modernity, and the Consumer's World: A Rational Framework for Geographical Analysis*. Baltimore: Johns Hopkins University Press.

Walker, Peter, and Louise Fortmann. 2003. "Whose Landscape? A Political Ecology of the 'Exurban' Sierra." *Cultural Geographies* 10: 469-491.

Williams, Raymond. 1976. *Keywords: A Vocabulary of Culture and Society*. New York: Oxford University Press.

Zonn, Leo (ed.). 1990. *Place Images in Media: Portrayal, Experience, and Meaning*. Savage, Md.: Rowman & Littlefield.

Ten Recommended Books on Geographies of Cultural Difference

(For additional suggested readings, see the *Contemporary Human Geography* LaunchPad: http://www.macmillanhighered.com/launchpad/DomoshCHG1e.)

Aitken, Stuart. 2001. *Geographies of Young People: The Morally Contested Spaces of Identity*. London and New York: Routledge. The author presents youth as a fundamental category of cultural difference rather than a developmental stage toward adulthood in this study of the interaction of space, place, and childhood identity.

Burgess, Jacquelin A., and John R. Gold (eds.). 1985. *Geography, the Media, and Popular Culture*. New York: St. Martin's Press. The geography of popular culture is linked in diverse ways to the communications media, and this collection of essays explores facets of that relationship.

Harris, Dianne. 2013. *Little White Houses: How the Postwar Home Constructed Race in America*. Minneapolis: University of Minnesota Press. The author shows how, following the Second World War, marketing and popular media portrayed the new suburban developments as exclusively White. The study demonstrates that racial segregation and racial identities were constructed and reinforced through inequalities in the housing market.

Jackson, Peter, and Jan Penrose (eds.). 1993. *Constructions of Race, Place, and Nation*. Minneapolis: University of Minnesota Press. An edited collection that examines the way in which the ideas of racial and national identity vary from place to place; rich in empirical research.

Johnston, Lynda, and Robyn Longhurst. 2010. *Space, Place, and Sex: Geographies of Sexualities*. Lanham, Md.: Rowman & Littlefield. Employing feminist and queer theories, the authors examine the geography of sex and sexuality from the scale of the body to the globe. They highlight throughout the interactive effects between sexuality and place.

Jordan, Terry G., Jon T. Kilpinen, and Charles F. Gritzner. 1997. *The Mountain West: Interpreting the Folk Landscape*. Baltimore: Johns Hopkins University Press. Reading the folk landscapes of the American West, three geographers reach conclusions about the regional culture and how it evolved.

Price, Patricia. 2004. *Dry Place: Landscapes of Belonging and Exclusion*. Minneapolis: University of Minnesota Press. Price explores the narratives that have sought to establish claims to the dry lands along the U.S.–Mexico border, demonstrating how stories can become vehicles for reshaping places and cultural identities.

Roberts, Les (ed.). 2012. *Mapping Cultures: Place, Practice, Performance*. New York: Palgrave. The individual chapter authors explore ways to map a range of cultural expression such as music, collective memories, and political identity.

Warf, Barney (ed.). 2012. *Encounters and Engagements between Economic and Cultural Geography*. New York: Springer. A diverse set of chapters based on up-to-date literature reviews on the themes of the "cultural turn" in economic geography and the "rematerialization" of cultural geography.

Zelinsky, Wilbur. 2011 *Not Yet a Placeless Land: Tracking an Evolving American Geography*. Amherst: University of Massachusetts Press. Zelinsky brings a lifetime's worth of knowledge of North American cultural geography to bear on the question of whether the country's land and citizens are homogenizing. To his own observations he adds a survey of literature on a range of American landscapes, regions, and cities. Arguing there are multiple countervailing forces at work, he concludes that standardization is occurring simultaneously with diversification.

Journals in Geographies of Cultural Difference

Indigenous Affairs. A quarterly journal published by the International Working Group for Indigenous Affairs thematically focused on issues of indigenous cultures. Volume 1 was published in 1976.

Journal of Popular Culture. Published by the Popular Culture Association since 1967, this journal focuses on the role of popular culture in the making of contemporary society. See, in particular, Volume 11, No. 4, 1978, a special issue on cultural geography and popular culture.

Material Culture: Journal of the Pioneer America Society. Published twice annually, this leading periodical specializes in the subject of the American rural material culture of the past. Volume 1 was published in 1969, and prior to 1984 the journal was called *Pioneer America*.

CHAPTER 3

Would you feel comfortable walking
here? If not, why not?

A street scene in the large city of Kolkata, India.
(National Geographic Image Collection/Alamy.)

Go to "Seeing Geography" on page 133 to learn more about this image.

POPULATION GEOGRAPHY

Shaping the Human Mosaic

The nineteenth-century French philosopher **Auguste Comte** famously asserted that "demography is destiny." Though this might be a bit of an exaggeration, it is true that one of the most important aspects of the world's human population is its demographic characteristics, such as age, gender, health, mortality, density, and mobility. In fact, many cultural geographers argue that familiarity with the spatial dimensions of demography provides a baseline for the discipline. Thus, **population geography** provides an ideal topic to launch our substantive discussion of the human mosaic.

population geography The study of the spatial and ecological aspects of population, including distribution, density per unit of land area, fertility, gender, health, age, mortality, and migration.

The most essential demographic fact is that more than 7 billion people inhabit the Earth today. But, numbers alone tell only part of the story of the delicate balance between human populations and the resources on which we depend for our survival, comfort, and enjoyment. Think of the sort of lifestyle you may now enjoy or aspire to in the future. Does it involve driving a car? Eating meat regularly? Owning a spacious house with central heating and air conditioning? If so, you are not alone. In fact, these "Western" consumption habits have become so widespread that they may now be more properly regarded as universal. Satisfying these lifestyle demands requires using a wide range of nonrenewable resources—fossil fuels, extensive farmlands, and fresh water among them—whose consumption ultimately limits the number of people the Earth can support. Indeed, it has been argued that if Western lifestyles are adopted by a significant number of the globe's inhabitants, then our current population of 7 billion is already excessive and will soon deplete or contaminate

LEARNING OBJECTIVES

3.1 Describe the regional patterns of population characteristics and how these are distributed spatially and change over time.

3.2 Identify patterns, causes, and consequences of population migrations and the things—such as diseases—that accompany them.

3.3 Discuss theories of population growth and control and how these have changed over time.

3.4 Recognize the ways in which the natural world shapes population characteristics, and how population characteristics in turn shape the natural world.

3.5 Analyze the imprint of demographic factors on the cultural landscape.

the Earth's life-support systems: the air, soil, and water we depend on for our very survival. Although we may think of our geodemographic choices, such as how many children we will have or what to eat for dinner, as highly individual ones, when aggregated across whole groups, they can have truly global repercussions.

As we do throughout this book, we approach our study using the five themes of cultural geography. The demographics of human populations—their size, age, gender compositions, and spatial distribution—are discussed under the regional theme. Population mobility and the related movement of diseases that affect human populations are discussed next. Population debates are considered under the theme of globalization. Although population is experienced locally and policies affecting population are typically set at the national level, the size and impact of human populations involve a debate that is most commonly pitched at the global scale. Next, the theme of nature–culture reveals that the ways in which we interpret the natural world and adapt it to our needs are deeply entwined with demographic practices. We close with a consideration of the geodemographic cultural landscape, which highlights the surprisingly diverse ways that places across the world respond to changing population dynamics.

REGION

3.1 LEARNING OBJECTIVE
Describe the regional patterns of population characteristics and how these are distributed spatially and change over time.

The principal characteristics of human populations—their densities, spatial distributions, age and gender structures, the ways they increase and decrease, and how rapidly population numbers change—vary enormously from place to place. Understanding the demographic characteristics of populations, and why and how they change over time, gives us important clues to their cultural characteristics.

POPULATION DISTRIBUTION AND DENSITY

population density
A measurement of population per unit area (e.g., per square mile).

If the 7,100,000,000 inhabitants of the Earth were evenly distributed across the land area, the **population density** would be about 125 persons per square mile (48 per square kilometer). However, people are very unevenly distributed, creating huge disparities in density. Mongolia, for example, has 5.2 persons per square mile (2 per square kilometer), whereas Bangladesh has 2897 persons per square mile (1118 per square kilometer) (**Figure 3.1**).

If we consider the distribution of people by continents, we find that 70.2 percent of the human race lives in Eurasia—Europe and Asia. The continent of North America is home to only 7.7 percent of all people, Africa to 15.2 percent, South America to 5.6 percent, and Australia and the Pacific islands to 0.5 percent. When we consider population distribution by country, we find that 19 percent of all humans reside in China, 17 percent in India, and only 4.4 percent in the third-largest nation in the world, the United States (**Table 3.1**).

For analyzing data, it is convenient to divide population density into categories. For example, in Figure 3.1, one end of the spectrum contains thickly settled areas having 250 or more persons per square mile (100 or more per square kilometer); on the other end, largely unpopulated areas have fewer than 2 persons per square mile (less than 1 per square kilometer). Moderately settled areas, with 60 to 250 persons per square mile (25 to 100 per square kilometer), and thinly settled areas, inhabited

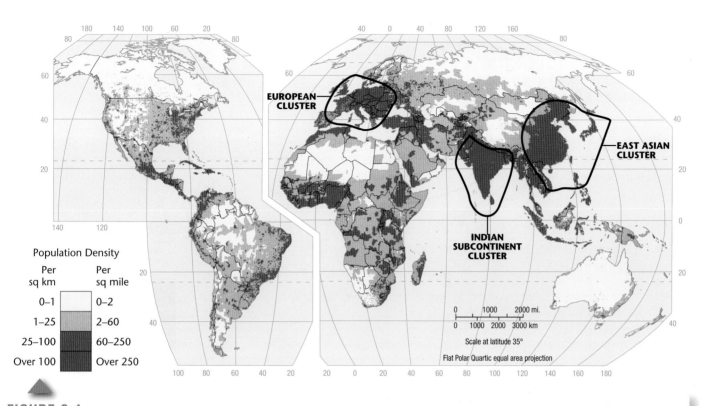

FIGURE 3.1   **Population density in the world.** Try to imagine the diverse causal forces—physical, environmental, and cultural—that have been at work over the centuries to produce this complicated spatial pattern. It represents the most basic cultural geographical distribution of all. *(Source: Population Reference Bureau.)*

by 2 to 60 persons per square mile (1 to 25 per square kilometer), fall between these two extremes. These categories create formal demographic regions based on the single trait of population density. As Figure 3.1 shows, a fragmented crescent of densely settled areas stretches along the western, southern, and eastern

TABLE 3.1   The World's 10 Most Populous Countries, 2012 and 2050

Largest Countries in 2012	Population in 2012 (in millions)	Largest Countries in 2050	Population in 2050 (estimated, in millions)
China	1,343	India	1,691
India	1,205	China	1,311
United States	314	United States	423
Indonesia	249	Nigeria	402
Brazil	199	Pakistan	314
Pakistan	190	Indonesia	309
Nigeria	170	Bangladesh	226
Bangladesh	161	Brazil	213
Russia	143	Democratic Republic of Congo	194
Japan	127	Ehiopia	166

Source: Population Reference Bureau (2012).

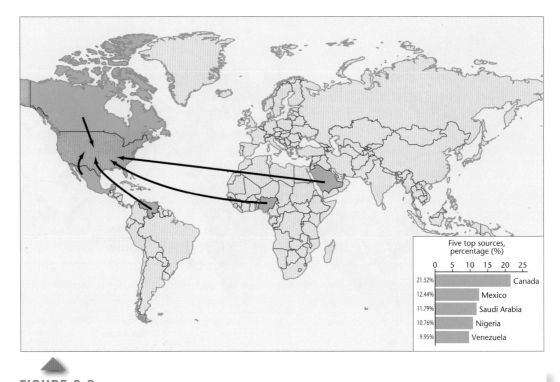

FIGURE 3.2   **Where does U.S. oil come from?** The United States has domestic oil resources, but they cannot fully supply the domestic demand, so oil is imported from other areas of the world. The United States consumes 26 percent of all the oil produced in the world. *(Source: U.S. Department of Energy, 2010.)*

edges of the huge Eurasian continent. Two-thirds of the human race is concentrated in this crescent, which contains three major population clusters: eastern Asia, the Indian subcontinent, and Europe. Outside of Eurasia, only scattered districts are so densely settled. Despite the image of a crowded world, thinly settled regions are much more extensive than thickly settled ones, and they appear on every continent. Thin settlement characterizes the northern sections of Eurasia and North America, the interior of South America, most of Australia, and a desert belt through North Africa and the Arabian Peninsula into the heart of Eurasia.

There is more we as geographers want to know about population geography than simply population density. For example, what are people's standards of living and are they related to population density? Some of the most thickly populated areas in the world have the highest standards of living—and even suffer from labor shortages (for example, the major industrial areas of western Europe and Japan). The

small principality of Monaco in southern Europe, for instance, is the most densely populated nation in the world; it also ranks among the highest incomes per capita, highest average life expectancies, and lowest unemployment rates. In other cases, thinly settled regions may actually be severely overpopulated relative to their ability to support their populations, a situation usually associated with marginal agricultural lands. Although 1000 persons per square mile (400 per square kilometer) is a "sparse" population for an industrial district, it is "dense" for a rural area. For this reason, **carrying capacity**—the population beyond which a given environment cannot provide support without becoming significantly damaged—provides a far more meaningful index of overpopulation than density alone. Often, however, it is difficult to determine carrying capacity until the region under study is near or over the limit.

> ■ **carrying capacity** The maximum number of people that can be supported in a given area.

Sometimes the carrying capacity of one place can be expanded by drawing on the resources of another place. Americans, for example, consume far more food, products, and natural resources than do most other people in the world: 26 percent of the entire world's petroleum, for instance, is consumed in the United States (**Figure 3.2**). The carrying capacity of the United States would be exceeded if it did not annex the resources—including the labor—of much of the rest of the world.

A critical feature of population geography is the demographic changes that occur over time. Analyzing these gives us a dynamic perspective from which we can glean insights into cultural changes occurring at local, regional, and global scales. Populations change primarily in two ways: people are born and others die in a particular place, and people move into and out of that place. The latter refers to migration, which we will consider later in this chapter. For now we discuss births and deaths, which can be thought of as additions to and subtractions from a population. They provide what demographers refer to as natural increases and natural decreases.

PATTERNS OF NATALITY

Births can be measured by several methods. The older way was simply to calculate the birth rate: the number of births per year per thousand people (**Figure 3.3**).

More revealing is the **total fertility rate (TFR)**, which is measured as the average number of children born per woman during her reproductive lifetime, considered to be from 15 to 49 years of age (with some countries having a younger upper-age limit of

> **total fertility rate (TFR)** The number of children the average woman will bear during her reproductive lifetime (15–44 or 15–49 years of age). A TFR of less than 2.1, if maintained, will cause a natural decline of population.

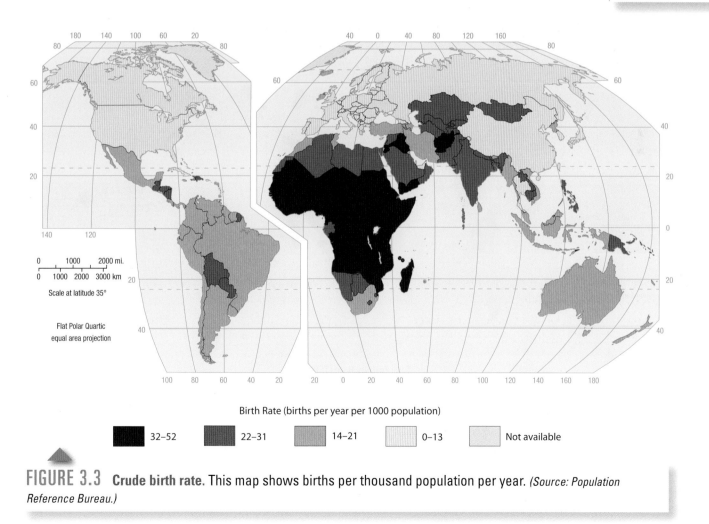

Birth Rate (births per year per 1000 population)

32–52	22–31	14–21	0–13	Not available

FIGURE 3.3 **Crude birth rate.** This map shows births per thousand population per year. *(Source: Population Reference Bureau.)*

44 years of age). The TFR is a more useful measure than the birth rate because it focuses on the female segment of the population, reveals average family size, and gives an indication of future changes in the population structure. A TFR of 2.1 is needed to produce a stabilized population over time, one that does not increase or decrease. Once achieved, this condition is called **zero population growth**.

> **zero population growth** A stabilized population created when an average of only two children per couple survive to adulthood. Eventually, the number of deaths equals the number of births.

The TFR varies markedly from one part of the world to another, revealing a vivid geographical pattern (**Figure 3.4**). In southern and eastern Europe, the average TFR is only 1.3. Every country with a TFR of 2.0 or lower will eventually experience population decline. Bulgaria, for example, has a TFR of 1.2 and is expected to lose 38 percent of its population by 2050. Taiwan and Latvia are tied at 1.1 for the lowest total fertility rate in the world in 2012.

By contrast, sub-Saharan Africa has the highest TFR of any sizable part of the world, led by Niger with 7.1 as of 2014. Elsewhere in the world, only Afghanistan can rival the sub-Saharan African rates. However, according to the World Bank, during the past two decades TFRs have fallen in all sub-Saharan African nations.

THE GEOGRAPHY OF MORTALITY

Another way to assess demographic change is to analyze **death rates**: the number of deaths per year per 1000 people (**Figure 3.5**). Of course,

> **death rate** The number of deaths per year per 1000 people.

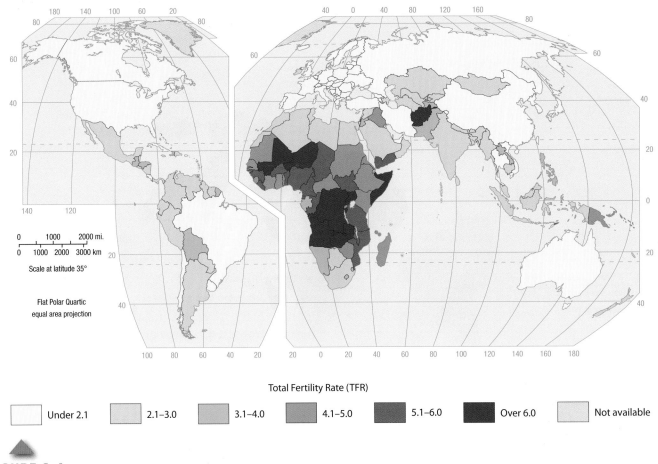

Total Fertility Rate (TFR)

| Under 2.1 | 2.1–3.0 | 3.1–4.0 | 4.1–5.0 | 5.1–6.0 | Over 6.0 | Not available |

FIGURE 3.4   The total fertility rate (TFR) in the world. The TFR indicates the average number of children born to women over their lifetimes. A rate of 2.1 is needed to produce a stable population over the long run; below that, population will decline. Fast growth is associated with a TFR of 5.0 or higher. *(Source: Population Reference Bureau.)*

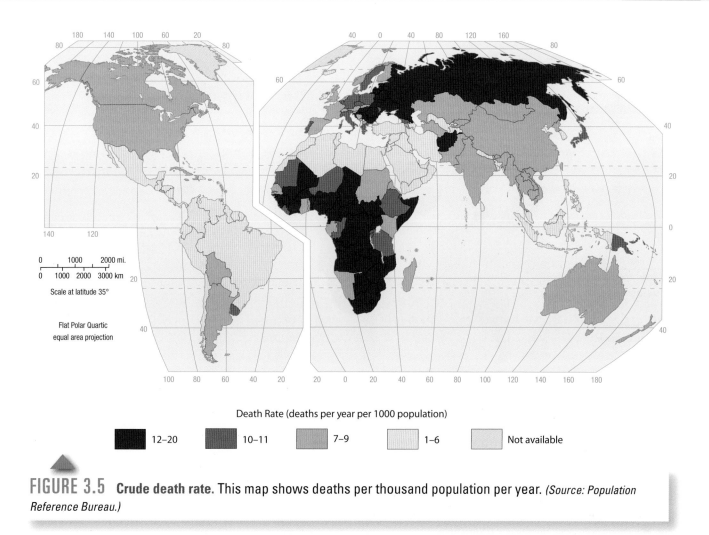

Death Rate (deaths per year per 1000 population)

■ 12–20	■ 10–11	■ 7–9	□ 1–6	□ Not available

FIGURE 3.5 **Crude death rate.** This map shows deaths per thousand population per year. *(Source: Population Reference Bureau.)*

death is a natural part of the life cycle, and there is no way to achieve a death rate of zero. But geographically speaking, death comes in different forms. In the developed world, most people die of age-induced degenerative conditions, such as heart disease, or from maladies caused by industrial pollution of the environment. Many types of cancer fall into the latter category. By contrast, contagious diseases such as malaria, HIV/AIDS, and diarrheal diseases are a leading cause of death in poorer countries. Civil warfare, inadequate health services, and the age structure of a country's population will also affect its death rate.

Some of the world's highest death rates occur in sub-Saharan Africa, the poorest world region and most afflicted by life-threatening diseases and civil strife (**Figure 3.6** illustrates the geography of HIV/AIDS). In general, death rates of more than 20 per 1000 people

are uncommon today. The world's highest death rate as of 2013—just over 17 per 1000 people—was found in South Africa, and is primarily the result of a very high prevalence of HIV/AIDS. South Africa is the country with the highest HIV/AIDS prevalence in the world. High death rates are also found in eastern European nations—with Ukraine, Bulgaria, and Russia, for instance, ranked second, eighth, and tenth in the world in 2013—thanks to a collapsing public health care system in the post-Soviet era, environmental contamination and increased cancer incidence, poor health choices including smoking and alcohol consumption, and very high rates of diseases such as tuberculosis and HIV/AIDS.

By contrast, the American tropics generally have rather low death rates, as does the desert belt across North Africa, the Middle East, and central Asia. In these regions, the predominantly young population

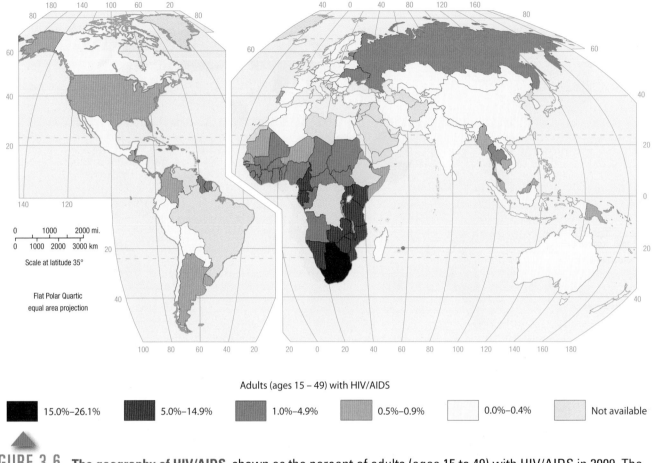

Adults (ages 15 – 49) with HIV/AIDS

■ 15.0%–26.1%	■ 5.0%–14.9%	■ 1.0%–4.9%	■ 0.5%–0.9%	□ 0.0%–0.4%	□ Not available

FIGURE 3.6 **The geography of HIV/AIDS,** shown as the percent of adults (ages 15 to 49) with HIV/AIDS in 2009. The quality of data gathering varies widely from one country to another and is particularly poor in Africa and most of Asia. *(Sources: Population Reference Bureau and UNAIDS.)*

suppresses the death rate. Compared to South Africa, Paraguay's death rate of only 4.6 per 1000 seems quite low. Because of its older population structure, the average death rate in the European Union is 10.5 per 1000. Australia, Canada, and the United States, which continue to attract young immigrants, have lower death rates than most of Europe. Canada's death rate, for instance, is slightly less than 8 people per 1000.

THE DEMOGRAPHIC TRANSITION

demographic transition Describes the movement from high birth and death rates to low birth and death rates.

All industrialized, technologically advanced countries have low fertility rates and stabilized or declining populations, having passed through what is called the **demographic transition**

(Figure 3.7). In preindustrial societies, birth and death rates were both high, resulting in almost no population growth. Because these were agrarian societies that depended on family labor, many children meant larger workforces, thus the high birth rates. But low levels of public health and limited access to health care, particularly for the very young, also meant high death rates. With the coming of the industrial era, medical advances and improvements in diet set the stage for a drop in death rates. Human life expectancy in industrialized countries soared from an average of 35 years in the eighteenth century to 75 years or more at present. Yet, birth rates did not fall so quickly, leading to a population explosion as fertility outpaced mortality. In Figure 3.7, this is shown in late stage 2 and early stage 3 of the model. Eventually, a decline in the birth rate is followed the decline in the death rate,

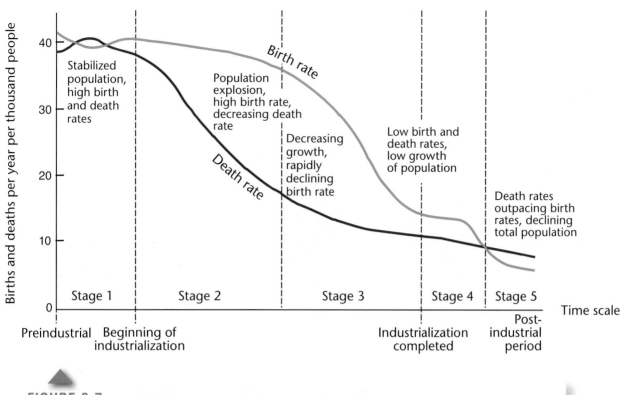

FIGURE 3.7 **The demographic transition as a graph.** The "transition" occurs in several stages as the industrialization of a country progresses. In stage 2, the death rate declines rapidly, causing a population explosion as the gap between the number of births and deaths widens. Then, in stage 3, the birth rate begins a sharp decline. The transition ends when, in stage 4, both birth and death rates have reached low levels, by which time the total population is many times greater than at the beginning of the transformation. In the postindustrial stage, population decline eventually begins.

slowing population growth. An important reason leading to lower fertility levels involves the high cost of children in industrial societies, particularly because childhood itself becomes a prolonged period of economic dependence on parents. Finally, in the postindustrial period, the demographic transition produced zero population growth or actual population decline (Figure 3.7 and **Figure 3.8**).

Achieving lower death rates is relatively cost effective, historically requiring little more than the provision of safe drinking water and vaccinations against common infectious diseases. Lowering death rates tends to be uncontroversial and quickly achieved, demographically speaking. Getting birth rates to fall, however, can be far more difficult, especially for a government official who wants to be reelected. Birth control, abortion, and challenging

long-held beliefs about family size can prove quite controversial, and political leaders may be reluctant to legislate them. Indeed, the Chinese implementation of its one-child-per-couple policy (see Subject to Debate, page 105) probably would not have been possible in a country with a democratically elected government. In addition, because it involves changing a cultural norm, the idea of smaller families can take three or four generations to become a reality. Increased educational levels for women are closely associated with falling fertility levels, as is access to various contraceptive devices (**Figure 3.9**).

The demographic transition is a model that predicts trends in birth rates, death rates, and overall population levels in the abstract. Yet, like many models used by demographers, it is based on the historical experience of western Europe: it is

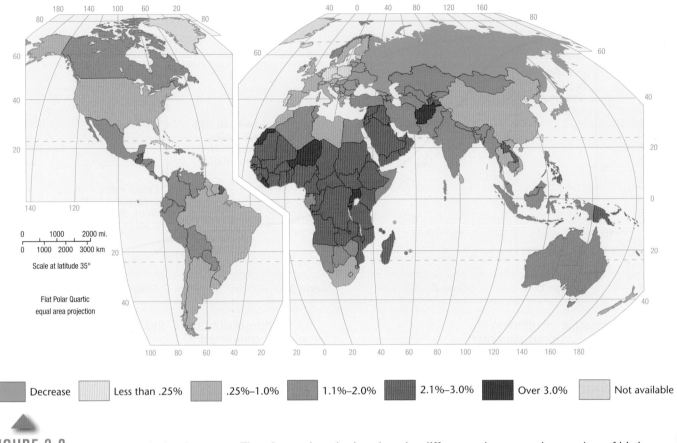

| Decrease | Less than .25% | .25%–1.0% | 1.1%–2.0% | 2.1%–3.0% | Over 3.0% | Not available |

FIGURE 3.8 **Annual population increase.** The change is calculated as the difference between the number of births and deaths in a year, taken as a percentage of total population. Migration is not considered. Note the contrast between tropical areas and the middle and upper latitudes. In several places, countries with a very slow increase border areas with extremely high growth. *(Source: Population Reference Bureau.)*

Eurocentric Using the historical experience of Europe as the benchmark for all cases.

Eurocentric. It does a good job of describing changes in population patterns over time in Europe, as well as in other wealthy regions. However, it has several shortcomings. First is the inexorable stage-by-stage progression implicit in the model. Have countries or regions ever skipped a stage or regressed? Certainly. The case of China shows how policy, in this case government-imposed restrictions on births, can fast-forward an entire nation to stage 4 (again, see the Subject to Debate feature). War, too, can cause a return to an earlier stage in the model by increasing death rates. For instance, Angola and Afghanistan are two countries with recent histories of conflict; both

had topped the world's death rate list until recently. In other cases, wealth has not led to declining fertility. Thanks to oil exports, residents of Oman enjoy relatively high average incomes. But fertility, too, remains relatively high at nearly 3 children per woman in 2013. Indeed, the Population Reference Bureau has pointed to a "demographic divide" between countries where the demographic transition model applies well, and others—mostly poorer countries or those experiencing widespread conflict or disease—where birth and death rates do not necessarily follow the model's predictions. Even in Europe, there are countries where fertility has dropped precipitously, while at the same time death rates have escalated. In countries such as Russia,

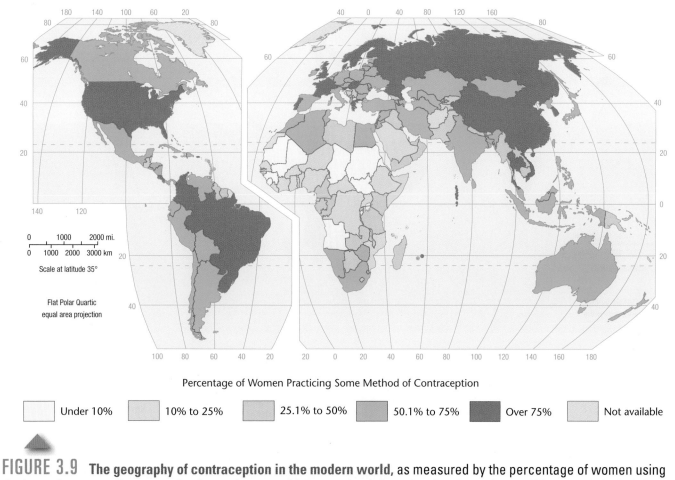

Percentage of Women Practicing Some Method of Contraception

| | Under 10% | | 10% to 25% | | 25.1% to 50% | | 50.1% to 75% | | Over 75% | | Not available |

FIGURE 3.9 **The geography of contraception in the modern world,** as measured by the percentage of women using devices of any sort. Contraception is much more widely practiced than abortion, but cultures differ greatly in their level of acceptance. Several different devices are included. *(Source: World Population Data Sheet.)*

which, according to Peter Coclanis, has "somehow managed to reverse the so-called demographic transition," the model itself must be questioned.

AGE DISTRIBUTIONS

Some countries have overwhelmingly young populations. In a majority of countries in Africa, as well as some countries in Latin America and tropical Asia, close to half the population is younger than 15 years of age (**Figure 3.10**). In Uganda, for example, 51 percent of the population is younger than 15 years of age. In sub-Saharan Africa, 44 percent of all people are younger than 15. Other countries, generally those

that industrialized early, have a preponderance of middle-aged people in the over 15–under 65 age bracket. A growing number of affluent countries have remarkably aged populations. In Germany, for example, fully 21 percent of the people have now passed the traditional retirement age of 65. Many other European countries are not far behind. A sharp contrast emerges when Europe is compared with Africa, Latin America, or parts of Asia, where the average person never even lives to age 65. In Mauritania, Niger, Afghanistan, Guatemala, and many other countries, only 2 to 3 percent of the population have reached that age.

Countries with disproportionate numbers of old or young people often address these imbalances in

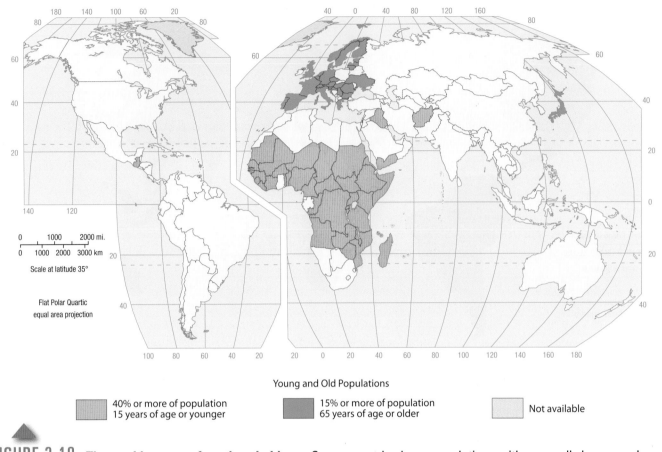

Young and Old Populations

40% or more of population 15 years of age or younger	15% or more of population 65 years of age or older	Not available

FIGURE 3.10   The world pattern of youth and old age. Some countries have populations with unusually large numbers of elderly people; others have preponderantly young populations. ***What issues might be associated with either situation?*** *(Source: Population Reference Bureau.)*

innovative ways. Italy, for example, has one of the lowest birth rates in the world, with a TFR of only 1.4. In addition, Italy's population is one of the oldest in the world: 20.5 percent of its population is age 65 or older. As a result of both its TFR and age distribution, Italy's population is projected to shrink by 10 percent between 2010 and 2050. Of course, immigrants to Italy from other countries can and no doubt will counteract this trend, but that will lead to contentious political and cultural debates.

Given that the Italian culture does not embrace the institutionalization of the growing ranks of its elderly, and faced with the reality that more Italian women than ever work outside the home and thus few adult women are willing or able to stay at home full time to care for their elders, Italians have gotten creative. Elderly Italians can apply for adoption by

families in need of grandfathers or grandmothers. Frances D'Emilio recounts the experience of one such man, Giorgio Angelozzi, who moved in with the Rivas, a Roman family with two teenagers. Angelozzi said that Marlena Riva's voice reminded him of his deceased wife, Lucia, and this is what convinced him to choose the Riva family. Dagmara Riva, the family's teenage daughter, said that Mr. Angelozzi has helped her with Latin studies and that "Grandpa is a person of great experience, an affectionate person. We're very happy we invited him to live with us."

Age structure also differs spatially within individual countries. For example, rural populations in the United States and many other countries are usually older than those in urban areas. The flight of young people to the cities has left some rural counties in the midsection of the United States with

FEMALE: An Endangered Gender?

Does the simple fact of being female expose a person to demographic peril? In most societies, women are viewed as valuable, even powerful, particularly as mothers, nurturers, teachers, and spiritual leaders. Yet in other important ways, to be female is to be endangered. We will consider this controversial idea with an eye to how demographics and culture closely shape each other.

Many cultures demonstrate a marked preference for males. The academic term describing this is *androcentrism;* you may know it as *patriarchy, male bias,* or simply *sexism.* Whether a preference for males is a feature of all societies has been disputed. Some societies pass along forms of their wealth, property, and prestige from mother to daughter, rather than from father to son. This is rare, however, and it is clear that the roots of the cultural preference for males are historically far-reaching and widespread. In most societies, positions of economic, political, social, and cultural prestige and power are held largely by men. Men typically are considered to be the heads of households. Family names tend to pass from father to son, and with them, family honor and wealth. In traditional societies, when sons marry, they usually bring their wives to live in their parents' house and are expected to assume economic responsibility for aging parents.

For all of these reasons, in many places a cultural premium is placed on producing male children. Because couples often can choose to have more children, having a girl first is usually not a problem. However, in countries that have enacted strict population control programs, couples may not be given the chance to try again for a male child. This has resulted in severe pressures on couples to have boys in some countries, particularly in China and India. In both these nations, female-specific infanticide or abortion has resulted in a growing gender imbalance. Ultrasound devices that allow gender identification of fetuses are now available even to rural peasants in China. About 100,000 such devices were in use as early as 1990 in China, and by the middle of that decade there were 121 males for every 100 females among children two years of age or younger. The sex ratio in China is radically changing, and a profound gender imbalance exists there: in 2011 there were 833 girls for every 1000 boys. The naturally occurring sex ratio is 952 girls for every 1000 boys. In India, too, there were only 914 girls per 1000 boys in 2011. In some states the imbalance is even worse: in Punjab, for instance, it is 830 to 1000.

The cultural ramifications of such male-heavy populations are potentially profound. Men of marriageable age are increasingly unable to find female partners. Social analysts speculate that this will lead to human trafficking and violence against women, and this is already the case in India where girls are kidnapped or trafficked in from neighboring Nepal and Myanmar. In China, the policy of one child per couple has resulted in the so-called four-two-one problem. This refers to the fact that the generational structure of many families now reflects four grandparents, two parents, and a single male child. This places enormous pressures on the shoulders of the male child to care for aging parents and grandparents. It also encourages parents and grandparents to lavish all their attention, wealth, and hopes on the only child. For some families, this has led to the "little emperor syndrome," whereby the male heir becomes spoiled, unable to cope independently, and even obese.

Continuing the Debate

As noted, most societies value females and males equally. For a number of reasons, however, some societies show a clear preference for males. Keeping all this in mind, consider the following questions:

- Are Chinese families somewhat justified in emphasizing the birth of a son at all costs?

- According to a recent report, Americans using technology to select their baby's gender, unlike the Chinese, more often choose to have a girl. Why do you think there is a difference between male and female preference in these two societies?

This "little emperor" poses with his grandparents.
(AStock/Corbis.)

FIGURE 3.11 **Residents of Sun City, Arizona,** enjoy the many recreational opportunities provided in this planned retirement community, where the average age is 75. *(A. Ramey/PhotoEdit Inc.)*

also show that future growth will rely on the momentum of all those young people growing into their reproductive years and having their own families, regardless of how small those families may be in contrast to earlier generations. Those population pyramids with more of a cylindrical shape represent countries approaching population stability or those in demographic decline. A quick look at a country's population pyramid can tell volumes about its past as well as its future. How many dependent people—the very old and the very young—live there? Has the country suffered the demographic effects of genocide or a massive disease epidemic (**Figure 3.13**)? Are significantly more boys than girls being born? These questions and more can be explored at a glance using a population pyramid.

THE GEOGRAPHY OF GENDER

populations whose median age is 45 or older. Some warm areas of the United States have become retirement havens for the elderly; parts of Arizona and Florida, for example, have populations far above the average age. Communities such as Sun City near Phoenix, Arizona, legally restrict residence to the elderly (**Figure 3.11**). In Great Britain, coastal districts have a much higher proportion of elderly than does the interior, suggesting that the aged often migrate to seaside locations when they retire. (Figure 3.11 with caption)

> **population pyramid** A graph used to show the age and sex composition of a population.

A very useful graphic device for comparing age characteristics is the **population pyramid** (**Figure 3.12**). Careful study of such pyramids not only reveals the past progress of birth control but also allows geographers to predict future population trends. Youth-weighted pyramids, those that are broad at the base, suggest the rapid growth typical of the population explosion. What's more, broad-based population pyramids not only reflect past births but

Although the human race is divided almost evenly between females and males, geographical differences do occur in the **sex ratio**: the ratio between men and women in a population (**Figure 3.14**). Slightly more boys than girls are born, but infant boys have slightly higher mortality rates than do infant girls. Recently settled areas typically have more males than females, as is evident in parts of Alaska, northern Canada, and tropical Australia. According to the U.S. Census Bureau, in 2010 there were 108 males for every 100 of Alaska's female inhabitants. Some poverty-stricken parts of South Africa are as much as 59 percent female. Prolonged wars reduce the male population. And, in general, women tend to outlive men. The population pyramid is also useful in showing gender ratios. Note, for instance, the larger female populations in the upper bars for the United States in Figure 3.12.

> **sex ratio** The numerical ratio of females to males in a population.

Beyond such general patterns, gender often influences demographic traits in specific ways. Often

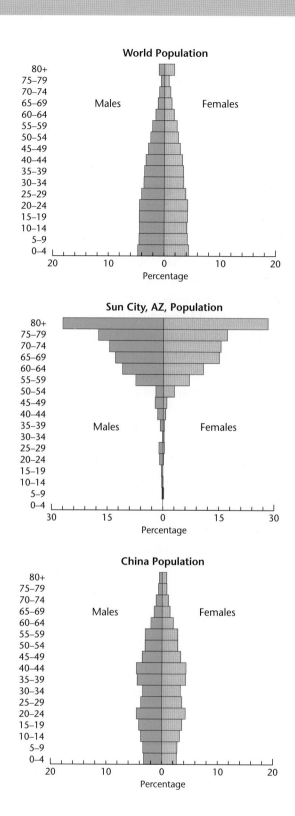

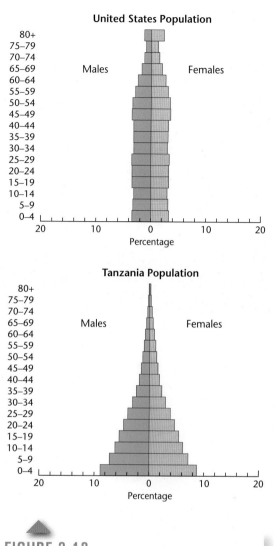

FIGURE 3.12 **Population pyramids for the world and selected countries and communities.** Tanzania displays the classic stepped pyramid of a rapidly expanding population, whereas the U.S. pyramid looks more like a precariously balanced pillar. China's population pyramid reflects the lowered numbers of young people as a result of that country's one-child policy. *How do these pyramids help predict future population growth?* (Source: Population Reference Bureau.)

gender roles What it means to be a man or a woman in different cultural and historical contexts.

gender roles—culturally specific notions of what it means to be a man and what it means to be a woman—are closely tied to how many children are produced by couples. In many cultures, women are considered more womanly when they produce many offspring. By the same token, men are seen as more manly when they father many

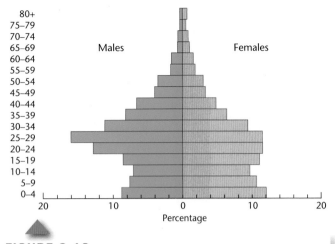

FIGURE 3.13 **Oman's youth bulge.** Sixty-five percent of the population of the Middle East is under 30 years of age. If provided educations and job opportunities, these individuals can constitute a valuable pool of human capital. If not, they can be the drivers of political unrest. *(Source: Population Reference Bureau.)*

children. Because the raising of children often falls to women, the spaces that many cultures associate with women tend to be the private family spaces of the home. Public spaces such as streets, plazas, and the workplace, by contrast, are often associated with men. Some cultures go so far as to restrict where women and men may and may not go, resulting in a distinctive geography of gender, as **Figure 3.15** illustrates (take a look at Seeing Geography, page 133). Falling fertility levels that coincide with higher levels of education for women, however, have resulted in numerous challenges to these cultural ideas of male and female spaces. As more and more women enter the workplace, for instance, ideas of where women should and should not go slowly become modified. Clearly, gender is an important factor to consider in our exploration of population geography (see also Subject to Debate, page 105, and Patricia's Notebook, page 110).

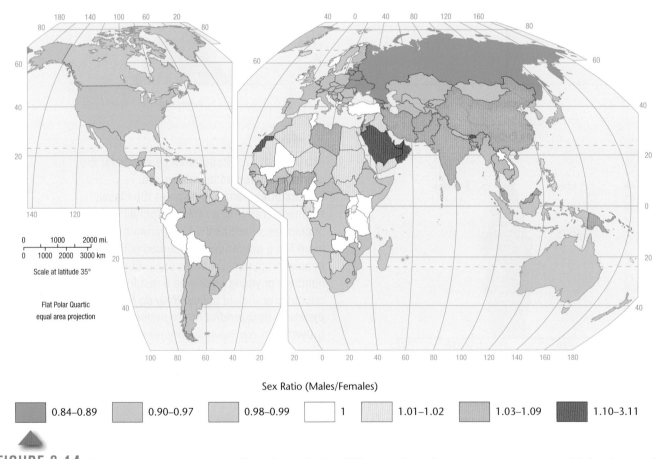

Sex Ratio (Males/Females)

| 0.84–0.89 | 0.90–0.97 | 0.98–0.99 | 1 | 1.01–1.02 | 1.03–1.09 | 1.10–3.11 |

FIGURE 3.14 **Females as a percentage of total population.** *What explanations can you come up with for the gender imbalances shown on this map?* *(Source: Population Reference Bureau.)*

FIGURE 3.15 **Segregated beach.** This portion of the beachfront along the Israeli city of Tel Aviv allows women and men to swim on separate days. Young children of both genders accompany their mothers, whereas older boys visit with their fathers. Gender-based segregation is important for Israel's orthodox Jews. *(Courtesy of Patricia L. Price.)*

STANDARD OF LIVING

Various demographic traits can be used to assess standard of living and analyze it geographically. **Figure 3.16** is a simple attempt to map living standards using the **infant mortality rate**: a measure of how many children per 1000 live births die before reaching one year of age. Many experts believe that the infant mortality rate is the best single index of living standards because it is affected by many different factors: health, nutrition, sanitation, access to doctors, availability of clinics, education, ability to obtain medicines, and adequacy of housing. A striking geographical pattern is revealed by the infant mortality rate.

Another good measure of quality of life is the United Nations Human Development Index (HDI), which combines measures of literacy, life expectancy, education, and wealth (see Figure 1.11, pages 22–23). The highest possible score is 1.000, and the three top-ranked countries in 2012 were Norway, Australia,

> **infant mortality rate** The number of infants per 1000 live births who die before reaching one year of age.

THE VIDEO CONNECTION

Great Expectations for Female Lawyers
http://www.tinyurl.com/ogfxqje

In this video, five women who work at the law firm Debevoise and Plimpton are interviewed. These same women were profiled in *The New York Times* Magazine 12 years earlier, when they started their jobs at the firm. In the interviews, they reflect on their careers and how being a woman has affected their professional success. They talk about how other aspects of their lives, such as their race, age, and motherhood, have combined in ways that have both advantaged and disadvantaged them.

Thinking Geographically:

1. Chapter 3 talks about a geography of gender, wherein women who work outside of the home are seen to be "out of place." What specifics do these women mention regarding feeling out of place, or not, in their jobs?

2. In Chapter 3 we learned that gender roles change when fertility levels fall and women become more highly educated. In what ways have these women experienced changing gender roles, and in what ways have the expectations and norms around being a woman not changed for them?

3. Though women make up roughly half of the human population, women are underrepresented relative to men in some professions, such as law. Women also tend to make less than men in equal positions: for the top executive positions, women are paid about $0.82 for every $1.00 that a man in the same job earns. At the end of the video, it is noted that only 4% of top U.S. law firms are led by a woman. What do you think the social as well as the individual consequences of these labor market imbalances may be?

Patricia's Notebook

Demystifying the Sunday Crowds in Hong Kong

Patricia Price *(American Council on Education)*

During a trip to Hong Kong, I was struck by the dense crowds of women that can be seen on Sundays. Some of these gatherings are in public spaces, such as parks. Others are found in city spaces that are usually heavily trafficked during workdays, such as the steps outside large office buildings. The largest crowd that I observed was under the covered walkway of a subway station. Initially, I thought that these women were gathered for a political demonstration of some kind. In Latin America—the world region I'm most familiar with—large gatherings of people almost always mean that a protest march, political rally, or labor strike is about to begin. But there were no men to be seen anywhere in these Hong Kong throngs! In addition, they didn't seem to be protesting anything. Rather, they happily chatted, ate, styled one another's hair, and played cards.

Later, I learned that these women were in fact maids enjoying their day off. For several decades now, women from the Philippines have migrated to Hong Kong to work as domestic servants, doing chores for and looking after the children of the families who hire them. Wages in Hong Kong are much higher than back home in the Philippines: a major pull factor. However, working conditions can be far from ideal, involving hard physical labor and long hours. Reports of abuse of Filipina servants by their employers are numerous. If a servant is fired, she will be deported if she cannot find another job quickly. The stress of long separations from husbands and children back in the Philippines has led to the

breakup of families. In addition, some of these women find themselves coerced into Asia's booming sex industry. Knowing this makes me think twice about these happy-looking women. Though being a domestic servant isn't a pleasant or highly paid job anywhere in the world, Filipina maids in Hong Kong certainly seem to face a number of pressures, ranging from labor conditions, family circumstances, and the lack of good job alternatives. While the news coverage of domestic servants in the United States seems to be mostly about their status as documented or undocumented workers, there are obviously additional factors to keep in mind. These factors complicated my picture of these women, who, at first glance, seemed happy and carefree as they enjoyed their day off.

Filipina domestic servants in Hong Kong. On Sunday, their day off, these women congregate in covered walkways and other public spaces. They gossip, play cards, cut each other's hair, and relax. *(Courtesy of Patricia L. Price.)*

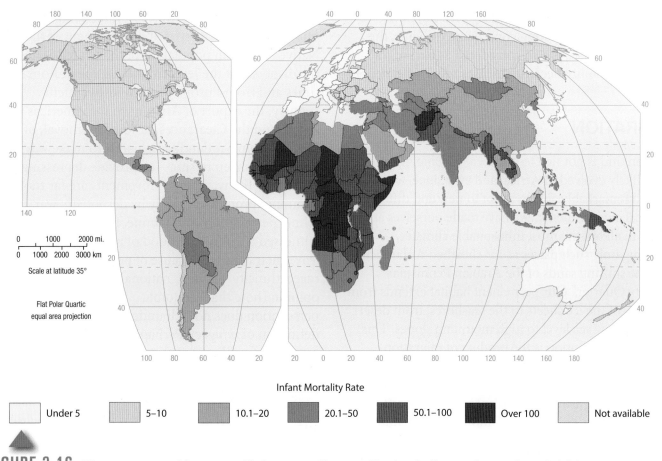

Infant Mortality Rate

| Under 5 | 5–10 | 10.1–20 | 20.1–50 | 50.1–100 | Over 100 | Not available |

FIGURE 3.16 **The present world pattern of infant mortality rate.** The key indicates the number of children per 1000 born who die before reaching one year of age. The world's infant mortality rate is 44. Many experts believe that this rate is the best single measure of living standards. *(Source: Population Reference Bureau.)*

and the United States. Examination of the HDI reveals some surprises. If all countries spent equally on those things that improve their HDI rankings, such as education and health care, then we would expect the wealthiest countries to place first on the list. According to the International Monetary Fund, Luxembourg ranked first among 183 ranked nations in wealth as measured by the gross domestic product (GDP) per capita in 2012. Yet, it ranked 26th in the world on the HDI. Compare this to Barbados, which ranked 24th by GDP per capita but came in much higher on the HDI at 6th. Why did Barbados rank higher in living standards than its monetary wealth would indicate, whereas Luxembourg ranked lower? We would have to conclude that the government of Barbados places a relative priority on spending for education and health care, whereas the government of Luxembourg does not.

MOBILITY

3.2 **LEARNING OBJECTIVE**
Identify patterns, causes, and consequences of population migrations and things—such as diseases—that accompany them.

After natural increases and decreases in populations, the second of the two basic ways in which population numbers are altered is the migration of people from one place to another, and it offers a straightforward illustration of relocation diffusion. When people migrate, they sometimes bring more than simply their culture; they can also bring disease. The introduction of diseases to new places can have dramatic and devastating consequences. Thus, the spread of diseases is also considered as part of the theme of mobility because diseases move with people.

MIGRATION

Humankind is not tied to one locale. *Homo sapiens* most likely evolved in Africa, and ever since, we have proved remarkably able to adapt to new and different physical environments. We have made ourselves at home in all but the most inhospitable climates, avoiding only such places as ice-sheathed Antarctica and the shifting sands of the Arabian Peninsula's "Empty Quarter." Our permanent habitat extends from the edge of the ice sheets to the seashores, from desert valleys below sea level to high mountain slopes. This far-flung distribution is the product of migration.

Early human groups moved in response to the migration of the animals they hunted for food and the ripening seasons of the plants they gathered. Indeed, the agricultural revolution, whereby humans domesticated crops and animals and accumulated surplus food supplies, allowed human groups to stop their seasonal migrations. But, why did certain groups opt for long-distance relocation? Some migrated in response to environmental collapse, others in response to religious or ethnic persecution. Still others—probably the majority, in fact—migrated in search of better opportunities. For those who migrate, the process generally ranks as one of the most significant events of their lives.

The Decision to Move Migration takes place when people decide that moving is preferable to staying and when the difficulties of moving seem to be more than offset by the expected rewards. In the nineteenth century, more than 50 million European emigrants left their homelands in search of better lives. Today, migration patterns are very different (**Figure 3.17**). Europe, for example, now predominantly receives immigrants rather than sending out emigrants. International migration stands at an all-time high, much of it labor migration associated with the process of globalization. About 215 million people today live outside the country of their birth.

Historically, and to this day, forced migration also often occurs. The westward displacement of the Native American population of the United States; the dispersal of the Jews from Palestine in Roman times and from Europe in the mid-twentieth century; the export of Africans to the Americas as slaves; and the Clearances, or forced removal, of farmers from Scotland's Highlands to make way for large-scale sheep raising—all provide depressing examples. Today, refugee movements are far too common, prompted mainly by despotism, war, ethnic persecution, and famine. Recent decades have witnessed a worldwide flood of **refugees**: people who leave their country because of persecution based on race, ethnicity, religion, nationality, or political opinion (note that economic persecution does not fall under the definition of *refugee*). Perhaps as many as 16 million people who live outside their native country are refugees (see also Chapter 5).

Every migration, from the ancient dispersal of humankind out of Africa to the present-day movement toward urban areas, is governed by a host of **push-and-pull factors** that act to make the old home unattractive or unlivable and the new land attractive. Generally, push factors are the most central. After all, a basic dissatisfaction with the homeland is prerequisite to voluntary migration. The most important factor prompting migration throughout the thousands of years of human existence has been economic. More often than not, migrating people seek greater prosperity through better access to resources, especially land. Both forced migrations and refugee movements, however, challenge the basic assumption of the push-and-pull model, which posits that human movement is the result of choices and is primarily driven by economic factors.

> **refugees** Those fleeing from persecution in their country of nationality. The persecution can be religious, political, racial, or ethnic.
>
> **push-and-pull factors** Unfavorable, repelling conditions (push factors) and favorable, attractive conditions (pull factors) that interact to affect migration and other elements of diffusion.

DISEASES ON THE MOVE

Throughout history, infectious disease has periodically decimated human populations. One has

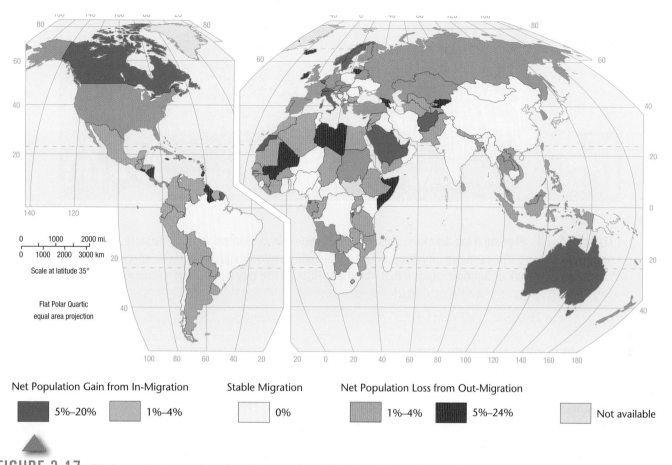

Net Population Gain from In-Migration

■ 5%–20%   ■ 1%–4%

Stable Migration

□ 0%

Net Population Loss from Out-Migration

■ 1%–4%   ■ 5%–24%

□ Not available

FIGURE 3.17 **Major and minor migration flows today.** Why have these flows changed so profoundly in the past hundred years? *(Source: Population Reference Bureau.)*

only to think of the vivid accounts of the Black Death episodes that together killed one-third of medieval Europe's population to get a sense of how devastating disease can be. Importantly for cultural geographers, diseases both move in spatially specific ways and result in responses that are spatial in nature.

The spread of disease provides classic illustrations of both expansion and relocation diffusion. Some diseases are noted for their tendency to expand outward from their points of origin. Commonly borne by air or water, the viral pathogens for contagious diseases such as influenza and cholera spread from person to person throughout an affected area. Some diseases spread in a hierarchical diffusion fashion, whereby only certain social strata are exposed. The poor, for example, have always been much more exposed to the unsanitary conditions—rats, fleas, excrement, and crowding—that have led to disease

epidemics. Other illnesses, particularly airborne diseases, affect all social and economic classes without regard for human hierarchies: their diffusion pattern is literally contagious. Indeed, disease and migration have long enjoyed a close relationship. Diseases have spread and relocated thanks to human movement. Likewise, widespread human migrations have occurred as the result of disease outbreaks.

When humans engage in long-distance mobility, their diseases move along with them. Thus, cholera—a waterborne disease that had for centuries been endemic to the Ganges River region of India—broke out in the early 1800s in Calcutta (now called Kolkata; see Figure 4.21 on page 166 and Seeing Geography on page 133). Because Calcutta was an important node in Britain's colonial empire, cholera quickly began to relocate far beyond India's borders. Soldiers, pilgrims, traders, and travelers spread

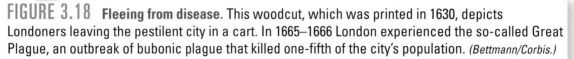
FIGURE 3.18 **Fleeing from disease**. This woodcut, which was printed in 1630, depicts Londoners leaving the pestilent city in a cart. In 1665–1666 London experienced the so-called Great Plague, an outbreak of bubonic plague that killed one-fifth of the city's population. *(Bettmann/Corbis.)*

cholera throughout Europe, then from port to port across the world through relocation diffusion, making cholera the first disease of global proportions.

Early responses to contagious diseases were often spatial in nature. Isolation of infected people from the healthy population—known as quarantine—was practiced. Sometimes only the sick individuals were quarantined, whereas at other times households or entire villages were shut off from outside contact. Whole shiploads of eastern European immigrants to New York City were routinely quarantined in the late nineteenth century. Another early spatial response was to flee the area where infection had occurred. During the time of the plague in fourteenth-century Europe, some healthy individuals abandoned their ill neighbors, spouses, and even children. Some of them formed altogether separate communities, whereas others sought merely to escape the city walls for the countryside (**Figure 3.18**).

Targeted spatial strategies could be implemented once it became known that some diseases spread through specific means. The best-known example is John Snow's 1854 mapping of cholera outbreaks in London, illustrated in **Figure 3.19**, which allowed him to trace the source of the infection to one water pump. Thanks to his detective work, and

FIGURE 3.19 **Mapping disease**. This map was constructed by London physician John Snow. Snow was skeptical of the notion that "bad air" somehow carried disease. He interviewed residents of the Soho neighborhood stricken by cholera to construct this map of cholera cases and used it to trace the outbreak to the contaminated Broad Street pump. Snow is considered to be a founder of modern epidemiology. *(Source: Courtesy of the John Snow Archive.)*

the development of a broader medical understanding of the role of germs in the spread of disease, cholera was discovered to be a waterborne disease that could be controlled by increasing the sanitary conditions of water delivery.

The threat of deadly disease is hardly a thing of the past. Today, geographers play vital roles in understanding the diffusion of HIV/AIDS, SARS (severe acute respiratory syndrome), and the so-called swine flu and bird flu in order to better address outbreaks and halt the spread of these global scourges. By mapping the spread of these contemporary diseases, understanding the pathways traveled by their carriers, and developing appropriate spatial responses, it is hoped that mass epidemics and disease-related panics can be avoided.

GLOBALIZATION

3.3 LEARNING OBJECTIVE
Discuss theories of population growth and control and how these have changed over time.

One of the great debates today involves the population of the Earth. How many people can the Earth support? Should humans limit their reproduction, and who should decide? There are no precise numbers involved in this debate, but these questions are so hotly contested because addressing them in a conscientious way may well determine the long-term fate of the human race.

POPULATION EXPLOSION?

One of the fundamental issues of the modern age is the **population explosion**: a dramatic increase in world population since 1900 (**Figure 3.20**). The crucial element triggering this explosion has been a steep decline in the death rate, particularly for infants and children, in most of the world, without an accompanying universal decline in

> **population explosion** The rapid, accelerating increase in world population since about 1650 and especially since 1900.

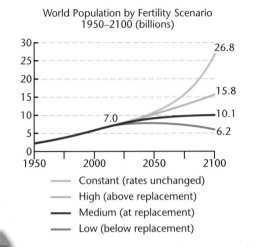

FIGURE 3.20 **World population under various fertility scenarios.** This prediction, put together by the United Nations, shows various total world populations that could be reached by the year 2100. The high number is based on current fertility patterns remaining unchanged. Notice what a difference it makes if fertility is at or below replacement level.

fertility. At one time in traditional cultures, only two or three offspring in a family of six to eight children might live to adulthood, but when improved health conditions allowed more children to survive, the cultural norm encouraging large families persisted.

Until very recently, the number of people in the world has been increasing geometrically, doubling in shorter and shorter periods of time. It took from the beginning of human history until 1800 C.E. for the Earth's population to grow to 1 billion people. But from 1800 to 1930, it grew to 2 billion, and in only 45 more years it doubled again (**Figure 3.21**). The overall effect of even a few population doublings is astonishing. As an illustration of a simple geometric progression, consider the following legend. A king was willing to grant any wish to the person who could supply a grain of wheat for the first square of his chessboard, two grains for the second square, four for the third, eight for the fourth, and so on. To cover all 64 squares and win, the candidate would have had to present a cache of wheat larger than today's worldwide wheat crop. Looking at the population explosion in another way, it is estimated that 107 billion humans have lived in the entire 200,000-year period since *Homo sapiens* originated. Of these, 7 billion (roughly 7 percent) are alive today.

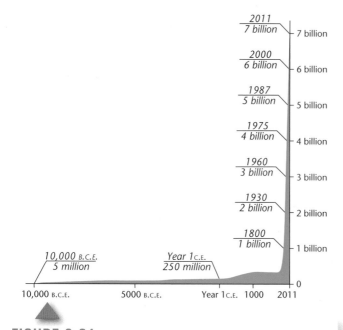

FIGURE 3.21 **World population doubling timeline.**
This graph illustrates the ever-faster doubling times of the world's population. Whereas accumulating the first billion people took all of human history until about 1800 C.E., the next billion took slightly more than a century to add, the third billion took only 30 years, and the fourth took only 15 years. (*Source: Adapted from Sustainablescale.org.*)

Some scholars foresaw long ago that an ever-increasing global population would eventually present difficulties. The most famous pioneer observer of population growth, the English economist and cleric Thomas Malthus, published *An Essay on the Principle of Population*—known as the "dismal essay"—in 1798. He believed that the human ability to multiply far exceeded our ability to increase food production. Consequently, Malthus maintained that "a strong and constantly operating check on population" would necessarily act as a natural control on numbers. Malthus regarded famine, disease, and war as the inevitable outcome of the human population's outstripping the food supply (**Figure 3.22**). He wrote, "Population, when un-checked, increases in a geometrical ratio. Subsistence only increases in an arithmetical ratio. A slight acquaintance with numbers will show the immensity of the first power in comparison of the second."

The adjective **Malthusian** entered the English language to describe the dismal future

> **Malthusian** Those who hold the views of Thomas Malthus, who believed that overpopulation is the root cause of poverty, illness, and warfare.

Malthus foresaw. Being a cleric as well as an economist, however, Malthus believed that if humans could voluntarily restrain the "passion between the sexes," they might avoid their otherwise miserable fate.

OR CREATIVITY IN THE FACE OF SCARCITY?

But, was Malthus right? From the very first, his ideas were controversial. The founders of communism, Karl Marx and Friedrich Engels, blamed poverty and starvation on the evils of capitalist society. Taking this latter view might lead one to believe that the miseries of starvation, warfare, and disease are more the result of maldistribution of the world's wealth than of overpopulation. Indeed, severe food shortages in several regions point to precisely this issue of inequitable food distribution and its catastrophic consequences.

Malthus did not consider that when faced with complex problems such as scarce food supplies, human beings are highly creative. This has led critics of Malthus and his modern-day followers to point out that although the global population has doubled three

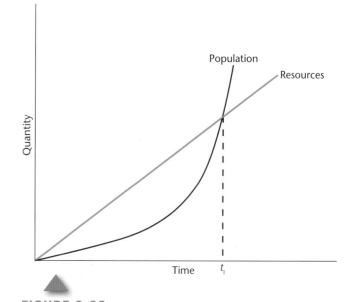

FIGURE 3.22 **Malthus's dismal equation.** Thomas Malthus based his theory of population growth on the simple notion that resources (food) grow in a linear fashion, whereas population grows geometrically. Beyond time t_1, population outstrips resources, resulting in famine, conflict, and disease.

times since Malthus wrote his essay, food supplies have doubled five times. Scientific innovations such as the green revolution have led to food increases that have far outpaced population growth (see Chapter 8). Other measures of well-being, including life expectancy, air quality, and average education levels, have all improved, too. Some of Malthus's critics, known as **cornucopians**, argue that human beings are, in fact, our greatest resource and that attempts to curb our numbers misguidedly cheat us out of geniuses who could devise creative solutions to our resource shortages. Modern-day followers of Malthus, known as **neo-Malthusians**, counter that the Earth's support systems are being strained beyond their capacity by the widespread adoption of wasteful Western lifestyles.

> **cornucopians** Those who believe that science and technology can solve resource shortages. In this view, human beings are our greatest resource rather than a burden to be limited.
>
> **neo-Malthusians** Modern-day followers of Thomas Malthus.

The fact is that the world's population is growing more slowly than before. The world's TFR has fallen to 2.47; one demographer has declared that "the population explosion is over"; and Figure 3.21 depicts a leveling-off of the global population at about 11 billion by the year 2100, suggesting that the world's current "population explosion" is merely a stage in a global demographic transition. Relatively stable population totals worldwide, however, mask the population declines and aging of the population structure already under way in some regions of the world.

It is difficult to speculate about the state of the world's population beyond the year 2050. What is clear, however, is that the lifestyles we adopt will affect how many people the Earth can ultimately support. In particular, whether wealthy countries continue to use the amount of resources they currently do, and whether developing countries decide to follow a Western-style route toward more and more resource consumption, will significantly determine whether we have already overpopulated the planet.

THE RULE OF 72

A handy tool for calculating the doubling time of a population is called the Rule of 72: take a country's rate of annual increase, expressed as a percent, and divide it into the number 72. The result is the number of years a population, growing at a given rate, will take to double.

For example, the natural annual growth of the United States in 2013 was 0.6 percent. Dividing 72 by 0.6 yields 120, which means that the population of the United States will double every 120 years. This does not, however, factor in the relatively high levels of immigration experienced by the United States, which will cause its population to double faster than every 120 years.

What about countries with faster rates of growth? Consider Guatemala, which is growing at 2.1 percent per year. Doing the math, we find that Guatemala's population is doubling every 34 years!

Percentages such as 0.6 or 2.1 don't sound like such high rates. If we were discussing your bank account rather than the populations of countries, you would hope that your money would double more quickly than every 34 years! (Incidentally, you can apply the Rule of 72 to your bank account or to any other figure that grows at a steady annual rate.) You may be tempted to say, "Look, there are only 14 million people in Guatemala, so it doesn't really matter if its population is doubling quickly. What really matters is that at an annual increase of 1.3 percent, India's 1.2 billion people will double to 2.4 billion in 55 years, and China's 1.35 billion will double almost as fast as the U.S. population—every 146 and 137 years, respectively—but that will add another 1.35 billion to the world's population, more than four times what the United States will add by doubling!"

This assessment is partly right and partly wrong, depending on your vantage point. Viewed at the global scale, it does indeed make a significant difference when China's or India's population doubles. But if you are a resident of Guatemala or an official of the Guatemalan government, a doubling of your country's population every 34 years means that health care, education, jobs, fresh water, and housing must be supplied to twice as many people every 34 years. As mentioned in the introduction to this chapter, the scale at which population questions are asked is vital for the answers that are given.

POPULATION CONTROL PROGRAMS

Though the debates may occur at a global scale, most population control programs are devised and

FIGURE 3.23 China's One Child Per Couple Policy.
This Chinese billboard encourages families to use family planning in order to achieve the goal of one child per couple. *(Iain Masterson/Alamy.)*

implemented at the national level. When faced with perceived national security threats, some governments respond by supporting pronatalist programs designed to increase the population. Labor shortages in Nordic countries have led to incentives for couples to have larger families. Most population programs, however, are antinatalist: they seek to reduce fertility. Needless to say, this is an easier task for a nonelected government to carry out because limiting fertility challenges traditional gender roles and norms about family size.

Although China certainly is not the only country that has sought to limit its population growth, its so-called one-child policy provides the best-known modern example. Mao Zedong, the longtime leader of the People's Republic of China (1949–1976), did not initially discourage large families in the belief that "every mouth comes with two hands." In other words, he believed that a large population would strengthen the country in the face of external political pressures. He reversed his position in the 1970s when it became clear that China faced resource shortages as a result of its burgeoning population. In 1980 the one-child-per-couple policy was adopted. With it, Chinese authorities sought not merely to halt population growth but, ultimately, to decrease the national population. All over China today one sees billboards and posters admonishing the citizens that "one couple, one child" is the ideal family (**Figure 3.23**).

Violators face huge monetary fines, cannot request new housing, lose the rather generous benefits provided to the elderly by the government, forfeit their children's access to higher education, and may even lose their jobs. Late marriages are encouraged. In response, between 1970 and 1980, the TFR in China plummeted from 5.9 births per woman to only 2.7, then to 2.2 by 1990, 2.0 by 1994, 1.7 by 2007, and 1.5 by 2013 (see Figure 3.3, page 97). The Chinese government claims that 400 million fewer births have occurred over the lifetime of the policy.

China achieved one of the greatest short-term reductions of birth rates ever recorded, thus proving that cultural changes can be imposed from above, rather than simply waiting for them to diffuse organically. In recent years, the Chinese population control program has been less rigidly enforced as economic growth has eroded the government's control over the people. This relaxation has allowed more couples to have two children instead of one; however, the increase in economic opportunity and migration to cities have led some couples to have smaller families voluntarily. Because of China's population size, its actions will greatly affect the global context as well as its own national one (see Subject to Debate on page 105 for a discussion of the gender implications of strict population control policies).

NATURE–CULTURE

3.4 LEARNING OBJECTIVE
Recognize the ways in which the natural world shapes population characteristics, and how population characteristics in turn shape the natural world.

The reasons for human populations forming such diverse patterns are often cultural. Though the differences may have started out as adaptations to given physical conditions, when repeated from generation to generation, these patterns become woven tightly into the cultural fabric of places. Thus, demographic practices such as living in crowded settlements or having large families may well have deep roots in both nature and culture.

ENVIRONMENTAL INFLUENCE

Local population characteristics are often influenced by the availability of resources. In the middle latitudes, population densities tend to be greatest where the terrain is level, the climate is mild and humid, the soil is fertile, mineral resources are abundant, and the sea is accessible. Conversely, population tends to thin out with excessive elevation, aridity, coldness, ruggedness of terrain, and distance from the coast.

Climatic factors influence where people settle. Most of the sparsely populated zones in the world have, in some respect, difficult climates from the viewpoint of human habitation (see Figure 3.1, page 95). The thinly populated northern edges of Eurasia and North America are excessively cold, and the belt of sparse population extending from North Africa into the heart of Eurasia matches the pattern in major desert zones of the Eastern Hemisphere. Humans remain largely creatures of the humid and subhumid tropics, subtropics, or midlatitudes and have not fared well in excessively cold or dry areas. Small populations of Inuit (Eskimo), Sami (Lapps), and other peoples live in some of the less hospitable areas of the Earth, but these regions do not support large populations, while Antarctica is home only to a handful of researchers. Humans have proven remarkably adaptable, and our cultures include strategies that allow us to live in many different physical environments. As a species, however, perhaps we have not entirely moved beyond the **adaptive strategies** that suited us so well to the climatic features of sub-Saharan Africa, where we began.

> **adaptive strategies** The unique way in which each culture uses its particular physical environment; those aspects of culture that serve to provide the necessities of life—food, clothing, shelter, and defense.

Humankind's preference for lower elevations is especially true for the middle and higher latitudes. Most mountain ranges in those latitudes stand out as sparsely populated regions. By contrast, inhabitants of the tropics often prefer to live at higher elevations, concentrating in dense clusters in mountain valleys and basins (see Figure 3.1). For example, in tropical portions of South America, more people live in the Andes Mountains than in the nearby Amazon lowlands. The capital cities of many tropical and subtropical nations lie in mountain areas above 3000 feet (900 meters) in elevation. Living at higher elevations allows residents to escape the hot, humid climate and diseases of the tropic lowlands. In addition, these areas were settled because the fertile volcanic soils of these mountain valleys and basins were able to support larger populations in agrarian societies.

Humans often live near the sea. The continents of Eurasia, Australia, and South America resemble hollow shells, with the majority of the population clustered around the rim of each continent (see Figure 3.1). In Australia, half the total population lives in only five port cities, and most of the remainder is spread out over nearby coastal areas. This preference for living by the sea stems partly from the trade and fishing opportunities the sea offers. At the same time, continental interiors tend to be regions of climatic extremes. For example, Australians speak of the "dead heart" of their continent, an interior land of excessive dryness and heat (**Figure 3.24**).

Disease also affects population distribution. Some diseases attack valuable domestic animals, depriving people of food and clothing resources. Such diseases

FIGURE 3.24 Australia's "Dead Heart." This shot was taken west of Alice Springs in Australia's Northern Territory, which is just about at the center of the continent. The climate here does not lend itself to dense human settlements. *(Auscape/UIG/Getty Images.)*

have an indirect effect on population density. For example, in parts of East Africa, a form of sleeping sickness attacks livestock. This particular disease is almost invariably fatal to cattle but not humans. The people in this part of East Africa depend heavily on cattle, which provide food, represent wealth, and serve a religious function in some tribes. The spread of a disease fatal to cattle has caused entire tribes to migrate away from infested areas, leaving those areas unpopulated.

ENVIRONMENTAL PERCEPTION, SETTLEMENT, AND MIGRATION

Perception of the physical environment plays a major role in a group's decision about where to settle and live. Different cultural groups often "see" the same physical environment in different ways. These varied responses to a single environment influence the distribution of people. A good example appears in a part of the European Alps shared by German- and Italian-speaking peoples. The mountain ridges in that area—near the point where Switzerland, Italy, and Austria border one another—run in an east–west direction, so that each ridge has a sunny, south-facing slope and a shady, north-facing one. German-speaking people, who rely on dairy farming, long ago established permanent settlements some 650 feet (200 meters) higher on the shady slopes than the settlements of Italians, who are culturally tied to warmth-loving crops, on the sunny slopes. This example demonstrates how contrasting cultural attitudes toward the physical environment and land use affect settlement patterns.

Sometimes the same cultural group changes its perception of an environment over time, with a resulting redistribution of its population. The coalfields of western Europe provide a good case in point. Before the industrial age, many coal-rich areas—such as the Midlands of England, southern Wales, and the lands between the headwaters of the Oder (or Odra) and Vistula rivers in Poland—were only sparsely or moderately settled. The development of steam-powered engines and the increased use of coal in the iron-smelting process, however, created a tremendous demand. Industries grew up near the

European coalfields, and people flocked to these areas to take advantage of the new jobs. In other words, once a technological development gave a new cultural value to coal, many sparsely populated areas containing that resource acquired large concentrations of people.

Recent studies indicate that much of the migration within the United States today is prompted by a desire for a pleasant climate and other desirable physical environmental traits, such as beautiful scenery. Surveys of immigrants to Arizona revealed that its sunny, warm climate is a major reason for migration. An attractive environment provided the dominant factor in the growth of the population and economy of Florida. The most desirable environmental traits that serve as stimulants for American migration include (1) mild winter climate and a mild summer climate with low humidity, (2) mountainous terrain, (3) diverse natural vegetation that includes forests, (4) the presence of lakes and rivers, and (5) nearness to the seacoast. Different age and cultural groups often express different preferences, but all are influenced by their perceptions of the physical environment in making decisions about migration.

POPULATION DENSITY AND ENVIRONMENTAL ALTERATION

People modify their habitats through their adaptive strategies. Particularly in areas where population density is high, radical alterations often occur. This can also happen in fragile environments even at relatively low population densities because, as discussed earlier in this chapter, the Earth's carrying capacity varies greatly from one place to another and from one culture to another.

Many of our adaptive strategies are not sustainable. Population pressures and local ecological crises are closely related. For example, in Haiti, where rural population pressures have become particularly severe, the trees in previously forested areas have been felled for fuel, leaving the surrounding fields and pastures increasingly denuded and vulnerable to erosion (Figure 3.25). In short, overpopulation relative to resource availability can precipitate

FIGURE 3.25 **Aerial view of Hispaniola.** The border between Haiti, on the left, and the Dominican Republic, to the right, is clearly demarcated by the absence of forest cover on the Haitian side. *(James P. Blair/National Geographic Creative.)*

consume appears to be limited only by the ability to pay for it. Indeed, in mid-2010, China surpassed the United States to become the world's largest consumer of energy resources.

ENVIRONMENTAL REFUGEES

Sometimes human adaptive strategies stress the natural environment past the breaking point and previous population densities can no longer be sustained. When that happens, people are forced to migrate. Sudden environmental disasters, such as floods, tornadoes, or forest fires, give rise to a massive human exodus from a destroyed place (**Figure 3.27**). People who are displaced from their homes due to severe environmental disruption are called **environmental refugees**.

> **environmental refugees** People who are displaced from their homes due to severe environmental disruption.

environmental destruction—which, in turn, results in a downward cycle of worsening poverty, with an eventual catastrophe that is both ecological and demographic. Thus, many cultural ecologists believe that attempts to restore the balance of nature will not succeed until we halt or even reverse population growth, although they recognize that other causes are also at work in ecological crises (**Figure 3.26**).

The worldwide ecological crisis is not solely a function of overpopulation. A relatively small percentage of the Earth's population controls much of the industrial technology and consumes a disproportionate percentage of the world's resources each year. Americans, who make up less than 5 percent of the global population, account for about 25 percent of the natural resources consumed globally each year. New houses built in the United States in 2002 were, on average, 38 percent bigger than those built in 1975, despite a shrinking average household size. If everyone in the world had an average American standard of living, the Earth could support only about 500 million people—only 8 percent of the present population. As the economies of large countries such as India and China continue to surge, the resource consumption of their populations is likely to rise as well because the human desire to

FIGURE 3.26 **Landslide on Whidbey Island, Washington State.** This massive landslide in 2013 changed the topography of Whidbey Island by moving nearly 5.3 million square feet of earth and raising the elevation of the beach some 30 feet above its previous height. This area is geologically vulnerable to landslides. *(AP Photo/Ted S. Warren, File.)*

FIGURE 3.27 **Environmental refugees from Hurricane Sandy.** This image is from Breezy Point, Queens, New York. This neighborhood was devastated by Hurricane Sandy and an ensuing fire in 2012. More than 100 homes were destroyed here. *(Peter van Agtmael/Magnum Photos.)*

FIGURE 3.28 **Sinking building in Mexico City.** This line of rooftops was initially straight, but has shifted as the ground beneath became unstable over the years. ***Can you imagine living or working in one of these units?*** *(National Geographic/Getty Images.)*

Environmental Changes in Mexico City. Sometimes environmental changes occur slowly over time. Depletion of the underground water table, for instance, has caused Mexico City to sink, on average, 2.5 inches each year. Some of the larger, heavier buildings in Mexico City, such as the city's main cathedral, have been sinking even more quickly than that. The Palace of Fine Arts, for instance, has sunk 13 feet in the last century, and its original ground floor has now become the building's basement.

Mexico City was originally the Aztec capital of Tenochtitlan. Built in the middle of a lake, the Aztecs made use of a series of canals and floating gardens to transport themselves and grow their food. The natural environment was thus minimally altered by its Aztec inhabitants. When the Spaniards conquered Mexico in the sixteenth century, they built Mexico City on top of the ruins of Tenochtitlan and drained the lake in an effort to control flooding. Since then, Mexico City has grown to one of the world's largest urban areas. Its 21 million inhabitants get their water from an aquifer—a natural underground reservoir. As the water is drained out over time, the underlying layers of soft clay and volcanic silt dry out and collapse unevenly.

Cracked foundations, window frames that are no longer square, and buildings leaning on each other present multiple inconveniences, costs, and hazards (**Figure 3.28**). According to one Mexico City–based engineer, "When a building tilts more than 1 degree, then I think it becomes very uncomfortable." He notes that people living in crooked buildings are unable to comfortably lie in bed or even wash dishes, with tap water flowing oddly. "Tables aren't stable. Liquids don't look right when they are in big containers. Window panes can break. Doors don't close right." Vertigo is a common complaint of people who live and work in tilted structures. Other urban features are also affected. Subway lines and sidewalks need extensive repairs due to sinking, tilting, and buckling. Liquids in drainage canals and pipelines flow uphill.

Some structures are in imminent danger of collapse, and 50 buildings have been condemned by the city since 2006, with another 5000 or so known to be unstable. These structures become vulnerable to natural disasters. Because Mexico City is located in an earthquake-prone area, frequent tremors further damage buildings and a larger earthquake could prove disastrous. An earthquake that struck in

1985 caused 800 buildings to collapse and left thousands dead.

The adaptive strategy of draining Mexico City's underground aquifer to provide water for its population has resulted in a slow collapse of the city. At what point will residents begin to migrate to other places because of the hazards posed? Another large earthquake like the one that struck in 1985 might bring about significant population movements out of the damaged city.

A number of environmental catastrophes are closely related to population pressures. Violent hurricanes and sea level rise are associated with global warming; land degradation occurs due to local pressures such as overfarming, contamination, or broader climate change resulting in desertification; and landslides result from deforestation and other effects of human settlement expansion. These are just some of the catastrophes resulting from unsustainable adaptive strategies (**Figure 3.29**). The Internal

Displacement Monitoring Centre estimated that some 32 million people worldwide were forced to flee their homes due to weather-related events alone in 2012. These disasters are more man-made than they are "natural," and they are often the trigger for large-scale migrations of human populations in search of safer environments.

CULTURAL LANDSCAPE

3.5 LEARNING OBJECTIVE
Analyze the imprint of demographic factors on the cultural landscape.

Population geographies are visible in the landscapes around us. The varied densities of human settlements and the shapes these take in different places provide

FIGURE 3.29 **Global climate change refugees.** Climate change can lead to environmental hazards that make places uninhabitable. This map depicts the types of hazards that are likely to occur. People in the developing world are most likely to become environmental refugees, because climate change is coupled with the effects of poverty and war.

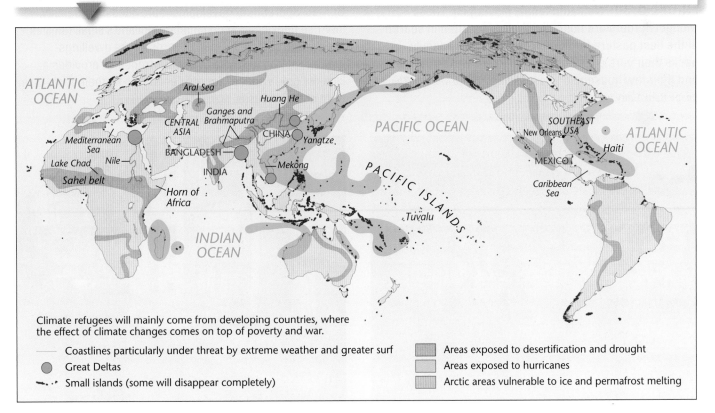

Climate refugees will mainly come from developing countries, where the effect of climate changes comes on top of poverty and war.

— Coastlines particularly under threat by extreme weather and greater surf

● Great Deltas

-•-• Small islands (some will disappear completely)

▨ Areas exposed to desertification and drought

▨ Areas exposed to hurricanes

▨ Arctic areas vulnerable to ice and permafrost melting

clues to the intertwined cultural and demographic strategies implemented in response to diverse local factors. These clues are evident wherever people have settled, be they urban, suburban, or rural locations. Less visible, but no less important, factors are also at work on and throughout the cultural landscape. In this final part of the chapter, we examine the diverse, and often creative, ways in which different places are organized spatially in order to accommodate the populations that live there.

DIVERSE SETTLEMENT TYPES

Human settlements range in density from the isolated farmsteads found in some rural areas to the teeming streets of megacities such as Kolkata, India (see Seeing Geography, page 133). The size and pattern of human settlements depend, in part, on the size and needs of the population to be accommodated. Small rural populations and those engaged in subsistence farming don't inhabit large, dense cities. Nomadic peoples require dwellings that can be easily packed up and moved, and they tend to have few household possessions (**Figure 3.30**). On the other end of the spectrum, dense cities make more sense for populations employed in the service and information sectors. These people benefit from being as close as possible to their jobs, and higher densities—at least in theory—reduce commuting distances. People in postindustrial societies also tend to have smaller families, which can be more easily accommodated in apartments (**Figure 3.31**).

But, there is also a significant cultural aspect at work, such that similar populations may live in settlements that look and feel vastly different from each other. Take rural settlements, for instance. In many parts of the world, farming people group themselves together in clustered settlements called **farm villages**. These tightly bunched settlements vary in size from a few dozen inhabitants to several thousand. Contained in the village **farmstead** are the house, barn, sheds,

> **farm villages** Clustered rural settlements of moderate size, inhabited by people who are engaged in farming.
>
> **farmstead** The center of farm operations, containing the house, barn, sheds, and livestock pens.

FIGURE 3.30 **Nomadic family moving on.** These Mongol herders are nomadic; they move often in search of the best pasture for their livestock. Here, a family packs their yurt, a traditional Mongolian tent structure, and their few household possessions. *(Bruno Morandi/The Image Bank/Getty Images.)*

FIGURE 3.31 **High-density dwelling in Amsterdam.** The Borneo-Sporenburg development provides high-density, low-rise housing to accommodate Holland's small families in an urban setting. Instead of yards, these dwellings feature rooftop terraces. The waterfront also provides a kind of "green space" for residents, who tie their boats up in front of their houses. *(© Iain Masterton/Alamy.)*

FIGURE 3.32 **A truly isolated farmstead, in the Vesturland region of western Iceland.** This type of rural settlement dominates almost all lands colonized by Europeans who migrated overseas. Iceland was settled by Norse Vikings a thousand years ago. *(© Jay Dickman/Corbis.)*

pens, and garden. The fields, pastures, and meadows lie out in the country beyond the limits of the village, and farmers must journey out from the village each day to work the land.

Farm villages are the most common form of agricultural settlement in much of Europe, in many parts of Latin America, in the densely settled farming regions of Asia (including much of India, China, and Japan), and among the sedentary farming peoples of Africa and the Middle East.

In many other parts of the world, the rural population lives on dispersed, isolated farmsteads, often some distance from the nearest neighbors (**Figure 3.32**). These dispersed rural settlements grew up mainly in Anglo America, Australia, New Zealand, and South Africa—that is, in the lands colonized by emigrating Europeans.

Why do so many farm people settle together in villages? Historically, the countryside was unsafe, threatened by roving bands of outlaws and raiders. Farmers could better defend themselves against such dangers by grouping together in villages. In many

parts of the world, the populations of villages have grown larger during periods of insecurity and shrunk again when peace returned. Many farm villages occupy the most easily defended sites in their vicinity.

Various communal ties strongly bind villagers together. Farmers linked to one another by blood relationships, religious customs, communal landownership, or other similar bonds usually form clustered villages. Mormon farm villages in the United States provide an excellent example of the clustering force of religion. Communal or state ownership of the land—as in China and parts of Israel—encourages the formation of farm villages.

The conditions encouraging dispersed settlement are precisely the opposite of those favoring village development. These include peace and security in the countryside, eliminating the need for defense; colonization by individual pioneer families rather than by socially cohesive groups; private agricultural enterprise, as opposed to some form of communalism; and well-drained land where water is readily available. Most dispersed farmsteads originated rather recently, dating primarily from the colonization of new farmland in the past two or three centuries.

LANDSCAPES AND DEMOGRAPHIC CHANGE

Population change in a place can occur rapidly or more slowly over time. Rapid **depopulation** can come about as the result of sudden catastrophic events, such as natural disasters, disease epidemics, and warfare (**Figure 3.33**). Or depopulation can take place at a slower pace. For example, places may lose population over time because of the gradual out-migration of people in search of opportunities elsewhere, as a cumulative result of declining fertility rates, or as we saw in the previous section in response to climate change.

depopulation A decrease in population that sometimes occurs as the result of sudden catastrophic events, such as natural disasters, disease epidemics, and warfare.

FIGURE 3.33 **Refugees flee fighting in eastern Congo.** This area of the Democratic Republic of Congo, near the Rwandan border, has experienced the repeated movement of people displaced by civil strife. *(Stephen Morrison/epa/Corbis.)*

Populations can also grow in more or less rapid fashions. A place might experience an abrupt influx of people who have been displaced from other areas. Population increases commonly occur more gradually, as the result of demographic improvements such as longer life spans or lower infant mortality rates, or the accumulation of steady streams of immigrants to attractive areas over time.

Depopulation in one area and population increase in another are often linked processes. For instance, you could easily envision a scenario in which the environmental refugees exiting one place suddenly flood into a neighboring area. As a longer-term example, you could equally well imagine how broad economic changes or technological innovations could lead to regional population shifts whereby some areas become depopulated and others gain population.

Indeed, shifting economic and technological conditions have influenced the demographic landscape in profound ways. The process of industrialization during the past 200 years has caused the greatest voluntary relocation of people in world history. Within industrial nations, people moved from rural areas to cluster in manufacturing regions (see Chapter 10). Agricultural changes have also influenced population density. For example, the complete mechanization of cotton and wheat cultivation in mid-twentieth-century America allowed those crops to be raised by a much smaller labor force (see Chapter 8). As a result, profound

depopulation occurred, to the extent that many small towns serving these rural inhabitants ceased to exist.

Regardless of the reasons, population changes must be accommodated, and these adaptations are invariably reflected in the cultural landscape.

Depopulation in History: Ancient Rome There are as many theories accounting for the decline of the Roman Empire as there are historians theorizing about it. Lead poisoning, overextension of the empire's reach, conquest by Germanic tribes, and at least 200 other theories have been proposed.

Rome, the empire's capital city, suffered this process acutely. For nearly a millennium, Rome had been the world's wealthiest, most powerful, and most populous city (**Figure 3.34**). At the close of the first century C.E., Rome's population surpassed 1 million. As the Roman Empire's economic, political, and military might waned across the third, fourth, and fifth centuries, the demographic balance shifted as well. Under the emperor Constantine in 330 C.E., the empire was split into eastern and western halves, and its capital was relocated to Constantinople (now Istanbul, in modern-day Turkey).

As barbarian invasions further weakened the divided empire, Rome's physical infrastructure—its famed roads, bridges, aqueducts, and monumental buildings—crumbled, as did its impressive administrative infrastructure. With no access to former amenities, from running water to educational opportunities, Rome's population had shrunk to around 20,000 by 550 C.E. The Roman landscape had become a ghost of its former glory: dilapidated, uninhabited, and in ruins (see Figure 3.34).

Bogotá Rising Dilapidated-looking landscapes can result from population decline, as the previous example of ancient Rome illustrates, but they can also arise from rapid population influx. Many of the world's shantytowns exemplify the sort of chaotic landscape that can result from rapid population increases. Los Altos de Cazucá, a neighborhood in the Colombian capital of Bogotá, is but one example. In this case, the population influx comes mostly from people arriving in the capital after being displaced by armed conflict in the countryside.

The United Nations High Commissioner for Refugees (UNHCR) estimates that, by 2012, there

FIGURE 3.34 **Rome at its height and Rome in ruins.** The image on the left, from a nineteenth-century engraving, shows Rome at its height, as seen from Mount Palatine, one of Rome's seven hills. The image on the right, from a nineteenth-century painting, shows the ruins of the Roman Forum from the Capitoline Hill, another of Rome's seven hills. *(Both photos: The Granger Collection, New York/The Granger Collection.)*

were 28.8 million internally displaced persons (IDPs) due to armed conflict worldwide. Colombia is one of the globe's hotspots for IDPs and has the highest number of IDPs of any country in the world. Colombia's estimated 5.8 million IDPs have been uprooted largely as the result of ongoing civil warfare between the paramilitary group known as the FARC (Revolutionary Armed Forces of Colombia) and government forces bent on eradicating them. The UNHCR has identified this as the "worst humanitarian crisis in the western hemisphere."

Colombia's armed conflict has depopulated the rural areas where it occurs, forcing its victims into the cities. Terrorized, landless, and impoverished, the mainly indigenous and Afro-Colombian IDPs settle in the outskirts of cities like Colombia's capital, Bogotá.

Los Altos de Cazucá is one of the settlements inhabited by Colombia's internally displaced persons (**Figure 3.35**). Known as shantytowns, areas like Los Altos de Cazucá arise for different reasons, but they exist in all large cities throughout the developing world.

FIGURE 3.35 **Los Altos de Cazucá.** This is a shantytown on the outskirts of Colombia's capital city, Bogotá. Many of its approximately 50,000 residents are displaced people from other parts of the country. *(imagebroker.net/SuperStock.)*

World Heritage Site

Pico Island is the second largest island in the Azores, a chain of nine volcanic islands. Politically a part of Portugal, the islands are located some 950 miles west of Portugal, in the middle of the Atlantic Ocean. Pico Island is noted for its vineyard culture, established in the 1500s, which exemplifies the isolated farmstead settlement.

■ Pico Island consists of 987 hectares (40,515 square feet) of volcanic rock whose man-made landscape has been shaped entirely for small-scale agriculture; namely, growing grapes and processing them into wine (viniculture).

■ The island remained uninhabited until 1439, when the first Portuguese settlers arrived. Initially, they were cattle herders, but winemaking was introduced from the mainland a few decades later, with Portuguese Franciscan and Carmelite religious orders improving its techniques in the 1500s. Winemaking continued to be the island's mainstay until the mid-1800s, when the grape vines became afflicted by grape mildew and phyllorexa (plant diseases). Today, wine production has experienced a resurgence, although the sweet Verdelho wine exported to European nobility in past centuries is hard to find because the island's small size limits total production.

■ With a population of just over 15,000 people, Pico Island counts some cities—the capital Madalena being the largest with 6000 inhabitants—but the majority of its population lives in small farm villages and port towns.

(Thomas Stankiewicz/Getty Images.)

Pico Island Vineyard

THE WALLS: The viniculture landscape of Pico Island is unique and ingenious. The island has mild year-round temperatures and adequate rain, both of which are conducive to growing wine grapes. But the island is constantly buffeted by strong winds and sea surges that are harmful to grape vines. In response, islanders used volcanic basalt rock to construct walls along the coastline, which enclose small plots of land used to grow grapes. The walls protect the fragile plants from the wind and sea water, and support them as they grow. The resulting landscape is one of many small interlocking rectangular plots, called currais, bounded by dark-colored basalt rock walls. UNESCO has deemed this an "extraordinarily beautiful man-made landscape of small, stone-walled fields."

(Yulia_B/Shutterstock.)

■ Vine cultivation is done entirely by hand, rather than using farm machinery. This is possible because of the small scale of wine production on the island. In addition, the rocks used to construct the walls were stacked meticulously by hand: no mortar was used.

■ Buildings constructed to support winemaking and the people engaged in it include cellars, warehouses, churches, ports, houses, and wells. The settlements are small and the architecture distinctive.

TOURISM: The island's natural features as well as its historic industries attract a small but growing number of tourists.

(imageBROKER/Alamy.)

■ Pico is "the mountain island" of the Azores, dominated by a volcanic cone rising nearly 8000 feet above sea level. Hiking to the crater at the top of the dormant volcano takes a full day and is quite challenging.

■ Whaling, officially banned by the European Union in the 1980s, was Pico Island's main industry for a period of time. Today, whale watching is one of the island's primary tourist activities.

■ Pico Island has received an international award for its efforts to promote sustainable tourism (tourism that results in low impact to the environment and local culture).

■ Pico Island is a beautiful expression of the balance forged between nature and man in isolated conditions.

■ The small, geometric basalt-walled plots are ingenious constructions that allow for the successful cultivation of wine grapes in a harsh environment.

■ Placed on the list of World Heritage Sites in 2004, Pico Island has a small but growing tourism industry.

www.whc.unesco.org/en/list/1117

FIGURE 3.36 **Landscapes of poverty and wealth.** This scene from Bogotá is a common Latin American urban landscape, with wealthy high-rises located next to impoverished shantytowns. *(Jacques Jangoux/Getty Images.)*

Housing is constructed by the residents themselves, using found materials like cardboard, tin panels, and old tires. Some shantytowns are located far away from downtown areas where wealthy people reside and where jobs are, while others are literally pressed up against wealthier neighborhoods (Figure 3.36).

Most shantytowns arise spontaneously, in order to address population influx to cities that lack the resources to plan systematically for rapid growth. But, are shantytowns temporary, disappearing as their residents become incorporated into city life? Hardly. Instead, shantytowns gradually become part of the urban fabric. Indeed, most of the spatial expansion of Latin American cities like Bogotá occurs precisely thanks to growth of the shantytowns surrounding them. Over time, the dwellings are constructed of more-permanent materials such as concrete blocks. Roads are paved and running water installed. Power and phone lines are extended. The once-temporary areas slowly become visually, economically, and culturally integrated into the permanent fabric of the city. New shantytowns then arise beyond their borders, to accommodate recent arrivals.

CONCLUSION

In our study of population geography, we have seen that humankind is unevenly distributed over the Earth. Spatial variations in fertility, death rates, rates of population change, age groups, gender ratios, and standards of living also exist: these patterns can be depicted as demographic culture regions. The principles of mobility prove useful in analyzing human migration and also help explain the spread of factors influencing demographic characteristics such as disease.

Population geography proves particularly intriguing when scale is taken into account. Although most geodemographic issues are experienced locally, and policies shaping them are usually set at the national level, the important debates about the world's population are truly global in nature.

The theme of nature–culture shows how the natural environment and people's perception of and engagement with it influence demographic factors. Nature–culture interactions shape the spatial distribution of people and sometimes help guide migrations. In addition, population density is linked to the level of environmental alteration, and overpopulation can have a destructive impact on the environment and give rise to environmental refugees.

The cultural landscape visually expresses the varied ways in which societies accommodate their populations. How people distribute themselves over the Earth's surface finds a vivid expression in the cultural landscape. The look and feel of places are constantly adapting to demographic change.

GEOGRAPHY @ WORK

Aaron Hoard. *(Jennifer Boyle Photography.)*

Aaron Hoard
Deputy Director,
Office of Regional and Community
Relations, University of Washington
Education: BA Geography
University of Washington

Q. *Why did you major in geography and decide to pursue a career in the field?*

A. I was interested in urban planning as a career, but my university didn't offer an undergraduate planning degree at the time. The mix of subjects and technical skills taught in the geography department seemed closely aligned to what I wanted to do as a career.

Q. *Please describe your job.*

A. I act as a liaison between the university and local governments, businesses, and neighbors. In this capacity, I help manage projects related to transportation, land use, public safety, housing, and city regulations. I have helped organize the university's response to design and mitigation issues related to the construction of major transportation projects. I have worked on affordable housing strategies for university employees including a home mortgage program and an affordable housing project. I also work closely with the community and city to address ongoing concerns about public safety and student behavior in the surrounding neighborhoods. And, I represent the university on a variety of boards and committees in the community.

Q. *How does your geographical background help you in your day-to-day work?*

A. My projects cut across multiple fields of study such as housing, transportation, public safety, social services, and economic development. Geography has a rich tradition of being an interdisciplinary degree and having that experience working across different disciplines helped prepare me to work on complex challenges in my career. Many of my projects are also spatial in nature. Where do people live? Where can we develop new buildings? Where does crime occur? The ability to identify and answer these spatial questions helps to focus solutions.

Q. *What advice do you have for college students considering a career in geography?*

A. Develop a "tool box" of analytical skills offered through your geography program. Skills in GIS, programming, cartography, statistics, demographics, remote sensing, and other areas can help make you a competitive candidate for your first job. The interdisciplinary nature of geography will serve you well throughout your career. Learning to fuse together different subjects to answer spatial questions in geography will help you to develop the intellectual flexibility you'll need to tackle a variety of complex problems.

DOING GEOGRAPHY

Public Space, Personal Space: Too Close for Comfort?

Culture can condition people to accept or reject crowding. Personal space—the amount of space that individuals feel "belongs" to them as they move about their everyday business—varies from one cultural group to another. As the Seeing Geography section for this chapter notes, different people seem to require different amounts of personal space. One's comfort zone varies with social class, gender, ethnicity, the situation at hand, and what one has grown accustomed to over one's life. Some Arabs, for example, consider it appropriate and even polite to be close enough for another to smell his or her breath during conversation. Those from cultures that have not developed a high tolerance for personal contact might experience such closeness as intrusive. Americans conducting business in Japan are often surprised at the level of physical closeness expected in their dealings. Such closeness might well be interpreted as overstepping one's bounds, literally, in the United States!

In this exercise, you will gather some data on the amount of personal space needed by those around you. Observe and record your findings, and discuss them in class as a group.

Steps to Understanding Indigenous Culture

Step 1: Observe your professors as they lecture in class. Is your class a large one that meets in a lecture hall? If so, where does your professor sit or stand in relation to the students? Does the professor have his or her own designated space in the classroom? Where is it located, and how big is it? Does the professor ever step outside of it? How does this professor's use of space compare to that of your other professors, and why do you think this is so? If you have a smaller class, compare the use of space by that professor. Is it different from the behavior of the professors in large lecture halls? Under what circumstances, if any, do your professors get close to students, and how close do they get?

Step 2: How close can you get to friends? Strike up a conversation with a same-gender friend standing next to you. Discreetly move closer and closer to your friend until he or she moves away or says something about your proximity. How much space separated you when this happened? What do you think would happen if you tried this with a stranger? With a friend or a stranger of a different gender? With a friend or a stranger from a different culture?

Step 3: Discuss your findings. Did all your classmates have similar experiences, or were your findings notably different? Are all students in your class from similar economic or ethnic backgrounds? If not, that may explain some of the differences that emerge.

There has been some talk lately about the future of the ever-huger "McMansions" and sprawling suburban developments that for several decades have characterized middle-class life in the United States. Clearly, both waste resources. McMansions are expensive to heat and cool, and their landscaping is often out-of-step with the local environment. Living in a suburb often means commuting long distances to jobs or school, in cars occupied by just one or two people. But think, too, about the proxemics involved and consider these questions:

- Is there such a thing as too much space?

- Do you know or have you heard of people who have opted out of living large in the 'burbs? Why do you think they made this decision?

- In what ways do planners and architects take culturally diverse preferences about personal space into account when they design cities, streets, buildings, homes, and classrooms?

Population densities in parks. These two images depict similarly designed public playgrounds featuring a large sandbox. The park on the top is located in Minnesota in the United States; the park on the bottom is located in Taipei, Taiwan. *Where would you rather play? Why?* (Top: James Shaffer/ PhotoEdit; Bottom: Christian Klein/Alamy.)

SEEING GEOGRAPHY Kolkata, India
Would you feel comfortable walking here? If not, why not?

A street scene in the large city of Kolkata, India.
(National Geographic Image Collection/Alamy.)

Do you need your "personal space"? Most Americans and Canadians do. If so, Kolkata (formerly Calcutta), India, is a place you might want to avoid. West Bengal state, where Kolkata is located, has the highest population density in the country.

Why do people form such dense clusters? The theme of cultural interaction would tell us of push factors that encourage people to leave their farms and move to the city. Some can no longer make a living or feed their families with the food provided by the tiny plots of land they work. Others are forced off the land by landlords who want to convert their farms to use mechanized Western methods of agriculture that use far less labor (see Chapter 8). But cultural interaction also tells us of pull factors

exerted by cities such as Kolkata— the hope or promise of better-paying jobs, the encouragement of friends and relatives who came to the city earlier, or the greater availability of government services. And so, pushed and pulled, they come to the teeming, crowded city, to jostle and elbow their way through the streets.

But, it is not only cities in the developing world that become so dense. If you have ever visited, or lived in, Manhattan, New York, you are all too familiar with dense crowds of people. In fact, there are some who grow up in these environments and find that the relative solitude of rural areas verges on terrifying. They prefer the bustle of activity and the sounds of the city, and they feel at home in a crowd.

Being a woman is another reason that you might feel uncomfortable in this Kolkata street environment. Notice the relative absence of women in this crowd. Many societies have strict norms that dictate where women, and men, may and may not go. Harassment or even violence may be the result of violating these norms.

The study of the size and shape of people's envelopes of personal space is called **proxemics.** Anthropologist Edward T. Hall, whose book *The Hidden Dimension* is listed in Ten Recommended Books on Population Geography at the end of the chapter, is the founder of this science. Urban planners, architects, psychologists, and sociologists, as well as geographers, use proxemics to explain why some people need more space than others and how this varies culturally.

> ■ **proxemics**
> The study of the size and shape of people's envelopes of personal space.

Chapter 3
LEARNING OBJECTIVES REEXAMINED

3.1 Describe the regional patterns of population characteristics and how these are distributed spatially and change over time.
See pages 94–111. How do birth and death rates shape populations in different regions of the world?

3.2 Identify patterns, causes, and consequences of population migrations and things—such as diseases—that accompany them.
See pages 111–115. In what ways does disease play a role in population migration?

3.3 Discuss theories of population growth and control and how these have changed over time.
See pages 115–118. How does the Malthusian theory of population growth differ from that of cornucopians?

3.4 Recognize the ways in which the natural world shapes population characteristics, and how population characteristics in turn shape the natural world.
See pages 118–123. What role does the environment play in a group's decision on where to settle?

3.5 Analyze the imprint of demographic factors on the cultural landscape.
See pages 123–130. Name two ways that populations provide visual imprints on the cultural landscape.

KEY TERMS

Population Geography on the Internet

You can learn more about population geography on the Internet at the following web sites:

Population Reference Bureau, Washington, D.C.
http://www.prb.org
This organization is concerned principally with overpopulation and standard of living. The annual "World Population Data Sheet" provides up-to-date basic demographic information at a glance, and the "Datafinder" section contains a wealth of images on all aspects of global population for use in presentations and reports. Some of the maps in this chapter were adapted from PRB maps.

United Nations High Commissioner for Refugees
http://www.unhcr.org
This United Nations web site provides basic information about refugee situations worldwide. The site includes regularly updated maps showing refugee locations and populations as well as photos of refugee life.

U.S. Census Bureau Population Clocks
http://www.census.gov/main/www/popclock.html
Check real-time figures here for the population of the United States and the population of the world. The main web site, http://www.census.gov, hosts the most important and comprehensive data sets available on the U.S. population.

World Health Organization, Geneva, Switzerland
http://www.who.int/en
Learn about the group that distributes information on health, mortality, and epidemics as it seeks to improve health conditions around the globe. The "Global Health Atlas" allows you to create detailed maps from WHO data.

Worldwatch Institute, Washington, D.C.
http://www.worldwatch.org
This organization is concerned with the ecological consequences of overpopulation and the wasteful use of resources. It seeks sustainable ways to support the world's population and brings attention to ecological crises.

Sources

Coclanis, Peter. 2010. "Russia's Demographic Crisis and Gloomy Future." *The Chronicle of Higher Education,* 19 February, pp. B9 – B11.

D'Emilio, Frances. 2004. "Italy's Seniors Finding Comfort with Strangers." *Miami Herald,* 30 October, pp. 1A–2A.

Howden, Lindsay M., and Meyer, Julie A. 2011. *Age and Sex Composition: 2010.* Washington, D.C.: U.S. Census Bureau.

Johnson, Tim. 2011. "Mexico City Copes With That Sinking Feeling." *The Seattle Times,* 24 September. http://seattletimes.com/html/nationworld/2016310507_mexicosinking25.html.

Marsh, Viv. 2012. "China to Overhaul 'Threatening' One-Child Slogans." *BBC News,* 27 February. http://www.bbc.co.uk/news/world-asia-17181951.

Population Reference Bureau. 2007. *2007 World Population Data Sheet.* Washington, D.C.: Population Reference Bureau. http://www.prb.org/Publications/Datasheets/2007/2007WorldPopulationDataSheet.aspx.

Population Reference Bureau. 2012. *2012 World Population Data Sheet.* Washington, D.C.: Population Reference Bureau. http://www.prb.org/pdf12/2012-population-data-sheet_eng.pdf.

Rosin, Hanna. 2010. "The End of Men." *Atlantic Monthly,* July/August. http://www.theatlantic.com/magazine/archive/2010/07/the-end-of-men/8135/.

United Nations High Commissioner for Refugees. 2012. *Global Overview 2011: People Internally Displaced by Conflict and Violence.* New York: United Nations.

World Bank. *Statistics in Africa.* http://www.worldbank.org/afr/stats.

Yonetani, Michelle. 2013. *Global Estimates 2012: People Displaced by Disasters.* Geneva: Internal Displacement Monitoring Centre and Norwegian Refugee Council.

Ten Recommended Books on Population Geography

(For additional suggested readings, see the *Contemporary Human Geography* LaunchPad: http://www.macmillanhighered.com/launchpad/DomoshCHG1e)

Diamond, Jared. 2005. *Collapse: How Societies Choose to Fail or Succeed.* New York: Penguin. Why do thriving civilizations die out? Ecology, biology, and geography all play a role in Diamond's analysis.

Hall, Edward T. 1966. *The Hidden Dimension.* Garden City, N.Y.: Doubleday. This is the classic study of proxemics conducted by an anthropologist. Hall argues that culture, above all else, shapes our criteria for defining, organizing, and using space.

Harrison, Mark. 2012. *Contagion: How Commerce Has Spread Disease.* New Haven, Conn.: Yale University Press. Contagious diseases that have shaped the course of world history—the bubonic plague (Black Death), yellow fever, influenza, and cholera—have spread along trade routes. This fascinating account traces the close relation between mobility necessitated by commercial interactions and devastating illness.

Johnson, Steven. 2006. *The Ghost Map: The Story of London's Most Terrifying Epidemic—And How It Changed Science, Cities, and the Modern World.* New York: Riverhead Books. In this gripping historical narrative, physician John Snow's tracing of the Soho cholera outbreak of 1854 is placed in the larger context of the history of science and urbanization.

Livi Bacci, Massimo. 2012. *A Short History of Migration.* Cambridge, U.K.: Polity Press. Examines past and present migrations the world over, with an eye toward those factors that have helped and hindered specific groups of migrants.

Mann, Charles C. 2006. *1491.* New York: Vintage. This highly readable account of the American landscape on the eve of European conquest challenges long-held notions. Contending that the indigenous population of the Americas was, in fact, much larger and well-off than assumed, Mann paints a picture of a preconquest landscape that was culturally rich, technologically sophisticated, and environmentally compromised.

Meade, Melinda S., and Michael Emch. 2010. *Medical Geography.* 3rd ed. New York: Guilford. Surveys the perspectives, theories, and methodologies that geographers use in studying human health; a primary text that undergraduates can readily understand.

Newbold, K. Bruce. 2007. *Six Billion Plus: World Population in the Twenty-First Century.* 2nd ed. Lanham, Md.: Rowman & Littlefield. The impact of increased global population levels across the next century is assessed with respect to interaction with environmental, epidemiological, mobility, and security issues.

Seager, Joni, and Mona Domosh. 2001. *Putting Women in Place: Feminist Geographers Make Sense of the World.* New York: Guilford. A highly readable account of why paying attention to gender is crucial to understanding the spaces in which we live and work.

Tone, Andrea. 2002. *Devices and Desires: A History of Contraceptives in America.* New York: Hill & Wang. This social history of birth control in the United States details the fascinating relationship between the state and the long-standing attempts of men and women to limit their fertility.

Journals in Population Geography

Gender, Place and Culture: A Journal of Feminist Geography. Published by the Carfax Publishing Co., P.O. Box 2025, Dunnellon, Fla. 34430. Volume 1 appeared in 1994.

Population and Environment: A Journal of Interdisciplinary Studies. Volume 1 was published in 1996.

AN ARMY OF ONE

Aquí se habla Spanglish. What does this sign tell you about who lives in this city?

U.S. Army veterans at a recruiting table in Miami, Florida.
(Jeff Greenberg/AgeFotostock.)

Go to "Seeing Geography" on page 170 to learn more about this image.

YO SOY EL ARM

THE GEOGRAPHY OF LANGUAGE
Building the Spoken Word

anguage is one of the primary features that distinguishes humans from other animals. Many animals, including dolphins, whales, and birds, do indeed communicate with one another through patterned systems of sounds, movements, or scents and other chemicals. Some nonhuman primates have been taught to use sign language to communicate with humans. In turn, we have attempted to translate animal noises into words in human languages, with results that vary across cultures. Although, for instance, a cat obviously sounds the same in any part of the world, the speakers of various human languages render that sound in diverse ways. Table 4.1 shows how the sounds of various animals "translate" into different human languages.

However, the complexity of human language, its ability to convey nuanced emotions and ideas, and its importance for our existence as social beings set it apart from the communication systems used by other animals. In many ways, language is the essence of culture. It provides the single most common variable by which different cultural groups are identified and by which groups assert their unique identity. Language not only facilitates the cultural diffusion of innovations but it also helps to shape the way we think about, perceive, and name our environment. **Language**, a mutually agreed-on system of symbolic

language A mutually agreed-on system of symbolic communication that has a spoken and usually a written expression.

LEARNING OBJECTIVES

4.1 Understand the geographical patterning of languages.

4.2 Analyze how languages and dialects have come to exist.

4.3 Describe the relationship between technology and language.

4.4 Explain the relationships between language and the physical environment.

4.5 Identify the ways languages are visibly part of the cultural landscape.

TABLE 4.1 **Translating Animal Sounds**

	English	Indonesian	Japanese	Greek
Dog	*bow-wow*	*gonggong*	*wanwan*	*gav*
Cat	*meow*	*ngeong*	*nyaa*	*niaou*
Bird	*tweet-tweet*	*kicau*	*chunchun*	*tsiou tsiou*
Rooster	*cock-a-doodle-doo*	*kikeriku*	*kokikokkoo*	*ki-kiriki*

Source: Compiled using information from the Sounds of the World's Animals project conducted by Dr. Catherine N. Ball, Department of Linguistics, Georgetown University.

communication, offers the main means by which learned belief systems, customs, and skills pass from one generation to the next.

One of our first and most long-lasting ties to place is forged through language. The ability to name, or rename, a place is a key step in claiming a place as one's own, as shown, for example, by **Figure 4.1**, a political map of Antarctica. Notice how many places are named after Antarctic explorers and monarchs of countries that claim portions of the continent: the Ross Sea and Ross Ice Shelf are named after James Clark Ross, the English explorer who charted much of Antarctica's coastline, and Queen Maud Land is named after Norway's Queen Maud. Interestingly, Roald Amundsen, the Norwegian explorer who gave Queen Maud Land its name in 1939, took part in an earlier trip that highlights the importance of language. In 1898 Amundsen was part of an ill-fated expedition to locate the magnetic south pole when his ship became stuck in the ice for more than one year. Several of Amundsen's shipmates, including men from Poland, Romania, Norway, the United States, and Belgium, went insane. This was not only because of the long winter nights and the cold weather; in great part, language differences among the men became their biggest problem. They simply could not understand one another!

REGION

4.1 LEARNING OBJECTIVE
Understand the geographical patterning of languages.

The spatial variation of speech is remarkably complicated, adding intricate patterns to the human mosaic (**Figure 4.2**). Because language is such a central component of culture, understanding the spatiality of language, and how and why its patterns change over time, provides a particularly valuable window into cultural geography more generally. The logical place to begin our geographical study of language is with the regional theme.

Separate languages are those that cannot be mutually understood. In other words, a monolingual speaker of one language cannot comprehend the speaker of another. **Dialects,** by contrast, are variant forms of a language where mutual comprehension is possible. A speaker of English, for example, can generally understand that language's various dialects,

> **dialect** A distinctive local or regional variant of a language that remains mutually intelligible to speakers of other dialects of that language; a subtype of a language.

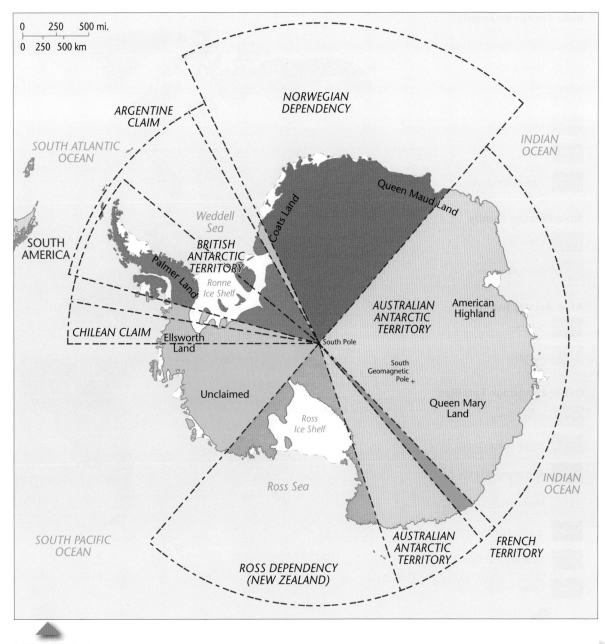

FIGURE 4.1 **Naming place is closely related to claiming place.** This map shows the pie-shaped land claims of various nations. Notice how place-names reflect the names of Antarctic explorers or political rulers on the political map of this continent. The green portions are unclaimed. *(Source: Adapted from Latrimer Clarke Corporation Pty Ltd.)*

regardless of whether the speaker comes from Australia, Scotland, or Mississippi. Nevertheless, a dialect is distinctive enough in vocabulary and pronunciation to label its speaker as hailing from one place or another, or even from a particular city. About 7000 languages and many more dialects are spoken in the world today.

When different linguistic groups come into contact, a **pidgin** language, characterized by a very small vocabulary derived from the languages of the groups in contact, often results. Pidgins primarily serve the purposes of trade and commerce: they facilitate exchange at a basic

pidgin A composite language consisting of a small vocabulary borrowed from the linguistic groups involved in commerce.

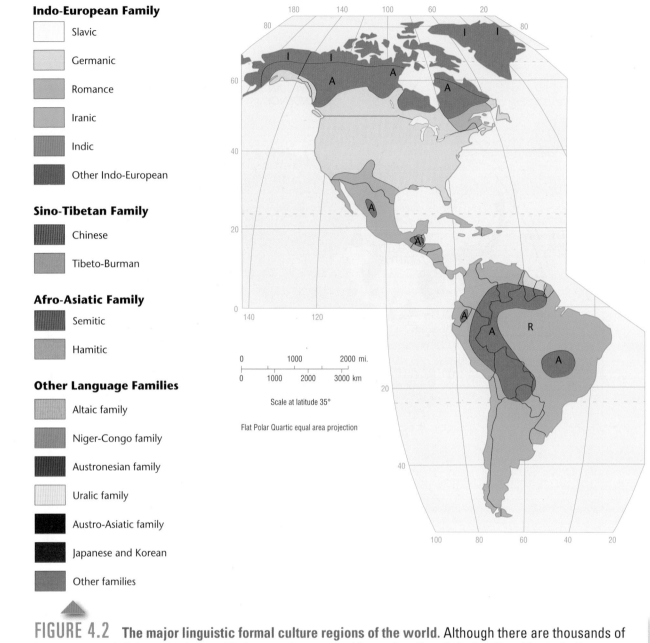

Indo-European Family
- Slavic
- Germanic
- Romance
- Iranic
- Indic
- Other Indo-European

Sino-Tibetan Family
- Chinese
- Tibeto-Burman

Afro-Asiatic Family
- Semitic
- Hamitic

Other Language Families
- Altaic family
- Niger-Congo family
- Austronesian family
- Uralic family
- Austro-Asiatic family
- Japanese and Korean
- Other families

0   1000   2000 mi.
0   1000   2000   3000 km

Scale at latitude 35°

Flat Polar Quartic equal area projection

FIGURE 4.2 **The major linguistic formal culture regions of the world.** Although there are thousands of languages and dialects in the world, they can be grouped into a few linguistic families. The Indo-European language family represents about half of the world's population. It spread throughout the world, in part, through Europe's empire-building efforts.

level but do not have complex vocabularies or grammatical structures. An example is Tok Pisin, meaning "talk business." Tok Pisin is a largely English-derived pidgin spoken in Papua New Guinea, where it has become the official national language in a country where many native Papuan tongues are spoken. Although New Guinea pidgin is not readily intelligible to a speaker of Standard English, certain common words such as *gut bai* ("good-bye"), *tenkyu* ("thank you"), and *haumas* ("how much") reflect the influence of English. When pidgin languages acquire fuller vocabularies and become native languages of their speakers, they are called **creole** languages. Obviously, deciding

creole A language derived from a pidgin language that has acquired a fuller vocabulary and become the native language of its speakers.

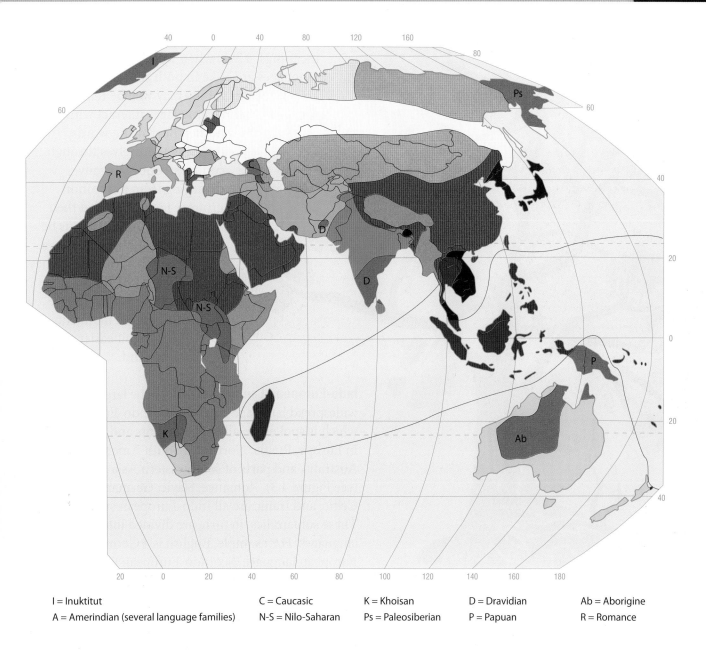

I = Inuktitut
A = Amerindian (several language families)

C = Caucasic
N-S = Nilo-Saharan

K = Khoisan
Ps = Paleosiberian

D = Dravidian
P = Papuan

Ab = Aborigine
R = Romance

precisely *when* a pidgin becomes a creole language has at least as much to do with a group's political and social recognition as it does with what are, in practice, fuzzy boundaries between language forms.

Another response to the need for speakers of different languages to communicate with one another is the elevation of one existing language to the status of a **lingua franca.** A lingua franca is a language of communication and commerce spoken across a wide area where it is not a mother tongue. The Swahili language enjoys lingua franca status in much of East Africa, where inhabitants speak a number of other regional languages and dialects. English is fast becoming a global lingua franca. Finally, regions that have linguistically mixed populations may be characterized by **bilingualism,** which is the ability to speak two languages with fluency. For example, along the U.S.-Mexico border, so many residents speak both English and Spanish (with varying degrees of fluidity) that

lingua franca An existing, well-established language of communication and commerce used widely where it is not a mother tongue.

bilingualism The ability to speak two languages fluently.

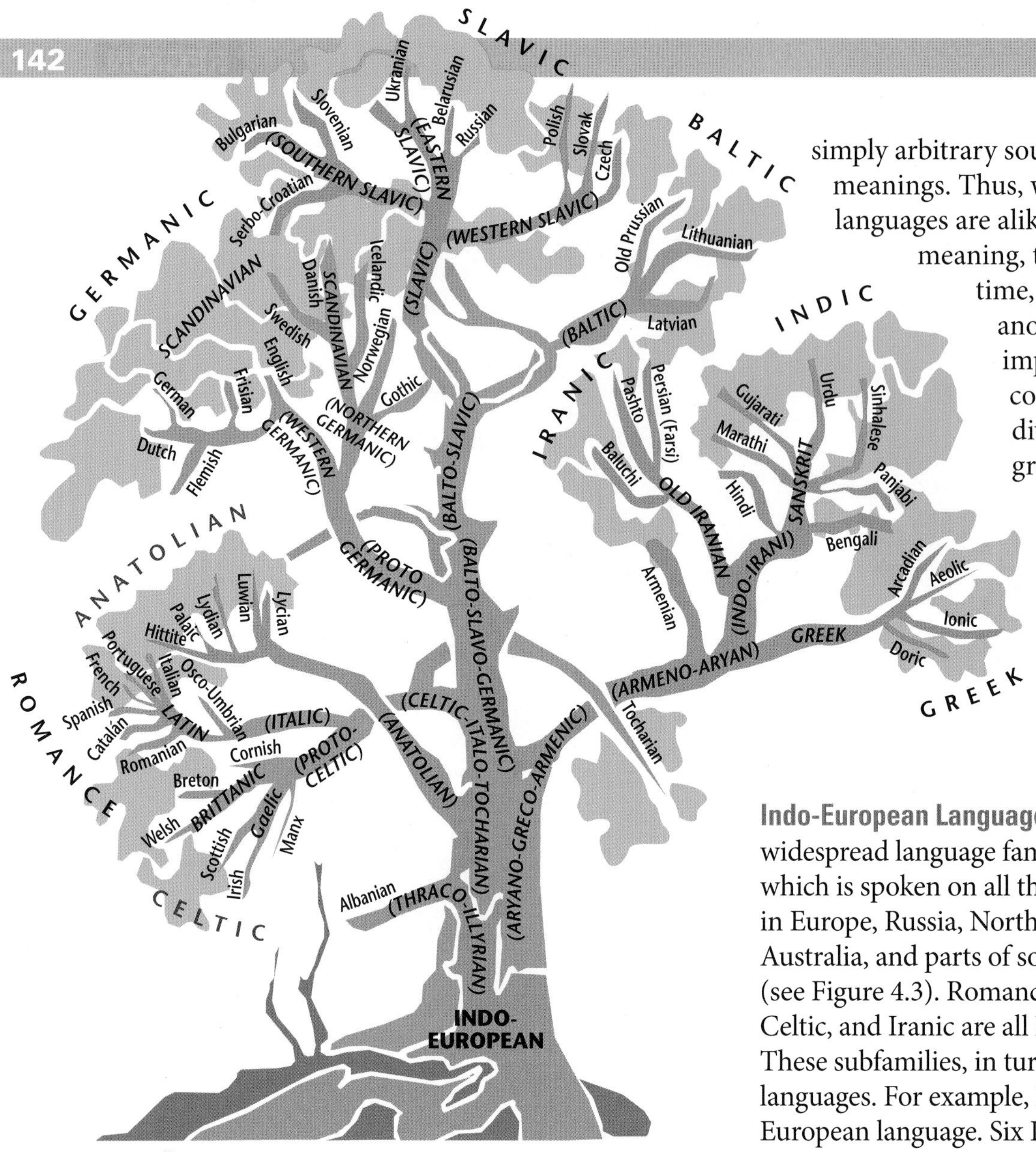

FIGURE 4.3 **Linguistic family tree.** Shown here is a detailed image of one branch of the linguistic family tree. *(Source: Adapted from Ford.)*

simply arbitrary sounds associated with certain meanings. Thus, when words in different languages are alike in both sound and meaning, they may well be related. Over time, languages interact with one another, borrowing words, imposing themselves through conquest, or organically diverging from a common ground. Languages and their interrelations can thus be graphically depicted as a tree with various branches (**Figure 4.3**). This classification makes the complicated linguistic mosaic a bit easier to comprehend.

Indo-European Language Family The largest and most widespread language family is the Indo-European, which is spoken on all the continents and is dominant in Europe, Russia, North and South America, Australia, and parts of southwestern Asia and India (see Figure 4.3). Romance, Slavic, Germanic, Indic, Celtic, and Iranic are all Indo-European subfamilies. These subfamilies, in turn, are divided into individual languages. For example, English is a Germanic Indo-European language. Six Indo-European tongues, including English, are among the 10 most spoken languages in the world as classified by the number of native speakers (**Table 4.2**).

Comparing the vocabularies of various Indo-European tongues reveals their kinship. For example, the English word *mother* is similar to the Polish *matka*, the Greek *meter*, the Spanish *madre*, the Farsi *madar* in Iran, and the Sinhalese *mava* in Sri Lanka. Such similarities demonstrate that these languages have a common ancestral tongue.

Sino-Tibetan Family Sino-Tibetan is another of the major language families of the world and is second only to Indo-European in numbers of native speakers. The Sino-Tibetan region extends throughout most of China and Southeast Asia (see Figure 4.2). The two language branches that make up this group, Sino- and Tibeto-Burman, are believed to have had a common

bilingualism in practice—even if not in policy—means there is no need for a lingua franca.

LANGUAGE FAMILIES

> **language family** A group of related languages derived from a common ancestor.

One way in which geolinguists often simplify the mapping of languages is by grouping them into **language families:** tongues that are related and share a common ancestor. Words are

TABLE 4.2 The 10 Leading Languages in Numbers of Native Speakers*

Language	Family	Speakers (in millions)	Main Areas Where Spoken
Chinese	Sino-Tibetan	1197	China, Taiwan, Singapore
Spanish	Indo-European	406	Spain, Latin America, southwestern United States
English	Indo-European	335	British Isles, United States, Caribbean, Australia, New Zealand, South Africa, Philippines, former British colonies in tropical Asia and Africa
Hindi	Indo-European	260	Northern India, Pakistan
Arabic	Afro-Asiatic	223	Middle East, North Africa
Portuguese	Indo-European	202	Portugal, Brazil, southern Africa
Bengali	Indo-European	193	Bangladesh, eastern India
Russian	Indo-European	162	Russia, Kazakhstan, parts of Ukraine and other former Soviet Republics
Japanese	Japanese and Korean	122	Japan
Javanese	Austronesian	84.3	Indonesia

*"Native speakers" means mother tongue.
Source: Ethnologue 2013, http://www.ethnologue.com/statistics/size.

origin some 6000 years ago in the Himalayan Plateau; speakers of the two language groups subsequently moved along the great Asian rivers that originate in this area. "Sino" refers to China and in this context indicates the various languages spoken by more than 1.3 billion people in China. Han Chinese (Mandarin) is spoken in a variety of dialects and serves as the official language of China. The nearly 400 languages and dialects that make up the Burmese and Tibetan branch of this language family border the Chinese language region on the south and west. Other East Asian languages, such as Vietnamese, have been heavily influenced by contact with the Chinese and their languages, although it is not clear that they are linguistically related to Chinese at all.

Afro-Asiatic Family The third major language family is the Afro-Asiatic. It consists of two major divisions: Semitic and Hamitic. The Semitic languages cover the area from the Arabian Peninsula and the Tigris-Euphrates river valley of Iraq westward through Syria and North Africa to the Atlantic Ocean. Despite the considerable size of this region, there are fewer speakers of the Semitic languages than you might expect because most of the areas that Semites inhabit are sparsely populated deserts. Arabic is by far the most widespread Semitic language and has the greatest number of native speakers, about 223 million. Although many different dialects of Arabic are spoken, there is only one written form.

Hebrew, which is closely related to Arabic, is another Semitic tongue. For many centuries, Hebrew was a "dead" language, used only in religious ceremonies by millions of Jews throughout the world. With the creation of the state of Israel in 1948, a common language was needed to unite the immigrant Jews, who spoke the languages of their many different countries of origin. Hebrew was revived as the official national language of what otherwise would have been a **polyglot,** or multilanguage, state (**Figure 4.4**). Amharic, a third major Semitic tongue, is spoken today by 21 million in the mountains of East Africa.

Smaller numbers of people who speak Hamitic languages share North and East Africa with the speakers of Semitic languages. Like the Semitic

polyglot A mixture of different languages.

FIGURE 4.4 **Sign in Arab quarter of Nazareth.** Many of Israel's cities are home to diverse populations. This sign at a child-care center reflects Israel's polyglot population, with its English, Arabic, and Hebrew wording. *(Courtesy of Patricia L. Price.)*

languages, these tongues originated in Asia but today are spoken almost exclusively in Africa by the Berbers of Morocco and Algeria, the Tuaregs of the Sahara, and the Cushites of East Africa.

Other Major Language Families Most of the rest of the world's population speak languages belonging to one of six remaining major families. The Niger-Congo language family, also called Niger-Kordofanian, which is spoken by about 430 million people, dominates Africa south of the Sahara Desert. The greater part of the Niger-Congo culture region belongs to the Bantu subgroup. Both Niger-Congo and its Bantu constituent are fragmented into a great many different languages and dialects, including Swahili. The Bantu and their many related languages spread from what is now southeastern Nigeria about 4000 years ago, first west and then south in response to climate change and new agricultural techniques (**Figure 4.5**).

Flanking the Slavic Indo-Europeans on the north and south in Asia are the speakers of the Altaic language family, including Turkic, Mongolic, and several other subgroups. The Altaic homeland lies largely in the inhospitable deserts, tundra, and

coniferous forests of northern and central Asia. Also occupying tundra and grassland areas adjacent to the Slavs is the Uralic family. Finnish and Hungarian are the two more widely spoken Uralic tongues, and both enjoy the status of official languages in their respective countries.

One of the most remarkable language families in terms of distribution is the Austronesian. Representatives of this group live mainly on tropical islands stretching from Madagascar, off the east coast of Africa, through Indonesia and the Pacific islands, to Hawaii and Easter Island. This longitudinal span is

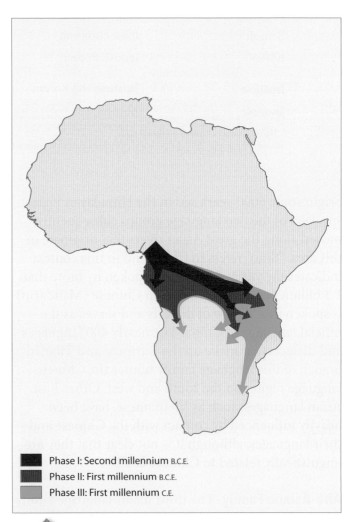

	Phase I: Second millennium B.C.E.
	Phase II: First millennium B.C.E.
	Phase III: First millennium C.E.

FIGURE 4.5 **Bantu expansionism.** The Bantu and their many languages spread from what is today southeastern Nigeria to the west and south. *(Source: Adapted from Dingemanse/Wikipedia.)*

more than half the distance around the world. The north–south, or latitudinal, range of this language area is bounded by Hawaii and Taiwan in the north and New Zealand in the south. The largest single language in this family is Javanese, with 84.3 million native speakers, but the most geographically widespread is Polynesian.

Japanese and Korean, with about 188.4 million speakers combined, probably form another Asian language family. The two perhaps have some link to the Altaic family, but even their kinship to each other remains controversial and unproven.

In Southeast Asia, the Vietnamese, Cambodians, Thais, and some tribal peoples of Malaya and parts of India speak languages that constitute the Austro-Asiatic family. They occupy an area into which Sino-Tibetan, Indo-European, and Austronesian languages have all encroached.

MOBILITY

4.2 LEARNING OBJECTIVE
Analyze how languages and dialects have come to exist.

Different types of cultural diffusion have helped shape the linguistic map. Relocation diffusion has been extremely important because languages spread when groups, in whole or in part, migrate from one area to another. Some individual tongues or entire language families are no longer spoken in the regions where they originated, and in certain other cases the linguistic hearth is peripheral to the present distribution (compare Figures 4.2 and 4.6). Today, languages continue to evolve and change based on the shifting locations of peoples and on their needs as well as on outside forces.

INDO-EUROPEAN DIFFUSION

How did Indo-European languages arise and spread, to become the largest language family on Earth? One theory suggests that the earliest speakers of the Indo-

European languages lived in southern and southeastern Turkey, a region known as Anatolia, about 9000 years ago. According to the **Anatolian hypothesis,** the initial diffusion of these Indo-European speakers was facilitated by the innovation of plant domestication. As sedentary farming became adopted throughout Europe, a gradual and peaceful expansion diffusion of Indo-European languages occurred. As these people dispersed and lost contact with one another, different Indo-European groups gradually developed variant forms of the language, causing fragmentation of the language family.

The Anatolian hypothesis has been criticized by scholars who note that specific terms related to animals (particularly horses), as opposed to agriculture, appear to link Indo-European languages to a common origin. The so-called **Kurgan hypothesis,** which is more widely accepted than the Anatolian hypothesis, places the rise of Indo-European languages in the central Asian steppes only 6000 years ago (**Figure 4.6**). It asserts that the spread of Indo-European languages was both swifter and less peaceful than is maintained by those who subscribe to the Anatolian hypothesis. The domestication of horses by militaristic Kurgans allowed them to overtake the more peaceful agricultural societies and to rapidly spread their languages through imposition. No one theory has been definitively proven.

What is more certain is that in later millennia, the diffusion of certain Indo-European languages—in particular, Latin, English, and Russian—occurred in conjunction with the territorial spread of great political empires. In such cases of imperial conquest, relocation and expansion diffusion were not mutually exclusive. Relocation diffusion occurred as a small number of conquering elites came to rule an area. The language of the conqueror, implanted by relocation diffusion, often gained wider acceptance through expansion diffusion. Typically, the conqueror's language spread hierarchically—adopted

> **Anatolian hypothesis** A theory of language diffusion holding that the movement of Indo-European languages from the area in contemporary Turkey known as Anatolia followed the spread of plant domestication technologies.
>
> **Kurgan hypothesis** A theory of language diffusion holding that the spread of Indo-European languages originated with animal domestication in the central Asian steppes and grew more aggressively and swiftly than proponents of the Anatolian hypothesis maintain.

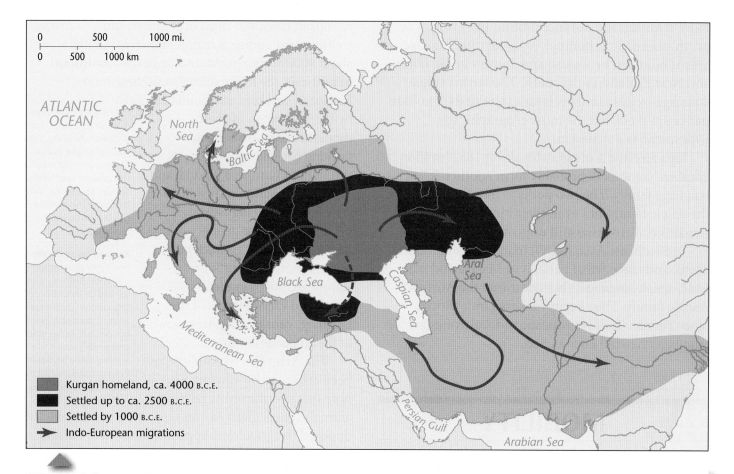

FIGURE 4.6 **The spread of Indo-European language.** This map depicts the so-called Kurgan hypothesis, named after the burial mounds (*kurgan*) characteristic of the warrior pastoralists who inhabited the area north of the Black and Caspian seas. Around 6000 B.C.E., they began to spread outward, conquering and imposing their language across Europe, Central Asia, India, the Balkans, and Anatolia (modern-day Turkey).

first by the more important and influential persons and by city dwellers. The diffusion of Latin with Roman conquests, and Spanish with the conquest of Latin America, occurred in this manner.

MIGRATION AND THE SURVIVAL OF LANGUAGE

As we have seen above, conquest can lead to the imposition of a new language and the abandonment or suppression of native tongues. However, these threatened languages may reappear and thrive in new places, as their speakers migrate for reasons of economic or cultural survival.

New York City is thought to be home to as many as 800 languages, making it the most linguistically dense place in the world. There are more speakers of Vlashki in Queens than in the Croatian mountain villages where the language originated, and a roughly equal number of Garifuna speakers in the Bronx and Brooklyn as in Honduras and Belize. These are but two of "a remarkable trove of endangered tongues that have taken root in New York" (Roberts, 2010).

How did New York City become home to such linguistic riches? These languages relocated there through the migration of their native speakers. While populations in the language source region may fall victim to ethnic conflict, disease, starvation, compulsory schooling, or merely assimilation into dominant language groups, thereby losing their ability to speak their native language, migrant speakers from these places may better manage to keep the tongue alive and well in their new homes.

Yet, many of these relocated languages find themselves under new pressure from the dominant English language in the United States. For this reason, the City University of New York started the Endangered Language Alliance, whose members canvass city neighborhoods in search of immigrant speakers of vulnerable languages. The speakers are videotaped, and the Alliance encourages the teaching and use of these languages. The reality, however, is that many of these languages will vanish from New York City's linguistic landscape when the children of these immigrants cease to speak them regularly or stop teaching them to their children.

RELIGION AND LINGUISTIC MOBILITY

Cultural interaction creates situations in which language is linked to a particular religious faith or denomination, a linkage that greatly heightens cultural identity. Perhaps Arabic provides the best example of this cultural link. It spread from a core area on the Arabian Peninsula with the expansion of Islam. Had it not been for the evangelical success of the Muslims, Arabic would not have diffused so widely. The other Semitic languages also correspond to particular religious groups. Hebrew-speaking people are of the Jewish faith, and the Amharic speakers in Ethiopia tend to be Coptic, or Eastern, Christians. Indeed, we can attribute the preservation and recent revival of Hebrew to its active promotion by Jewish nationalists who believe that teaching and promoting Hebrew to diasporic Jews facilitate unity.

Certain languages have even acquired a religious status. Latin survived mainly as the ceremonial language of the Roman Catholic Church and Vatican City. In non-Arabic Muslim lands, such as Iran, where people consider themselves Persians and speak Farsi, Arabic is still used in religious ceremonies. Great religious books can also shape languages by providing them with a standard form. Martin Luther's translation of the Bible led to the standardization of the German language, and the Qur'an is the model for written

Arabic. Because they act as common points of frequent cultural reference and interaction, great religious books can also aid in the survival of languages that would otherwise become extinct. The early appearance of a hymnal and the Bible in the Welsh language aided the survival of that Celtic tongue, and Christian missionaries in diverse countries have translated the Bible into local languages, helping to preserve them. In Fiji, the appearance of the Bible in one of the 15 local dialects elevated that dialect to the dominant native language of the islands.

LANGUAGE'S SHIFTING BOUNDARIES

Dialects, as well as the language families discussed previously, reveal a vivid geography. Their boundaries—what separates them from other dialects and languages—shift over time, both spatially and in terms of what elements they contain or discard.

Geolinguists map dialects by using **isoglosses,** which indicate the spatial borders of individual words or pronunciations. The dialect boundaries between Latin American Spanish speakers using *tú* and those using *vos* are clearly defined in some areas, as shown in **Figure 4.7**. The choice of *tú* or *vos* for the second-person singular carries with it a cultural indication. *Vos* represents a usage closer to the original Spanish but is considered by many in the Spanish-speaking world to be rather archaic and, in fact, has died out in Spain itself. In other regions, particularly throughout Argentina, *vos* has long been used by the media; often, it is considered to reflect the "standard" dialect and usage for the area in which it is used. Because certain words or dialects can fall out of fashion or simply become overwhelmed by an influx of new speakers, isogloss boundaries are rarely clear or stable over time. Indeed, in Central America, the media are increasingly using *vos*—long used in conversation in the region—thus elevating *vos* to a more official status covering a larger territory. Because of this, geolinguists often disagree about how many dialects are present in an area or exactly where isogloss borders should be drawn. The language map of any place is a constantly shifting kaleidoscope.

> **isogloss** The border of usage of an individual word or pronunciation.

FIGURE 4.7 **Dialect boundaries in Latin America.** Spanish speakers in the Americas use either *vos* or *tú* as the second-person singular verb form. They represent dialects of Spanish: both are correct, linguistically speaking, but the *vos* form is older. Some regions use *vos* and *tú* interchangeably. *(Source: Adapted from Pountain, 2005.)*

The dialects of American English provide another good example. At least three major dialects, corresponding to major culture regions, had developed in the eastern United States by the time of the American Revolution: the Northern, Midland, and Southern dialects (**Figure 4.8**). As the three subcultures expanded westward, their dialects spread and fragmented. Nevertheless, they retained much of their basic character, even beyond the Mississippi River. These culture regions have unusually stable boundaries. Even today, the "*r*-less" pronunciation of words such as *car* ("cah") and *storm* ("stohm"), characteristic of the East Coast Midland regions, is readily discernible in the speech of its inhabitants.

Although we are sometimes led to believe that Americans are becoming more alike, as a national culture overwhelms regional ones, the current status of American English dialects suggests otherwise. Linguistic divergence is still under way, and dialects continue to mutate on a regional level, just as they always have. Local variations in grammar and pronunciation proliferate, confounding the proponents of standardized speech and defying the homogenizing influence of the Internet, television, and other mass media.

Shifting language boundaries involve content as well as spatial reach, and this, too, changes over time. Today, for example, some of the unique vocabulary of American English dialects is becoming old-fashioned. For instance, the term *icebox,* which was literally a wooden box with a compartment for ice that was utilized to cool food, was used widely throughout the United States to refer to the precursor of the refrigerator. Although the modern electric refrigerator is ubiquitous in the United States today, some people,

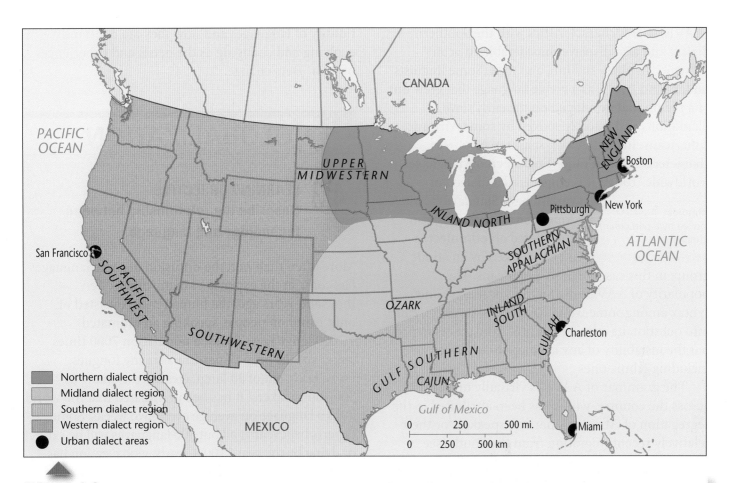

FIGURE 4.8 **Major dialects of North American English, with a few selected subdialects.** These dialects had developed by the time of the American Revolution and have remained remarkably stable over time.

particularly those of older generations and in the South, still use the term *icebox.* Many young people, by contrast, no longer pause to say the entire word *refrigerator,* shortening it instead to *fridge.*

As illustrated by the birth of the new word *fridge,* slang terms are quite common in most languages, and American English is no exception. **Slang** refers to words and phrases that are not a part of a standard, recognized vocabulary for a given language but are nonetheless used and understood by some or most of its speakers. Often, subcultures—for example, youth, drug dealers, and nightclub-goers—have their own

> **slang** Words and phrases that are not part of a standard, recognized vocabulary for a given language but that are nonetheless used and understood by some of its speakers.

slang that is used within that community but is not readily understood by nonmembers. Slang words tend to be used for a period of time and then discarded as newer terms replace them. For example, *fresh, the bomb,* and *phat* were used to refer to desirable, attractive, or fashionable things in the 1990s. Although many people today still recognize these words, a new generation of young people is much more likely to use a new set of words. Their children, in turn, will more than likely use yet another set of words. Slang illustrates another way in which American English changes over time.

Some African Americans speak a distinctive form of English. African American Vernacular English

(AAVE) shares characteristics with the Southern dialect and also displays considerable African influence in pitch, rhythm, and tone. Some linguists understand AAVE as a creole language that grew out of a pidgin that developed on the early slave plantations and is today spoken by some African Americans. Indeed, AAVE shares many characteristics with other English creole languages worldwide. Today, it is considered a dialect, or variation, of Standard American English. It is also considered an **ethnolect,** a dialect spoken by an ethnic group, in this case, African Americans. The popularity of AAVE's distinctive vocabulary and syntax among some white Americans, however, calls into question its ethnic exclusivity and serves to point out the instability of any language boundaries, including ethnic ones.

> **ethnolect** A dialect spoken by a particular ethnic group.

The grammar of AAVE is virtually uniform across the country, which has been attributed to the segregation of African American speakers or their relatively recent migration from the South (see Chapter 5, pages 196–197). Some distinguishing characteristics of AAVE's speech patterns include the use of double negatives ("She don't like nothing"), omission of forms of the verb *to be* ("He my friend"), and nonconjugation of verbs ("She give him her paper yesterday"). There is some controversy over the place of AAVE in the U.S. educational system. Does AAVE constitute a distinctive language, rather than simply a dialect of English? Should it be taught—with its attendant grammar, structure, and literature—to American schoolchildren? Are those who speak AAVE and Standard English technically bilingual? This has led some linguists to refer to AAVE as African American Language (AAL), indicating the status of a separate language.

In the United States today, many descendants of Spanish speakers have adapted their speech to include words and variants of words in both Spanish and English, in a dialect known as "Spanglish" (see Seeing Geography, page 170). Although acceptance of AAVE and Spanglish as legitimate language forms is hardly without conflict (see Patricia's Notebook, page 151, and Subject to Debate, page 152), they illustrate the fluidity of languages and how they are constantly evolving and changing as the needs and experiences of their users change.

GLOBALIZATION

4.3 LEARNING OBJECTIVE
Describe the relationship between technology and language.

More often than not, the diffusion of some languages has come at the expense of many others. Ten thousand years ago, the human race consisted of only 1 million people, speaking an estimated 15,000 languages. Today, a population 7000 times larger speaks only 47 percent as many tongues. Only 1 percent of all languages have as many as 500,000 speakers. It has been estimated that the world loses a language on average every two weeks. Some experts believe that all but 300 languages will be extinct or dying by the year 2100. Clearly, globalization has worked to favor some languages and eliminate others. According to linguist Suzanne Romaine, there are no children today who are learning any of California's nearly 100 native languages. Languages die out when their speakers do; often the entire cultural world associated with a language vanishes as well. Thus, globalization both presents the opportunity for more people to communicate directly with one another and, at the same time, threatens to extinguish the cultural diversity that goes hand-in-hand with linguistic diversity.

TECHNOLOGY, LANGUAGE, AND EMPIRE

Technological innovations affecting language range from the basic practice of writing down spoken languages to the sophisticated information superhighway provided by the Internet. Technological innovations have in the past facilitated the spread and proliferation of multiple languages, but more recently they have encouraged the tendency of only a few languages—especially English, but also Chinese and

Patricia's Notebook

Miami State of Mind

Patricia Price. *(American Council on Education.)*

Geographically speaking, Miami—located in the state of Florida—is undoubtedly a part of the United States of America. However, many aspects of life here make visitors and residents alike feel more like they are somewhere else. As novelist Joan Didion wrote, Miami is "a settlement of considerable interest, not exactly an American city as American cities have until recently been understood but a tropical capital: long on rumor, short on memory . . . referring not to New York or Boston or Los Angeles or Atlanta but to Caracas and Mexico, to Havana and to Bogotá and to Paris and Madrid." The climate is subtropical, with the only real seasonal variation being between the rainy (summer) and dry (winter) months. Even after 19 years, I still miss autumn! Demographically, about 65 percent of residents are Latino, while the rest are roughly evenly split between African- and Anglo-Americans. And the United Nations has ranked Miami number one in the world in terms of the proportion of foreign-born residents (just over 50 percent).

Perhaps the biggest transition for a new arrival to Miami from another U.S. city, however, involves language. Spanish is the lingua franca of Miami. You can get by here just fine without speaking a word of English, though the reverse is not true: monolingual English speakers will struggle to understand and be understood in many areas of their daily lives. In other words, not speaking Spanish is an impediment here, whether in one's personal transactions or even in the ability to get a job, since many require applicants to understand and speak some Spanish.

I wonder whether Miami will remain this way or whether—as with other so-called immigrant gateways such as New York, San Francisco, and Chicago—immigrants will eventually acculturate, and the lingua franca of this town will transition yet again to English. I can already see this happening with my students, most of whose parents or grandparents are immigrants. Already their Spanish (or Portuguese, or Haitian Kreyol) is limited to what I call "kitchen" status. In other words, their vocabularies are limited, and they speak Spanish or other languages mostly with non-English-speaking relatives. Their children might not fully learn any language beyond English.

Little Havana commercial landscape. The lingua franca of Miami's Little Havana neighborhood is Spanish. This is reflected in the names and descriptions of local businesses. English is seldom seen (or heard). *(Jeff Greenberg 2 of 6/Alamy.)*

SUBJECT TO *Debate*

IMPOSING ENGLISH

English-only laws are nothing new in the United States. Its history as a nation of immigrants has led to a population that, at any one point in time, has spoken a variety of languages besides English. In its early days as a colony, one could hear German, Dutch, French, and a multitude of Native American languages spoken alongside English. This prompted both Benjamin Franklin and John Adams to propose enforcing English as the sole acceptable language, and Theodore Roosevelt once said, "The one absolutely certain way of bringing this nation to ruin or preventing all possibility of its continuing as a nation at all would be to permit it to become a tangle of squabbling nationalities. We have but one flag. We must also learn one language, and that language is English."

Citing concerns that providing official documents and services in multiple languages would simply be too expensive, contemporary advocates of English-only legislation claim that mandating one language is one way to reduce the cost of government. Some proponents also believe that English-only laws encourage immigrants to assimilate by learning the official language of the United States. Opponents accuse the laws of being racist and suggest that supporters of English-only legislation are threatened by cultural diversity. Linguistic unity, they say, does not lead to political or cultural unity. Furthermore, providing official documents and services only in English in effect denies these services and information to those who do not understand English. Debates such as these bring up questions of the legal, social, and political status of minority groups and their languages, debates that exist in many countries besides the United States.

Today, most of those who wish to legislate English as the official language of the country target Spanish-speaking immigrants as the object of their concern. Language is often at the heart of the immigration debates that occur so often on the nightly news. Anxieties about being culturally "overwhelmed" by Spanish speakers who refuse to learn English culminate in claims that Latino immigrants are dividing the nation in two: one English-speaking and culturally "American," and the other Spanish-speaking and unable to assimilate into the mainstream (see Patricia's Notebook, page 151).

Historically, most immigrants to the United States sooner or later abandon their native tongues. As late as 1910, one out of every four Americans could speak some language other than English with the skill of a native; today that number is significantly lower. This was a result of the mass immigrations from Germany, Poland, Italy, Russia, China, and many other foreign lands. Much of this linguistic diversity has given way to English, partly because these other languages lacked legal status, partly because of the monolingual educational system in the United States, and partly because of social pressures.

Continuing the Debate

As this discussion illustrates, many cultures do not want to assimilate into English-speaking society, choosing instead to actively assert pride in their language. Keeping this in mind, consider these questions:

- Do you think the wave of immigrants today, with their pride in language, is different from earlier waves? How so?

- Will monoglot English speakers become a dwindling minority as more and more people become bilingual through either choice or necessity?

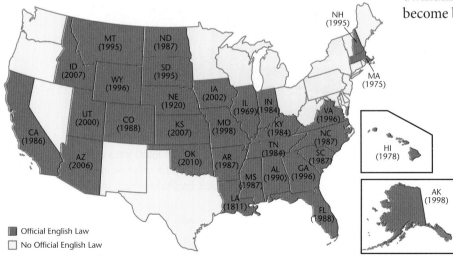

Official English Law
No Official English Law

States that have some form of official English-only laws. Dates show the years English-only laws were enacted. In the 1980s, the influx of immigrants to the United States from Asia and Latin America prompted many of these laws. Typically, English-only laws require that state documents be published in English only. Some states' laws, however, also prohibit the state from doing business in a language other than English or providing services such as multilingual emergency medical hotlines. (http://www.us-english.org/view/13.)

FIGURE 4.9 **Multilingual restroom sign.** This sign is located in the Cleveland, Ohio airport. Six languages— English, Braille, Spanish, Arabic, French, and Chinese—in addition to the visual images are used to convey to a variety of international travelers that this is the women's restroom. *(Courtesy of Patricia L. Price.)*

Britain, France, the Netherlands, Belgium, Portugal, Spain, and the United States across the globe altered the linguistic practices of millions of people. This empire building superimposed Indo-European tongues on the map of the tropics and subtropics. The areas most affected were Asia, Africa, and the Austronesian island world. A parallel case from the ancient world is China, also a formidable imperial power that spread its language to those it conquered. During the Tang dynasty (618–907 C.E.), Chinese control extended to Tibet, Mongolia, Manchuria (in contemporary northeastern China), and Korea. The 4000-year-old written Chinese language proved essential for the cohesion and maintenance of its far-flung empire. Although people throughout the empire spoke different dialects or even different languages, a common writing system lent a measure of mutual intelligibility at the level of the written word. Today, however, the Chinese are concerned about the "Romanization" of their language, as Chinese schools prioritize learning English and youth are not familiar with as many Chinese characters as their parents or grandparents.

Spanish—to dominate all others (**Figure 4.9**). Particular language groups achieve cultural dominance over neighboring groups in a variety of ways, often with profound results for the linguistic map of the world. Technological superiority is usually involved (see the cartoon at right). Earlier, we saw how plant and animal domestication—the technology of the "agricultural revolution"—aided the early diffusion of the Indo-European language family.

An even more basic technology was the invention of writing, which appears to have developed as early as 5300 years ago in several hearth areas, including in Egypt, among the Sumerians in what is today Iraq, and in China. Writing helped civilizations develop and spread, giving written languages a major advantage over those that remained spoken only. Written languages can be published and distributed widely, and they carry with them the status of standard, official, and legal communication.

Written language facilitates record keeping, allowing governments and bureaucracies to develop. Thus, the languages of conquerors tend to spread with imperial expansion. The imperial expansion of

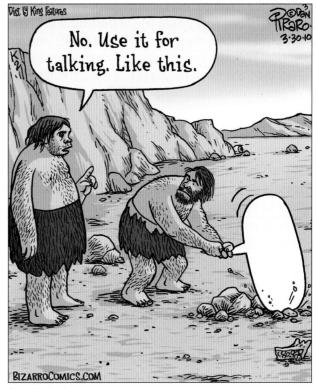

(Dan Piraro.)

ILE PROPRE - VIE PROPRE

Bora - Bora, île propre, c'est notre affaire à tous,
Bora - Bora, Fenua ma, E ohipa ia na te Taatoaraa.

Ensemble, gardons notre île propre.
Tatou paatoa, ia vai ma noa to tatou Fenua.

FIGURE 4.10 **French, the colonial language of the empire, shares this sign on the isle of Bora Bora** in French Polynesia with the native variant of the Polynesian tongue. Until recently, French rulers allowed no public display of the Polynesian language and tried to make the natives adopt French. *(Courtesy of Terry G. Jordan-Bychkov.)*

Even though imperial nations have, for the most part, given up their colonial empires, the languages they transplanted overseas survive. As a result, English still has a foothold in much of Africa, South Asia, the Philippines, and the Pacific islands. French persists in former French and Belgian colonies, especially in northern, western, and central Africa; Madagascar; and Polynesia (**Figure 4.10**). In most of these areas, English and French function as the languages of the educated elite, often holding official legal status. They are also used as a lingua franca of government, commerce, and higher education, helping hold together states with multiple native languages.

Transportation technology also profoundly affects the geography of languages. Ships, railroads, and highways all serve to spread the languages of the culture groups that build them, sometimes spelling doom for the speech of less technologically advanced peoples whose lands are suddenly opened to outside contacts. The Trans-Siberian Railroad, built about a century ago, spread the Russian language eastward to the Pacific Ocean. The Alaska Highway, which runs through Canada, carried English into Native American refuges. The construction of highways in Brazil's remote Amazonian interior threatens the native languages of that region.

Another example is the predominance of English on the Internet, which can be understood as a contemporary information highway (**Figure 4.11**). What will happen when other languages begin to challenge the dominance of English on the Internet? This will inevitably happen sooner or later, although exactly when English will be surpassed by another language is anyone's guess. For example, from 2000 to 2010 there was a truly impressive 1277 percent growth in the number of Chinese speakers on the Internet. If this trend continues, and when—not if—the 71 percent of Chinese speakers who do not now use the Internet begin to log on, we can expect Chinese to surpass English as the most popular language on the Internet. Speakers of other languages grew even more impressively over this time period. For instance, Arabic-speaking Internet users grew 2501 percent from 2000 to 2010, while speakers of Russian grew 1826 percent over the same period.

Yet, technology can also be employed to preserve and revive endangered languages. Native American groups, for instance, are working with software developers to create language apps for the iPhone and iPad. They are also working to produce toys that speak in Native languages and video games. All of these are intended to preserve and teach Native languages, especially to younger people. Most speakers of Native American languages—around 90 percent—are middle-aged and older, so capturing the interest of younger speakers is vitally important to the survival of these languages. The U.S. government provides federal funds in the form of competitive grants to support language preservation efforts, while some tribes can utilize casino earnings for this purpose. Regardless of the funding source, most Native American groups are engaged in language preservation and instruction. Besides the apps and

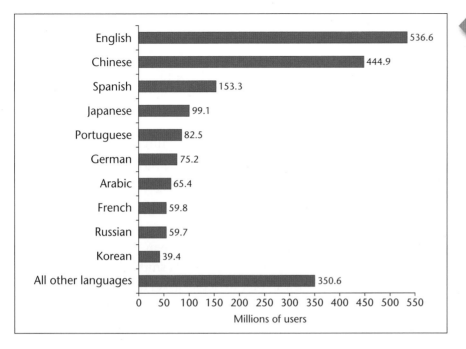

English — 536.6
Chinese — 444.9
Spanish — 153.3
Japanese — 99.1
Portuguese — 82.5
German — 75.2
Arabic — 65.4
French — 59.8
Russian — 59.7
Korean — 39.4
All other languages — 350.6

Millions of users

FIGURE 4.11 **The 10 most prevalent languages on the Internet,** measured as a percentage of total users. English is the second most widely spoken language on Earth, after Mandarin Chinese. And, English is the most widely spoken second language in the world. It dominates the Internet as well. *(Source: Adapted from http://www.internetworldstats.com/stats7.htm.)*

video games, language immersion schools have been established (**Figure 4.12**), with mentor relationships forming between elders and youth, YouTube videos of Native speakers are also available, and Google Hangout video chats facilitate live conversations among remotely located individuals.

FIGURE 4.12 **Learning Cherokee with technology.** These two fifth-graders attend the Cherokee Nation Immersion School in Tahlequah, Oklahoma. They are working with their teacher to learn the Cherokee language with the assistance of an app. *(AP Photo/Sue Ogrocki.)*

TEXTING AND LANGUAGE MODIFICATION

Though English may dominate the Internet (see Figure 4.11), much of what comes across our computer and cell phone screens isn't a readily recognizable form of English. As with the diffusion of spoken English to far-flung regions of the British Empire, the English language that is spread via electronic correspondence is subject to significant modification. E-mailing, instant messaging, and text messaging Standard English on cell phones requires a lot of typing, and text is notoriously deficient in conveying emotions when compared to the spoken word. For these reasons, users often use abbreviations and symbols to shorten the number of keystrokes used, to add emotional punctuation to their correspondence, and to make electronic communication difficult to monitor by those who don't understand the language—particularly parents and teachers! (See the cartoon on the following page.)

English is an alphabetic writing system, where letters represent discrete sounds that must be strung together to form a word's complete sound. A second major writing system is syllabic, where characters represent blocks of word sounds. This type of writing, prevalent throughout the Middle East and Southeast

(Sidney Harris/ScienceCartoonsPlus.com.)

"FIRST PAPER I'VE HAD TO GRADE WRITTEN ENIRELY IN EMOTICONS."

Asia, includes Arabic, Hebrew, Hindi, and Thai. The third form of writing is logographic, where characters represent entire words. Chinese is the only major language in this category.

English-speaking texters quickly learn to take shortcuts around the lengthiness inherent to alphabetic writing systems. The simplest and most commonly used shortcut involves using acronyms instead of whole words to convey common phrases. *POS* (parent over shoulder), *VBG* (very big grin), *LOL* (laughing out loud), and *GMTA* (great minds think alike) are some examples of this technique. Another shortcut involves using characters that sound like words. For example, the number "8" can substitute for the word or sound "ate," so "h8" is "hate," and "i 8" is "I ate." In the first example, a syllabic approach to writing is used because "8" condenses the syllable "ate" into one character. Another technique involves using symbols to substitute for entire words. Using "8" for the word "ate" is an example of this technique, called rebus writing, where a symbol is used for what it sounds like as opposed to what it stands for. As your instructor for this class will likely confirm, such abbreviations and symbols have even worked their way into term papers at the college level, much to the consternation or delight of language scholars.

The use of symbols such as ☺ and ♥, which are called pictograms, or word pictures, is similar to

Chinese writing. Their popularity has resulted in a vast lexicon of pictograms as well as simpler symbol groupings, called emoticons, used to convey entire words, ideas, and emotions in a compact and often humorous form. Here are some examples:

:)   (user is smiling or joking)

:@   (user is screaming or cursing)

:- #   (well, shut my mouth!)

Although the meaning of these symbols is understood by speakers of many languages because of their ubiquity, non-English languages also employ their own symbol combinations. In Korean, for instance, ^^ is used instead of :) to convey a smiling face and -_- is used instead of :(to depict a sad face. In Chinese, the number "5" is pronounced in a way that resembles crying, so "555" is the Chinese texter's way of conveying sadness.

Spoken and written English are quickly picking up these cyberspeech patterns with expressions such as *PITA,* which is used to indicate that someone is a "pain in the ass" without actually saying so, and *G2G,* which is uttered instead of the marginally longer sounds "got to go" for which this acronym stands. As speakers of non-English languages become more heavily involved in texting-based activities, their spoken and written languages also doubtlessly will become modified.

LANGUAGE PROLIFERATION: ONE OR MANY?

Could all the world's languages have derived from one single mother tongue? It may seem a large leap from the primordial tongue to a consideration of globalization and languages, but, in fact, the two are related. If we humans began with one language, why shouldn't we return to that condition? If one language became 15,000 and the 7000 or so that remain will dwindle to 300 within a century, then why not end up with one again?

Are the forces of modernization working to produce, through cultural diffusion, a single world language? And if so, what will that language be? English? Worldwide, about 335 million people speak English as their mother tongue and perhaps another

350 million speak it well as a second, learned language. Adding other reasonably competent speakers who can "get by" in English, the world total reaches about 1.5 billion, more than for any other language. What's more, the Internet is one of the most potent agents of diffusion, and its language, overwhelmingly, is English.

English earlier diffused widely with the British Empire and U.S. imperialism, and today it has become the de facto language of globalization. Consider the case of India, where the English language imposed by British rulers was retained (after independence) as the country's language of business, government, and education. It provided some linguistic unity for India, which had 800 indigenous languages and dialects. This is why today many of India's 1.2 billion people speak English well enough to provide customer support services over the telephone for clients in the United States. Even so, many people resent its use and wish India to be rid of this hated linguistic colonial legacy once and for all (see Figure 4.21). Although English is not likely to be driven out of India any time soon, it is true that the spoken English of India has drifted away from Standard British English. The same holds true for the English of Singapore, which is now a separate language called Singlish. Many other regional, English-based languages have developed—languages that could not be readily understood in London or Chicago.

But, is the diffusion of English to the entire world population likely? Will globalization and cultural diffusion produce one world language? Probably not. More likely, the world will ultimately be divided largely among 5 to 10 major languages.

LANGUAGE AND CULTURAL SURVIVAL

Because language is the primary way of expressing culture, if a language dies out, there is a good chance that the culture of its speakers will, too. Languages, like animal species, are classified as endangered or extinct. Endangered languages are those that are not being taught to children by their parents and are not being used actively in everyday matters. Some

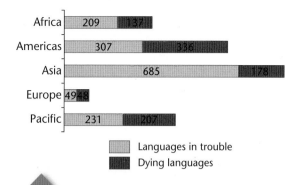

FIGURE 4.13 **Distribution of the world's nearly extinct languages.** This chart shows the regional distribution of the world's nearly extinct languages, or those languages for which only a few elderly speakers are still alive. In 2013 "dying" languages accounted for 12.7 percent of the world's total living languages, whereas 20.8 percent were considered to be "in trouble." *(Source: http://www.ethnologue.com/world.)*

linguists believe that more than half of the world's roughly 7000 languages are endangered. Ethnologue, an online language resource (see Linguistic Geography on the Internet, page 173, for this web site address), considers 377 languages to have become extinct since 1950, and another 906 to be on the "dying" list. Languages that have only a few elderly speakers still living fall into this category. The Americas and the Pacific regions together account for more than three-quarters of the world's current nearly extinct languages, thanks to their many and varied indigenous language groups (**Figure 4.13**). For example, as of 2010, only five families in Argentina spoke Vilela; today, the language is classified as "dead," with no remaining living speakers. When speakers die, it is most likely that their language will die out with them.

As Figure 4.13 shows, one-quarter of the world's in trouble and dying languages may be found in the Americas. They represent a wealth of Native American languages slowly becoming suffocated by English, Spanish, and Portuguese. Other **language hotspots**—places with the most unique, misunderstood, or endangered tongues—are located

> **language hotspots**
> Those places on Earth that are home to the most unique, misunderstood, or endangered languages.

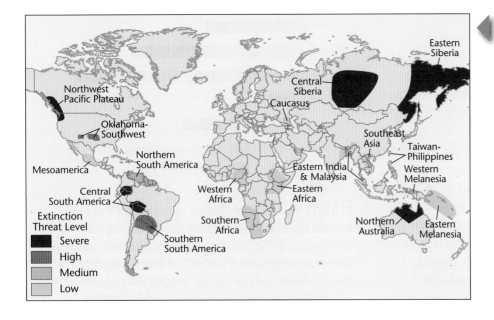

FIGURE 4.14 **Global "language hotspots."** The Enduring Voices Project and National Geographic Society have teamed up to document endangered languages and thereby attempt to prevent language extinction. *(Source: http:// travel.nationalgeographic.com/travel/ enduring-voices/.)*

around the globe (**Figure 4.14**). Three of the world's most vulnerable regions are located in the United States.

Languages can also be used to keep cultural traditions alive by speaking them repeatedly. Keith Basso, an anthropologist who has written an intriguing book titled *Wisdom Sits in Places,* discusses the landscape of distinctive place-names used by the Western Apache of New Mexico. The people Basso studied use place-names to invoke stories that help the Western Apache remember their collective history. According to Nick Thompson, one of Basso's interviewees, "White men need paper maps. . . . We have maps in our minds." Thompson goes on to assert that calling up the names of places can guard against forgetting the correct way of living, or adopting the bad habits of white men, once Western Apaches move to other areas. "The names of all these places are good. They make you remember how to live right, so you want to replace yourself again." One of the places Basso heard about is called Shades of Shit. Here is what he was told:

> *It happened here at Shades of Shit.*
> *They had much corn, those people who lived here, and their relatives had only a little. They refused to share it. Their relatives begged them but still they refused to share it.*
> *Then their relatives got angry and forced them to stay at home. They wouldn't let them go anywhere, not even to defecate. So they had to do it at home. Their shades*

> *[shelters] filled up with it. There was more and more of it! It was very bad! Those people got sick and nearly died.*
> *Then their relatives said, "You have brought this on yourselves. Now you live in shades of shit!" Finally, they agreed to share their corn.*
> *It happened at Shades of Shit.* (1996, p. 24)

Today, merely standing at this place or speaking its graphic name reminds the Western Apache that stinginess is a vice that can threaten the survival of the entire community.

NATURE–CULTURE

4.4 **LEARNING OBJECTIVE**
Explain the relationships between language and the physical environment.

Language interacts with the environment in two basic ways. First, the specific physical habitats in which languages evolve help shape their vocabularies. Second, the environment can guide the migrations of linguistic groups or provide refuges for languages in retreat. The following section, from the viewpoint of possibilism—the notion that the physical environment shapes, but does not fully determine, cultural phenomena—illustrates how the physical environment influences vocabulary and the distribution of language.

HABITAT AND VOCABULARY

Humankind's relationship to the land played a strong role in the emergence of linguistic differences, even at the level of vocabulary. For example, the Spanish language—which originated in Castile, Spain, a dry and relatively barren land rimmed by hills and high mountains—is especially rich in words describing rough terrain, allowing speakers of this tongue to distinguish even subtle differences in the shape and configuration of mountains, as Table 4.3 reveals. Similarly, Scottish Gaelic possesses a rich vocabulary to describe types of topography; this terrain-focused vocabulary is a common attribute of all the Celtic languages spoken by hill peoples. In the Romanian tongue, also born of a rugged landscape, words relating to mountainous features emphasize use of that terrain for livestock herding. English, by contrast, which developed in the temperate wet coastal plains of northern Europe, is relatively deficient in words describing mountainous terrain (Figure 4.15). However, English abounds with words describing flowing streams and wetlands: typical physical features found in northern Europe. This vocabulary transferred well to the temperate East Coast of the United States. In the rural American South alone, one finds *river, creek, branch, fork, prong, run, bayou,* and

TABLE 4.3 Some Spanish Words Describing Mountains and Hills

Spanish Word	English Meaning
Candelas	Literally "candles"; a collection of *peñas*
Ceja	Steep-sided breaks or escarpments separating two plains of different elevations
Cejita	A low escarpment
Cerrillo or *cerrito*	A small *cerro*; a hill
Cerro	A single eminence, intermediate in size between English *hill* and *mountain*
Chiquito	Literally "small," describing minor secondary fringing elevations at the base of and parallel to a *sierra* or *cordillera*
Cordillera	A mass of mountains, as distinguished from a single mountain summit
Cuchilla	Literally "knife"; the comblike secondary crests that project at right angles from the sides of a *sierra*
Cumber	The highest elevation or peak within a *sierra* or *cordillera*; a summit
Eminencia	A mountainous or hilly protuberance
Loma	A hill in the midst of a plain
Lomita	A small hill in the midst of a plain
Mesa	Literally "table"; a flat-topped eminence
Montaña	Equivalent to English *mountain*
Pelado	A barren, treeless mountain
Pelon	A bare conical eminence
Peloncilla	A small *pelon*
Peña	A needlelike eminence
Picacho	A peaked or pointed eminence
Pico	A summit point; English *peak*
Sandia	Literally "watermelon"; an oblong, rounded eminencee
Sierra	An elongated mountain mass with a serrated crest
Tinaja	A solitary, hemispheric mountain shaped like an inverted bowl

Source: Hill, 1986

FIGURE 4.15   **A scene in the desert of the western United States.** "Mountains," yes, but what kind of mountains? (See Table 4.3.) The English language cannot describe such a place adequately because it is the product of a very different, humid and cool, physical landscape. As a result, the ability of English speakers to name dryland environmental features will be less precise in such places. *(David Muench/Corbis.)*

slough. This vocabulary indicates that the area is a well-watered land with a dense network of streams.

Clearly, then, language serves an adaptive strategy. Vocabularies are highly developed for those features of the environment that involve livelihood. Without such detailed vocabularies, it would be difficult to communicate sophisticated information relevant to the community's livelihood, which in most places is closely bound to the physical landscape.

THE HABITAT HELPS SHAPE LANGUAGE AREAS

Environmental barriers and natural routes have often guided linguistic groups onto certain paths. The wide distribution of the Austronesian language group, for instance, was profoundly affected by prevailing winds and water currents in the Pacific and Indian oceans. The Himalayas and the barren Deccan Plateau deflected migrating Indo-Europeans entering the Indian subcontinent into the rich Ganges-Indus River plain. The northern and southern dialect boundaries

in the United States are loosely limited by the Mississippi River.

Because such physical barriers as rivers and mountain ridges can discourage groups from migrating from one area to another, they often serve as linguistic borders as well. In parts of the Alps, speakers of German and Italian live on opposite sides of a major mountain ridge. Portions of the mountain rim along the northern edge of the Fertile Crescent in the Middle East form the border between Semitic and Indo-European tongues. Linguistic borders that follow such physical features generally tend to be stable, and they often endure for thousands of years. By contrast, language borders that cross plains and major routes of communication are often unstable.

THE HABITAT PROVIDES REFUGE

The environment also influences language insofar as inhospitable areas provide protection and isolation. Such areas often provide minority linguistic groups refuge from aggressive neighbors and are, accordingly, referred to as **linguistic refuge areas.** Rugged hilly and mountainous areas, excessively cold or dry climates, dense forests, remote islands, and extensive marshes and swamps can all offer protection to minority language groups. For one thing, unpleasant environments rarely attract conquerors. Also, mountains tend to isolate the inhabitants of one valley from those in adjacent ones, discouraging contact that might lead to linguistic diffusion.

> **linguistic refuge area** An area protected by isolation or inhospitable environmental conditions in which a language or dialect has survived.

Examples of these linguistic refuge areas are numerous. The rugged Caucasus Mountains and nearby ranges in central Eurasia are populated by a large variety of peoples and languages (**Figure 4.16**). In the Rocky Mountains of northern New Mexico, an

FIGURE 4.16   **The environment provides a linguistic refuge in the Caucasus Mountains.** The rugged mountainous region between the Black and Caspian seas—including parts of Armenia, Russia, Georgia, and Azerbaijan—is peopled by a great variety of linguistic groups, representing three major language families. Mountain areas are often linguistic mosaics because the rough terrain provides refuge and isolation.

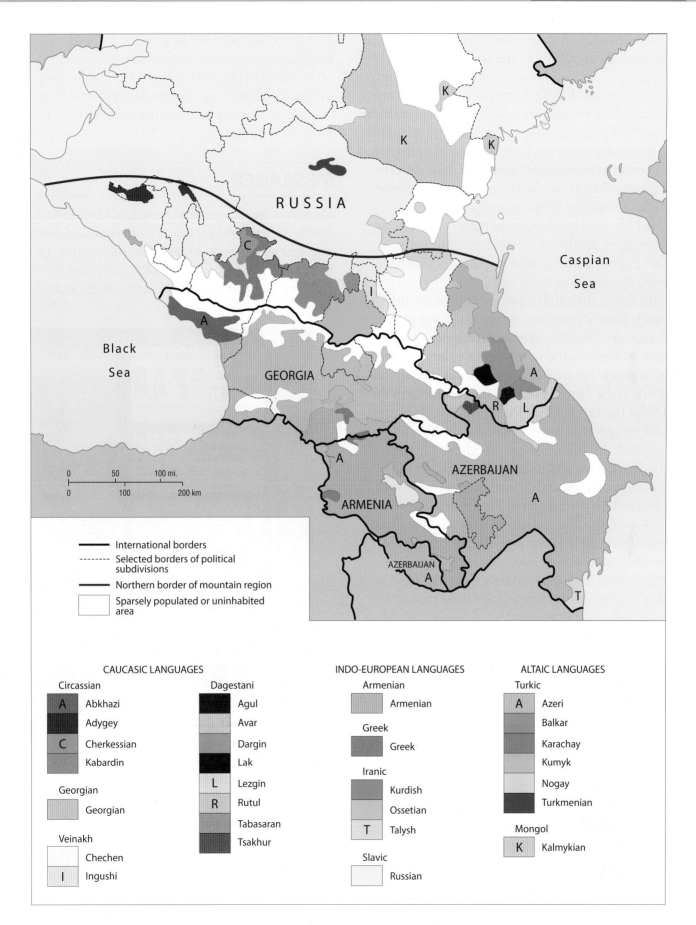

International borders
Selected borders of political subdivisions
Northern border of mountain region
Sparsely populated or uninhabited area

CAUCASIC LANGUAGES

Circassian
A Abkhazi
Adygey
C Cherkessian
Kabardin

Georgian
Georgian

Veinakh
Chechen
I Ingushi

Dagestani
Agul
Avar
Dargin
Lak
L Lezgin
R Rutul
Tabasaran
Tsakhur

INDO-EUROPEAN LANGUAGES

Armenian
Armenian

Greek
Greek

Iranic
Kurdish
Ossetian
T Talysh

Slavic
Russian

ALTAIC LANGUAGES

Turkic
A Azeri
Balkar
Karachay
Kumyk
Nogay
Turkmenian

Mongol
K Kalmykian

archaic form of Spanish survives, largely as a result of isolation that ended only in the early 1900s. Similarly, the Alps, the Himalayas, and the highlands of Mexico form fine-grained linguistic mosaics, thanks to the mountains that provide both isolation and protection for multitudinous languages. The Dhofar, a mountain tribe in Oman, preserves Hamitic speech, a language family otherwise vanished from all of Asia. Bitterly cold tundra climates of the far north have sheltered Uralic and Inuktitut speakers, and a desert has shielded Khoisan speakers from Bantu invaders. In short, rugged, hostile, or isolated environments protect linguistic groups that might otherwise be eclipsed by more dominant languages.

Still, environmental isolation is no longer the vital linguistic force it once was. Fewer and fewer places are so isolated that they remain little touched by outside influences. Today, inhospitable lands may offer linguistic refuge, but it is no longer certain that they will in the future. Even an island situated in the middle of the vast Pacific Ocean does not offer reliable refuge in an age of airplanes, satellite-transmitted communications, and global tourism. Similarly, marshes and forests provide refuge only if they are not drained and cleared by those who wish to use the land more intensively. The nearly 10,000 Gullah-speaking descendants of African slaves have long nurtured their distinctive African-influenced culture and language, in part because they reside on the Sea Islands of South Carolina, Georgia, and North Florida. Today, the development of these islands for tourism and housing for wealthy nonlocals, as well as the out-migration of Gullah youth in search of better economic opportunities, threatens the survival of the Gullah culture and language. The reality of the world is no longer isolation, but contact.

CULTURAL LANDSCAPE

4.5 LEARNING OBJECTIVE
Identify the ways languages are visibly part of the cultural landscape.

Road signs, billboards, graffiti, placards, and other publicly displayed writings not only reveal the locally dominant language but also can be a visual index to bilingualism, linguistic oppression of minorities, and other facets of linguistic geography. Furthermore, differences in writing systems render some linguistic landscapes illegible to those not familiar with these forms of writing (**Figure 4.17**).

MESSAGES

Linguistic landscapes send messages, both friendly and hostile. Often these messages have a political content and deal with power, domination, subjugation, or freedom. In Turkey, for example, until recently Kurdish-speaking minorities were not allowed to

FIGURE 4.17 **Linguistic landscapes.** This image, from Los Angeles's Koreatown, is difficult for non-Korean speakers to read. The Korean characters used are not the Latin alphabet with which most English speakers are familiar. *(Nik Wheeler/Corbis.)*

FIGURE 4.18 **Road sign in Dublin, Ireland,** shows directions in Irish Gaelic on top and English underneath. *(JoeFoxDublin/Alamy.)*

reinstate them in the landscape offer an indication of the social and political status of minority populations more generally.

Other types of writing, such as gang-related graffiti, can denote ownership of territory or send messages to others that they are not welcome (**Figure 4.19**). Only those who understand the specific gang symbols used will be able to decipher the message. Misreading such writing can have dangerous consequences for those who stray into unfriendly territory. In this way, gang symbols can be understood as a dialect that is particular to a subculture and transmitted through symbols or a highly stylized script.

TOPONYMS

Language and culture also intersect in the names that people place on the land, whether they are given to settlements, terrain features, streams, or various other aspects of their surroundings. These place-names, or **toponyms,** often directly reflect the spatial patterns of language, dialect, and ethnicity. Toponyms become part of the cultural

> **toponym** A place-name usually consisting of two parts: the generic and the specific.

broadcast music or television programs in Kurdish, to publish books in Kurdish, or even to give their children Kurdish names. Because Turkey wishes to join the European Union, these minority language restrictions have come under intense outside scrutiny. In 2002 Turkey reformed its legal restrictions to allow the Kurdish language to be used in daily life but not in public education. In 2012 Kurdish language instruction became an elective subject in Turkey's public schools. The Canadian province of Québec, similarly, has tried to eliminate English-language signs. French-speaking immigrants settled Québec, and its official language is French, in contrast to Canada's policy elsewhere of bilingualism in English and French. As **Figure 4.18** indicates, there is a movement in Ireland to replace English-language place-name signs with signs depicting the original Gaelic place-names. The suppression of minority languages and attempts to

FIGURE 4.19 **Graffiti is used to mark gang territory.** This wall in the Polanco neighborhood of Guadalajara, Mexico, is covered with graffiti. Gangs use stylized scripts often unintelligible to nonmembers to mark their territory. *(Courtesy of Patricia L. Price.)*

landscape when they appear on signs and placards. Many place-names consist of two parts—the generic and the specific. For example, in the American place-names Huntsville, Harrisburg, Ohio River, Newfound Gap, and Cape Hatteras, the specific segments are *Hunts-, Harris-, Ohio, Newfound,* and *Hatteras.* The generic parts, which tell what *kind* of place is being described, are *-ville, -burg, River, Gap,* and *Cape.*

Generic toponyms are of greater potential value to the cultural geographer than specific names because they appear again and again throughout a culture region. There are literally thousands of generic place-names,

> **generic toponym** The descriptive part of many place-names, often repeated throughout a culture area.

and every culture or subculture has its own distinctive set of them. They are particularly valuable both in tracing the spread of a culture and in reconstructing culture regions of the past. Sometimes generic toponyms provide information on changes people brought about long ago in their physical surroundings.

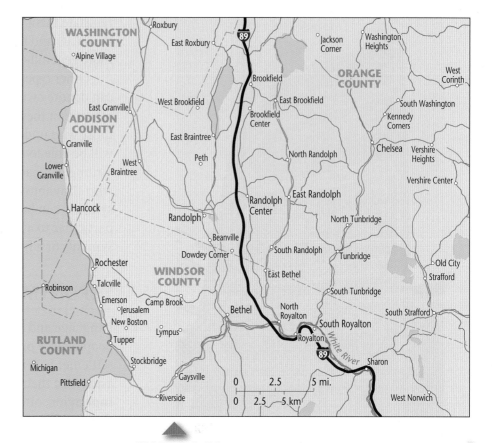

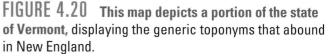

FIGURE 4.20 **This map depicts a portion of the state of Vermont,** displaying the generic toponyms that abound in New England.

GENERIC TOPONYMS OF THE UNITED STATES

The three dialects of the eastern United States (see Figure 4.8)—Northern, Midland, and Southern—illustrate the value of generic toponyms in cultural geographical detective work. For example, New Englanders, speakers of the Northern dialect, often used the terms *Center* and *Corners* in the names of the towns or hamlets. Outlying settlements frequently bear the prefix *East, West, North,* or *South,* with the specific name of the township as the suffix. Thus, in Randolph Township, Orange County, Vermont, we find settlements named Randolph Center, South Randolph, East Randolph, and North Randolph (**Figure 4.20**). A few miles away lies Hewitts Corners.

These generic usages and duplications are peculiar to New England, and we can locate areas settled by New Englanders as they migrated westward by looking for such place-names in other parts of the country. A trail of "Centers" and name duplications extending westward from New England through upstate New York and Ontario and into the upper Midwest clearly indicates their path of migration and settlement. Toponymic evidence of New England exists in areas as far away as Walworth County, Wisconsin, where Troy, Troy Center, East Troy, and Abels Corners are clustered; in Dufferin County, Ontario, where one finds places such as Mono Centre; and even in distant Alberta, near Edmonton, where the toponym Michigan Centre doubly suggests a particular cultural diffusion. Similarly, we can identify Midland American areas by such terms as

Gap, Cove, Hollow, Knob (a low, rounded hill), and *-burg,* as in Stone Gap, Cades Cove, Stillhouse Hollow, Bald Knob, and Fredericksburg. We can recognize southern speech by such names as *Bayou, Gully,* and *Store* (for rural hamlets), as in Cypress Bayou, Gum Gully, and Halls Store.

TOPONYMS AND CULTURES OF THE PAST

Place-names often survive long after the culture that produced them vanishes from an area, thereby preserving traces of the past. Australia abounds in Aborigine toponyms, even in areas from which the native peoples disappeared long ago. No toponyms are more permanently established than those identifying physical geographical features, such as rivers and mountains. Even the most absolute conquest, exterminating an aboriginal people, usually does not entirely destroy such names. Quite the contrary, in fact. Geographer R. D. K. Herman speaks of anticonquest, in which the defeated people finds its toponyms venerated and perpetuated by the conqueror, who at the same time denies the people any real power or cultural influence. The abundance of Native American toponyms in the United States provides an example (see Doing Geography on page 172). India, however, has recently decided to revert to traditional toponyms, after many Indian place-names had been Anglicized under British colonial rule (**Figure 4.21**).

In Spain and Portugal, seven centuries of Moorish rule left behind a great many Arabic place-names. An example is the prefix *guada-* on river names (as in Guadalquivir and Guadalupejo). The prefix is a corruption of the Arabic *wadi,* meaning "river" or "stream." Thus, Guadalquivir, corrupted from Wadi-al-Kabir, means "the great river." The frequent occurrence of Arabic names in any particular region or province of Spain reveals the remnants of Moorish cultural influence in that area, rather than anticonquest. Many such names were brought to the Americas through Iberian conquest, so that Guadalajara, for example, appears on the map as an important Mexican city.

THE VIDEO CONNECTION

Vigilante Copyeditor
http://www.tinyurl.com/l5mxlpk

This video depicts vandalism of the Sculpture Park at Brooklyn's Pratt Institute. The vandalism was not done by your run-of-the-mill vandal bent on destroying or defacing the sculptures. Instead, Pratt's vandal has made it his (or her) job to correct the grammar on the sculpture placards. The narrator discusses how the edits to the text of the placards interact with the sculptures in a way that demonstrates critical thinking. Because the corrections are not signed, the identity of the "vigilante copyeditor" remains unknown.

Thinking Geographically:

1. This chapter discusses toponyms, or place-names, and notes that there are sometimes very contentious battles about what name a place should be given, and who should do the naming. Can you view what happened with the placards at the Sculpture Park as a battle over toponyms? How (or how not)?

2. What is the relationship between art and language in a public space such as Pratt's Sculpture Park?

3. The video claims that most graffiti artists in Brooklyn "tattoo ... egocentric message[s]" on public spaces. Is this a fair assessment? In what ways was the work of the "vigilante copyeditor" different from—and the same as—that of other urban graffiti artists? What's the difference between a graffiti artist and a vandal? What label would you assign to the individual (or individuals) who corrected the grammar on the placards at the Pratt Institute Sculpture Park, and why?

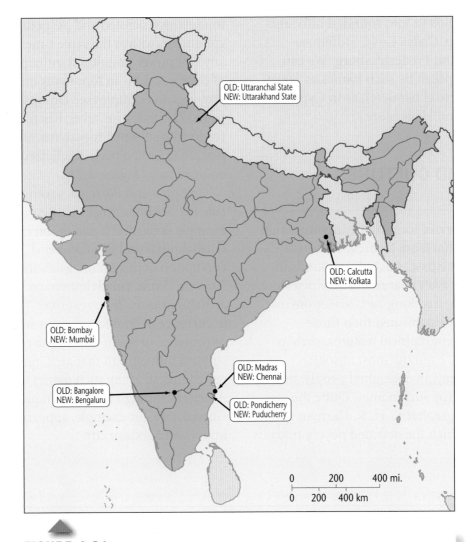

FIGURE 4.21   India's postcolonial toponym shift. More than 50 years after the English colonizers "quit" India, their colonial place-names are being swept from the map, too. *(Source: Adapted from Sappenfield, 2006.)*

THE POLITICAL ECONOMY OF TOPONYMS

Without a doubt, you are familiar with places that bear the names of wealthy and influential individuals, politicians, or corporate sponsors. Sports stadiums, campus buildings, museums—all provide specific spaces that can be named. The owners of these spaces typically use naming opportunities as a way to raise funds. For instance, universities provide naming rights for donors contributing anywhere from $5 million to $20 million (Figure 4.22).

But, what happens when a place that has been built using public monies—say, a train station that was constructed using tax revenues—is given a corporate name (Figure 4.23)? Geographers Reuben Rose-Redwood and Derek Alderman examine this practice, using the example of the building that was constructed to replace the World Trade Center's Twin Towers, which were destroyed in the terrorist attacks of 9/11. The building was initially dubbed "The Freedom Tower," but once it acquired corporate tenants, the building began to be referred to by its legal address, One World Trade Center. According to Rose-Redwood and Alderman, this illustrates "how

FIGURE 4.22   **University of Louisville sports stadium.** The home of the Cardinals is named after a pizza chain and supported by other corporate sponsors noted on the sign. Many universities are turning to corporate sponsorship to build sports facilities such as this one. *(Andy Lyons/Getty Images.)*

the naming of places has become . . . one of the next 'frontiers' in the neoliberalization of urban spaces" (2011, p. 2).

What happens when the corporate sponsor of a place has a less-than-acceptable reputation? Florida Atlantic University, for instance, sold the naming rights for its sports stadium to GEO Group for $6 million in February 2013. GEO Group operates private prisons and has been implicated in the inhuman treatment of inmates at its facilities. Less than two months later, the agreement was cancelled.

In this instance, the negative place image cast by a shady corporate sponsor simply wasn't worth the price paid.

Is no place safe from corporate ownership? Apparently not. People are paid to allow companies to use their cars as mobile billboards. What about the items of clothing and accessories that prominently bear the names of their designers? Is what poet Adrienne Rich called "the geography closest-in" —our bodies—a place that is also open for corporate naming opportunities?

FIGURE 4.23   **Barclays Center Station.** The New York City subway system is public. Yet, this stop—just outside the Barclays Center arena—is named after the arena's corporate sponsor: a bank. *(Spencer Platt/Getty Images.)*

World Heritage Site

The Alhambra fortress-palace complex, gardens and rural estates of the Generalife, and the Albayzín residential quarter together form this World Heritage Site. All three are located in the city of Granada, in southern Spain. The Alhambra and the Albayzín are situated on hills above the modern lower city. Hills were important defense sites in medieval European cities (see Chapter 10). And, the irrigated gardens of the Generalife were part of the rulers' rural estates.

■ From the eleventh through fourteenth centuries, Muslim rulers, or *emirs,* oversaw the construction of the Alhambra fortress and residences, gardens of the Generalife, and the Albayzín residential quarter. From the eighth through fifteenth centuries, Spain was ruled by the Moors, Muslims of North African descent.

■ These sites are exceptional examples of Moorish imperial architecture. Though Spain became Christianized after the Moors were expelled in 1492, Spanish monarchs greatly admired and sought to safeguard the architectural achievements of the Moors in the Andalusian region of southern Spain. Through careful restoration, the architecture of these sites remains true to its Spanish-Moorish roots. The urban layout, colors, and materials look today much as they did over 700 years ago.

■ The Albayzín residential quarter occupies a hill that, according to archaeological evidence, has been continuously occupied since early Roman times (about 2500 years).

(Karol Kozlowski/Alamy))

Alhambra, Generalife, and Albayzín

WALL WRITING IN THE ALHAMBRA:

The original Alhambra fortress-palace was built by Samuel Ha-Nagid, an eleventh-century Jewish grand vizier (a prime minister of sorts) to the Zirid sultans of Granada. In the thirteenth and fourteenth centuries, the Nazrid emirs constructed an entire complex of palaces and irrigated gardens at the site. The buildings are adorned with what appear at first glance to be elaborate intricate designs. In reality, the walls, ceilings, and other architectural elements are covered with words. Over 10,000 Arabic inscriptions adorn the interior of the Alhambra complex. "There is probably no other place in the world where studying walls, columns and fountains is so similar to turning the pages of a book," according to Spanish researcher Juan Castilla.

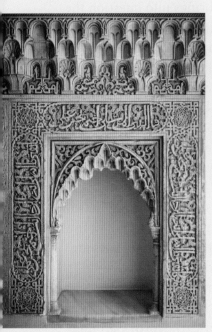

(caracterdesign/iStockphoto/Getty Images.)

■ Islamic architecture is typically adorned with decorative writing called calligraphy and other geometric patterns, rather than representations of living beings. According to some Muslims, figural representation was a form of idolatry (the worship of a physical object as a God), while others saw the body as an imperfect covering for the soul.

■ Because calligraphy was commonly regarded as the preeminent expression of the visual arts, calligraphers typically had a higher status than other artists and were housed in the Ministry of Writing.

■ Efforts to digitally archive and transcribe the wall writing reveal that fewer than 10 percent of the inscriptions are Qur'anic (religious) verses or poetry. The rest offer praise to the Nasrid emirs or consist of other popular sayings, with "There is no victor but Allah" being the most common inscription.

TOURISM:

This site attracts a huge number of international tourists, most of whom arrive in the summer months.

(Geoff A Howard/Alamy.)

■ The number of tourists to the Alhambra is limited to a daily quota of 6600.

■ Unfortunately, the decorative surfaces of the site's buildings are not well protected, so they continue to be worn away through the touch of so many tourists.

■ The picturesque city of Granada is situated on a series of hills against the backdrop of the Sierra Mountains. Tourists enjoy the mix of Mediterranean cultures, cuisines, and traditions there.

■ This World Heritage Site was built by and for the Moorish rulers of southern Spain, and exemplifies the pinnacle of Andalusian architecture, a synthesis of classical Arabic and southern Spanish styles.

■ The interior structures of the Alhambra palace are covered with more than 10,000 intricate inscriptions in Arabic calligraphy.

■ The Alhambra and the Generalife were placed on the list of World Heritage Sites in 1984, and in 1994 the site was expanded to include the Albayzín quarter.

www.whc.unesco.org/en/list/314

SEEING GEOGRAPHY   Aquí se habla Spanglish
What does this sign tell you about who lives in this city?

U.S. Army veterans at a recruiting table in Miami, Florida.
(Jeff Greenberg/AgeFotostock.)

Languages are fluid, always being altered and reinvented as the needs and experiences of their users change. Thanks to relocation diffusion resulting from the conquest of the East Coast of North America by the British in the seventeenth century, the primary language of the United States is English. The language then expanded westward as English-speaking peoples conquered more and more of the continent's territory. But, the English spoken in Britain's overseas colonies has never been "the Queen's English." Rather, British colonies in North America, Australia, the Caribbean, Africa, and eastern and southern Asia all developed their own distinctive dialects. Indigenous words have been incorporated into English vocabularies, as the

sections on toponyms in this chapter show. Pidgins, creoles, and distinctive dialects have resulted in places of high multilingual exposure, such as Singapore and the Anglophone Caribbean islands. Waves of immigrants have added further to the linguistic richness of English-speaking areas.

In 2002, Hispanics surpassed African Americans as the nation's numerically most significant minority group. In some U.S. cities, Hispanics constitute more than half of the population, a fact that brings into question the designation "minority." For example, the population of Miami, Florida, is two-thirds Hispanic, and more than three-fourths of the residents of El Paso and San Antonio, Texas, are Hispanic. In the United States today, Spanish-speaking peoples from Latin America provide the largest flow of immigrants into the United States. Even midsize and smaller towns in the midwestern and southern United States are becoming destinations for Spanish-speaking immigrants, sharply changing the ethnic composition of cities such as Shelbyville, Tennessee; Dubuque, Iowa; and Siler City, North Carolina.

It is logical to assume that the Spanish language spoken by these new immigrants, and by the families of Hispanic Americans, will have a growing impact on American English. As the opening photograph for this chapter illustrates, Spanish words have become common sights in U.S. cities, appearing frequently on street signs. But, Spanish and English have combined in a rich, complex fashion as well, to produce a hybrid language called Spanglish. The phrase "Vámonos al downtown a tomar una bironga after work hoy" is an excellent example. It translates into Standard English as "Let's go downtown and have a beer after work today." *Vámonos* ("let's go"), *tomar* ("to drink"), and *hoy* ("today") are Spanish words that are combined in the same sentence with the English words *downtown* and *after work*. Linguists refer to this tendency to shift between languages in the same sentence, a very common practice among bilingual speakers, as code-switching. However, the noun *bironga,* which means "beer" in English, is a Spanglish invention: it exists in neither English nor Spanish. This is quite common in Spanglish, and neologisms such as *hanguear* ("to hang out"), *deioff* ("day off"), and *parquear* ("to park" a vehicle) abound.

In the image that opens this chapter, the phrase "Yo soy el Army" employs both Spanish and English to recruit. This image also underscores that people of Hispanic descent can be of any race.

Spanglish reflects the growing Spanish-English bilingualism of many U.S. residents, the flexibility of language, and the enduring creativity of human beings as we attempt to communicate with one another. Ilan Stavans, author of *Spanglish: The Making of a New American Language,* likens Spanglish to jazz. "Yes," Stavans writes, "it is the tongue of the uneducated. Yes, it's a hodgepodge. . . . But its creativity astonished me. In many ways, I see in it the beauties and achievements of jazz, a musical style that sprung up [sic] among African-Americans as a result of improvisation and lack of education. Eventually, though, it became a major force in America, a state of mind breaching out of the ghetto into the middle class and beyond. Will Spanglish follow a similar route?"

CONCLUSION

Language, then, is an essential part of culture that can be studied using the five themes of cultural geography. Language is firmly enmeshed in the cultural whole. Its families, dialects, vocabulary, pronunciation, and toponyms display distinct spatial variations that are shown on maps of linguistic culture regions. Languages are mobile entities, ebbing and flowing across geographical areas. Relocation and expansion diffusion, both hierarchical and contagious, are apparent in the movement of language. They are also shaped and reshaped with the changing needs of their users.

The globalization of language, through the expansion of ancient empires as well as today's interlinked global exchanges, underscores the fact that human interactions are primarily language based. The progression of languages used from one, to many, and back again to a few—perhaps even to one—shows that the number of languages in existence is variable. The trend toward the dominance of a few "big" languages may afford opportunities to communicate on a global scale, but it also may signal the demise of much of the cultural richness across the Earth. Efforts to preserve and teach dying languages draw on the same cultural innovations that in other ways act to threaten the most vulnerable of tongues.

Language and physical environment interact in a nature–culture dynamic, with the physical environment helping to shape linguistic elements, such as vocabulary, and language shaping our use and perception of the environment. Finally, we can see language in the landscapes created by literate societies. The public signs, generic toponyms, and the sponsorship of the wealthy together create a linguistic landscape that can be read using one's "geographic eyes." Dominance of one group over another is often expressed in the latter's exclusion from the linguistic cultural landscape.

DOING GEOGRAPHY

Toponyms and Roots of Place

As you recall from this chapter, toponyms can give us important clues about the historical, social, political, and physical geography of a place. One example of this is the prevalence of indigenous place-names throughout the Americas, from Canada to Chile. You may say a place-name on a daily basis without being aware of its roots in an indigenous language. According to Charles Cutler, European settlers simply appropriated many of the Native American words for plants, animals, foods, and places with little or no modification in their pronunciation. These words are known as *loanwords*. For example, *Milwaukee* comes from an Algonquin word meaning "good spot or place," and *Chicago,* also Algonquin in origin, probably means "garlic field." The commonly used derogatory place-name *Podunk* is also indigenous in origin, from the Natick word for "swampy place." In fact, the names of more than half of the states in the United States are of Native American origin.

　　For this exercise, you will explore toponyms in more detail.

Steps to Exploring Toponyms

Step 1: Choose (or your instructor will assign) a state in the United States or a province in Canada on which to focus.

Step 2: Find a map of your chosen or assigned state or province. It should be a map that is detailed enough to show the names of political and physical features, such as cities, towns, counties or parishes, rivers, mountains, lakes, and so on. You can use an online map program like Google Maps, or look for maps in the reference section of your library; in printed or CD-ROM atlases; and online at sites such as the University of Texas's Perry-Castañeda map collection, located at http://www.lib.utexas.edu/maps/. Many maps in atlases also provide an index of place-names that can be useful to you.

Step 3: Examine your map with an eye toward the different categories of toponyms. Make a list of at least five place-names for each category:

- Historical people or events

- Non-English place-names (excluding Native American names)

- Native American place-names

- Place-names transplanted from elsewhere (e.g., "New" York)

- Descriptions of physical features (landforms, elevation, etc.)

- Descriptions of natural resources

Based on what you learned, answer the following questions:

- Did you find at least five examples for each category? If not, why do you think you didn't?

- For which category of toponyms did you find the most examples? Why?

- Were toponyms of one or more categories clustered spatially on the map? If so, where and why?

- What do the names say about the history, culture, and physical geography of the state or province you examined?

Yosemite National Park in California. The word *Yosemite* comes from the name of the indigenous group that inhabited this area. Notice the arrowhead symbol, which underscores the park's Native American history. *(Adamsmith/Taxi/Getty Images.)*

Chapter 4
LEARNING OBJECTIVES REEXAMINED

4.1 Understand the geographical patterning of languages.

See pages 138–145. Name three major language families and the regions in which those languages are spoken.

4.2 Analyze how languages and dialects have come to exist.

See pages 145–150. What is the Kurgan hypothesis and how does it differ from the Anatolian hypothesis?

4.3 Describe the relationship between technology and language.

See pages 150–156. How is technology being used to preserve and revive endangered languages?

4.4 Explain the relationships between language and the physical environment.

See pages 158–162. Name at least one instance where the physical environment of a region provided a linguistic border.

4.5 Identify the ways languages are visibly part of the cultural landscape.

See pages 162–167. What are toponyms and how do they reflect the cultural landscape?

KEY TERMS

Anatolian hypothesis	p. 145	language family	p. 142
bilingualism	p. 141	language hotspots	p. 157
creole	p. 140	lingua franca	p. 141
dialect	p. 138	linguistic refuge areas	p. 160
ethnolect	p. 150	pidgin	p. 139
generic toponym	p. 164	polyglot	p. 143
isogloss	p. 147	slang	p. 149
Kurgan hypothesis	p. 145	toponym	p. 163
language	p. 137		

Linguistic Geography on the Internet

You can learn more about linguistic geography on the Internet at the following web sites:

Dictionary of American Regional English

http://dare.wisc.edu/

Discover a reference web site that describes regional vocabulary contrasts of the English language in the United States and includes numerous maps.

Endangered Language Alliance

http://endangeredlanguagealliance.org/main/

A poet, a professor, and a field linguist have joined forces to find, record, and promote New York City's vulnerable languages

Enduring Voices

http://www.nationalgeographic.com/mission/enduringvoices/

This flash map allows you to explore the world's "language hotspots," those regions that are home to the most linguistic diversity, the highest levels of linguistic endangerment, and the least studied tongues.

Ethnologue

http://www.ethnologue.com

This site provides information on how languages change over time as well as on endangered and nearly extinct languages. The interactive map allows you to browse different parts of the world for more information on their vulnerable languages.

Language Log

http://languagelog.ldc.upenn.edu/nll/

Search the fascinating posts on this language-themed blog, run by University of Pennsylvania phonetician Mark Liberman and featuring guest posters. Open up the "Categories" menu to find topics ranging from animal behavior, linguistics in the news, and the ever-entertaining "lost in translation" examples from around the world.

Sources

Anderson, Greg, and David Harrison. 2007. "Language Hotspots." Model developed at the Living Tongues Institute for Endangered Languages. http://www.nationalgeographic.com/mission/enduringvoices/.

Basso, Keith H. 1996. *Wisdom Sits in Places: Landscape and Language Among the Western Apache.* Albuquerque: University of New Mexico Press.

Bird, Steven. 2013. "Cyberlinguistics: Recording the World's Vanishing Voices." *Phys.org. 12 March.* http://phys.org/news/2013-03-cyberlinguistics-world-voices.html.

Cutler, Charles. 1994. *O Brave New Words! Native American Loanwords in Current English.* Norman: University of Oklahoma Press.

Didion, Joan. 1987. *Miami.* New York: Simon & Schuster.

Dingemanse, Mark. "Bantu Expansion." *Wikipedia Commons.* http://en.wikipedia.org/wiki/Image:Bantu_expansion.png.

Ford, Clark. "Early World History: Indo-Europeans to the Middle Ages." http://www.public.iastate.edu/~cfford/342worldhistoryearly.html.

Herman, R. D. K. 1999. "The Aloha State: Place Names and the Anti-Conquest of Hawaii." *Annals of the Association of American Geographers* 89: 76–102.

Hill, Robert T. 1986. "Descriptive Topographic Terms of Spanish America." *National Geographic* 7: 292–297.

Latrimer Clarke Corporation Pty Ltd. http://www.altapedia.com.

Pereltsvaig, Aysa. 2012. "Re-mapping the Languages of the Caucasus." *GeoCurrents*, 1 June. http://geocurrents.info/place/russia-ukraine-and-caucasus/caucasus-series/re-mapping-languages-of-the-caucasus.

Pountain, Chris. 2005. "Varieties of Spanish." Queen Mary School of Modern Languages. http://www.qmul.ac.uk/~mlw058/varspan/varspanla.pdf.

Roberts, Sam. 2010. Listening to (and Saving) the World's Languages. *The New York Times,* 28 April. http://www.nytimes.com/2010/04/29/nyregion/29lost.html?pagewanted=all.

Romaine, Suzanne. 2007. "Preserving Endangered Languages." *Language and Linguistics Compass,* 1 (1–2): 115–132.

Sappenfield, Mark. 2006. "Tear up the Maps: India's Cities Shed Colonial Names." *Christian Science Monitor,* 7 September. http://www.csmonitor.com/2006/0907/p01s02-wosc.html.

Stavans, Ilan. 2003. *Spanglish: The Making of a New American Language.* New York: Rayo.

U.S. English. http://www.us-english.org/view/13.

Ten Recommended Books and Special Issues on the Geography of Language

(For additional suggested readings, see the *Contemporary Human Geography* LaunchPad: http://www.macmillanhighered.com/launchpad/DomoshCHG1e.)

Abley, Mark. 2005. *Spoken Here: Travels among Threatened Languages.* New York: Mariner Books. A fascinating journey across the global map of dwindling languages, exploring their diversity, their speakers, and their vulnerability to eclipse by the ever-more-dominant major tongues, especially English.

Comrie, Bernard, Stephen Matthews, and Maria Polinsky. 2003. *The Atlas of Languages: The Origin and Development of Languages throughout the World.* New York: Checkmark Books. Utilizes maps and illustrations to depict the origin and spread of the world's 7000-plus languages.

Deutscher, Guy. 2010. *Through the Language Glass: Why the World Looks Different in Other Languages.* New York: Metropolitan Books–Henry Holt. The author makes a convincing case that one's native tongue substantively shapes a speaker's experience of the world, in ways that are profoundly different from the experiences of speakers of other languages.

Kenneally, Christine. 2007. *The First Word: The Search for the Origins of Language.* New York: Penguin. Do languages really "evolve," increasing in a linear fashion from the first grunts of cavemen millennia ago to today's cacophony of 7000 different tongues? Kenneally traces both the science of language evolution and the history of its study.

MacNeil, Robert, and William Cran. 2005. *Do You Speak American?* New York: Mariner Books. Spoken English in the United States displays a wealth of regional dialects, urban and rural differences, and historical as well as contemporary ties to languages spoken in other countries. The authors traveled the country documenting this fascinating language mosaic. A three-volume DVD of the same title is also available.

Ostler, Nicholas. 2005. *Empires of the Word: A Language History of the World.* New York: HarperCollins. This fascinating book explores the spread and evolution of languages through conquest, with many maps accompanying the text.

Pinker, Steven. 2007. *The Language Instinct: How the Mind Creates Language.* New York: Harper Perennial. Ever notice how linguistically inventive the average toddler is? This best seller looks at many angles of language development in humans, from its origins, to its social functions, to the instinct humans share to learn, understand, and speak language.

Rose-Redwood, Reuben, and Derek Alderman (eds.). 2011. Special Thematic Interventions Section: *New Directions in Political Toponymy. ACME: An International E-Journal for Critical Geographies* 10(1): 1–41. http://www.acme-journal.org/Volume10-1.htm. A collection of the latest geographic scholarship on political place-names.

Stewart, George R. 2008. *Names on the Land: A Historical Account of Place-Naming in the United States.* New York: New York Review of Books Classics. Originally written during World War II as a tribute to the diversity of American culture, Stewart's exploration of the history of American place-names is as informative today as it ever was.

Wolfram, Walt, and Ben Ward. 2006. *American Voices: How Dialects Differ from Coast to Coast.* Malden, Mass.: Wiley-Blackwell. A lively account of the rich linguistic topography of the United States.

Journals in Population Geography

World Englishes. Published by the International Association for World Englishes, the journal documents the fragmentation of English into separate languages around the world. Edited by Margie Berns and Daniel R. Davis.

When does cuisine cease to be ethnic and become simply "American"? What roles do mobility and globalization play in the process?

Ethnic restaurants abound in Manhattan's Greenwich Village neighborhood.
(Steve Lewis Stock/Getty Images.)

Go to "Seeing Geography" on page 216 to learn more about this image.

GEOGRAPHIES OF RACE AND ETHNICITY

Melting Pot or Tapestry?

One of the enduring stories that the people of the United States proudly tell themselves is that "ours is an immigrant nation." This story is on display during annual festivals celebrating the variety of ethnic traditions in countless cities, towns, and villages across the nation. For example, the midwestern town of Wilber, settled by Bohemian immigrants beginning about 1865, bills itself as "The Czech Capital of Nebraska" and annually invites visitors to attend a "National Czech Festival." Celebrants are promised Czech foods, such as *koláce, jaternice,* poppy seed cake, and *jelita;* Czech folk dancing; Czech postcards and souvenirs imported from Europe; and handicraft items made by Nebraska Czechs (bearing an official seal and trademark to prove their authenticity). Thousands of visitors attend the festival each year. Without leaving Nebraska, these tourists can move on to "Norwegian Days" at Newman Grove, the "Greek Festival" at Bridgeport, the Danish "Grundlovs Fest" in Dannebrog, "German Heritage Days" at McCook, the "Swedish Festival" at Stromsburg, the "St. Patrick's Day Celebration" at O'Neill, several Native American powwows, and assorted other ethnic celebrations (**Figure 5.1**).

Today, Nebraska is still a magnet for immigrants, but since the 1990s, the state's new arrivals have been overwhelmingly non-European. In particular, Mexican immigrants employed in Nebraska's meat-processing industry find destinations such as Nebraska and other

LEARNING OBJECTIVES

5.1 Describe how ethnic groups are distributed geographically, and discuss how and why these distributions have changed over time.

5.2 Define the different types of migration.

5.3 Discuss how globalization might variously deepen, reshape, or erase ethnicity, race, and racism.

5.4 Analyze the diverse ways in which ethnic and racial groups interact with the natural environment, and discuss how these groups may be protected, or made more vulnerable.

5.5 Recognize the different ways that ethnicity and race leave imprints on the cultural landscape.

FIGURE 5.1 **The town of Lindsborg, Kansas (*left*).** Proud of its Swedish heritage, Lindsborg holds a Swedish Festival each year in June. **Hispanic Heritage Festival in Nebraska (*right*).** Hispanic immigrants have expanded the range of ethnic pride festivals throughout the Midwest. *(Left: Betts Anderson Loman/PhotoEdit. All rights reserved.; Right: Steve Skjold/Alamy.)*

upper midwestern states attractive. In general, immigrants to the United States today are more likely to come from Asia or Latin America than from Europe, and they are changing the face of ethnicity in the United States (**Figure 5.2**). Indeed, ethnicity is a central aspect of the cultural geography of most places, forming one of the brightest motifs in the human tapestry.

RACE OR ETHNICITY: WHAT'S THE DIFFERENCE?

Race is often used interchangeably with *ethnicity,* but the two terms have very different meanings, and one must be careful in choosing between them. **Race** can be understood as a genetically significant difference among human populations. A few biologists today support the view that human populations do form racially distinct groupings, arguing that race explains phenomena such as the susceptibility to certain

> **race** A classification system that is sometimes understood as arising from genetically significant differences among human populations, or from visible differences in human physiognomy, or as a social construction that varies across time and space.

diseases. In contrast, many social scientists (and many biologists as well) have noted the fluidity of definitions of *race* across time and space, suggesting that race is a social construct rather than a biological fact (**Figures 5.3** and **5.4**, page 180).

Because race is a social construction, it takes different forms in different places and times. In the United States, for example, the so-called one-drop rule meant that anyone with African American ancestry at all was considered black. This law was intended to prevent interracial marriage. It also meant that moving out of the category "black" was, and still is today, extremely difficult because one's racial status is determined by ancestry. Yet, the notion that "black blood" somehow makes a person completely black is challenged by the growing numbers of young people with diverse racial

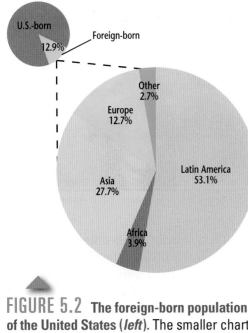

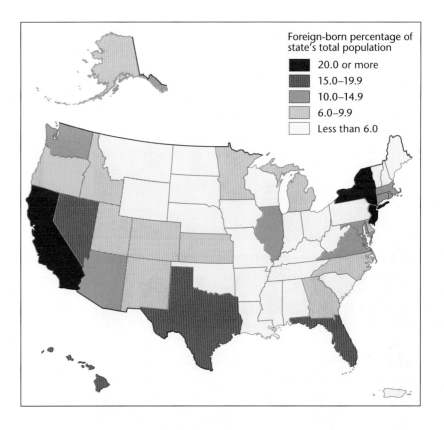

FIGURE 5.2 **The foreign-born population of the United States (*left*).** The smaller chart shows that 12.9 percent, or 40 million, of the total U.S. population in 2010 was born abroad. The larger chart shows that the majority of these people came from Latin America and Asia. **The foreign-born population by state and Puerto Rico (*right*).** This map shows where foreign-born people reside throughout the United States and Puerto Rico. *(Sources: Grieco and Trevelyan, 2010; U.S. Census Bureau, American Community Survey, 2009; Puerto Rico Community Survey, 2009.)*

backgrounds. The rise in interracial marriages in the United States, along with the fact that—for the first time in 2000—one could declare multiple races on the census form, means that more and more people identify themselves as "racially mixed" or "biracial" instead of feeling they must choose only one facet of their ancestry as their sole identity. Thus, golfer Tiger Woods calls himself a "Cablinasian" to describe his mixed Caucasian, black, American Indian, and Asian background. Race and ethnicity arise from multiple sources: your own definition of yourself; the way others see you; and the way society treats you, particularly in a legal context. All three of these are subject to change over time. Still, President Barack Obama, whose mixed-race ancestry has led him to identify with both his black and his white relatives,

has struggled during his two terms in office with claims that he is at once "too black" and "not black enough" to be a representative president.

In Brazil, by contrast, a range of physiognomic features, such as skin pigmentation, eye color, and hair texture, are used to identify a person racially, with the result that many racial categories exist in Brazil. Siblings are frequently classified in significantly different racial terms depending on their appearance. The same person can be put into multiple racial categories by different people, and individuals' own racial self-designations can depend on such variables as their mood at the time they are asked. Increased economic or educational status can "whiten" individuals formerly classified as black. However, one must keep in mind that Brazil was also the last country in the Americas to abolish slavery, and discrimination against darker-skinned Brazilians is still common today.

Studies of genetic variation have demonstrated that there is far more variability within so-called racial groups than between them, which has led most scholars to believe that all human beings are, genetically speaking, members of only one race:

White   Black   Hispanic   Asian

FIGURE 5.4 **Mug shots of people from different races.** Phenotype variations—visible bodily differences such as facial features, skin color, and hair texture—are considered indicative of "race" by the U.S. Federal Bureau of Investigation (FBI). "Race" is one of several visible characteristics—which also include tattoos, scars, height, and weight—and is often used by law enforcement to identify suspects. These are photos from the FBI's 10 Most Wanted list. *How is this idea of race as defined by visible differences different from race as it is commonly understood in the United States? Do you see any similarities between this image and Figure 5.3?* *(Courtesy of the Federal Bureau of Investigation.)*

FIGURE 5.3 **Castas painting.** These paintings were common in colonial Mexico. They depict the myriad racial and ethnic combinations perceived to arise from intermixing among Europeans, indigenous peoples, and African slaves in the New World. They provided a way for those in power to keep track of the confusing racial hierarchy they had created. *(Schalkwijk/Art Resource, New York.)*

Homo sapiens sapiens. In fact, some social scientists have dropped the term race altogether in favor of *ethnicity.* This is not to say, however, that **racism,** the belief that human capabilities are determined by racial classification and that some races are superior to others, does not exist. In this chapter, we use the verb *racialize* to refer to the processes whereby these socially constructed differences are understood— usually, but not always, by the

> **racism** The belief that human capabilities are determined by racial classification and that some races are superior to others.

powerful majority—to be impervious to assimilation. Across the world, hatreds based in racism are at the root of the most incendiary conflicts imaginable (see Subject to Debate, page 181).

What exactly is an ethnic group? The word *ethnic* is derived from the Greek word *ethnos,* meaning a "people" or "nation," but that definition is too broad. For our purposes, an **ethnic group** consists of people of common ancestry and cultural tradition. A strong feeling of group identity characterizes ethnicity. Membership in an ethnic group is largely involuntary, in the sense that a person cannot simply decide to join; instead, he or she must be born into the group. In some cases, outsiders can join an ethnic group by marriage or adoption.

> **ethnic group** A group of people who share a common ancestry and cultural tradition, often living as a minority group in a larger society.

Different ethnic groups may base their identities on different traits. For some, such as Jews, ethnicity

SUBJECT TO *Debate*

RACISM: An Embarrassment of the Past, or Here to Stay?

Most modern societies are quick to claim that they have moved beyond the scourge of racism. Brazil, for example, bills itself as a "racial democracy." Swedes see theirs as a nation that is particularly tolerant in matters of race. And in the "land of opportunity"—the United States—privileges are supposedly extended to all, regardless of race, creed, color, and so forth.

Yet, we know through our day-to-day experiences of studying, working, and socializing in specific places that although "post-racism" may be the ideal of modern societies, it is not always the fact in real life. In France in 2005, for instance, working-class Paris suburbs such as Clichy-sous-Bois—home to mostly Muslim immigrants from North Africa—became the site of rioting by youths who felt persecuted and excluded from French society because of their religion, culture, skin color, and immigrant status. A group of teenage boys who had been playing soccer were returning to the housing projects where they lived when they spotted the police. Thinking that they were being chased in relation to a reported burglary at a construction site, three of the teenagers hid in a power substation; two of them were accidentally electrocuted. This incident tapped into a well of unrest over what many see as the systematic exclusion in France of those with Arabic- or African-sounding names and darker skins, exclusion that has led to high levels of unemployment among *banlieue* (the French equivalent of a ghetto) residents. Months of rioting ensued throughout France.

The following year in the United States, widespread immigration protests filled city streets across the nation. In the spring of 2006, an estimated half a million people hit the streets of Los Angeles, Chicago, and Washington, D.C., and groups numbering in the tens of thousands marched in Denver, Milwaukee, Charlotte, Atlanta, Phoenix, and many other cities across the nation. Hispanics and their supporters took to the streets to protest legislation that threatened to deport undocumented workers and to classify anyone who helped an undocumented individual as a felon. As in the French incidents, Hispanic immigrants to the United States said they felt discriminated against and excluded from society. Even those who merely "look" or "sound" Hispanic are often the targets of racism. Rather than being violent terrorists or drains on society, protesters argued that Hispanics are, in fact, the backbone of low-wage labor in the United States.

These incidents highlight the fact that race-related tensions are about not only ancestry or skin pigmentation but also religious differences, immigration, and social opportunities. As anxieties arise over ever-higher levels of contact in a globalizing world, racialized intolerance is all too often the response. Indeed, A. Sivanandan has argued that "poverty is the new black" in a globalizing world that has become increasingly hostile toward involuntary migrants regardless of their skin color.

Continuing the Debate

Possible responses to racism lie along a continuum, from violent protest to waiting for time to erase racism. Consider these questions:

- Is racism really a serious problem, or are minority groups simply agitating as they always have?

- Will racism finally go away by itself as societies modernize and move beyond intolerance?

- Can legislation adequately and appropriately address the fundamental causes of racism, or are more radical measures—protests or work stoppages, for example—necessary given the deeply rooted nature of race-based discrimination?

Protesters numbering around half a million rally for immigrant rights in Los Angeles, part of a wave of protests occurring throughout U.S. cities in the spring of 2006. *(J. Emilio Flores/Getty Images.)*

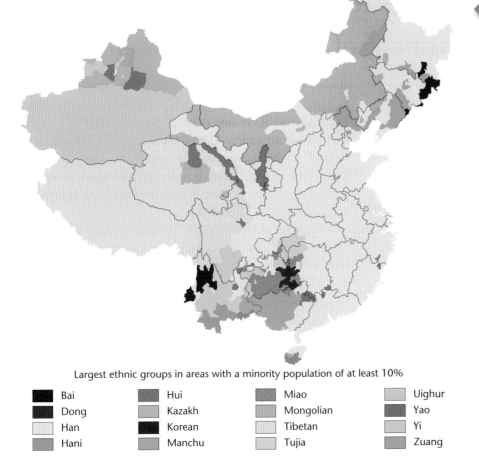

FIGURE 5.5 **Ethnic minorities in China.** Han Chinese are the dominant ethnic group, comprising about 92 percent of China's population. In fact, 1 out of 5 of Earth's inhabitants are Han Chinese, making this the world's largest ethnic group. *Why are China's ethnic minority populations concentrated in sparsely populated peripheries of the country? (Source: Adapted from Hsieh and Hsieh, 1995.)*

Largest ethnic groups in areas with a minority population of at least 10%

■	Bai	■	Hui	■	Miao	■	Uighur
■	Dong	■	Kazakh	■	Mongolian	■	Yao
■	Han	■	Korean	■	Tibetan	■	Yi
■	Hani	■	Manchu	■	Tujia	■	Zuang

primarily means religion; for the Amish, it is both folk culture and religion; for Swiss Americans, it is national origin; for German Americans, it is ancestral language; for African Americans, it is a shared history stemming from slavery. Religion, language, folk culture, history, and place of origin can all help provide the basis of the sense of "we-ness" that underlies ethnicity.

As with race, ethnicity is a notion that is at once vexingly vague yet hugely powerful (see Patricia's Notebook, page 183). The boundaries of ethnicity are often fuzzy and shift over time. Moreover, ethnicity often serves to mark minority groups as different, yet majority groups also have an ethnicity, which may be based on a common heritage, language, religion, or culture. Indeed, an important aspect of being a member of the majority is the ability to decide how, when, and even if one's ethnicity forms an overt aspect of one's identity. Finally, some scholars question whether ethnicity even exists as intrinsic qualities of a group, suggesting instead that the

perception of ethnic difference arises only through contact and interaction.

Apropos of this last point is the distinction between *immigrant* and *indigenous* (sometimes called *aboriginal*) groups. Many, if not most, ethnic groups around the world originated when they migrated from their native lands and settled in a new country. In their old home, they often belonged to the host culture and were not ethnic, but when they were transplanted by relocation diffusion to a foreign land, they simultaneously became a minority and ethnic. Han Chinese are not considered ethnic in China (**Figure 5.5**), but if they come to North America, they are. Indigenous ethnic groups that continue to live in their ancient homes become ethnic when they are absorbed into larger political states. The Navajo, for example, reside on their traditional and ancient lands and became ethnic only when the United States annexed their territory. The same is true of Mexicans, who, long resident in what is today the southwestern United States, found themselves labeled ethnic

Patricia's Notebook

Teaching Race and Ethnicity

Patricia Price. *(American Council on Education.)*

In my own research, I explore the changing geographies of race in the United States. In particular, I examine new immigration flows and cultural debates about immigration, race, and Latinos that have arisen in response. There is no question that Latino immigrants, particularly those who are poorest and brownest, are treated as a racial group in this country. True, the U.S. Census recognizes that "Latino" or "Hispanic" is an ethnicity and that those who identify as Latino or Hispanic can be of any (or mixed) race. But in practice most Americans, in their day-to-day engagement with race, view Latinos as one "point" in what David Hollinger has termed the ethno-racial pentagon: black, white, red, yellow, and brown.

We academics tend to teach what we research, and so I teach a great deal on race and ethnicity. Despite my affection for the topic, however, I'm the first to admit that it's hard to think of a subject that is more controversial or more difficult to teach. Students are often angry or confused about race, based on their own, often bitter, experiences. Race and ethnicity are hard to define, and their borders seem to be perennially under siege. Are Hispanics a separate race? Can I change my ethnicity if I wish to do so? Is a meaningful conversation about race even possible? These are just some of the questions my students ask.

Regardless of these questions' difficulty, it is important for students and professors to keep learning about and researching race and ethnicity. Racism is one of the most important social justice issues of our time. Avoiding candid discussions about racial and ethnic difference may make for a more congenial classroom climate in the short run, but it almost certainly stymies engagement that might constructively approach and transform the injustices of racism on our campuses, in our communities, and across the country.

Geography itself has been transformed thanks in part to our long and sometimes difficult conversations about race and other differences: sexuality, gender, and (dis)ability among them. Today, these differences are the subject of some of the most exciting research in the discipline, whereas 20 years ago there was little research if any at all on sexuality, gender, race and racism, and (dis)ability. Who geographers are has changed as well, and there are more "minority"—racial, sexual, physical

Anti-immigrant rally in Washington, D.C. This protest was organized by the Minutemen Project. Named after the American Revolutionary militiamen, this activist group attempts to police the border between the United States and Mexico. The Minutemen and their supporters view this border as a source of drugs, crime, and undocumented immigration into the United States. ***Critics claim that Mexicans and Mexican Americans are stigmatized, discriminated against, and persecuted by groups like this. What do you think?*** *(Paul J. Richards/AFP/Getty Images.)*

minorities when the border was moved northward after the 1848 U.S.–Mexican War. "We did not jump the border, the border jumped us!" is a common local comeback to this sudden shift in ethnic status.

This is not to say that ethnic minorities remain unchanged by their host culture. **Acculturation** often occurs, meaning that the ethnic group adopts enough of the ways of the host society to be able to function economically and socially. Stronger still is **assimilation,** which implies a complete blending with the host culture and may involve the loss of many distinctive ethnic traits. Intermarriage is perhaps the most effective way of encouraging assimilation. Many students of American culture long assumed that all ethnic groups would eventually be assimilated into the American melting pot, but relatively few have been; instead, they use acculturation as their way of survival. The past three decades, in fact, have witnessed a resurgence of ethnic identity across the globe (see Subject to Debate, page 181). Perhaps then **transculturation,** which is the notion that people adopt elements of other cultures as well as contributing elements of their own culture, thereby transforming both cultures, is a better way to view the interaction among cultures.

Ethnic geography is the study of the spatial aspects of ethnicity. Ethnic groups are the keepers of distinctive cultural traditions and the focal points of various kinds of social interaction. They are the basis of not only group identity but also friendships, marriage partners, recreational outlets, business success, and political power bases. These interactions can offer cultural security and reinforcement of tradition. Ethnic groups often practice unique adaptive strategies and usually occupy clearly defined areas, whether rural or urban. In other words, the study of ethnicity has built-in geographical dimensions. The geography of race is a related field of study that focuses on the spatial aspects of how race is socially constructed and negotiated. Cultural geographers who study race and ethnicity tend to always have an eye on the larger economic, political,

> **acculturation**
> An ethnic group's adoption of enough of the ways of the host society to be able to function economically and socially.
>
> **assimilation**
> The complete blending of an ethnic group into the host society, resulting in the loss of all distinctive ethnic traits.
>
> **transculturation**
> The notion that people adopt elements of other cultures as well as contributing elements of their own culture, thereby transforming both cultures.

environmental, and social power relations at work. This chapter draws on insights and examples from both ethnic geography and the geography of race.

REGION

5.1 LEARNING OBJECTIVE
Describe how ethnic groups are distributed geographically, and discuss how and why these distributions have changed over time.

Formal ethnic culture regions exist in most countries (see Figure 5.5, page 182). Ethnic and racialized minority populations tend to cluster in particular spatial patterns. These tend to shift over time in response to changes in policies, attitudes, and the changing cultural boundaries of racial and ethnic difference.

ETHNIC HOMELANDS AND ISLANDS

There are four types of ethnic culture regions: the rural ethnic homelands and islands, and urban ethnic neighborhoods and ghettos. The difference between ethnic homelands and islands is their size, in terms of both area and population. Rural **ethnic homelands** cover large areas, often have overlapping municipal borders, and include sizable populations. Because of their size, the age of their inhabitants, and their geographical segregation, they tend to reinforce ethnic identities. The residents of homelands typically seek or enjoy some measure of political autonomy or self-rule. Homeland populations usually exhibit a strong sense of attachment to the region. Most homelands belong to indigenous ethnic groups and include special, venerated places that serve to symbolize and celebrate the region—shrines to the special identity of the ethnic group. In its fully developed form, the homeland represents that most powerful of geographical entities, one combining the attributes of both formal and functional culture

> **ethnic homelands**
> Sizable areas inhabited by an ethnic minority that exhibits a strong sense of attachment to the region and often exercises some measure of political and social control over it.

ethnic islands Small
ethnic areas in the rural
countryside; sometimes
called folk islands.

regions. By contrast, **ethnic islands** are small dots in the countryside, usually occupying an area smaller than a county and serving as home to several hundred to several thousand people (at most). Because of their small size and isolation, they do not exert as powerful an influence as homelands do.

North America includes a number of viable ethnic homelands (**Figure 5.6**), including Acadiana,

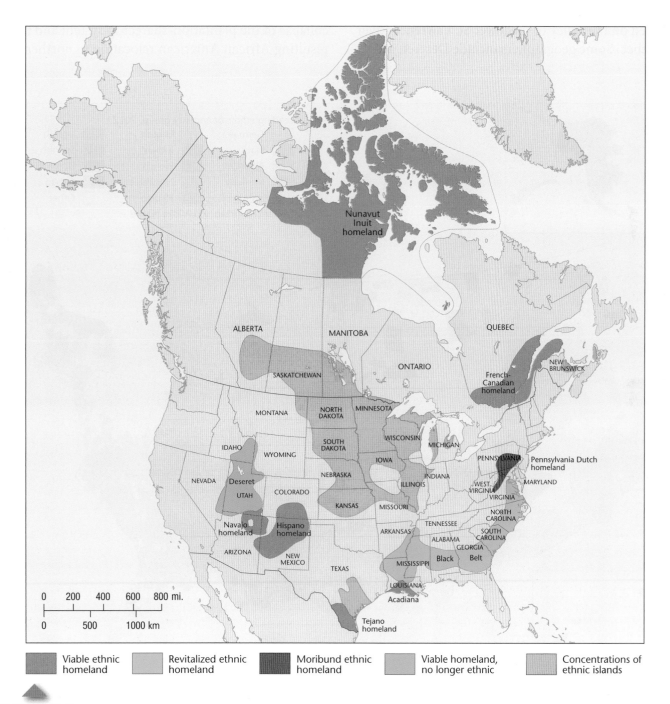

FIGURE 5.6 **Selected ethnic homelands in North America, past and present, and concentrations of rural ethnic islands.**
The Hispano homeland is also referred to as the Spanish American homeland. With the return migration of African Americans from northern industrial cities such as Chicago to rural southern areas, the long-moribund Black Belt homeland appears to be undergoing a revitalization. *(Sources: Arreola, 2002; Carlson, 1990; Nostrand and Estaville, 2001; U.S. Bureau of the Census, 2010.)*

the Louisiana French homeland now increasingly identified with the Cajun people and also recognized as a vernacular region; the Hispano or Spanish American homeland of highland New Mexico and Colorado; the Tejano homeland of south Texas; the Navajo reservation homeland in Arizona, Utah, and New Mexico; and the French Canadian homeland centered on the valley of the lower St. Lawrence River in Québec. Some geographers include Deseret, a Mormon homeland in the Great Basin of the intermontane West. Some ethnic homelands have experienced decline and decay. These include the Pennsylvania Dutch homeland, which has been weakened to the point of extinction by assimilation. Other homelands, notably the southern Black Belt, was once moribund, having been diminished by the collapse of the plantation-sharecrop system and the resulting African American relocation to northern

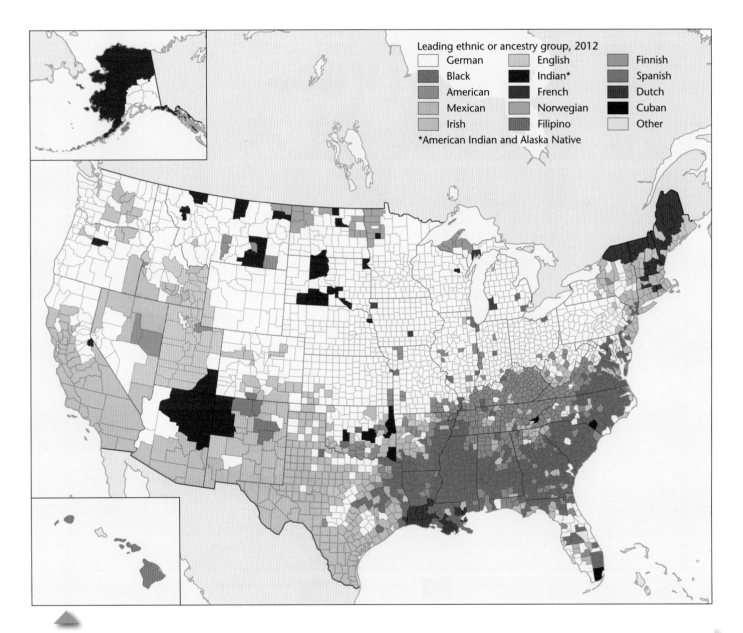

FIGURE 5.7 **Ethnic and national-origin groups in the United States.** Several ethnic homelands appear on this map, as do many ethnic islands. *(U.S. Census Bureau, 2008–2012 American Community Survey 5-Year Estimates (B02001, Race; B02006, Asian Alone by Selected Groups; B03001, Hispanic or Latino Origin by Specific Origin; B04003, Total Ancestry Reported). Copyright © 2014 by Jon T. Kilpinen. Kilpinen, Jon T., 2014. "Leading Ethnic or Ancestry Group, 2012." United States Map Gallery. Map 01.01. http://scholar.valpo.edu/usmaps/1.)*

urban areas. But, it has experienced a revitalization through the recent return migration of African Americans to the South (see also page 196). Nonethnic immigration has diluted the Hispano homeland. At present, the most vigorous ethnic homelands are those of the French Canadians and Mexican Americans in south Texas.

If residents of ethnic homelands assimilate and are absorbed into the host culture, then at the very least, a geographical residue, or **ethnic substrate,** remains. The resulting culture region, though no longer ethnic, nevertheless retains some distinctiveness, whether in local cuisine, dialect, or traditions.

> **ethnic substrate**
> Regional cultural distinctiveness that remains following the assimilation of an ethnic homeland.

Thus, it differs from surrounding regions in a variety of ways. In seeking to explain its distinctiveness, geographers often discover an ancient, vanished ethnicity. For example, the Italian province of Tuscany owes both its name and some of its uniqueness to the Etruscan people, who ceased to be an ethnic group 2000 years ago, when they were absorbed into the Latin-speaking Roman Empire. More recently, the massive German presence in the American heartland (**Figure 5.7**), now largely nonethnic, helped shape the cultural character of the Midwest, which can be said to have a German ethnic substrate.

ETHNIC NEIGHBORHOODS AND RACIALIZED GHETTOS

Formal ethnic culture regions also occur in cities throughout the world, as minority populations initially create, or are consigned to, separate ethnic residential quarters. Two types of urban ethnic culture regions exist. An **ethnic neighborhood** is a voluntary community where people of common ethnicity reside by choice. An ethnic neighborhood has many benefits: common use of a language other than that of the majority culture, nearby kin, stores and services specially tailored to a certain group's tastes, the presence of employment that relies on an ethnically based division of labor, and institutions important to the group—such as churches and

> **ethnic neighborhood**
> A voluntary community where people of like origin reside by choice.

FIGURE 5.8 Miami Beach, Florida, is home to a large community of Orthodox Jews, who constitute a visible ethnic presence in this neighborhood. *(Jeff Greenberg 2 of 6/Alamy.)*

lodges—that remain viable only when a number of people live close enough to participate frequently in their activities. Miami's orthodox Jewish population clusters in Miami Beach–area neighborhoods (**Figure 5.8**) in part because the proximity of synagogues and kosher food establishments makes religious observance far easier than it would be in a neighborhood that did not have a sizable orthodox Jewish population.

The second type of urban ethnic region is a **ghetto.** Historically, the

> **ghetto** Traditionally, an area within a city where an ethnic group lives, either by choice or by force. Today in the United States, the term typically indicates an impoverished African American urban neighborhood.

term dates from thirteenth-century medieval Europe, when Jews lived in segregated, walled communities called ghettos (**Figure 5.9**). Ethnic residential quarters have, in fact, long been a part of urban cultural geography. In ancient times, conquerors often forced the vanquished native population to live in ghettos. Religious minorities usually received similar treatment. Islamic cities, for example, had Christian districts. If one abides by the origin of the term, ghettos need be neither composed of racialized minorities nor impoverished. Typically, however, the term *ghetto* is commonly used in the United States today to signal an impoverished, urban, African American neighborhood. A related term, *barrio*, refers to an impoverished, urban, Hispanic neighborhood. Ghettos and barrios are as much functional culture regions as formal ones.

Coinciding with the urbanization and industrialization of North America, ethnic neighborhoods became typical in the northern United States and in Canada about 1840. Instead of dispersing throughout the residential areas of the city, immigrant groups clustered together in a spatial expression of diaspora cultures, discussed in Chapter 2. To some degree, ethnic groups that migrated to cities came from different parts of Europe than those who settled in rural areas. Whereas Germany and Scandinavia supplied most of the rural settlers, the cities attracted those from Ireland and eastern and southern Europe. Catholic Irish, Italians, and Poles, along with Jews from eastern Europe, became the main urban ethnic groups, although lesser numbers of virtually every nationality in Europe also came to the cities of North America. These groups were later joined by French Canadians, southern blacks, Puerto Ricans, Appalachian whites, Native Americans, Asians, and other non-European groups.

Regardless of their particular history, the neighborhoods created by ethnic migrants tend to be transitory. As a rule, urban ethnic groups remain in neighborhoods while "learning the ropes" as sociologists Alejandro Portes and Rubén Rumbaut put it. As a result, their central-city ethnic neighborhoods experience a life cycle in which one group is replaced by another, later-arriving one. We can see this process in action in the succession of

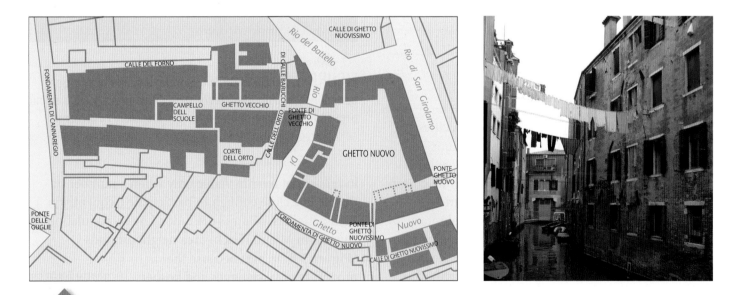

FIGURE 5.9 **Venetian ghetto.** In the sixteenth century, Venice's Jewish population lived in a segregated, walled neighborhood called a ghetto. On the left is a map of this early ghetto. Though most of Venice's Jews do not reside in the ghetto today (*right*), many attend religious services there, and the ghetto continues to lie at the heart of Venetian Jewish life. *(Left: Adapted from the Jewish Museum of Venice; Right: LusoItaly/Alamy.)*

FIGURE 5.10 **Two Chinese ethnic neighborhoods.** The image on the left depicts Los Angeles's traditional urban Chinatown, which is a popular tourist destination as well. The image on the right illustrates the suburban location and flavor of new ethnoburbs. This image is from the San Gabriel Valley, outside Los Angeles. *(Left: Richard Cummins/Robert Harding World Imagery/Corbis; Right: Ron Lim.)*

groups that resided in certain neighborhoods and then moved on to more desirable areas. The list of groups that passed through one Chicago neighborhood from the nineteenth century to the present provides an almost complete history of American migratory patterns. First came the Germans and Irish, who were succeeded by the Greeks, Poles, French Canadians, Czechs, and Russian Jews, who were soon replaced by the Italians. The Italians, in turn, were replaced by Chicanos and a small group of Puerto Ricans. As this succession occurred, the established groups had often attained enough economic and cultural capital to move to new areas of the city. In many cities, established ethnic groups moved to the suburbs. Even when ethnic groups relocated from inner-city neighborhoods to the suburbs, ethnic residential clustering survived (**Figure 5.10**). The San Gabriel Valley, about 20 miles (32 kilometers) from downtown Los Angeles, has developed as a major Chinese suburb. These suburban ethnic neighborhoods, called **ethnoburbs**, can house relatively affluent populations.

> **ethnoburbs** A suburban ethnic neighborhood, sometimes home to relatively affluent immigration populations.

RECENT SHIFTS IN ETHNIC MOSAICS

In the United States, immigration laws have changed during the past half century, shifting in 1965 from the quota system based on national origins to one that allowed a certain number of immigrants from the Eastern and Western hemispheres, as well as giving preference to certain categories of migrants, such as family members of those already residing in the United States. These changes, and the rising levels of undocumented immigration, have led to a growing ethnic variety in North American cities.

Asia and Latin America, rather than Europe, are now the principal source of immigrants to North America. In 2010 Mexico alone supplied 23.7 percent of immigrants to the United States; China, the Philippines, India, and Vietnam were next, with a combined total of 15.9 percent. Asian and Pacific Islanders are now the fastest-growing group in the United States, projected to grow from only 5.6 percent of the U.S. population today to 10 percent by 2050. People of Japanese ancestry form the largest national-origin group in Hawaii, and many West Coast cities, from Vancouver to San Diego, have very sizable Asian populations. Vancouver, already 11 percent Asian in 1981, has since absorbed many more Asian immigrants, particularly from Hong Kong, which again became part of China in 1997.

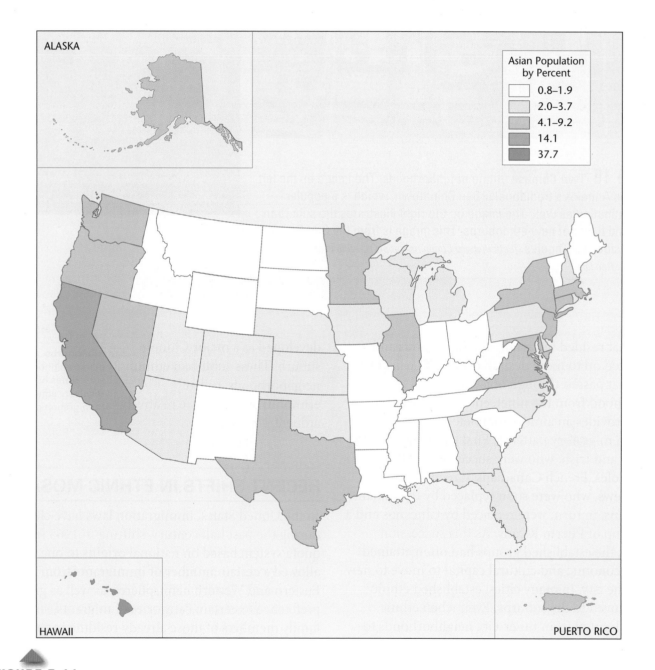

FIGURE 5.11 **Asian population by state.** People indicating "Asian" alone as a percentage of the total population by state. (*Source: U.S. Bureau of the Census, 2010.*)

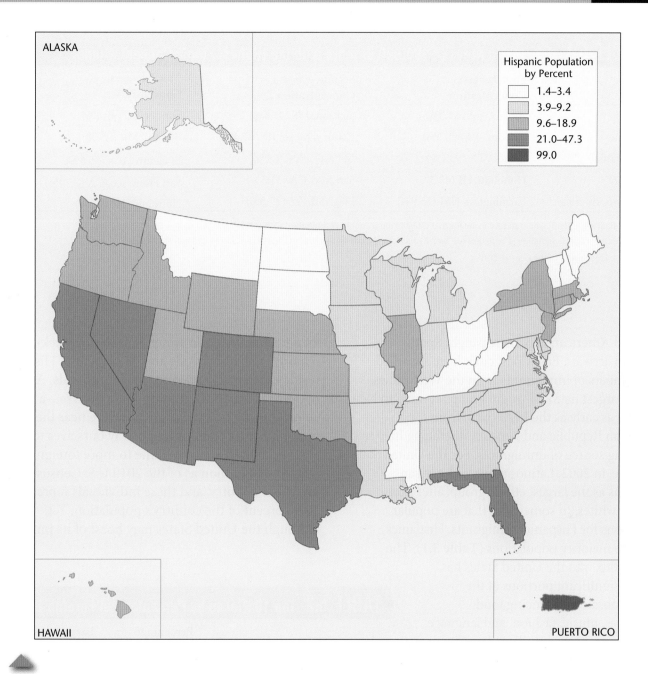

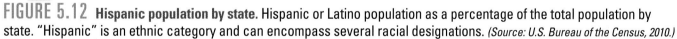

FIGURE 5.12 **Hispanic population by state.** Hispanic or Latino population as a percentage of the total population by state. "Hispanic" is an ethnic category and can encompass several racial designations. *(Source: U.S. Bureau of the Census, 2010.)*

Canadian statistical data show that, in 2012, Chinese were by far the largest non-European ethnic group in Vancouver, accounting for 1 out of every 5 Vancouverites. In the United States, the West Coast is home to about 40 percent of the Asian population, mostly in California, whereas the urban corridor that stretches from New York City to Boston houses another concentration (Figure 5.11). Spatially speaking, Asians are less segregated than African American or Hispanic populations.

Latin America, including the Caribbean countries, has also surpassed Europe as a source of immigrants to North America (Figure 5.12). The two largest national-origin groups of Hispanic immigrants—Mexicans and Cubans—are still the most prevalent, but they have been supplemented since the 1980s by waves

TABLE 5.1 U.S. Metro Areas of 100,000 or more with the Highest Ethnic Concentrations*

	First-Largest Concentration	Second-Largest Concentration	Third-Largest Concentration
Black	Albany, GA (53.3%)	Jackson, MS (47.4%)	Sumter, SC (47.2%)
White	Altoona, PA (96.3%)	Coeur d'Alene, ID (95.4%)	Bangor, ME (95.3%)
Hispanic	Laredo, TX (93.1%)	El Paso, TX (79.9%)	El Centro, CA (66.6%)
Asian	Honolulu, HI (43.5%)	San Jose, CA (31.3%)	San Francisco, CA (23.4%)
Native American	Farmington, NM (35.9%)	Flagstaff, AZ (27.5%)	Honolulu, HI (9.3%)

*The figures for black, white, and Asian are for that category alone, not in combination with other races. The figures for Hispanic can include any racial designation. The figures for Native American include Native American, Alaskan Native, and native Hawaiian.

Source: U.S. Bureau of the Census, 2010.

of Central American and, increasingly, South American arrivals. East Coast cities have absorbed large numbers of immigrants from the West Indies. The two largest national-origin groups coming to New York City as early as the 1970s were from the Dominican Republic and Jamaica, displacing Italy as the leading source of immigrants. For the United States as a whole, in 2002, Latinos surpassed African Americans as the largest ethnic group, after non-Hispanic whites. In some cities that are popular destinations for Hispanic immigrants, Hispanics constitute majority populations (Table 5.1). The Latino influx into the United States has rewoven significant portions of the cultural mosaic, influencing food preferences, music, fashion, and language.

As demographer William Frey has noted, ethnic populations continue to concentrate in established ports of entry, most of them in coastal locations. There is a large swath of the United States that has remained largely non-Hispanic white (Figure 5.13). Figure 5.14 shows that black Americans, too, remain spatially concentrated (see also Figure 2.20 in Chapter 2 for a map of Native American and Alaska Native population distribution). The 2010 census reported that 79 percent of whites live in white-majority neighborhoods, while 44 percent of African Americans live in black-majority neighborhoods and 45 percent of Hispanics reside in Hispanic-majority neighborhoods. As a nation, we may be becoming more diverse—a more colorful mosaic—but we are nowhere near the melting pot we sometimes portray ourselves to be.

The United States is home to more foreign-born people—40 million as of the 2010 U.S. Census—than any other country, and these individuals represent 12.9 percent of the country's population. Yet, although the United States may boast of its immigrant

TABLE 5.2 Top 10 Places by Percentage of Foreign-Born

Place	Percentage Foreign-Born	Largest Source of Immigrants
Dubai, United Arab Emirates	82	India
Miami, United States	51	Cuba
Toronto, Canada	49	India
Queens, United States	48	China
Amsterdam, The Netherlands	47	Suriname
Muscat, Oman	45	India
Singapore	43	Malaysia
Vancouver, Canada	40	China
Geneva, Switzerland	39	Portugal
Auckland, New Zealand	39	England

Source: Adapted from http://en.wikipedia.org/wiki/Foreign_born.

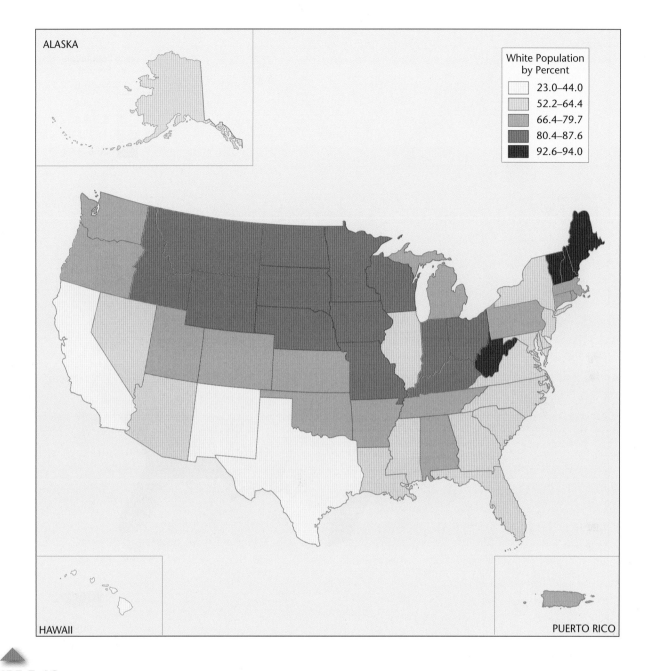

FIGURE 5.13 **White population by state.** People indicating "white" alone as a percentage of the total population by state. *(Source: U.S. Bureau of the Census, 2010.)*

background, other countries are home to higher percentages of the foreign-born. Nineteen percent of Canada's population is foreign-born, as is 38 percent of Israel's. Some of the world's small city-states have even higher percentages of foreign-born populations, with Singapore at 43 percent and the United Arab Emirates at 71 percent foreign-born. Only two U.S. cities—Miami, Florida, and Queens, New York—ranked in the top 10 list of the world's places that have significant proportions of foreign-born residents (Table 5.2). Two of the top 10 places are located in the Middle East, two in Europe, two in Canada, and one each in Asia and the Pacific. In addition, national-origin populations prevalent in the United

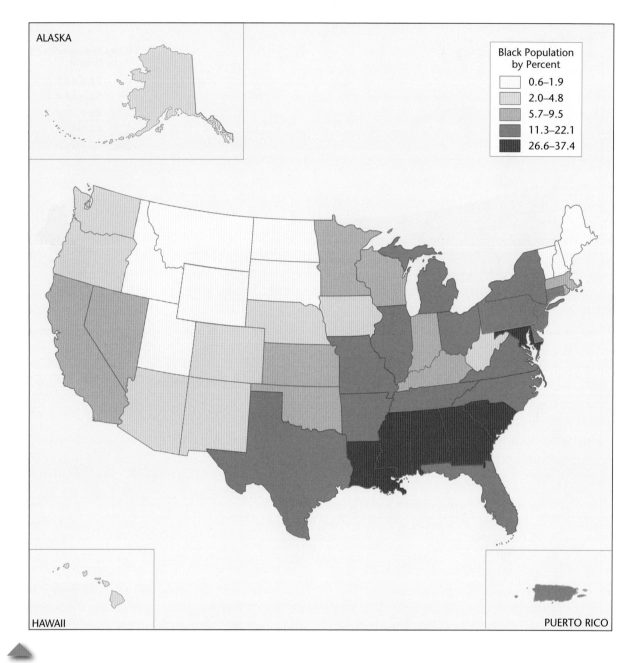

ALASKA

HAWAII

PUERTO RICO

Black Population
by Percent

	0.6–1.9
	2.0–4.8
	5.7–9.5
	11.3–22.1
	26.6–37.4

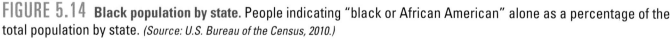

FIGURE 5.14 **Black population by state.** People indicating "black or African American" alone as a percentage of the total population by state. *(Source: U.S. Bureau of the Census, 2010.)*

States are even more prevalent in other countries. Fifty million ethnic Chinese reside outside China and Taiwan. Most of these overseas Chinese live not in North America but in Southeast Asian countries. Thailand counts more than 9 million, Indonesia has nearly that number, and Malaysia more than 7 million. Parts of East Africa have long been home to relatively affluent South Asian Indian populations, whereas Lebanese populations in West Africa enjoy a similarly privileged position. Even European countries such as Germany, the United Kingdom, Italy, and Spain—long known as sources rather than destinations of migrants—today are home to millions of Africans, Turks, and Asians. Immigration-based ethnicity is far from being a phenomenon limited to North America (see Subject to Debate, page 181).

MOBILITY

5.2 LEARNING OBJECTIVE
Define the different types of migration.

As noted in the introduction to this chapter, it is often through migration that a group formerly in the majority becomes, in a new land, different from the mainstream and thus labeled as ethnic or racialized. The many motives for mobility can have different results in terms of where a group chooses to go, which parts of its original culture relocate and which do not, and who may become its neighbors in its new home.

MIGRATION AND ETHNICITY

National Geographic's Genographic Project uses DNA samples from volunteers worldwide to substantiate the claim that all humans descended from a group of Africans who began to migrate out of Africa about 60,000 years ago. The contemporary global pattern of ethnic populations can be traced to their specific journeys. Thus, much of the ethnic pattern in many parts of the world is the result of migration. The migration process itself often creates ethnicity, as people leave places where they belonged to a nonethnic majority and become a minority in a new home. Voluntary migrations have produced much of the ethnic diversity in the United States and Canada, and the involuntary migration of political and economic refugees has always been an important factor in ethnicity worldwide and is becoming ever more so in North America.

chain migration The tendency of people to migrate along routes, over a period of time, from specific source areas to specific destinations.

Chain migration may be involved in relocation diffusion. In chain migration, an individual or small group decides to migrate to a foreign country. This decision typically arises from negative conditions in the home area, such as political persecution or lack of employment, and the perception of better conditions in the receiving country. Often ties between the sending and receiving areas are preexisting, such as those formed when military bases of receiving countries are established in sending countries. The first emigrants, or "innovators," may be natural leaders who influence others, particularly friends and relatives, to accompany them in the migration. The word spreads to nearby communities, and soon a sizable migration is under way from a fairly small district in the source country to a comparably small area or neighborhood in the destination country (**Figure 5.15**, page 196). In village after village, the first emigrants often rank high in the local social order, so that hierarchical diffusion also occurs. That is, the *decision* to migrate spreads by a mixture of hierarchical and contagious diffusion, whereas the actual migration itself represents relocation diffusion. This type of ethnic migration is also channelized. **Channelization** is a process whereby a specific source region becomes linked to a particular destination, so that neighbors in the old place became neighbors in the new place as well.

channelization A process whereby a specific source region becomes linked to a particular destination, so that neighbors in the old place became neighbors in the new place as well.

involuntary migration Also called forced migration, refers to the forced displacement of a population, whether by government policy (such as a resettlement program), warfare or other violence, ethnic cleansing, disease, natural disaster, or enslavement.

Involuntary migration also contributes to ethnic diffusion and the formation of ethnic culture regions. African slavery constituted the most demographically significant involuntary migration in human history (see the World Heritage Site on page 198) and has strongly shaped the ethnic mosaic of the Americas. Refugees from Cambodia and Vietnam created ethnic groups in North America, as did Guatemalans and Salvadorans fleeing political repression in Central America. Following forced migration and initial resettlement, the relocated group often engages in voluntary migration to concentrate in some new locality. Cuban political refugees, scattered widely throughout the United States in the 1960s, reconvened in south Florida, and Vietnamese refugees continue to gather in Southern California, Louisiana, and Texas. This is an example of **step migration** whereby a group proceeds to its final destination via a series of intermediate migrations. Outright warfare can also displace populations, both within the

step migration A process by which a group proceeds to its final destination via a series of intermediate migrations.

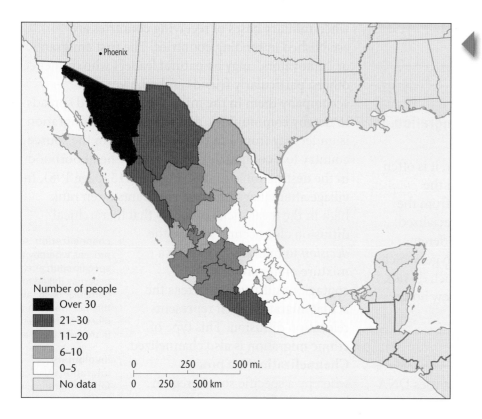

FIGURE 5.15 **Origins of migrants.** This map depicts the Mexican state of origin for the residents of Garfield, an ethnic neighborhood in Phoenix, Arizona, that is composed mostly of Mexican and Mexican American residents. Note that most of the residents trace their homeland to only a few northern Mexican states; immigrants from Mexico's more southern states, such as Oaxaca and Veracruz, are relatively rare. *(Source: Adapted from Oberle and Arreola, 2008.)*

Number of people

- Over 30
- 21–30
- 11–20
- 6–10
- 0–5
- No data

> **internally displaced persons** Persons or groups that have been forced to flee their homes due to conflict, natural disaster, or persecution but who remain within the borders of their home country.

> **refugees** Persons or groups that have been forced to flee their home country due to conflict, natural disaster, or persecution.

> **ethnic cleansing** The removal of unwanted ethnic minority populations from a nation-state through mass killing, deportation, or imprisonment.

country as **internally displaced persons (IDPs)** and beyond its borders, as **refugees**. Syria presents a recent example, with internally displaced persons far outnumbering Syrian refugees fleeing the country's borders (**Figure 5.16**).

Often, such involuntary migrations may result from policies of **ethnic cleansing,** whereby countries expel or massacre minorities in an attempt to produce cultural homogeneity in their populations. In addition, the culturally important sites of the targeted group, such as houses of worship, graveyards, and homes, are destroyed or desecrated. Notable past ethnic cleansings include Spain's expulsion of its Sephardic Jewish population in 1492, several forced removals of Native Americans to reservations in the United States over the course of the eighteenth and nineteenth

centuries, and the murder or expulsion of 1.5 million Kosovar Albanians by the Serbian ethnic majority in the Balkans in the first half of the 1990s.

Return migration represents another type of ethnic mobility and involves the voluntary movement of a group back to its ancestral homeland or native country. The large-scale return since 1975 of African Americans from the cities of the northern and western United States to the Black Belt ethnic homeland in the South is one of the most notable such movements now under way. The 2010 census found that, thanks to migration from cities like Los Angeles and Philadelphia, the South is home to 57 percent of the black population. This proportion is higher than that at any time in the preceding 50 years. Black Americans are returning to the South, particularly to cities like Atlanta, because of greater economic opportunities there than elsewhere in the country. Cultural factors play a role as well, particularly the historic attachment to place and the perception that racial prejudice has lessened in the region. Return migration has recently

> **return migration** A type of ethnic diffusion that involves the voluntary movement of a group of migrants back to their ancestral or native country or homeland.

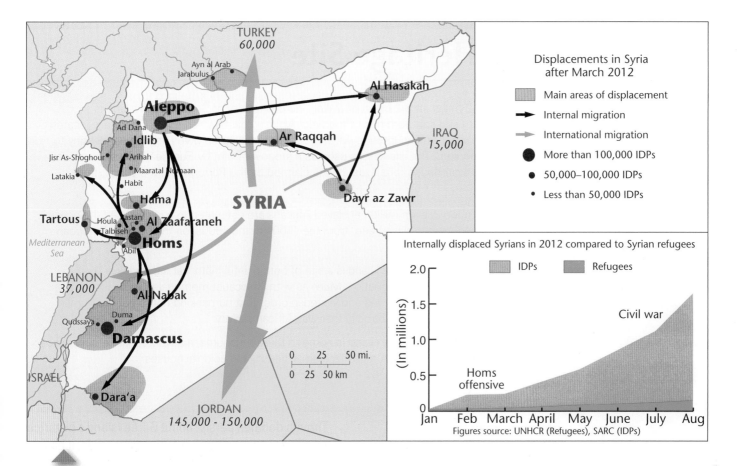

FIGURE 5.16 Syrians displaced by conflict. This map shows the civil war–related displacement of more than 1.5 million Syrians, who have constituted involuntary migrants since March 2012. Some have left Syria and are refugees in neighboring countries. There, they may face discrimination and poverty as unwanted minority populations. Far more are internally displaced peoples (IDPs) who have moved from one place to another within Syria, often ending up in so-called buffer zones established on the borders to protect neighboring states from large flows of Syrian refugees. *(Source: Adapted from Internal Displacement Monitoring Centre, 2012.)*

revitalized the once-moribund Black homeland (see Figure 5.6, page 185).

SIMPLIFICATION AND ISOLATION

When groups migrate and become ethnic in a new land, they have, in theory at least, the potential to introduce the totality of their culture by relocation diffusion. Conceivably, they could reestablish every facet of their traditional way of life in the area where they settle. However, ethnic immigrants never successfully introduce the totality of their culture. Rather, profound **cultural simplification** occurs. This happens, in part, because of chain migration:

only localized fragments of a national culture diffuse overseas, borne by groups from particular places migrating in particular eras. In other words, some simplification occurs at the point of departure. Moreover, far more cultural traits are implanted in the new home than actually survive. Only selected traits are successfully introduced, and others undergo considerable modification before becoming established in the new homeland. In other words, absorbing barriers prevent the diffusion of many traits, and permeable barriers cause changes in many other traits, greatly simplifying the migrant

> **cultural simplification** The process by which immigrant ethnic groups lose certain aspects of their traditional culture in the process of settling overseas, creating a new culture that is less complex than the old.

World Heritage Site

The small island of Gorée is located just 2 miles off the coast of Dakar, the capital city of Senegal. The island is positioned midway between the northern and southern tips of the west African coast, which made it centrally located for the slave trade funneling in from the west African interior. As UNESCO notes, "Gorée owes its singular destiny to the extreme centrality of its geographic positioning." The island was used as a stopover point by Europeans on the long sea voyage down the west African coast and as a warehouse for slaves. It was claimed by the Portuguese, the Dutch, the English, and the French.

▥ Mobility is a central dimension of the slave trade. Eleven million enslaved Africans are estimated to have crossed the ocean over the three centuries encompassing the trans-Atlantic slave trade, from the 1500s to the 1800s, on their way to European colonies in the Americas.

▥ The island is a site of both painful historical memories and reconciliation. Much as with Holocaust memorials, Gorée Island is preserved and visited in order that humankind not forget the human rights atrocity that was African slavery.

▥ The island is home to three museums, a seventeenth-century police station, a castle, a small beach, and colonial houses.

▥ Though debates persist about Gorée Island's historical significance in the trans-Atlantic slave trade, it has become a symbolically important site of remembrance, particularly for African diaspora populations.

▥ The House of Slaves, with its Door of No Return, is a museum visited by over 200,000 tourists annually.

▥ Gorée Island was declared a World Heritage Site in 1978; before that, the colonial government of Senegal had declared it a National History Site in 1944.

www.whc.unesco.org/en/list/26

(Barry Lewis/Alamy.)

Island of Gorée

THE HOUSE OF SLAVES:

Built in the 1770s, this house later belonged to a Senegalese *métis* (mixed-ancestry) woman, Anna Colas Pépin. Scholars believe that though she may have sold small numbers of slaves and kept domestic slaves herself, the actual slave transshipment site was not the house itself, but a location nearby. The house was restored and opened as a museum in 1962. Boubacar Joseph Ndiaye oversaw and curated the museum for 40 years, until his death in 2009.

(SAUL LOEB/AFP/Getty Images.)

■ **The Door of No Return.** The corridor ending in a gate opening onto a wharf from which slave ships departed was known as "the trip from which no one returned." Ndiaye claimed that a total of over 1 million African slaves made their final exit through the doors of this house over the four centuries of the trans-Atlantic slave trade.

■ Some scholars have contested these numbers, claiming that they were, in fact, much smaller, and that Gorée itself was not central to the Senegalese slave trade. This controversy notwithstanding, the island is hugely symbolic and therefore remains meaningful to many.

TOURISM:

Gorée Island attracts so-called heritage tourists, who can be defined as people looking for significant traces of their own past. Indeed, UNESCO refers to it as a "memory island."

■ Gorée Island is a pilgrimage site (see also Chapter 7) for the global African diaspora. Approximately 200,000 tourists visit the House of Slaves annually.

■ Most visitors to Gorée Island, whether or not they identify as descendants of African slaves, report being extremely moved by their visit.

■ The island's only real source of income is tourism, which supports the fewer than 1500 local residents.

■ Because of Gorée Island's historical significance, it has been visited by presidents, prime ministers, and even the pope himself. The Door of No Return is a popular framing for photographs.

(Carlos Mora/Alamy.)

cultures. In addition, choices that did not exist in the old home become available to immigrant ethnic groups. They can borrow novel ways from those they encounter in the new land, invent new techniques better suited to the adopted place, or modify existing approaches as they see fit. Most immigrant ethnic groups resort to all these devices to varying degrees.

The displacement of a group and its relocation to a new homeland can have widely differing results. The degree of isolation an ethnic group experiences in a new home helps determine whether traditional traits will be retained, modified, abandoned, or even adopted by the host culture. If the new settlement area is remote and contacts with outsiders are few, diffusion of traits from the sending area is more likely. Because contacts with groups in the receiving area are rare, little borrowing of traits can occur. Isolated ethnic groups often preserve in archaic form cultural elements that disappear from their ancestral country; that is, they may, in some respects, change less than their kinfolk back in the mother country.

Language and dialects offer some good examples of this preservation of the archaic. Germans living in ethnic islands in the Balkan region of southeastern Europe preserve archaic South German dialects better than do Germans living in Germany itself, and some medieval elements survive in the Spanish spoken in the Hispano homeland of New Mexico. The highland location of Taiwanese aboriginal peoples helped maintain their archaic Formosan language, belonging to the Austronesian family, despite attempts by Chinese conquerors to acculturate them to the Han Chinese culture and language.

GLOBALIZATION

5.3 LEARNING OBJECTIVE
Discuss how globalization might variously deepen, reshape, or erase ethnicity, race, and racism.

The perception of differences among human groups, whether based in visible physical traits, cultural practices, language differences, or religion, is probably as old as human civilization itself. For at least a century, however, the demise of ethnic groups has been predicted. The idea of America as a melting pot has been used to describe the process by which the mixing of diverse peoples would eventually absorb everyone into one American mainstream culture. While the potent forces of globalization certainly play a role in this blending process, the fact that ethnic differences and racial strife still persist—and in some instances have worsened—points to their deeply rooted nature.

A LONG VIEW OF RACE AND ETHNICITY

In keeping with the notion that globalization is a process that has unfolded over several centuries, as opposed to only over the past 40 years or so, it is useful to look at global concepts of ethnicity and race in historical perspective. Long-standing cohesion through shared language, religion, or ethnicity provides group members with the perception of a common history and shared destiny—that deep feeling of "we-ness"—that is the basis of the modern nation-state. Israel's identity as a Jewish nation, Turkey's common language, and Japan's distinctive cultural traditions set the stage for these groups to cohere as political entities.

Yet, all three of these nation-states harbor long-standing tensions among groups that reside within the national borders but who are not considered (or do not consider themselves) part of the cultural mainstream. They may not practice the official religion of the country, as in the case of Israel's Palestinians, who are made up of Muslims and Christians. They may not speak the official language, as with Turkey's Kurdish-speaking minority. Or, they may follow cultural traditions distinct from the mainstream, as is the case with Japan's indigenous Ainu population.

Occupying a minority religious, linguistic, or ethnic position can expose a group to persecution. Violence against racial or ethnic minorities dates back to ancient times. The persecution of Jews, for example, began with pogroms, or widespread anti-Semitic rioting, during the Roman Empire some two millennia ago. Across medieval Europe, violence against Jews was episodic, occurring within Spain in the eleventh century, England in the twelfth century, Germany in the fourteenth century, and Russia in the nineteenth century. Anti-Semitic hate crimes still occur today throughout Europe and the United States. Yet, when Israel was founded in 1948 as the only nation-state to have Judaism as its official religion, non-Jewish minorities were, in turn, marginalized. Even minority Jewish populations, such as the Mizrahim (Arab) Jews residing in Israel, "are excluded and marginalized not only from social and political centers, but . . . from the very definition of what it is to be 'Israeli,'" claims Israeli anthropologist Pnina Motzafi-Haller. Thus, it appears that no place or time is free of tensions based on ethnic or racialized difference.

RACE AND EUROPEAN COLONIZATION

Europe's colonization of vast territories in Africa, Asia, and Latin America was predicated on the drawing of sharp distinctions between the colonizers and the colonized. These distinctions were often depicted in racial terms. Indeed, conquered peoples were viewed as so different from Europeans that their very status as human beings was debated. The famous exchanges in the mid-sixteenth century between Caribbean-based Spanish priests and intellectuals in Spain is a case in point. Observing the abuses of indigenous populations firsthand led to an outcry by these clerics, who argued that the subjugated peoples were not animals and should not be subject to such treatment. "Are these not men? Have they not rational souls? Must you not love them as you love yourselves?" asked Antonio de Montesinos in 1511, in a sermon delivered on the island of Hispaniola (**Figure 5.17**). These same clerics, however, saw no problem with owning black slaves, simply because they did not view them as human beings.

European colonialism frequently drew on existing ethnic and racial cleavages in colonized societies. Rwanda provides a tragic example of this practice. Rwanda is composed of two major ethnic groups: the Hutu, who, at 85 percent of the population, constitute the majority, and the minority Tutsi. Although differences in status have long existed between the

FIGURE 5.17 **La Leyenda Negra.** Illustrations such as this gruesome depiction of Spanish conquistadores hanging and roasting indigenous people, while throwing their babies against a wall, were used to construct La Leyenda Negra, or "The Black Legend." La Leyenda Negra was circulated in the sixteenth century by the Dutch and the British, Spain's religious and colonial rivals. They wished to depict the Spaniards as cruel barbarians who had no moral business building empires in the New World. *(akg-images.)*

two groups, it was not until the Belgian takeover of the territory in 1918 that the distinction between Hutu and Tutsi became racialized. Tutsis were given positions of power in the colonial government, educational system, and economic structure of colonial Rwanda under Belgian rule (**Figure 5.18**). Hutus' physical differences from Tutsis were examined under a European racial lens. The shorter and darker Hutus were considered ugly, intellectually inferior, and natural slaves, whereas the taller, lighter Tutsis were extolled for their physical beauty and virtues of cultural refinement and leadership.

In 1959 the oppressed Hutu majority rebelled, leading to the deaths of some 20,000 Tutsis and the displacement of many more. Rwanda became independent from Belgium in 1961. Displaced Tutsi

FIGURE 5.18 **Identification card of a Rwandan Tutsi.** In Rwanda, cards like this one were used to distinguish racialized Tutsis from Hutus. To show a card with "Tutsi" written on it was tantamount to a death sentence during the Rwandan genocide. Whether to instigate a mandatory national identification card is a hot topic in many nations faced with undocumented immigration today. Because these cards can serve to mark some groups as undesirable in a potentially dangerous way, you can understand why there is usually vigorous opposition to a national identification card. *(Antony Njuguna/Reuters/Landov.)*

refugees regrouped in neighboring Uganda and in 1990 launched an invasion of Rwanda, demanding an end to racial discrimination and a reinstatement of their citizenship. Rapid deterioration of Hutu–Tutsi relations ensued, and plans to exterminate Tutsis and their Hutu sympathizers were promoted as the only way to rid Rwanda of its problems. In 1994 this **genocide** was conducted, resulting in the massacre of as many as 1 million people in the span of a few months (Figure 5.19).

> **genocide** The systematic killing of a racial, ethnic, religious, or linguistic group.

As these examples illustrate, ethnic and racial distinctions provided central features of European colonization and the rise of nation-states. Yet, as we have seen, tensions based on race and ethnicity predate the modern era. Racial and ethnic conflict is far from erased from the face of today's global world (Table 5.3) and may, in fact, become aggravated as the bonds of the nation-state are loosened (see Subject to Debate, page 181).

INDIGENOUS AND MINORITY IDENTITIES IN THE FACE OF GLOBALIZATION

Differences do not always cause conflict, tension, and oppression. Indeed, pride in one's ethnic or racial distinctiveness is common. Yet as revolutions in communications and transportation bring diverse

FIGURE 5.19 **Rwandan genocide.** These skulls are the remains of some 500,00 to 1,000,000 Tutsis, and their Hutu sympathizers, who were massacred over a period of four months in 1994 by Hutu militias. *(Joe McNally/Contributor/Getty Images.)*

TABLE 5.3 10 Largest Ethnocides in the Twentieth Century

Location	Dates	Description	Deaths
Turkey	1915–1918	Armenian genocide	500,000 –1,500,000
Soviet Union	1941–1953	Deportation of nationalities	300,000–600,000
Germany	1939–1945	Genocide of Jews and other Nazi race enemies	5,400,000–6,800,000
Yugoslavia	1941–1945	Ustasha violence against Serbs	350,000–530,000
Eastern Europe	1945–1947	Post–World War II expulsion of ethnic Germans from Poland, Czechoslovakia, Yugoslavia, and elsewhere	2,000,000–2,300,000
India	1947–1948	Partition of India	500,000–1,000,000
Bangladesh	1971	Partition of East Pakistan	500,000–3,000,000
Burundi	1972	Hutu genocide	100,000–200,000
Bosnia-Herzogovina	1990–1995	Ethnic cleansing of Muslims from Bosnia	25,000–155,000
Rwanda	1994	Tutsi genocide	500,000–800,000

Source: Valentino, 2005, p. 77.

peoples into heightened contact with one another the world over, distinctive markers of ethnicity come under pressure. Exposure to seductive U.S.-based cultural practices—in music, cuisine, fashion, lifestyles, and values—is thought to encourage people to shed their distinctive ways in favor of adopting a homogeneous modern, Westernized culture.

Late in the twentieth century, however, an ethnic resurgence, especially among indigenous groups, became evident in many countries around the world. In a very real way, many ethnic groups and their geographical territories have become bulwarks of resistance to globalization. Yet, it is too simplistic to romanticize indigenous groups as simply the keepers of ethnic distinctiveness in a globalizing world (**Figure 5.20**).

Members of indigenous groups must interpret the challenges as well as the opportunities offered by

FIGURE 5.20 **Native Huichol girl.** Most contemporary indigenous peoples do not wear head-to-toe ethnic garb on a daily basis. Those who live in cities or regularly interact with the majority society may adopt Western-style dress full time. Others, perhaps those who live in smaller villages or those who do not interact often with the majority society, may blend indigenous and Western styles. Some indigenous peoples view indigenous clothing as a mark of ethnic pride and wear it daily regardless of their place of residence or level of interaction with the majority society. This Huichol girl blends a baseball cap with traditional dress, itself a blend of European and indigenous materials and styles. The Huicholes are residents of western Mexico. *(Susana Valadez/ Huichol Center.)*

globalization and can be expected to modify their identities in complex ways.

Ethnic minority groups face similar pressures with respect to how their identities are asserted in majority society. This is particularly true of young people. Regardless of ethnicity, this age group is typically concerned with fitting in, which often involves hiding those aspects that mark them as "different" from the mainstream in some way. These concerns are heightened for ethnic minorities and immigrants, whose differences are often markedly apparent in language, dress, religion, and cuisine.

Take the example of Brazil's Japanese population. Numbering around 1.5 million individuals, Brazil is home to the largest Japanese ancestry population outside of Japan. As we previously noted, Brazil was quite late in abolishing slavery in 1888—in fact, it was the last country in the Western Hemisphere to do so. Yet, the need for cheap agricultural labor remained, and in the first decade of the twentieth century, this labor began to be filled by immigrant Japanese. Fleeing poverty in their home country, Japanese immigrants worked Brazil's coffee plantations. This migration reached a crescendo in 1931–1935 when over 70,000 Japanese immigrated to Brazil as a result of the strife of World War I. Around this time the Brazilian government attempted to force its immigrant populations from Germany, Italy, and Japan, as well as Jews, to "whiten" by assimilating into the mainstream through intermarriage and acculturation. They found the Japanese particularly resistant to such efforts, and by the 1970s these immigrants and their Brazilian-born descendants (*Nikkeijin*) formed a distinctive and economically well-off minority group (**Figure 5.21**).

Takeyuki Tsuda's research on Brazil's Nikkejin youth reveals the pressures faced by ethnic minorities. For example, these youths are often stereotyped in racial ways, as being smarter, politer, and neater than non-Japanese Brazilians. One father remarked, "When my son gets good grades in school, they say, 'Of course, it's because he's Japanese.' . . . When he keeps his desk clean in the classroom, they say, 'Of

FIGURE 5.21 **Japanese Brazilians.** These girls, who are of Japanese ancestry, attend a traditional Japanese day of the dead ceremony in Registro, Brazil. Registro is home to one of the oldest Japanese Brazilian communities. *(Yasuyoshi Chiba/AFP/Getty Images.)*

course, he is Japanese.'" To complicate matters even further, many of these Brazilian-born youths and their parents have returned to Japan in order to take advantage of the economic opportunities now available there. When they arrive, these young people face considerable pressure to speak and act Japanese. According to one young man, "The Japanese make more demands on us than on whites or other foreigners in Japan because of our Japanese faces." Tsuda found that many of these young people react by wearing Brazilian clothing, speaking Portuguese loudly in public, and in general acting "more Brazilian" than they ever did in Brazil, as ways of asserting their ethnic identity.

Globalization is far from a straightforward, one-way process when it comes to race, ethnicity, or any other aspect of human geography for that matter. Rather, globalization and its associated movements of people and ideas lead to highly complex geographies of race and ethnicity.

NATURE–CULTURE

5.4 LEARNING OBJECTIVE
Analyze the diverse ways in which ethnic and racial groups interact with the natural environment, and discuss how these groups may be protected, or made more vulnerable.

Ethnicity is very closely linked to the nature–culture theme. The interplay between people and physical environment is often evident in the pattern of ethnic culture regions, in ethnic migration, and in the persistence or even survival of ethnic and racialized groups. Ethnicity and race are sometimes also linked to the environment in harmful ways, particularly when the places they inhabit become the targets of pollution.

CULTURAL PREADAPTATION

For those ethnic groups created by migration or relocation diffusion, the concept of cultural preadaptation provides an interesting approach.

cultural preadaptation A complex of adaptive traits and skills possessed in advance of migration by a group, giving it survival ability and competitive advantage in occupying the new environment.

Cultural preadaptation involves a set of adaptive traits possessed by a group in advance of migration that gives them the ability to survive and a competitive advantage in colonizing the new environment. Most often, preadaptation occurs in groups migrating to a place that is environmentally similar to the one they left behind. The adaptive strategy they had pursued before migration works reasonably well in the new home.

The preadaptation may be accidental, but in some cases, the immigrant ethnic group deliberately chooses a destination area that physically resembles their former home. In Africa, the Bantu expansion mentioned in Chapter 4 was probably initially driven by climate change and expansion of the Sahara (see Figure 4.5, page 144). The Bantu spread south and west in search of forested lands similar to those they had previously inhabited. But, their progress southward was finally inhibited because their agricultural techniques and cattle were not adapted to the drier Mediterranean climate of southern Africa.

Such ethnic niche-filling has continued to the present day. Cubans have clustered in southernmost Florida, the only part of the U.S. mainland to have a tropical savanna climate identical to that in Cuba, and many Vietnamese have settled as fishers on the Gulf of Mexico, especially in Texas and Louisiana, where they could continue their traditional livelihood. Yet, historical and political patterns, as well as the factors driving chain migration discussed earlier, are also at work in these contemporary patterns of ethnic clustering. They act to temper the influence of the physical environment on ethnic residential selection, making it only one of many considerations.

This deliberate site selection by ethnic immigrants represents a rather accurate environmental perception of the new land. As a rule, however, immigrants tend to perceive the ecosystem of their new home as more like that of their abandoned native land than is actually the case. Their perceptions of the new country emphasize the similarities and minimize the differences. Perhaps the search for similarity results from homesickness or an unwillingness to admit that migration has brought them to a largely alien land. Perhaps growing to adulthood in a particular kind of physical environment inhibits one's ability to perceive a different ecosystem accurately. Whatever the reason, the distorted perception occasionally caused problems for ethnic farming groups (for examples, see Chapter 8). A period of trial and error was often necessary to come to terms with the New World environment. Sometimes crops that thrived in the old homeland proved poorly suited to the new setting. In such cases, **cultural maladaptation** is said to occur.

cultural maladaptation Poor or inadequate adaptation that occurs when a group pursues an adaptive strategy that, in the short run, fails to provide the necessities of life or, in the long run, destroys the environment that nourishes it.

HABITAT AND THE PRESERVATION OF DIFFERENCE

Certain habitats may act to shelter and protect ethnically or racially distinct populations. The high altitudes and rugged terrain of many mountainous regions make these areas relatively inaccessible and

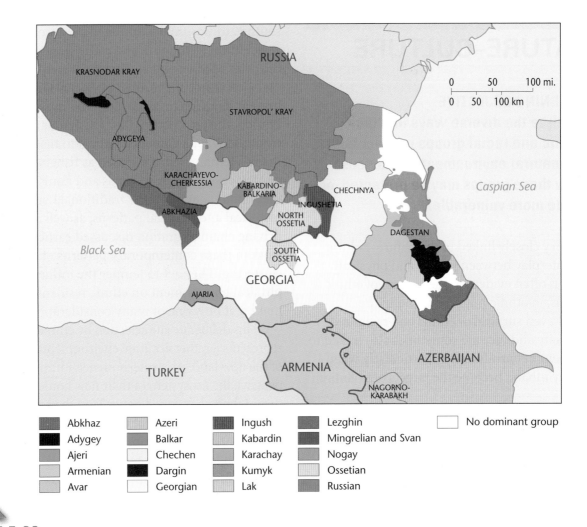

FIGURE 5.22 **Ethnic pluralities in the Caucasus.** The Caucasus Mountains, located in Southwest Asia, are home to one of the world's most ethnically diverse populations. They are also the site of much ethnically based conflict. Because ethnic territories often overlap, this map depicts pluralities. The populations shown comprise at least 40 percent of that place's ethnic population and in most cases also constitute a majority population, although some of the more heterogeneous areas have no dominant ethnic population. The racialized term *Caucasian* is derived from this area's name, although, in fact, it has little to do with the people actually living in the region. *(Source: Adapted from O'Loughlin et al., 2007.)*

thus can provide refuge for minority populations while providing a barrier to outside influences. As we saw in the previous chapter on language, mountain dwellers sometimes speak archaic dialects or preserve their unique tongues thanks to the refuge provided by their habitat. In more general terms, the ways of life—including language—associated with ethnic distinctiveness are often preserved by a mountainous location.

The ethnic patchwork in the Caucasus region, for example, persists thanks in no small measure to the mountainous terrain found there. As **Figure 5.22** shows, distinct groups occupy the rugged landscape of valleys, plateaus, peaks, foothills, steppes, and plains in this region. Religions and languages overlap in complex ways with ethnic identities and are made even more confounding by the geopolitical boundaries in the region. Thus, although this is one of the most ethnically diverse places on Earth, the Caucasus region has seen more than its fair share of conflict as well, much of it ethnically motivated.

Islands, too, can provide a measure of isolation and protection for ethnically or racially distinct groups. The Gullah, or Geechee, people are descendants of African slaves brought to the United States in the eighteenth and nineteenth centuries to work on plantations. Many of their nearly 10,000 descendants today inhabit the coastal islands of South Carolina, Georgia, and north Florida. Their island location has allowed much of the original African cultural roots to be preserved, so much so that the Gullah are sometimes called the most African American community in the United States. Today, as the tourist industry sets it sights on developing these coastal islands, and as Gullah youth migrate elsewhere in search of opportunity, the survival of this unique culture is in question.

ENVIRONMENTAL RACISM

Not only people but also the places where they dwell can be labeled by society as minority. And, it is a fact that, although belonging to a racial or ethnic minority group does not necessarily lead to poverty, the two often go hand-in-hand. In spatial terms, being poor frequently equals having the last and worst choice of where to live. In cities, that means impoverished and racialized minorities often reside in run-down, ecologically precarious, or peripheral places that no one else wants to inhabit (Figure 5.23). In rural areas, racialized and indigenous minorities work the smallest and least fertile lands, or—more commonly—they work the plots of others, having become dispossessed of their own lands.

environmental racism The targeting of areas where ethnic or racial minorities, immigrants, and/or poor people live with respect to environmental contamination or failure to enforce environmental regulations.

Environmental racism refers to the likelihood that a racialized minority population inhabits a polluted area. As Robert Bullard asserts, "Whether by conscious design or institutional neglect, communities of color in urban ghettos, in rural 'poverty pockets,' or on economically impoverished Native American reservations face some of the worst environmental devastation in the nation." In the United States, Hispanics, Native Americans, blacks, and poor people of all races are more likely than wealthier whites to

FIGURE 5.23 **Arab slum in Delhi, India.** This ethnic neighborhood is home to low-income Arabs, who are an ethnic minority population in India. Nearly 50 percent of Delhi's residents live in slums or unauthorized settlements. *(Viviane Dalles/REA/Redux.)*

reside in places where toxic wastes are dumped, polluting industries are located, or environmental legislation is not enforced. Inner-city populations, which in many U.S. cities are disproportionately made up of minorities and the poor, occupy older buildings contaminated with asbestos and toxic lead-based paints. Farmworkers, who in the United States are disproportionately Hispanic, are exposed to high levels of pesticides, fertilizers, and other agricultural chemicals. Native American reservation lands are targeted as possible sites for disposing of nuclear waste.

There is an overlap of nonwhite populations and toxic release facilities that is difficult to explain away as simply a random coincidence (Figure 5.24, page 208). Rather, areas such as the city of South Gate, located in Los Angeles County, have seen contaminating industries and activities purposely located in what is today a neighborhood composed of more than 90 percent Latino residents. In its early days, South Gate was inhabited mostly by white non-Hispanics. From its founding in 1923 until the 1960s, some of the largest U.S. polluters located their operations there: Firestone Tire and Rubber Company, A.R. Maas Chemical Company, Bethlehem Steel, Alcoa Aluminum, and

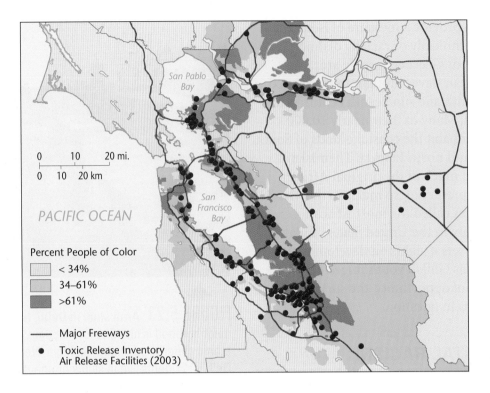

FIGURE 5.24 **Industrial air pollution and minority neighborhoods.** This figure maps the locations of TRI (Toxic Release Inventory) facilities with active air pollution releases in the San Francisco Bay Area, relative to the proportion of residences of people of color. The map clearly shows the close coincidence in space between activities harmful to human health and neighborhoods where one-third or more of the population is Black, Latino, or Asian. In this example, mapping is used to promote environmental social justice.

hundreds of smaller operations were allowed to set up shop in a period when few were aware of the health consequences of their activities. These industries attracted labor arriving from Mexico. During this period, the neighborhood was encircled by four major freeways and a flight path was routed overhead. Pollution resulting from the transportation-related fossil fuels added to the industrial waste in this community. As the deleterious health impacts of these environmental factors became more evident, white residents began to move away, being replaced by Hispanics, some 30 percent of whom are believed to be recent undocumented immigrants. Although many of the industries are now closed, their contaminants remain in the soil and water.

Today, as well, California's cities continue to contaminate in discriminatory ways. A study conducted by Manuel Pastor and his colleagues demonstrated that racialized minorities are far more likely to live near active toxic release facilities that produce air pollution in the San Francisco Bay area (**Figure 5.25**).

Racialized minority, immigrant, and poor populations are not powerless in the face of environmental racism. Organizations such as Communities for a Better Environment work with local residents, such as those living in South Gate, to promote research, legal assistance, and activism. *Environmental justice* is the general term for the movement to redress environmental discrimination.

Environmental racism extends to the global scale. Industrial countries often dispose of their garbage, industrial waste, and other contaminants in poor countries. Industries headquartered in wealthy nations often outsource the dangerous and polluting phases of production to countries that do not have—or do not enforce—strict environmental legislation. Whether areas inhabited by poor, ethnic, or racialized minorities are deliberately targeted for unhealthy conditions, or whether they are simply the

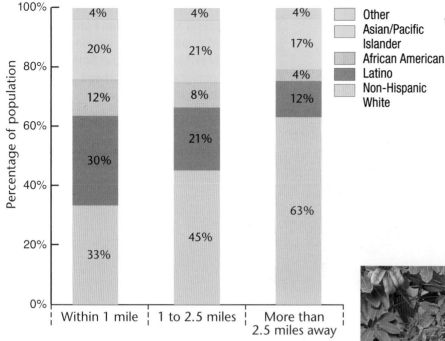

FIGURE 5.25 **Air pollution and environmental racism.** African American and Latino populations are far more likely to live less than 1 mile away from toxic release sites in the San Francisco Bay area. White and Asian populations, by contrast, are more likely to live 2.5 miles or more from these same sites of air pollution emissions. *(Source: Pastor, Morello-Frosch, and Sadd, 2010.)*

unfortunate recipients of the effects of larger detrimental processes, the connections among race, ethnicity, and the environment pose central questions for cultural geographers.

CULTURAL LANDSCAPE

5.5 **LEARNING OBJECTIVE**
Recognize the different ways that ethnicity and race leave imprints on the cultural landscape.

Ethnic and racialized landscapes often differ from mainstream landscapes in the styles of traditional architecture, in the patterns of surveying the land, in the distribution of houses and other buildings, and in the degree to which they "humanize" the land. Often the imprint is subtle, discernible only to those who

ethnic flag A readily visible marker of ethnicity on the landscape.

pause and look closely. Sometimes it is quite striking, flaunted as an **ethnic flag:** a readily visible marker of ethnicity on the landscape that strikes even the untrained eye (**Figure 5.26**).

FIGURE 5.26 **An "ethnic flag" in the cultural landscape.** This maize granary, called a *cuezcomatl*, is unique to the indigenous population of Tlaxcala state, Mexico. The structure holds shelled maize. In Mexico, even the cultivation of maize long remained an indigenous trait because the Spaniards preferred wheat. *(Ryan Watkins/ Photographer's Direct.)*

Sometimes the distinctive markers of race and ethnicity are not visible at all but rather are audible, tactile, olfactory, or—as we explore shortly with culinary landscapes—tasty.

URBAN ETHNIC LANDSCAPES

Ethnic cultural landscapes often appear in urban settings. A fine example is the brightly colored exterior mural typically found in Mexican American ethnic neighborhoods in the southwestern United States (Figure 5.27). These began to appear in the 1960s in Southern California, and they exhibit influences rooted in both the Spanish and the indigenous cultures of Mexico. A wide variety of wall surfaces, from apartment house and store exteriors to bridge abutments, provide the space for this ethnic expression. The subjects portrayed are also wide-ranging, from religious motifs to political ideology, from statements about historical wrongs to those about urban zoning disputes. Often they are specific to the site, incorporating well-known elements of the local landscape and thus heightening the sense of place and ethnic "turf." Inscriptions can be in either Spanish or English, but many Mexican murals do not contain a written message, relying instead on the sharpness of image and vividness of color to make an impression.

Usually, the visual ethnic expression is more subtle. Color alone can connote and reveal ethnicity to the trained eye. Red, for example, is a venerated

and auspicious color to the Chinese, and when they established Chinatowns in Canadian and American cities, red surfaces proliferated (Figure 5.28). Light

FIGURE 5.28 **House in Chinatown.** The Edward Mooney House in the Chinatown neighborhood of Manhattan is painted red: a sacred color for Chinese. *(Lonely Planet Images/Getty Images.)*

FIGURE 5.29 Ethnic storefront in Coney Island, New York. The liberal use of the color green in this storefront design might indicate an Islamic presence. Indeed, this is a Pakistani ethnic grocery. *(© David Grossman/Alamy.)*

FIGURE 5.30 The planned urban landscape of Brasília. As you can see from this aerial view, Brasília is a city of straight roadways, high-rise residential *superquadra,* and green spaces in between. It is a city that was literally built from scratch in an area of cleared jungle in Brazil's Amazonian interior. This allowed the planners to follow a modern urban design blueprint down to the last detail. Residents of Brasília complain that its larger-than-life public spaces, high-speed vehicle traffic, and impersonal housing blocks inhibit socialization and make Brasília a difficult place to live. *(Cassio Vasconcellos/SambaPhoto/Getty Images.)*

blue is a Greek ethnic color, derived from the flag of their ancestral country. Not only that, but Greeks also avoid red, which is perceived as the color of their ancient enemy, the Turks. Green, an Irish Catholic color, also finds favor in Muslim ethnic neighborhoods (**Figure 5.29**) in countries as far-flung as France and China because it is the sacred color of Islam (see Chapter 7).

Indeed, urban ethnic and racial landscapes are visible in cities across the world. This is true even when urban planners try their best to prevent the emergence of such landscapes. When Brazil's capital was moved from Rio de Janeiro to Brasília in 1960, for example, very few pedestrian-friendly public spaces were incorporated into the urban plan of architects Lúcio Costa and Oscar Niemeyer. Rather, residences were concentrated in high-rise apartments called *superquadra;* streets were designed for high-speed motor traffic only; and no smaller plazas, cafés, or sidewalks were included—all of which discouraged informal socialization (**Figure 5.30**). As historian James C. Scott notes in his discussion of Brasília, the lack of human-scale public gathering places was intentional. "Brasília was to be an exemplary city, a center that would transform the lives of the Brazilians who lived there—from their personal habits and

household organization to their social lives, leisure, and work. The goal of making over Brazil and Brazilians necessarily implied a disdain for what Brazil had been." But because the housing needs of those building the city and serving the government workers had not been planned, Brasília soon developed surrounding slums that did not follow an orderly layout. And because the desires of wealthier residents were not met by the uniform *superquadra* apartments, unplanned but luxurious residences and private clubs were also built. By 1980, 75 percent of Brasília's population lived in unanticipated settlements. And because, as previously mentioned, Brazil's social structure is, in part, based on skin pigmentation, it is the poorer and darker-skinned workers who live in the peripheral slums, whereas the wealthier and lighter-skinned residents tend to live in isolated enclaves for the rich. Thus, a racialized landscape of wealth and poverty emerged in Brasília despite all efforts to the contrary.

THE RE-CREATION OF ETHNIC CULTURAL LANDSCAPES

As mentioned previously in this chapter, migration is the principal way in which groups not ethnic in their homelands become ethnic or racialized, through relocation to a new place. As we have seen, ethnic groups may choose to relocate to places that remind them of their old homelands. Once there, they set about re-creating some—although, because of cultural simplification, not all—of their particular landscapes in their new homes.

Cuban immigrants, for example, have reconstructed in Miami aspects of their prerevolutionary Cuban homeland. In the early 1960s, the first wave of Cuban refugees came to the United States on the heels of Castro's communist revolution on the island. Some went north, to places such as Union City, New Jersey, where today there is a sizable Cuban American population. Others, however, came (or eventually relocated) to Miami. Once a predominantly Jewish neighborhood called Riverside, what is today called the Little Havana neighborhood became the heart of early Cuban immigration. Shops along the main thoroughfare, Calle Ocho (or Southwest Eighth Street), reflect the Cuban origins of the residents: grocers such as Sedanos and La Roca cater to the Spanish-inspired culinary traditions of Cuba; cafés selling small cups of strong, sweet *café Cubano* exist on every block; and famed restaurants such as Versailles are social gathering points for Cuban Americans.

Because ethnicity and its expression on the landscape are fluid and ever-changing, the landscape of Calle Ocho reflects the current demographic changes under way in Little Havana. Although the neighborhood is still a Latino enclave, with a Hispanic population of over 95 percent, only half identify as "Cuban." More than one-quarter of Little Havana's residents are recent arrivals from Central American countries, particularly Nicaragua and Honduras, whereas more and more Argentineans and Colombians are arriving—like the Cubans before

them—on the heels of political and economic chaos in their home countries. Today, you are just as likely to see a Nicaraguan *fritanga* restaurant as a Cuban coffee shop along Calle Ocho. This has prompted some to suggest that the neighborhood's name be changed from Little Havana to the Latin Quarter (**Figure 5.31**).

Geographer Christopher Airriess has studied Vietnamese refugees in the United States. These refugees were initially brought to one of four reception centers, and then further relocated to rural and urban locations throughout the country, on the theory that spatial dispersal would hasten their assimilation into mainstream U.S. society. Airriess's work reveals that a process of secondary, chain migration later led many of them to cluster in selected urban areas that offer warm weather and proximity to Vietnamese friends and relatives. Focusing on the Versailles neighborhood of New Orleans, Airriess found that the fact that the neighborhood is surrounded by swamps, canals, or bayous on three sides affords a degree of cultural isolation from other ethnic groups in the city. This has led to Versailles becoming an *ethnic neighborhood* where Vietnamese refugees re-create elements of their home landscape in the United States. The most prominent ethnic flag

FIGURE 5.31 **Restaurant in Little Havana, Miami.** *How many different national groups can you identify in this restaurant façade?* (Courtesy of Patricia L. Price.)

FIGURE 5.32 Lunch truck in New Orleans. *Loncheras,* or mobile lunch trucks, are a common sight in areas of New Orleans that suffered devastation during Hurricane Katrina in 2005. Mexican workers were attracted to New Orleans after 2005 because of opportunities to work in the construction sector as neighborhoods rebuilt. Lunch trucks cater to Mexican immigrant tastes and budgets. *(Russell McCulley/AFP/Getty Images.)*

of the Vietnamese in Versailles is the vegetable gardens found on the perimeter of the neighborhood. Cultural preadaptation has led to the reproduction of Vietnamese rural landscapes here, sometimes with plants sent directly from Vietnam. It is the older Vietnamese who plant and tend the gardens. Such gardening provides a therapeutic activity, allows them to retain traditional dietary habits, enables them to produce folk medicines from plants, and reduces household food expenditures. Airriess notes that "the image of conical-hat-wearing gardeners leaning over plants within a multi-textured and moisture-soaked, vibrant green agricultural scene, coupled with frequent flyovers of military helicopters, is an eerie scene of wartime Vietnam reproduced."

After Hurricane Katrina devastated sections of New Orleans—including Versailles (today more commonly called Village de L'est)—in 2005, the ethnic cultural landscape underwent a transformation. Many Vietnamese immigrants moved away from Versailles after the storm, whereas many Hispanics moved into the area, attracted by the many post-Katrina

construction jobs. A 2006 survey found nearly 7000 Asians in New Orleans after Katrina, compared to nearly 12,000 before the storm. By contrast, Latinos have grown from about 14,000 before the hurricane to 16,000 post-Katrina: the only ethnic group, in fact, to have increased after 2005. Today, the commercial landscape reflects this transition. The first arrivals on the scene were the *loncheras,* or mobile lunch trucks (**Figure 5.32**). Then came the Hispanic-oriented restaurants, supermarkets, and services. At present, the remaining Vietnamese American residents are learning to adapt to the newcomers. Sara Catania recounts how Mai Thi Nguyen, a business development director in Village de L'est, reports that her aunt, a market owner, "is learning Spanish. She's learning to say hello, how to tell customers how much something costs. It's wild. I love it. It's exciting."

ETHNIC CULINARY LANDSCAPES

"Tell me what you eat, and I'll tell you who you are." This often quoted statement arises from the connections between identity and **foodways,** or the customary behaviors associated with food preparation and consumption that vary from place to place and from ethnic group to ethnic group (**Figure 5.33**). As geographers Barbara and James Shortridge point out, "Food is a sensitive indicator of identity and change." Immigration, intermarriage, technological innovation, and the availability of certain ingredients mean that modifications and simplifications of traditional foods are inevitable over time. An examination of culinary cultural landscapes reveals that, although landscapes are typically understood to be visual entities, they can be constructed from the other four senses as well: touch, smell, sound, and—in this case—taste.

> **foodways** Customary behaviors associated with food preparation and consumption.

Although Singapore occupies the southern tip of the Malaysian Peninsula in extreme Southeast Asia, its cuisine draws heavily from southern Chinese cooking. This is because three-quarters of Singapore's multiethnic population is of Chinese ancestry. Intermarriage between Chinese men, who came to Singapore as traders and settled there, and local Malay women resulted in a distinctive spicy cuisine

▲ FIGURE 5.33 **Market in Chinatown, New York City.**
Grocery stores selling distinctive produce items are one of
the most visible landscape markers of an ethnic presence.
Here, spiny fruits called *durian,* also known as "stinky
fruit," hang from bags. *Durian* is sometimes banned from
public places in its native Asia because of its pungent
smell. *(AP Photo/Kathy Willens.)*

called *nonya.* As we have already seen, there are far
more ethnic Chinese in Asia than in other parts of the
world, principally because of China's proximity to
other Asian countries.

The corn tortilla remains a staple of central
Mexican foodways and is consumed at every meal by
some families. But, many middle-class urban women
in Mexico no longer have the time to soak, hull, and
grind corn by hand to make corn tortillas from
scratch. Rather, children are often sent every
afternoon to the corner *tortillería* to purchase *masa,*
or corn dough, or even hot stacks of the finished
product. Convenience notwithstanding, old-timers
complain that the uniform machine-made tortillas
will never come close to the flavor and texture of a
handmade tortilla. In some regions, notably in the
northern part of Mexico and southern Texas, the
labor-intensive corn tortilla was replaced altogether
by wheat tortillas, which are much easier to make.

In the rural, mountainous Appalachian region of
the eastern United States, distinct ethnic foodways
persist and literally flavor this region. Geographer
John Rehder observed that, during and after World
War II, Appalachians maintained their cultural
distinctiveness in the face of migration to northern
cities in part through "care packages" of lard, dried
beans, grits, cornmeal, and other foods unavailable in
the North. Grocery stores in the Little Appalachia
ethnic neighborhoods that formed in cities such as
Detroit, Cleveland, and Chicago soon began to cater
to the distinct food preferences of the population.
Thus, Appalachian foodways were maintained as
Appalachians moved north.

Rehder tells of sending students on a scavenger
hunt during a class field trip to Pikeville, Tennessee.
To one young man, he handed a card that read,
"What are cat head biscuits and sawmill gravy?"

> *About a half an hour later, my "biscuit man"*
> *returned with a scared look on his face and tears welling*
> *up in his eyes. I asked, "Where did you go and what*
> *happened?" He replied, "Well, . . . I . . . went to the*
> *feed store and asked an old man there, 'What are cat*
> *head biscuits and sawmill gravy?' just like you told me*
> *to do. Only the old man just growled at me and said,*
> *'Son, if you don't know, HELL, I ain't goin' to tell you!'*
> *And with that our lad left the feed store dejected and*
> *thoroughly upset. To calm him down, I said,"Why don't*
> *you just go on down to the café. Get a little something to*
> *drink and maybe calmly run your question by them down*
> *there." After a sufficient amount of time, I decided to go*
> *check on him. He was sitting on a red upholstered stool*
> *at the counter. As I drew closer, he looked up and, with*
> *fresh and quite different tears in his eyes, pointed to his*
> *plate: "This," he said proudly, "is cat head biscuits and*
> *sawmill gravy!" (p. 208)*

Rehder explains that cat head biscuits are flour
biscuits made in the exact size and shape of a cat's
head and baked golden in a wood stove. Sawmill
gravy is a southern-style white gravy made in a
cast-iron skillet. Legend has it that the cook at the
Little River Lumber Company logging camp ran out
of flour for the gravy and substituted cornmeal. The
loggers complained about the texture, referring to it
as "sawmill gravy." For further discussion of ethnicity
and food, see Doing Geography (page 215) and
Seeing Geography (page 216).

DOING GEOGRAPHY

Tracing Ethnic Foodways Through Recipes

At some point in our family histories, we all trace our roots back to migrants. Perhaps your ancestors walked here some 30,000 years ago. Perhaps they arrived on slave ships in the seventeenth century. Perhaps they were traders who settled and married locals. Were they part of the waves of Europeans in the eighteenth, nineteenth, and early twentieth centuries, or have you yourself only recently immigrated? More than likely, your ethnic inheritance results from a combination of different immigrant groups. As the geographer Doreen Massey wrote, "In one sense or another most places have been 'meeting places'; even their 'original inhabitants' usually came from somewhere else." In other words, if you dig into the history of any place, you'll find layers upon layers of people arriving from other places and bringing their cultural baggage—recipes and all—with them.

For this exercise, you will analyze one of the most commonplace, yet revealing, items of ethnic geography: a recipe. Certainly, you are what you eat, but you also eat where you are, and the foodways in which you take an active part are very revealing of both who you are and where you are. Even though you may not be conscious of it, the simple act of cooking a meal sets into motion all sorts of cultural geography elements: regional identity, ethnic heritage, place-specific agricultural traditions, and so on. Together, these form important components of identity and place.

Steps to Analyzing a Recipe

Step 1: Identify a recipe to analyze. Most of you grew up or are now living in households where meals are cooked on site at least some of the time. Choose a recipe that is used often and has been around for a while. The best candidate is a favorite family recipe that has been passed down through the generations (see the note below). If you don't have a copy of the recipe, you will need to interview a person who does—if necessary, by phone or e-mail.

Step 2: With a written recipe now in front of you, answer the following questions. You can interview someone in your family about this recipe, too.

- From where does this recipe come? With what country, or region, is it identified?

- Do any of the recipe's ingredients give clues about the origins of the recipe? Do the ingredients draw on particular animal or plant ingredients that are, or were, produced where the recipe originated?

- Has the recipe been modified to substitute ingredients that are no longer available, either because the person who used the recipe migrated or because the ingredients went out of production?

- Are there ingredients used in the recipe that are identified with particular ethnic groups and perhaps aren't consumed by others living in the same place?

- Do elements of the recipe's preparation give additional clues about the foodways of the people who developed the recipe?

- Are there special occasions, such as holidays, when this recipe is always used?

- Do any of the ingredients or methods of preparation have symbolic meanings or stories associated with them?

In sharing the results of the recipe analysis, the class can list the various places and ethnicities that together comprise the foodways of the students. The class might want to organize a potluck meal where students share the dishes they have analyzed. Students may also create a cookbook of their recipes and map the places from which they come.

Note: If you are attending school in a foreign country, use a family recipe from your native homeland. E-mail or call a family member to discuss this recipe using the guidelines for this exercise. Also, some of you may have grown up in an institutional setting, where you consumed food prepared in a cafeteria. Institutional foods can be very revealing of local ethnic influences, so for this exercise choose a dish commonly prepared for the evening meal.

Recipe box. Back when most Americans cooked their meals at home, a personalized recipe collection like this one—written on 3 × 5 cards and safeguarded in a recipe box—held treasured family recipes, some passed down for generations. *(Edward Pond/Masterfile.)*

SEEING GEOGRAPHY   America's Ethnic Foodscape

When does cuisine cease to be ethnic and become simply "American"? What roles do mobility and globalization play in the process?

Ethnic restaurants abound in Manhattan's Greenwich Village neighborhood.
(Steve Lewis Stock/Getty Images.)

This street scene was shot in Manhattan's Greenwich Village neighborhood, but it could have just as easily been taken in any number of major U.S. urban areas. The signs display a mixture of ethnic foods: Pakistani (Tastee Curritos), Indian (Kati Roll), and Italian (Monte's Trattoria) are the three most visible restaurants in the foreground of this photo.

As you know from Chapter 4, New York City is the most linguistically varied area on the face of the planet. This reflects the many and varied streams of immigration to this place. And with language comes food: two of the central aspects of ethnic identity. The ancestors of the owners of Monte's Trattoria may well have arrived in the United States between 1890 and the mid-1920s, the peak years for Italian immigrants. The owners of Tastee Curritos and Kati Roll probably arrived almost a century later. Changes in U.S. immigration laws in 1965 abolished the previous national quota system for immigrants, which had the effect of greatly diversifying the immigrant stream as well as increasing the total numbers of immigrants to America. As you know from reading this chapter, the major areas sending immigrants to the United States are no longer Europe; rather, they are Asia and Latin America.

Cultural geographer Richard Pillsbury famously quipped that America has no foreign food because we have accepted every possible foreign cuisine and made it our own. Americans are for the most part very willing to consume foods from all over the globe, as this image shows. In addition, Tex-Mex, Asian fusion, and many "Chinese" dishes—among them, chop suey, General Tso's chicken, and fortune cookies—are great examples of "foreign" cuisines that have been heavily Americanized, if not invented entirely in the United States.

So what exactly is "American" food, then? Perhaps the squash, beans, and corn consumed by Native Americans are our only example of a cuisine that is truly indigenous to the United States. Or, perhaps American fast food, which has itself diffused across the globe (along with the drive-in landscapes that accompany fast-food restaurants), is our only contemporary American food?

THE VIDEO CONNECTION

An Education in Equality
www.tinyurl.com/qy7nxqv

This video profiles a young African American man, Idris Brewster, and his family over the course of his thirteen years at the Dalton Academy, an elite private college preparatory school in Manhattan. His mother is an immigrant of mixed-race Caribbean ancestry, while his father describes his past as one of economic deprivation. Both utilized education as a ticket to an upper-class lifestyle, and they want their son to use education to excel as well. Idris moves between elite and not-so-elite spaces in his daily life, and in the process he is forging his identity as a black man with a bright future in a world where race still matters.

Thinking Geographically:

1. This video explores the role of education, and its power to elevate an individual's class status. What sorts of expectations, hopes, and pressures do Idris's parents place on his shoulders? How does race matter in this regard?

2. Geographers and others who study race note that so-called micro-aggressions—the slight, everyday indignities directed at people of color that question their legitimacy to occupy certain roles and spaces—are prevalent in the lives of people such as Idris Brewster. Can you find examples of racial micro-aggressions in the video?

3. Idris Brewster appears to lead two lives: one as a student at an elite college prep school, and one as a young black man in Brooklyn. How do those two lives overlap, and what disconnects between them can you identify from the video?

CONCLUSION

The inclusion of a critical focus on race into the long-standing field of ethnic geography provides contemporary, fruitful inroads into understanding human diversity. Here, the five themes of cultural geography have provided new perspectives on ethnicity and race. Ethnicity and race help shape the pattern of rural homelands and islands as well as urban ethnic neighborhoods and ghettos. Immigration is actively reshaping the ethnic mosaic of many places, including the United States. Human mobility routes—whether voluntary or involuntary, undertaken in response to persecution in a home region or opportunities beckoning beyond one's borders—shape the patterns that ethnicity and race assume in the new land. Sometimes similarities between home and host

habitats direct migrants to certain places rather than others. Globalization appears to intensify age-old tendencies to distinguish, and discriminate, among groups. Racism seems particularly resistant to change. Certain natural habitats can protect ethnically distinct groups. Yet, environmental racism illustrates how minority peoples and the places they inhabit can become the targets of undesirable practices, such as pollution. On the flip side, ethnic pride and the active construction of ethnically and racially distinct landscapes are evident throughout the world.

By now, we hope you are beginning to think and see as geographers do. The world is patterned in terms of culture. What these patterns are, why they change, how they change, and how these changes affect people living in places are the basic focus of cultural geography. Ethnicity and race, along with language and religion, lie at the heart of many of today's most pressing political geography questions.

GEOGRAPHY @ WORK

Vivian Gonzalez. *(WSVN Florida.)*

Vivian Gonzalez
Morning Meteorologist, WSVN Channel 7, Miami, Florida

Education:
BA Geography—Florida International University
BS Geoscience—Mississippi State University
Broadcast Meteorology Certificate—Mississippi State University

Q. *Why did you major in geography and decide to pursue a career in the field?*

A. Geography is a broad field of study that encompasses everything from the social, political, economic, and environmental issues applicable to the public and private sectors. It offers a wide array of opportunities that can be applied in any field. I chose the environmental sciences track because ever since I can remember, I was fascinated by the weather and natural elements. Geography helped give me a better understanding of the impact that climate change has on our society and environment. Climate change ultimately leads to weather pattern changes on a global scale and geography gave me the basic skills and understanding needed to work in the public safety sector.

Q. *Please describe your job.*

A. I am the weekday meteorologist for the number one–rated morning show in South Florida called *Today in Florida.*

My main responsibility is to keep the public safe and informed especially during the threat of severe weather. For example, in South Florida, during the summer months, the sea breeze each afternoon aids in the development of showers and storms, and these conditions can develop into severe weather capable of producing strong storms, tornadoes, flooding, damaging wind gusts and hail.

Among other responsibilities, I have to receive and perform analysis on data, specialize and produce around-the-clock live severe weather and hurricane coverage, give radio weather updates, build forecast pages and high-quality graphics for on-air use, write a blog about the current weather situation on a daily basis, use multiple social media platforms such as Facebook and Twitter to stay in constant communication with the public, manage real-time Doppler radar with street-level mapper and storm tracker, perform routine updates on all weather systems to keep them running at peak efficiency. Also, it is my responsibility to stay current with all new meteorological and graphic production software.

Q. *How does your geographical background help you in your day-to-day work?*

A. Having a geographical background has helped me become a better meteorologist. It has given me the basic knowledge necessary to deliver and explain all of the factors that influence our weather from season-to-season and has allowed me to give the public the information they need in an "easy to understand" format. I joined the WSVN 7 News team in the midst of the most active hurricane season on record in 2005. Thirty-one systems were spawned that year, with both Katrina and Wilma impacting South Florida directly. My background in geography was extremely valuable in tracking everything Mother Nature threw our way, as understanding topography played a key role in preparing the public for the intensity of the hurricanes. I feel that knowledgeable presentation and delivery of information are an art form, and to me, an exciting part of my job is coming up with different ways to make the story interesting. This is where my knowledge in geography shines, by combining what happens in the sky and how it affects what's on the ground, all through the visual medium of complex computer graphics fed by raw data.

Q. *What advice do you have for students considering a career in the geography field??*

A. The advice I would give anyone who wants to be a meteorologist or would like to attain a degree in the atmospheric science realm is to pursue a degree in geography. There are multiple career opportunities within the field of study that include job opportunities in the media and even federal, state, or local government. That being said, getting your big break or opportunity in the media industry isn't easy. There are only so many TV stations, and some small cities and rural clusters even outsource their weather programming to a regional service. Most of my colleagues tell me about their transition into the business, including some of the difficulties they encountered before getting their current opportunity. I often hear that rarely does someone start his or her career in a major market like Miami. The norm would be to start in a smaller market, gain experience, and then move your way up. Another great piece of advice is to "never stop learning." I attained my first degree in geography, but I continued to learn by earning my second degree in geosciences, a broadcast certificate in meteorology, and my AMS Seal of Approval.

Chapter 5
LEARNING OBJECTIVES REEXAMINED

5.1 Describe how ethnic groups are distributed geographically, and discuss how and why these distributions have changed over time.
See pages 184–194. Name two regions of the world that account for the main source of immigrants to North America today.

5.2 Define the different types of migration.
See pages 195–200. What are three types of migration?

5.3 Discuss how globalization might variously deepen, reshape, or erase ethnicity, race, and racism.
See pages 200–204. What are two ethnic groups that have been most affected by the forces of globalization?

5.4 Analyze the diverse ways in which ethnic and racial groups interact with the natural environment, and discuss how these groups may be protected, or made more vulnerable.

See pages 205–209. How has the natural environment helped to preserve ethnically or racially distinct populations?

5.5 Recognize the different ways that ethnicity and race leave imprints on the cultural landscape.
See pages 209–214. How do some ethnic groups visually express themselves within the cultural landscape?

KEY TERMS

acculturation	p. 184	ethnic substrate	p. 187
assimiliation	p. 184	ethnoburbs	p. 189
chain migration	p. 195	foodways	p. 213
channelization	p. 195	genocide	p. 202
cultural maladaptation	p. 205	ghetto	p. 187
cultural preadaptation	p. 205	internally displaced	
cultural simplification	p. 197	persons (IDPs)	p. 196
environmental racism	p. 207	involuntary migration	p. 195
ethnic cleansing	p. 196	race	p. 178
ethnic flag	p. 209	racism	p. 180
ethnic group	p. 180	refugees	p. 196
ethnic homelands	p. 184	return migration	p. 196
ethnic islands	p. 185	step migration	p. 195
ethnic neighborhood	p. 187	transculturation	p. 184

Geography of Ethnicity and Race on the Internet

You can learn more about ethnic geography and the geography of race on the Internet at the following web sites:

2010 Census of the United States
http://www.census.gov
Go to the American FactFinder section of the web page. Here, you will find a wonderful selection of maps that show themes (thematic maps) generated from census data. Some of the maps used in this chapter were found here. Data from the 2010 census were released through September 2013.

Guess My Race
This is a mobile device application ("app") for the iPhone, iPad, or iPod. Designed like a quiz, this game shows images of real people, and your task is to determine how each one answered the question "What race are you?" After you guess, you get to find out how each person really answered the question, as well as read a bit more on that person's experiences with race.

New America Media
http://newamericamedia.org/
This rich web site is dedicated to "expanding the news lens through ethnic media." You can select news stories by ethnicity, visit blogs, or view polls designed to capture the opinions of those typically excluded from mainstream surveys.

Racialicious
http://www.racialicious.com
Feel the need to know the top 10 trends in race and pop culture? Want to follow celebrity gaffes, politicians' missteps, and questionable media representations? Log on to Racialicious, a no-holds-barred blog about the intersection of race and pop culture.

Tyler Cowen's Ethnic Dining Guide
http://tylercowensethnicdiningguide.com
Starting with the premise that "all food is ethnic food," Professor Cowen asserts that ethnic restaurants bring together "entrepreneurship, international trade and migration, and cultural exchange." This video-heavy web site features ethnic food–related studies, book reviews, and—of course—dining recommendations.

Sources

Airriess, Christopher. 2002. "Creating Vietnamese Landscapes and Place in New Orleans." In Kate A. Berry and Martha L. Henderson (eds.), *Geographical Identities of Ethnic America: Race, Space, and Place*, pp. 228–254. Reno: University of Nevada Press.

Arreola, Daniel D. 2002. *Tejano South Texas: A Mexican American Cultural Province*. Austin: University of Texas Press.

Associated Press. 2006. "500,000 Rally Immigration Rights in Los Angeles." *MSNBC News,* March 25. http://oceanpark.com/webmuseum/2006/500000_immigration_rights_rally_los_angeles_.

Bhatty, Ayesha. 2012. "Canada Prepares for an Asian Future." *BBC News,* May 25. http://www.bbc.co.uk/news/world-radio-and-tv-18149316.

Bullard, Robert D. (ed.). 1993. *Confronting Environmental Racism: Voices from the Grassroots*. Boston: South End Press.

Carlson, Alvar W. 1990. *The Spanish American Homeland: Four Centuries in New Mexico's Río Arriba*. Baltimore: Johns Hopkins University Press.

Catania, Sara. 2006. "From Fish Sauce to Salsa—New Orleans Vietnamese Adapt to Influx of Latinos." *New American Media,* October 16. http://news.newamericamedia.org/news/view_article.html?article_id=3e0ffe22ee7a7bcbd9a2e1d4fb79f676.

Frey, William H. 2001. "Micro Melting Pots." *American Demographics* (June): 20–23.

Grieco, Elizabeth M., and Edward N. Trevelyan. 2010. *Place of Birth of the Foreign-Born Population: 2009*. American Community Survey Briefs. Washington, D.C.: U.S. Census Bureau. http://www.census.gov/prod/2010pubs/acsbr09-15.pdf.

Hollinger, David. 1995. *Postethnic America: Beyond Multiculturalism*. New York: Basic Books.

Hsieh, Chiao-min, and Jean Kan Hsieh. 1995. *China: A Provincial Atlas*. New York: Macmillan.

Massey, Doreen. 1994. *Space, Place, and Gender*. Minneapolis: University of Minnesota Press.

Motzafi-Haller, Pnina. 1997. "Writing Birthright: On Native Anthropologists and the Politics of Representation." In Deborah E. Reed-Danahay (ed.), *Auto/Ethnography: Rewriting the Self and the Social*, pp. 195–222. Oxford: Berg.

National Geographic. "Geno 2.0: The Greatest Journey Ever Told." *The Genographic Project.* https://genographic.nationalgeographic.com/.

Nostrand, Richard L., and Lawrence E. Estaville, Jr. (eds.). 2001. *Homelands: A Geography of Culture and Place Across America*. Baltimore: Johns Hopkins University Press.

Oberle, Alex P., and Daniel D. Arreola. 2008. "Resurgent Mexican Phoenix." *Geographical Review* 98(2): 171–196.

O'Loughlin, John, Frank Witmer, Thomas Dickinson, Nancy Thorwardson, and Edward Holland. 2007. "Preface and Map Supplement." *Special Issue: The Caucasus: Political, Population, and Economic Geographies. Eurasian Geography and Economics* 48(2): 127–134.

Pastor, Jr., Manuel, Rachel Morello-Frosch, and James Sadd. 2010. *Air Pollution and Environmental Justice: Integrating Indicators of Cumulative Impact and Socio-Economic Vulnerability into Regulatory Decision-Making*. California Air Resources Board. http://www.arb.ca.gov/research/apr/past/04-308.pdf.

Pillsbury, Richard. 1998. *No Foreign Food: American Diet in Time and Place*. Boulder, Colo.: Westview Press.

Portes, Alejandro, and Rubén G. Rumbaut. 2006. *Immigrant America: A Portrait*. 3rd ed. Berkeley and Los Angeles: University of California Press.

Rehder, John B. 2004. *Appalachian Folkways*. Baltimore: Johns Hopkins University Press.

Scott, James C. 1998. *Seeing Like a State: How Certain Schemes to Improve the Human Condition Have Failed*. New Haven, Conn.: Yale University Press.

Shortridge, Barbara G., and James R. Shortridge (eds.). 1998. *The Taste of American Place: A Reader on Regional and Ethnic Foods.* Lanham, Md.: Rowman & Littlefield.

Sivanandan, A. 2001. "Poverty Is the New Black." *The Guardian,* August 17. http://www.theguardian.com/politics/2001/aug/17/globalisation. race. Tsuda, Takeyuki. 2006. "When Minorities Migrate: The Racialization of Japanese Brazilians in Brazil and Japan." In Lola Romanucci, George De Vos, and Takeyuki Tsuda (eds.), *Ethnic Identity: Problems and Prospects for the Twenty-first Century,* pp. 208–232. Latham, Md.: AltaMira Press.

U.S. EPA. 1996. *Toxic Release Inventory.* http://www.inmotionmagazine .com/auto/mapofla.html.

Valentino, Benjamin A. 2005. *Final Solutions: Mass Killing and Genocide in the Twentieth Century.* Ithaca, N.Y.: Cornell University Press.

Ten Recommended Books on Ethnic Geography

(For additional suggested readings, see the *Contemporary Human Geography* LaunchPad: http://www.macmillanhighered.com/ launchpad/DomoshCHG1e.)

Alkon, Alison Hope. 2012. *Black, White, and Green: Farmers Markets, Race, and the Green Economy.* Athens, Ga.: University of Georgia Press. Often discussed as progressive sites for advancing environmentally progressive food agendas, farmers markets also incorporate problematic politics of race, class, and social justice.

Berry, Kate A., and Martha L. Henderson (eds.). 2002. *Geographical Identities of Ethnic America: Race, Space, and Place.* Reno: University of Nevada Press. Eighteen different experts give their views on American ethnic geography, explaining how place shapes ethnic/racial identities and, in turn, how these groups create distinctive spatial patterns and ethnic landscapes.

Conzen, Michael P. (ed.) 2010. *The Making of the American Landscape.* 2nd ed. New York and London: Routledge. A sweeping account of the last 500 years of human and environmental intersections that have shaped the look and feel of America's cultural landscapes today, with heavy emphasis on ethnicity.

Frazier, John W., Florence Margai, and Eugene Tetty-fio. 2003. *Race and Place: Equity Issues in Urban America.* Boulder, Colo.: Westview Press. Explores how the cultural landscapes of U.S. cities express geographies of racial and ethnic discrimination.

Lee, Jennifer 8. 2008. *The Fortune Cookie Chronicles: Adventures in the World of Chinese Food.* New York: Twelve. Many "ethnic" cuisines common in the United States are, in fact, largely invented and popularized, rather than in the home country of the immigrants in question. Such is the case with American "Chinese food" favorites, such as fortune cookies, General Tsao chicken, chop suey, and other dishes. Lee traces the fascinating cultural, nutritional, migratory, interethnic, and economic histories at work behind this American tradition.

Miyares, Ines M., and Christopher A. Airriess. 2007. *Contemporary Ethnic Geographies in America.* Lanham, Md.: Rowman & Littlefield. An edited collection featuring chapters authored by renowned contemporary ethnic geographers on a variety of topics ranging from Central American soccer leagues to Muslim immigrants from Lebanon and Iran; also includes a geographer's view of the ethnic festivals with which we began this chapter.

Nostrand, Richard L., and Lawrence E. Estaville, Jr. (eds.). 2001. *Homelands: A Geography of Culture and Place Across America.* Baltimore: Johns Hopkins University Press. A collection of essays on an array of North American ethnic homelands, together with in-depth treatment of the geographical concept of homeland.

Pulido, Laura. 2006. *Black, Brown, Yellow, and Left: Radical Activism in Los Angeles.* Berkeley: University of California Press. Analyzes radical activism by African Americans, Chicanos, and Japanese Americans in Southern California.

Rehder, John B. 2004. *Appalachian Folkways.* Baltimore: Johns Hopkins University Press. An exploration of the folk culture of the Appalachian region of the United States, including its distinctive settlement history, folk architecture, cuisine, speech, and belief systems.

Schein, Richard (ed.). 2006. *Landscape and Race in the United States.* London and New York: Routledge. Contributions by leading geographers studying the cultural geographies of race provide an updated and critical insight into this growing subfield.

Are these border fences or walls?

Post-9/11 border security: the boundary between Israeli and Arab-Palestinian lands (*top*) and the U.S.–Mexico border (*bottom*).
(Top: ImageBROKER/Alamy; Bottom: age fotostock/Getty Images.)

Go to "Seeing Geography" on page 265 to learn more about these images.

POLITICAL GEOGRAPHY

A Divided World

From the breakup of empires to regional differences in voting patterns, from the drawing of international boundaries to congressional redistricting in the U.S. electoral system, from the resurgence of nationalism to separatist violence, human political behavior is inherently geographical. As geographer Gearóid Ó Tuathail has said, **political geography** "is about power, an ever-changing map revealing the struggle over borders, space, and authority."

> **political geography**
> The geographic study of politics and political matters.

REGION

6.1 **LEARNING OBJECTIVE**
Identify the importance of region to political geography.

> **state** A centralized authority that enforces a single political, economic, and legal system within its territorial boundaries. Often used synonymously with "country."

The theme of region is essential to the study of political geography because an array of both formal and functional political regions exists. Among these, the most important and influential is the **state.**

A WORLD OF STATES

The fundamental political-geographical fact is that Earth is divided into roughly 200 independent countries or states, creating a diverse mosaic of functional regions (**Figure 6.1**). The state is a political institution that has taken a variety of forms over the centuries, ranging from Greek city-states, to Chinese dynastic states, to European feudal states. When we talk about states today, however, we mean something very specific and historically recent. States are independent political units with a centralized authority

LEARNING OBJECTIVES

6.1 Identify the importance of region to political geography.

6.2 Identify the various ways that mobility and political geography interact.

6.3 Describe the effects of globalization on political geography.

6.4 Recognize the importance of the physical environment to political geography.

6.5 Analyze the role of cultural landscape in political geography.

Abbreviations

BF	BURKINA FASO
BOTS	BOTSWANA
EG	EQUATORIAL GUINEA
GBI	GUINEA BISSAU
IC	IVORY COAST
RCA	CENTRAL AFRICAN REPUBLIC
RL	LEBANON
TC	TURKISH CYPRUS
TL	TIMOR-LESTE
UAE	UNITED ARAB EMIRATES
WAG	THE GAMBIA
WAL	SIERRA LEONE
ZW	ZIMBABWE

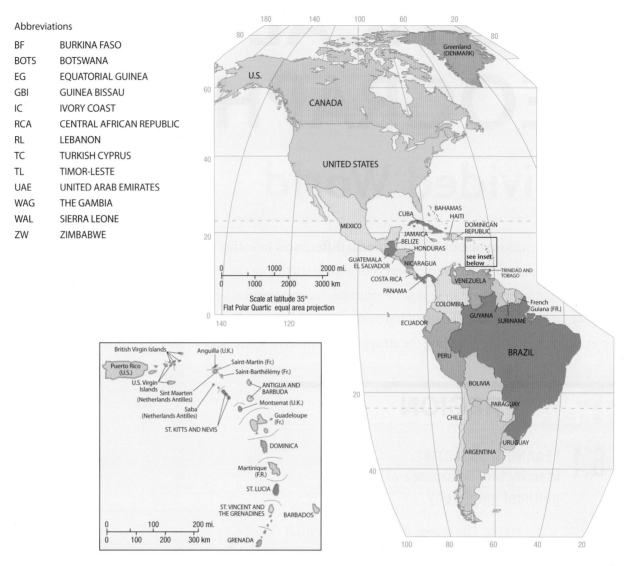

FIGURE 6.1 **The independent countries of the world.** In the twentieth century, the map was in rapid flux, with a proliferation of new countries. This process began after World War I with the breakup of such empires as those of Austria-Hungary and Turkey, then intensified after World War II when the overseas empires of the British, French, Italians, Dutch, Americans, and Belgians collapsed. More recently, the Russian-Soviet Empire disintegrated.

that makes claims to sole jurisdiction over a bounded territory. Within that territory, the central authority controls and enforces a single system of political and legal institutions. Importantly, in the modern international system, states recognize each other's **sovereignty.** That is, virtually every state recognizes every other state's right to exist and control its own affairs within its territorial boundaries.

sovereignty The right of individual states to control political and economic affairs within their territorial boundaries without external interference.

Closer inspection of Figure 6.1 reveals that some parts of the world are fragmented into many different states, whereas others exhibit much greater unity. The United States occupies about the same amount of territory as Europe, but the latter is divided into 47 independent countries. The continent of Australia is politically united, whereas South America has 12 independent entities and Africa has 55.

The modern state is a tangible geographical expression of one of the most common human

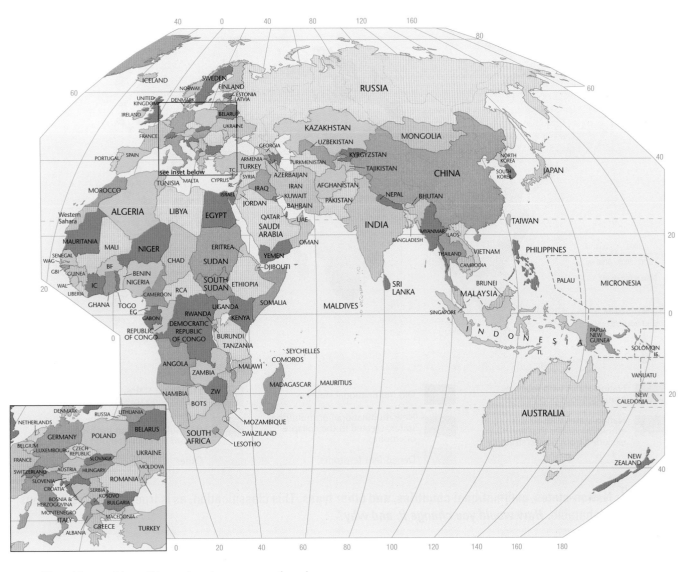

Most of the countries on this map have homepages on the web. Visit the Contemporary Human Geography *LaunchPad: http://www.macmillanhighered.com/launchpad/DomoshCHG1e to learn more about them.*

tendencies: the need to belong to a larger group that controls its own piece of Earth, its own territory. So universal is this trait that scholars coined the term **territoriality** to describe it. Most geographers view territoriality as a learned cultural response. Robert Sack, for example, regards territoriality as a cultural strategy that uses power to control an area and communicate that control, thereby subjugating the inhabitants and acquiring resources. He argues, for instance, that the precise marking of borders is a practice originally unique to modern Western culture. The modern territorial state, he claims, emerged rather recently in sixteenth-century Europe and diffused around the globe through European **colonialism.**

Political territoriality, then, is a thoroughly cultural-geographical phenomenon. The sense of

> ■ **territoriality** A learned cultural response, rooted in European history, that produced the external bounding and internal territorial organization characteristic of modern states.
>
> **colonialism** The building and maintaining of colonies in one territory by people based elsewhere.

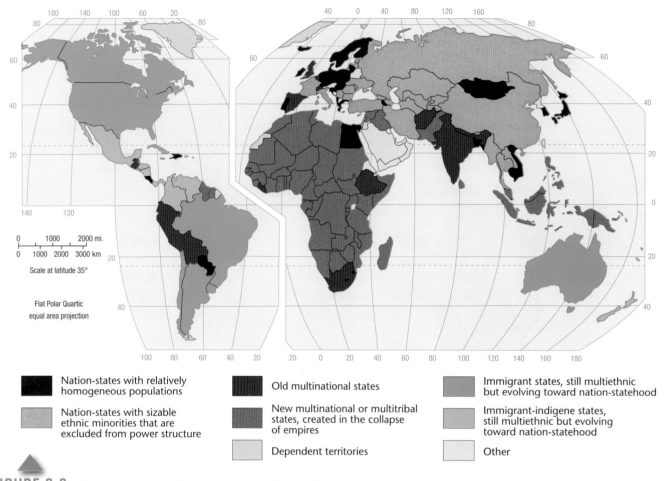

Nation-states with relatively homogeneous populations

Nation-states with sizable ethnic minorities that are excluded from power structure

Old multinational states

New multinational or multitribal states, created in the collapse of empires

Dependent territories

Immigrant states, still multiethnic but evolving toward nation-statehood

Immigrant-indigene states, still multiethnic but evolving toward nation-statehood

Other

FIGURE 6.2 **Nation-states, multinational countries, and other types.** This classification, as is true of all classifications, is arbitrary and debatable. *How would you change it, and why?*

nationalism The sense of belonging to and self-identifying with a national culture.

nation-state An independent country dominated by a relatively homogeneous cultural group.

collective identity that we call **nationalism** springs from a learned or acquired attachment to region and place. Nationalism is the idea that the individual derives a significant part of his or her social identity from a sense of belonging to a nation. We can trace the origins of modern nationalism to the late eighteenth century and the emergence of the **nation-state.** The nation-state was proposed as the ideal political geographical unit where the geographic boundaries of a nation (a people and its culture) would be identical to the territorial boundaries of the state. Nationality is culturally based in the nation-state, and the country's raison d'être lies in that cultural identity. The more the people have in

common culturally, the more stable and potent is the resultant nationalism. Examples of modern nation-states include Germany, Sweden, Japan, Greece, Armenia, and Finland (**Figure 6.2**). Geography and identity are thus tightly linked in the concepts of nationalism and the nation-state.

Distribution of National Territory An important geographical aspect of the modern state is the shape and configuration of the national territory. Theoretically, the more compact the territory, the easier is national governance. Circular or hexagonal forms maximize compactness, allow short communication lines, and minimize the amount of border to be defended. Of course, no country actually enjoys this ideal degree of compactness, although some—such as France and Brazil—come close (**Figure 6.3**).

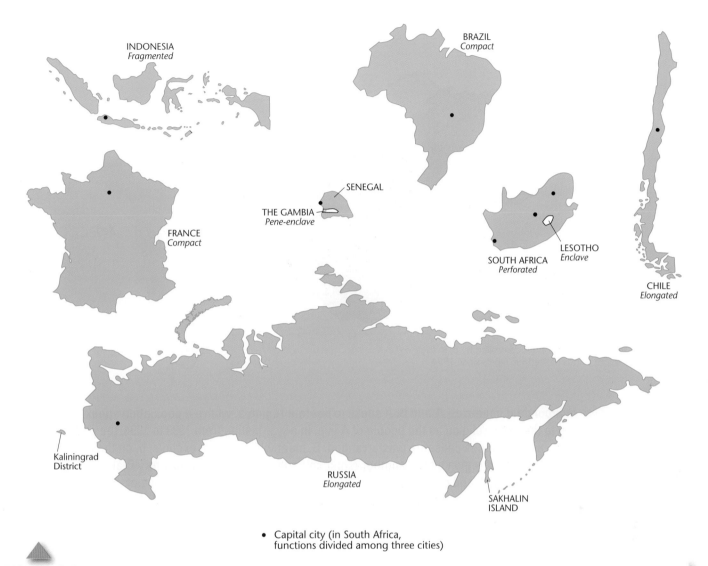

FIGURE 6.3 **Differences in the distribution of national territory.** The map, drawn from Eurasia, Africa, and South America, shows wide contrasts in territorial shape. France and, to a lesser extent, Brazil approach the ideal hexagonal shape, but Russia is elongated and has an exclave in the Kaliningrad District, whereas Indonesia is fragmented into a myriad of islands. The Gambia intrudes as a pene-enclave into the heart of Senegal, and South Africa has a foreign enclave, Lesotho. Chile must overcome extreme elongation. ***What problems can arise from elongation, enclaves, fragmentation, and exclaves?***

Any one of several unfavorable territorial distributions can inhibit national cohesiveness. Potentially most damaging to a country's stability are enclaves and exclaves. An **enclave** is a district surrounded by a country but not ruled by it. Enclaves can be either self-governing (e.g., Lesotho in Figure 6.3) or an exclave of another country. In either case, its presence can pose problems for the surrounding country. Potentially just as disruptive is the pene-enclave, an intrusive piece of territory with only the smallest of outlets (e.g., The Gambia in Figure 6.3).

Exclaves are parts of a national territory separated from the main body of the country to which they belong by the territory of another (e.g., Russia's Baltic seaport, Kaliningrad in

enclave A piece of territory surrounded by, but not part of, a country.

exclave A piece of national territory separated from the main body of a country by the territory of another country.

FIGURE 6.4 **Two independent countries, A and B.** A seeks to liberate region a, where a population speaking the same language and adhering to the same religion as the people of A live. In a war lasting from 1990 to 1994, the people of region a seceded from B, and a tenuous ceasefire was arranged. B, meanwhile, has an exclave, b, on the opposite, western side of A, and the people of region b form another ethnic minority, unrelated to B. Country B also possesses several much smaller enclaves—c, d, and e, the last of which is regarded as part of b (an exclave of an exclave!). A also has a tiny exclave, f. In other words, the distribution of national territories is troublesome to both A and B, particularly given the hostile relations between them. These are real countries. Using an atlas, try to identify them. If you fail, you can find the answer at the end of this chapter on page 269. *Have any recent events occurred here?* *(Sources: Office of the Geographer, U.S. Department of State, personal communication, 1997; Smith et al., 1997, p. 37.)*

Figure 6.3; **Figure 6.4**). Exclaves are particularly undesirable if a hostile power holds the intervening territory because defense of such an isolated area is difficult and makes substantial demands on national resources. Moreover, an exclave's inhabitants, isolated from their compatriots, may develop separatist feelings, thereby causing additional problems. Pakistan provides a good example of the national instability created by exclaves. Pakistan was created in 1947 as two main bodies of territory separated from each other by almost 1000 miles (1600 kilometers) of territory in northern India. West Pakistan held the capital and most of the territory, but East Pakistan was home to most of the people. West Pakistan hoarded the country's wealth, exploiting East Pakistan's resources but giving little in return. Ethnic

differences between the peoples of the two sectors further complicated matters. In 1971, a quarter of a century after its founding, Pakistan broke apart. The distant exclave seceded and became the independent country of Bangladesh (see Figure 6.1). Even when a national territory is geographically united, instability can develop if the shape of the state is awkward. Narrow, elongated countries, such as Chile, The Gambia, and Norway, can be difficult to administer, as can nations consisting of separate islands (see Figures 6.1 and 6.3). In these situations, transportation and communications are difficult, causing administrative problems. Several major secession movements have challenged the multi-island country of Indonesia; one of these—in East Timor—succeeded in 2002 in creating the first new sovereign state of the twenty-first century (see Figure 6.1). Similarly, the island nations of Japan, Mauritius, Comoros, and Philippines have all faced separatist movements in recent years (see Figure 6.1).

The shape and configuration of a country can make it harder or easier to administer, but they do not determine its stability. We can think of exceptions for all of the territorial forms in Figure 6.3. For example, Alaska is an exclave of the United States, but it is unlikely to produce a secessionist movement similar to the one in Bangladesh. Conversely, the Democratic Republic of the Congo and Nigeria both have compact territories and both have suffered through secessionist wars. Geography is an important factor in national governance, but it is rarely the determining factor.

Boundaries Political territories have different types of boundaries. Until fairly recent times, many boundaries were not sharp, clearly defined lines but instead were referred to as zones called **marchlands.** Today, the nearest equivalent to the marchland is the **buffer state,** an independent but small and weak country lying between two powerful, potentially belligerent countries. Mongolia, for example, is a buffer state between Russia and China; Nepal occupies a similar position between India and China (see Figure 6.1). If one of the neighboring countries assumes control of the buffer state, it loses much of its independence and becomes a **satellite state.**

Most modern boundaries are lines rather than zones, and we can distinguish several types. **Natural boundaries** follow some feature of the natural landscape, such as a river or mountain ridge. Examples of natural boundaries are numerous; for example, the Pyrenees lie between Spain and France, and the Rio Grande serves as part of the border between Mexico and the United States. **Ethnographic boundaries** are drawn on the basis of some cultural trait, usually a particular language spoken or religion practiced. The border that divides India from predominantly Islamic Pakistan is one case. **Geometric boundaries** are regular, often perfectly straight lines drawn without regard for physical or cultural features of the area. The U.S.–Canada border west of the Lake of the Woods (about 93° west longitude) is a geometric boundary, as are most county, state, and province borders in the central and western United States and Canada. Not all boundaries are easily categorized; some boundaries are of mixed type, composites of two or more of the types listed.

Finally, **relic boundaries** are those that no longer exist as international borders. Nevertheless, they often leave behind a trace in the local cultures. With the reunification of Germany in the autumn of 1990, the old Iron Curtain border between the former German Democratic Republic in the east and the Federal Republic of Germany in the west was quickly dismantled (**Figure 6.5**). In a remarkably short time span, measured in weeks, the Germans reopened severed transport lines and created new ones, knitting together the enlarged country. Even so, remnants and reminders of the old border remained, as it continued to function as provincial boundaries within Germany. Furthermore, it still separated two parts of the country with strikingly different levels of prosperity.

Spatial Organization of Territory States differ greatly in the way their territory is organized for purposes of administration. Political geographers recognize two basic types of spatial organization: the **unitary state** and the **federal state.** In unitary countries, power is concentrated centrally, with little or no provincial authority. All major decisions come from the central government, and policies are applied uniformly throughout the national territory. France and China

marchland A strip of territory, traditionally one day's march for infantry, that served as a boundary zone for independent countries in premodern times.

buffer state An independent but small and weak country lying between two powerful countries.

satellite state A small, weak country dominated by one powerful neighbor to the extent that some or much of its independence is lost.

natural boundary A political border that follows some feature of the natural environment, such as a river or mountain ridge.

ethnographic boundary A political boundary that follows some cultural border, such as a linguistic or religious border.

geometric boundary A political border drawn in a regular, geometric manner, often a straight line, without regard for environmental or cultural patterns.

relic boundary A former political border that no longer functions as a boundary.

unitary state An independent state that concentrates power in the central government and grants little authority to the provinces.

federal state An independent country that gives considerable powers and even autonomy to its constituent parts.

FIGURE 6.5 **A boundary disappears.** In Berlin, the view toward the Brandenburg Gate changed radically between 1989 and 1991, when the Berlin Wall was destroyed and Germany reunited. *(Courtesy of Terry G. Jordan-Bychkov.)*

are unitary in structure, even though one is democratic and the other authoritarian. A federal government, by contrast, is a more geographically expressive political system. That is, it acknowledges the existence of regional cultural differences and provides the mechanism by which the various regions can perpetuate their individual characters. Power is diffused, and the central government surrenders much authority to the individual provinces. The United States, Canada, Germany, Australia, and Switzerland, though exhibiting varying degrees of federalism, provide examples. The trend in the United States has been toward a more unitary, less federal government, with fewer states' rights. By contrast, federalism remains vital in Canada, representing an effort to counteract French Canadian demands for Québec's independence. That is, by emphasizing federalism, the central government allows more latitude for provincial self-rule, thus reducing public support for the more radical option of secession.

Whether federal or unitary, a country functions through some system of political subdivisions. In federal systems, these subdivisions sometimes overlap in authority, with confusing results. For example, the Native American reservations in the United States occupy a unique and ambiguous place in the federal system of political subdivisions. These semiautonomous enclaves are legally sanctioned political territories that, theoretically, only indigenous Americans can possess. In reality, ownership is often fragmented by non–Native American land holdings within the reservations. Although not completely sovereign, Native Americans do have certain rights to self-government that differ from those of surrounding local authorities. For example, many reservations can (and do) build casinos on their land even though gambling may be illegal in surrounding political jurisdictions. Reservations do not fit neatly into the American political system of states, counties, townships, precincts, and incorporated municipalities.

CENTRIFUGAL AND CENTRIPETAL FORCES

If there is one lesson to learn from the past few decades of global political geography it is that few nation-state boundaries are permanent. We have witnessed countless cycles of countries fragmenting into multiple states and scattered territories coalescing into larger political states. Many countries have, of course, remained more or less intact over the decades. Thus, we must ask, why do some countries break apart while others remain united?

Geographers refer to factors that promote national unity and solidarity as **centripetal forces.** By contrast, anything that disrupts internal order and furthers the breakup of the territory is called a **centrifugal force.** Many states encourage centripetal forces that help fuel nationalistic sentiment. Such factors as an official national language, national history museums, national parks and monuments, and sometimes even a national religion are actively promoted and supported by the state.

> ■ **centripetal force** Any factor that supports the internal unity of a country.
>
> **centrifugal force** Any factor that disrupts the internal unity of a country.

Although physical factors such as the spatial organization of territory, degree of compactness, and type of boundaries can influence an independent country's stability, other forces are also at work. In particular, cultural factors often make or break a country. The most viable independent countries, those least troubled by internal discord, have a strong feeling of group solidarity among their population. Strong group identity is often key to a state's territorial integrity.

In the case of the modern state, the primary source of group identity is nationalism. Political geographer John Agnew cautions us, however, that the meaning and form of nationalism are unstable, making it difficult to generalize. One version is state nationalism, wherein the nation-state is exalted and individuals are called to sacrifice for the good of the greater whole. The twentieth-century history of state nationalism in Europe is marked by two horrific world wars in which millions of lives were sacrificed. Referring to World War I, Agnew points out that the "war itself was also the outcome of a mentality in which the individual person had to sacrifice for the good of the greater whole: the nation-state."

When a specific ethnic group or ethnically based political party controls the state's military, we sometimes see forced deportations and even attempted genocides in efforts to create an ethnically homogeneous national citizenry (see Chapter 5). This happened in recent decades in Rwanda in 1994 and in the Balkans through the 1990s. Most recently, the military of Sudan has been accused of attempted genocide in the Darfur region of that country

FIGURE 6.6 **The war in Darfur.** The far west of Sudan is home to African cultures ethnically distinct from the Arab peoples that control the Sudanese government and military. Many international observers have labeled the Sudanese military's efforts to control Darfur as acts of genocide. *(AP Photo/Abd Raouf.)*

(**Figure 6.6**). The people of Darfur are ethnically distinct from the dominant culture controlling the Sudanese state. The most notorious modern example is Nazi Germany's attempt in the 1930s and 1940s to clear German territory of non-Germans, especially Jews and Roma (Gypsies) (see Rod's Notebook on page 234).

Another way to create a homogeneous citizenry is through **territorial fission.** This happens in cases of substate nationalism, in which ethnic or linguistic minority populations seek to secede from the state or to alter state territorial boundaries to promote cultural homogeneity and political autonomy. In such cases, a smaller, more culturally homogenous country may break off from the larger state. The world's newest nation-state, South Sudan, is a good example. We consider an array of centripetal and centrifugal forces later, in the sections on globalization and nature–culture.

> ■ **territorial fission** Occurs when an ethnic or linguistic minority population seeks to secede from the state or to alter state territorial boundaries to promote cultural homogeneity and political autonomy.

POLITICAL BOUNDARIES IN CYBERSPACE

What happens to political boundaries in cyberspace? E-mail, the Internet, and the World Wide Web can cross borders in ways previously not possible, although radio, telephone, and fax machines possess some of the same border-defying qualities. In Germany, for example, Nazi and neo-Nazi propaganda is prohibited, yet the First Amendment right of freedom of speech protects the dissemination of this material in the United States. What happens when this propaganda originates in the United States and is posted electronically to online bulletin boards worldwide? Can the German government prosecute the originator of the message—an American—for violating a German law?

How can countries impose their laws and boundaries in the computer age? One way is to restrict citizens' access to technological hardware such as satellite dishes, computers, phone lines, and cell phones. This is becoming increasingly difficult for even the most totalitarian regimes. Another tactic deployed by states is to allow citizens' access to the web but to control the flow of information they receive.

The most well-known example is the deal cut between the search engine company Google and the People's Republic of China in 2006. In order to gain access to China's rapidly expanding market for Internet services, estimated at 162 million customers in 2007, Google agreed to create a self-censoring search engine. Its Chinese-language version was specially designed to match the Chinese government's censorship laws. Web sites promoting ideas that the state views as threatening to its ideological dominance, such as free speech and the practice of unauthorized religions, are filtered out by Google's search engine. Google, based in Mountain View, California, has been widely criticized in the United States for this decision and has even been called to testify before Congress about its actions. In their own defense, Google officials argue that they can do more good for China's citizens by participating in their Internet business than by withdrawing completely.

As the Google case illustrates, states often rely on the cooperation of private service providers in their efforts to control Internet access. Twitter, an online social networking company, offers another example. In 2013 Twitter initiated a policy of compliance with state censorship laws. Beginning with Germany, Twitter began blocking a neo-Nazi group's tweets because the country has laws restricting the promotion of Nazi ideology. While we may agree that blocking discriminatory speech is a good thing, such censorship

FIGURE 6.7 **Cyber censorship during the Arab Spring.** Beginning in Tunisia in 2011, a wave of political protests swept across Arab countries with the aid of the Internet and various social media. Several governments responded by drastically restricting their citizens' access to the Internet. *(AFP/Getty Images.)*

in the United States would violate free speech rights. The Google and Twitter cases show that states still retain significant control over global information flows across borders, sometimes forcing private service providers to choose between complying with state censorship laws or being barred from doing business.

Governments often fear the power of the Internet to disseminate information that can threaten its authority. In such cases, they may take extreme action to halt information flow. In 2012, for example, China blocked its citizens' Internet access to the *New York Times* when the newspaper published an article unfavorable to the country's prime minister. During the political upheavals known as the "Arab Spring," several Middle Eastern and North African governments tried to limit the role of the Internet in antigovernment organizing. Egypt, Libya, and Syria severed their countries' Internet connections completely, while in Tunisia and Saudi Arabia governments used Internet censorship and surveillance to control information flow (**Figure 6.7**).

These cases illustrate that political boundaries are far from irrelevant in the age of the Internet. The Internet is, to a degree, eroding political boundaries by allowing information and ideas to diffuse more rapidly and completely. As a result, political barriers to cultural diffusion have become more fragile. However, countries still have some power to control what ideas are allowed inside their territorial boundaries.

SUPRANATIONAL POLITICAL BODIES

In addition to independent countries and their governmental subdivisions, the third major type of

political functional region is the **supranational organization** (**Figure 6.8**). **Supranationalism** exists when countries voluntarily give up some portion of their sovereignty to gain the advantages of a closer political, economic, and cultural association with their neighbors. Sometimes supranational organizations take the form of

supranational organization A group of independent countries joined together for purposes of mutual interest.

supranationalism A situation that occurs when states willingly relinquish some degree of sovereignty in order to gain the benefits of belonging to a larger political-economic entity.

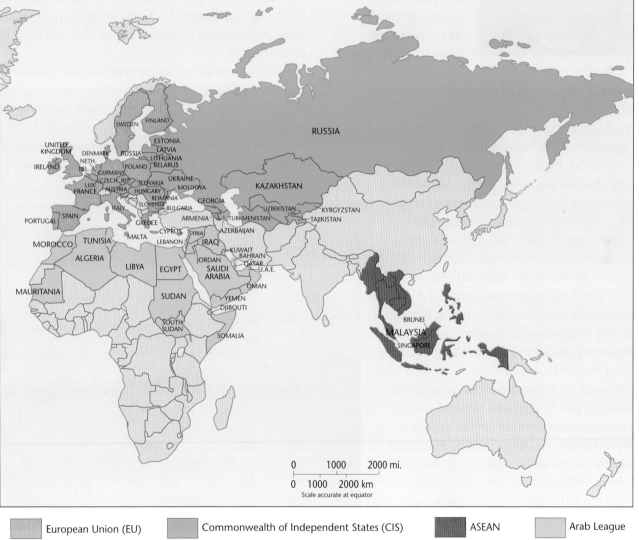

European Union (EU) Commonwealth of Independent States (CIS) ASEAN Arab League

FIGURE 6.8 **Some supranational political organizations in the Eastern Hemisphere.** These organizations vary greatly in purpose and cohesion. ASEAN stands for the Association of Southeast Asian Nations, and its purposes are both economic and political. ***What might this map indicate about globalization?***

Rod's Notebook

Places of a Genocidal State

Rod Neumann. *(Courtesy of Roderick Neumann.)*

For years I have known of the Holocaust in broad historical outline. I knew that 6 million Jews were systematically murdered by the Nazi-controlled German state. The ideology of the Nazi Party was a rabid form of German nationalism founded on racial hatred, especially anti-Semitism. For Nazi leader Adolf Hitler, the "final solution" to the "problem" of European Jews was their extermination. By the 1940s, Nazi Germany had become the world's first "genocidal state."

Although I had long been aware of these historical events, I did not fully grasp the nature of a genocidal state and how it functioned until I traveled to Poland and visited the Auschwitz-Birkenau State Museum. Auschwitz-Birkenau is a Nazi death camp complex where 1.1 million people were murdered during World War II. The abstraction of 6 million murders was suddenly made concrete as I saw firsthand how the German state organized the collection of Jews from as far away as the Netherlands and Greece and brought them to this concentration camp in southern Poland. I saw how the rail lines were built to efficiently unload thousands of prisoners and quickly send them to slave labor or quick death. I saw how the Birkenau camp had been built close to manufacturing facilities of German companies in order to more efficiently exploit prisoners' labor. Walking through the gas chambers where hundreds of thousands had died, I perceived the unspeakable horror of state-run mass murder.

I left Auschwitz-Birkenau thinking how important the experience of place is to knowledge. I could not have grasped fully the scale and magnitude of industrial mass genocide had I not looked out across the miles of rail lines and hundreds of hectares of barracks where people were stored until they could be murdered. I could never have understood how critical spatial relations were to the conduct of mass murder without physically moving through the spaces of the death camps.

I also came away realizing again how memory and place are so tightly bound. We memorialize places so that we do not forget. The United Nations declared Auschwitz-Birkenau a World Heritage Site in 1979 in hopes that the world will remember and prevent future atrocities. I know I will not forget my encounter with the places of a genocidal state.

Birkenau camp in the Auschwitz complex. These chimneys are all that remain of hundreds of prisoner barracks, which once stretched for many kilometers and held a total of approximately 90,000 people. *(Courtesy of Roderick Neumann.)*

regional trading blocs, such as the North American Free Trade Agreement (NAFTA, which comprises Canada, the United States, and Mexico), that promote the freer flow of goods and services across international borders.

> **regional trading blocs** Agreements made among geographically proximate countries that reduce trade barriers so the nations will be better able to compete with other regional markets.

In the twentieth century, supranational organizations grew in numbers and importance, coincident with and counterbalancing the proliferation of independent countries. Some represent the vestiges of collapsed empires, such as the British Commonwealth, French Community, and Commonwealth of Independent States (CIS)—the latter a shadow of the former Soviet Union. Most supranationals, such as the Arab League or the Association of Southeast Asian Nations (ASEAN), possess little cohesion.

The European Union, or EU, is by far the most powerful, ambitious, and successful supranational organization in the world (see Figure 6.8). It grew from a central core area of six countries in the 1950s to its present membership of 28. At first, the EU was merely a customs union whose purpose was to lower or remove tariffs that hindered trade, but it gradually took on more and more of a political and cultural role. An underlying motivation was to weaken the power of its member countries to the point that they could never again wage war against one another—a response to the devastation of Europe in two world wars.

The member countries have all sacrificed some of their sovereign powers to the EU administration. A single monetary currency, the euro, has been adopted by most EU members. Most international borders within the EU are now completely open, requiring no passport checks. More importantly for citizenship and national identity, the EU is standardizing a range of social norms related to the tolerance of religious and ethnic difference, human rights, and gender relations. In effect, the EU is pushing supranationalism to its logical conclusion. That is, at some future date nationalism as a focus of identity will be obsolete and people will think of themselves as European rather than as, say, Italian, or Latvian, or Polish. We will have to wait to learn whether centripetal or centrifugal forces will win out in the EU's grand experiment.

ELECTORAL GEOGRAPHICAL REGIONS

Voting in elections creates another set of political regions. A free vote of the people on some controversial issue provides one of the purest expressions of culture, revealing attitudes on religion, ethnicity, and ideology. Geographers can devise formal culture regions based on voting patterns, giving rise to the subspecialty known as **electoral geography.**

> **electoral geography** The study of the interactions among space, place, and region and the conduct and results of elections.

Electoral geographers also concern themselves with functional regions, in this case the voting district or precinct. Their interests are both scholarly and practical. For example, following each national census, political redistricting takes place in the United States. In redistricting, new boundaries are drawn for congressional districts to reflect the population changes since the previous census. The goal is to establish voting areas of more or less equal population and to increase or reduce the number of districts depending on the amount and direction of change in total population. These districts form the electoral basis for the U.S. House of Representatives. State legislatures are based on similar districts.

Geographers often assist in the redistricting process. Indeed, advances in GIS software make it possible to instantly draw and redraw congressional district boundaries to achieve a desired outcome. For example, Maptitude is a mapping software specially designed for professionals to use in redistricting. ProgressiveCongress.org supports the development and free distribution of software called Dave's Redistricting application. In all such cases, redistricting software offers a way to rapidly draw congressional maps based on the latest census data.

Gerrymandering the Vote The shape of district or precinct boundaries can significantly influences election results. Boundaries can be manipulated to favor one political party over another. If redistricting remains in the hands of legislators, then the majority

> **gerrymandering** The drawing of electoral district boundaries in an awkward pattern to enhance the voting impact of one constituency at the expense of another.

political party will often try to arrange the voting districts geographically to maximize and perpetuate its power. This practice is called **gerrymandering** and the resultant voting districts often have awkward, elongated shapes (**Figure 6.9**). A gerrymandered district in North Carolina was recently referred to as having the shape of a "fat squid," while a district in Georgia was labeled a "flat-cat roadkill."

Gerrymandering can be accomplished by one of two methods. One is to draw district boundaries so as to concentrate all of the opposition party into one district, thereby creating an unnecessarily large majority while also ensuring that it cannot win elsewhere. In the lingo of political operators, this is known as "packing." A second method is to draw the boundaries so as to divide opposition votes into many districts. This has the effect of diluting the

opposition's vote so that it does not form a simple majority in any district. This is known as "cracking."

Gerrymandering is nothing new. It has been a recognized, if sometimes illegal, political tactic in the United States since the early 1800s. However, recent demographic shifts, historic racial and ethnic party affiliations, and advances in redistricting software have combined to make gerrymandering an ever-more common practice in the United States. Texas provides a good example. Its population grew enough to gain an additional four congressional seats after the 2010 census. Hispanics and African Americans, historic supporters of the Democratic Party, accounted for the vast majority of the population growth. The Republican Party, however, controlled the state legislature, which is responsible for redrawing district boundaries. Seventy percent of the new districts drawn by the Republican legislature had white (traditionally Republican supporters in Texas) majorities even though whites' share of the state

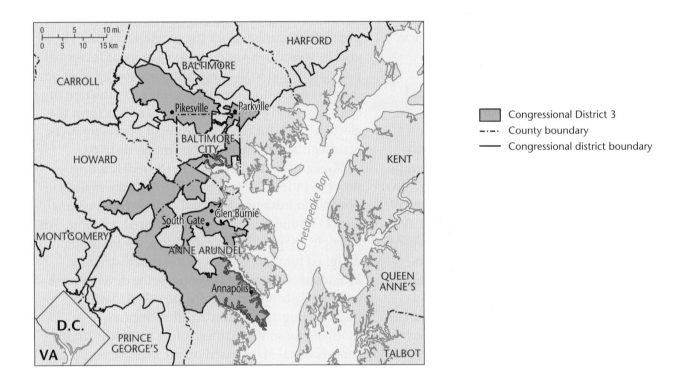

FIGURE 6.9 Gerrymandering of a congressional district in Maryland. Gerrymandering is often identifiable by the oddly shaped boundaries it produces. The shape of Maryland's Third District, likened to an "amoeba convention" and "a broken-winged pterodactyl," is a case in point. While gerrymandering is difficult to prove, such shapes raise suspicions. In this case, Democrats defended the shape, while Republicans wanted the district redrawn. Gerrymandering may be in the eye of the beholder. *(Source: NationalAtlas.gov.)*

population dropped from 52 percent to 45 percent. Republicans consequently won an additional four districts, while minority populations (now in the majority) gained none. Because the Texas redistricting was so blatantly discriminatory toward minority populations, it was repeatedly and successfully challenged in federal courts.

RED STATES, BLUE STATES

In recent years, various commentators in academia and the popular media have used maps of national election results to suggest that the United States is divided into distinct culture regions. We can trace this argument to the postelection analysis of the controversial 2000 presidential election, when, for the first time, news media adopted a universal color scheme for mapping voter preferences. States where the majority of voters favored the Democratic Party's presidential candidate were assigned the color blue, whereas those favoring the Republican candidate were assigned red. When cartographers mapped the state-by-state results of the 2000 and subsequent presidential elections, the solid blocks of starkly contrasting colors gave the appearance of a deeply divided country (**Figure 6.10**).

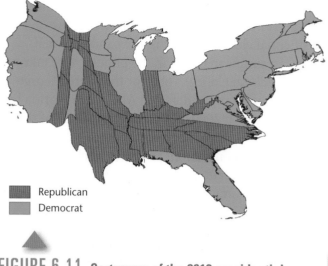

Republican
Democrat

FIGURE 6.11 **Cartogram of the 2012 presidential election results.** This map sizes America's 50 states in proportion to their respective populations. *How does this cartogram compare to the map in Figure 6.10 in its representation of the relative popularity of the Republican and Democratic candidates?* (Source: Courtesy of Mark Newman, http://www-personal.umich.edu/~mejn/election/2012/.)

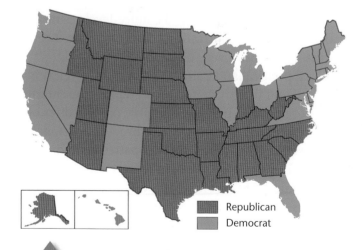

Republican
Democrat

FIGURE 6.10 **2012 presidential election results.** News media have adopted a standard color coding for political parties—red for Republican, blue for Democrat—in election result maps. It is now common to hear of the United States being politically and culturally divided into "red state" and "blue state" regions. (Source: Courtesy of Mark Newman, http://www-personal.umich.edu/~mejn/election/2012/.)

The terms *red state* and *blue state* entered the popular lexicon in reference to the division of the United States into regions of "conservative" and "liberal" cultures that these maps implied. Figure 6.10 suggests, for example, that the U.S. South is politically and culturally conservative, whereas the West Coast is liberal. These maps, so common in the popular media, are overly simplistic and, therefore, misleading. For example, the solid red coloring of a large portion of the country in Figure 6.10 implies that most of the United States is culturally and politically conservative. However, this map gives inordinate visual importance to states of larger areal extent, no matter what their population. A cartogram provides a better representation of both the true significance of states in the presidential election and the relative proportion of voters favoring Democrats or Republicans (**Figure 6.11**). When states' populations are taken into account, U.S. voters look much more favorable toward Democratic rather than Republican candidates.

The red state/blue state designation also exaggerates the appearance of sharp geographic divisions within the country. Whether a state is designated "red" or "blue" is based on the winner-take-all system that the United States uses for presidential elections. That is, the candidate that receives a simple majority in the popular vote takes all the Electoral College votes for that state. George W. Bush, for example, defeated Al Gore in Florida by only 537 votes in 2000, but he received all 25 of the state's electoral votes. This was an unusually close result, but it is common in presidential elections for only a small percentage of a state's popular vote to separate the winner and the loser. Designating a state such as Florida as "red" thus masks the narrow margin of the Republican victory and exaggerates the existence of regional polarization.

The limitations of the red state/blue state designation have led to the introduction of the term *purple state* to signify closely divided states. If we apply the purple state idea (i.e., shadings of color rather than stark contrasts) to county-level election results, we find that the simplistic division of the United States into liberal and conservative regions begins to break down (**Figure 6.12**). There is a great

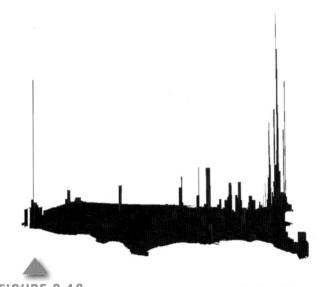

FIGURE 6.13 **Presidential election results in 3D.** In this map, height represents voters per square mile (for the 2004 presidential election), so urban areas stand out in contrast to suburban and rural areas. ***What political-geographic patterns are noticeable here?*** *(Source: Robert J. Vanderbei, Princeton University.)*

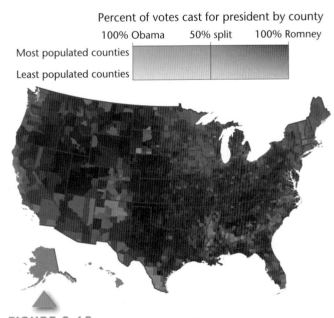

Percent of votes cast for president by county

100% Obama   50% split   100% Romney

Most populated counties

Least populated counties

FIGURE 6.12 **Purple America.** Rather than using stark color contrasts to represent the 2012 presidential election results, this map uses color shading. ***How does the use of shading affect our understanding of a sharply divided American electorate that appears so prominent in Figure 6.10? What other kinds of geographic voting patterns emerge?*** *(Source: Courtesy of Mark Newman, http://www-personal.umich.edu/~mejn/election/2012/.)*

deal of county-by-county variation within states across the United States. New geographic patterns emerge in this map that belie the idea of large regional divisions in the country.

Using shades of color and county-level results—rather than using contrasting colors and state-level results—provides a very different picture of U.S. regional political and cultural differences. Even this nuanced and finely grained cartographic presentation does not reveal all of the geographic complexities of U.S. presidential elections. With three-dimensional cartographic techniques, another pattern emerges. In the map in **Figure 6.13**, height represents voters per square mile, so that volume represents total number of voters. Here, the core geographic difference appears to exist between cities and rural areas, not between regions. The general pattern we see is that rural areas are predominantly Republican and cities are predominantly Democratic.

A closer analysis of the red state/blue state phenomenon reveals some of the complexities involved in electoral geography. Choices about the scale at which data are aggregated, the cartographic techniques employed, and even the map colors influence the interpretation of election results. This is something to keep in mind when you evaluate claims about a divided America.

MOBILITY

6.2 LEARNING OBJECTIVE
Identify the various ways that mobility and political geography interact.

Political events and developments can trigger massive human migrations, sometimes spanning continents. Moreover, the forces of globalization have accentuated the importance of mobility across political boundaries. However, political boundaries also act as barriers to the movement of people, ideas, knowledge, and resources.

MOVEMENT BETWEEN CORE AND PERIPHERY

Many independent states sprang fully formed into the world, but some expanded outward when powerful political entities emerged from a small nucleus called a **core area.** These core political powers enlarged their territories by annexing adjacent lands, often over many centuries. Generally, core areas possess a particularly attractive set of resources for human life and culture. Larger numbers of people cluster there than in surrounding districts, especially if the area has some measure of natural defense against aggressive neighboring political entities. This denser population, in turn, may produce enough wealth to support a large army, which then provides the base for further expansion outward from the core area. Resources, people, and capital investment begin to flow between the core and peripheral territories, usually resulting in a further consolidation of the core's wealth and power.

In states formed in this way, the core area typically remains the country's single most important

core area The territorial nucleus from which a country grows in area and over time, often containing the national capital and the main center of commerce, culture, and industry.

district, housing the capital city and the cultural and economic heart of the nation. The core area is the node of a functional region. France expanded to its present size from a small core area around the capital city of Paris. China grew from a nucleus in the northeast, and Russia originated in the small principality of Moscow (**Figure 6.14**). The United States grew westward from a core between Massachusetts and Virginia on the Atlantic coastal plain, an area that still includes the national capital and the densest population in the country.

The evolution of independent countries in this manner produces the core–periphery configuration, described in Chapter 1 as typical of both functional and formal regions. Although the core dominates the periphery, a certain amount of friction exists between the two. Peripheral areas generally display pronounced, self-conscious regionality and occasionally provide the settings for secession movements. Even so, countries that diffused from core areas are, as a rule, more stable than those created all at once to fill a political void. The absence of a core area can blur or weaken citizens' national identity and make it easier for various provinces to develop strong local or even foreign allegiances. Belgium and the Democratic Republic of the Congo offer examples of countries without political core areas. In the case of the Congo, this situation partly accounts for the history of secessionist conflicts and internecine wars since the country gained independence from colonial rule in 1960.

MOBILITY, DIFFUSION, AND POLITICAL INNOVATION

One of the consequences of colonialism was the creation of new patterns of mobility among the conquered populations. One pattern in particular contained the seeds of colonialism's demise. Colonial rule in Africa depended on the establishment of a relatively small cadre of educated African civil servants. The colonial rulers selected those whom they considered the best and the brightest (and most politically cooperative) among their African subjects and sent them to the best universities in Europe. In addition to learning the skills required of colonial functionaries, they absorbed the ideals of Western

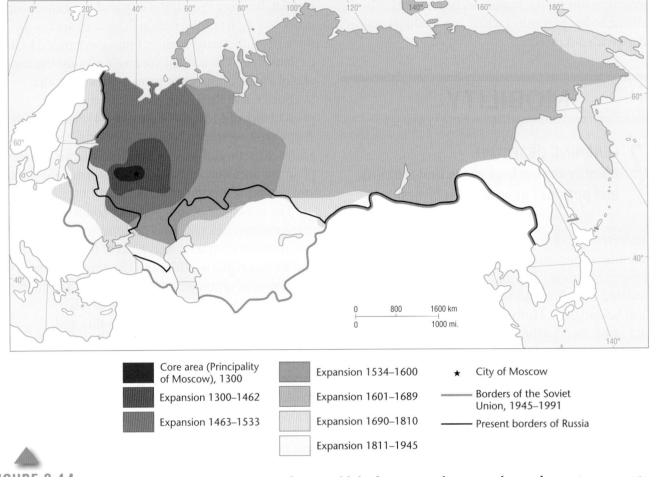

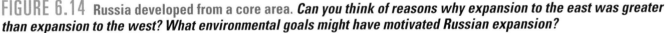

FIGURE 6.14 Russia developed from a core area. *Can you think of reasons why expansion to the east was greater than expansion to the west? What environmental goals might have motivated Russian expansion?*

political philosophy, such as the right of self-determination, independent self-rule, and individual rights. Educated Africans returned to their countries armed with these ideas and organized new nationalist movements to oppose colonial rule.

The consequences can be seen in the spread of political independence in Africa. In 1914 only two African countries—Liberia and Ethiopia—were fully independent of European colonial or white minority rule, and even Ethiopia later fell temporarily under Italian control. Influenced by developments in India and Pakistan, the Arabs of North Africa began a movement for independence. Their movement gained momentum in the 1950s and swept southward across most of the African continent between 1960 and 1965. Many of Africa's great nationalist leaders of this period had obtained university degrees in England, Scotland, France, and other European countries. By 1994 African

nationalists had helped spread the idea of independence into all remaining parts of the continent, eventually reaching the Republic of South Africa, formerly under white minority rule (**Figure 6.15**).

Despite its rapid spread, diffusion of African self-rule occasionally encountered barriers. Portugal, for example, clung tenaciously to its African colonies until 1975, when a change in government in Lisbon reversed a 500-year-old policy, allowing the colonies to become independent. In colonies containing large populations of European settlers, independence came slowly and usually with bloodshed. France, for example, sought to hold onto Algeria because many European colonists had settled there. The country nonetheless achieved independence in 1962, but only after years of violence. In Zimbabwe, a large population of European settlers refused London's orders to move toward majority rule, resulting in a

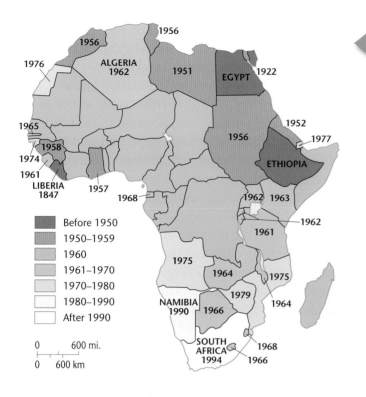

◀ **FIGURE 6.15** Independence from European colonial or white minority rule spread rapidly through Africa in the 1950s and 1960s. Prior to the 1950s, there were only three independent countries in Africa. Between 1951 and 1968, most attained self-rule. The remaining few gained freedom over the next three decades. By 1994 independence had spread from the Mediterranean to the Cape of Good Hope. Many of the ideas about national liberation were spread by African leaders educated in Europe. *Why do you think it took a few countries so much longer than most to gain independence?*

politics abounds with examples of cultural diffusion. A classic case is the spread of suffrage for women, a movement that culminated in 1920 with the ratification of a constitutional amendment (**Figure 6.16**). Opposition to women's suffrage was strongest in the U.S. South, a region that later exhibited the greatest resistance to ratification of the Equal Rights Amendment and displayed the most reluctance to elect women to public office.

Federal statutes permit, to some degree, laws to be adopted in the individual functional subdivisions. In the United States and Canada, for example, each state

bloody civil war and delaying the country's independence until 1979.

On a quite different scale, political innovations also spread within independent countries. American

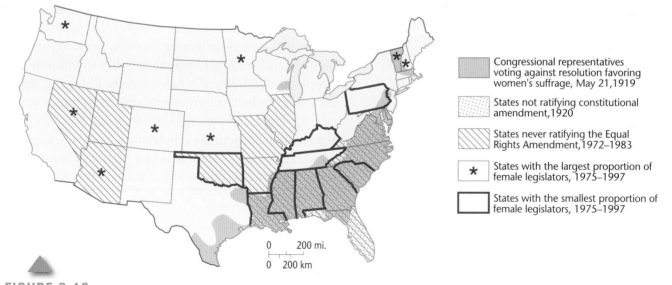

FIGURE 6.16 The diffusion of suffrage for women in the United States and of the Equal Rights Amendment. The suffrage movement achieved victory through a constitutional amendment in 1920. Both the suffrage movement and the campaign for an Equal Rights Amendment (ERA) for women failed to gain approval in the Deep South, an area that also lags behind most of the remainder of the country in the election of women to public office. *What might be the barriers to diffusion in the Deep South? The states failing to ratify the ERA lay mostly in the same area. The ERA movement did not succeed, in contrast to the earlier suffrage movement.* (Source: Adapted in part from Paulin and Wright, 1932.)

and province enjoys broad lawmaking powers, vested in the legislative bodies of these subdivisions. The result is often a patchwork legal pattern that reveals the processes of cultural diffusion at work. A good example is the clean air movement, which began in California with the initiation of state legislation regulating automotive and industrial emissions. It later spread to other states and provided the model for clean air legislation at the federal level.

THE FORCED MOBILITY OF VIOLENT CONFLICT

Since there have been states, there have been refugees. We learned in Chapter 3 that a refugee is a type of forced migrant, a definition that we expand upon here by looking to the United Nations. The UN definition of refugee pivots around people's relationship to the state. According to the extended definition in the UN 1951 Refugee Convention, "Refugees have to move if they are to save their lives or preserve their freedom. They have no protection from their own state—indeed it is often their own government that is threatening to persecute them." Often repression is driven by cultural difference—based on religion, race, or ethnicity—as much as by political disagreement. It seems that the very idea of the state contains both the promise of stability and protection and the threat of displacement and persecution. As geographers Jennifer Hyndman and Wenona Giles observe, "Displacement is the underbelly of mobility, a kind of movement that expresses the violent political relation of people to place."

Most refugees are fleeing some form of armed conflict including interstate wars, civil wars, insurgencies, and counterinsurgencies. Syria presents a tragic but all-too-common example of a twenty-first-century refugee crisis. Its citizenry is a complex assortment of religious and ethnic identities, including Arab-Sunni Muslim, Arab-Alawite (a Muslim Shi'ite sect), Kurd-Sunni, and Greek Orthodox Christian. The Baath Party controlled the government for decades, led by President Bashir al-Assad, an Alawite. When the Arab Spring swept through surrounding countries, overthrowing entrenched regimes, peaceful street protests broke out in Syria demanding a change in government. After violent government reprisals, they

soon developed into armed rebellion with multiple militias and ethnic, political, and religious factions from Syria and nearby countries taking sides in the fight. The protests subsequently escalated into a full-blown civil war. By 2013 Syria was the source of at least 1.6 million refugees who flowed into the nearby countries of Lebanon, Turkey, Egypt, Iraq, and Jordan at a rate of 250,000 per month. Jordan hosted the largest numbers of refugees, with its al-Zaatari refugee camp vying for the dubious distinction as the world's largest.

Worldwide, the number of refugees has been growing in the twenty-first century. Estimates vary greatly for political reasons (governments often impede accurate counting) as well as practical considerations (people on the move across borders are difficult to count). The United Nations High Commissioner for Refugees (UNHCR) is widely recognized as the most reliable source. As of 2012, the UN estimated that there were 15.2 million refugees worldwide, plus another 4.8 million semipermanent Palestinian refugees resulting from the establishment of the state of Israel in 1948 and the 1967 Arab-Israeli war (**Figure 6.17**). The great majority of refugees may be found in developing countries, many of which do not have the adequate infrastructure and resources to manage the flow of forced migrants. This is where the UNHCR comes in. Its primary mission is to safeguard the rights and well-being of refugees. In practical terms, this means delivering shelters, clothing, and medical and personal hygiene supplies to millions of people in hundreds of refugee camps. Refugee camps are meant to be temporary settlements to house and care for displaced persons until they can be safely repatriated. Sometimes repatriation takes years, in which case the UNHCR acts as a surrogate state, securing birth certificates, facilitating schooling, and improving sanitary infrastructure.

REFUGEE SPACES

There is thus a nearly hidden political geography of displaced persons worldwide. The UNHCR operates hundreds of camps and settlements housing millions of people in 124 countries. Although meant as temporary solutions, some camps last decades and become major population centers absent on most maps. Dadaab,

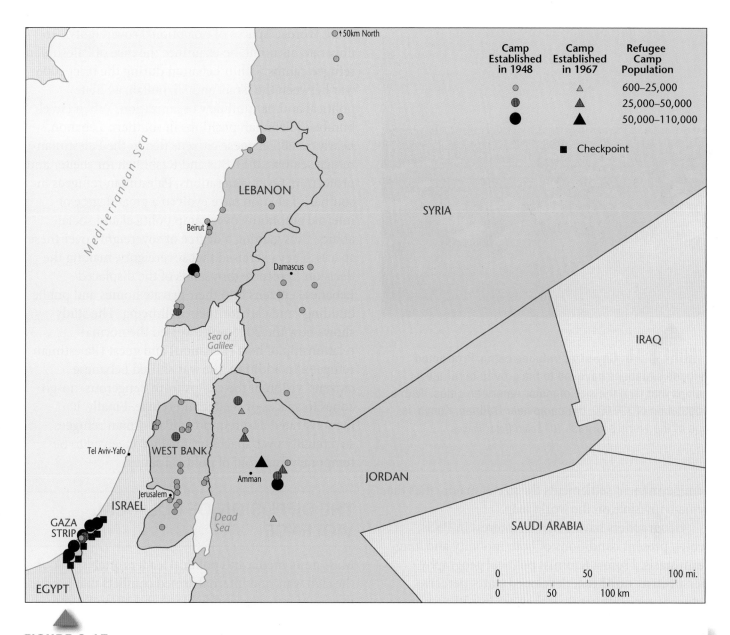

FIGURE 6.17 **Refugee spaces in the Middle East.** Civilians fleeing armed conflict that accompanied the establishment of the state of Israel in 1948 produced a new political geography of refugee settlements in the region. The United Nations continues to administer 58 semipermanent refugee camps in four countries. Civil and legal rights for the 1.4 million camp residents vary greatly from country to country. *(Source: http://www.unrwa.org/where-we-work.)*

Kenya, is the location of one of the oldest, established in 1991 and currently housing at least 320,000 people, probably many more (**Figure 6.18**). The camp and its population do not show up on standard world atlases. Imagine an atlas without St. Louis, Missouri, or Pittsburg, Pennsylvania, U.S. cities with equivalent population totals. Millions of people are born and come into adulthood in refugee camps with no promise of becoming citizens in a new country and little hope of returning to a home they never knew. The UN refers to such cases as a **protracted refugee situation (PRS).** An increasing number of refugee camps fall under this designation, thus creating a geography of

protracted refugee situation Results when people are born and mature into adulthood in refugee camps with no promise of becoming citizens in a new country and little hope of returning to their home country.

FIGURE 6.18 **City-sized refugee camp.** Protracted refugee situations have led to the growth of refugee camps that rival the size of major American cities. With a population of 320,000, the camp near Dadaab, Kenya, is one of the world's largest. *(Oli Scarff/Getty Images.)*

semipermanent settlements the size of major cities that remain invisible on the world map.

Geographers have taken an interest in PRSs, which present conditions of both mobility and long-term stasis. From a feminist political geography perspective, northern, developed states perceive refugees very differently if they are in a PRS or are actively on the move. Jennifer Hyndman and Wenona Giles show that refugees in PRSs are feminized based on their location and legal status. That is, refugees in PRSs are viewed as externalized, motionless, and politically passive and powerless, and thus are coded as nonthreatening/feminine. In contrast, refugees seeking asylum in northern, developed countries are viewed as internalized, on the move, and politically active and vocal, and thus are coded as threatening to both national security and the welfare state. Such feminization, Hyndman and Giles argue, can serve to reinforce the conditions that create PRSs by reducing urgency and lowering priorities for official action.

In another PRS study, geographer Adam Ramadan reveals the relative and contingent qualities of mobility/stasis. Ramadan categorizes the refugee camp as a distinctive political space. PRS camps can become,

in his words, "spaces of exceptional sovereignty." To illustrate his point, he examines the case of Palestinian refugee camps within Lebanon during the brief 2006 war between the Israel and Hizbullah, a Shi'ite political and paramilitary organization. When Israel bombed Hizbullah positions in southern Lebanon, nearly 5,000 Lebanese citizens fled to the Palestinian refugee camps of El-Bus and Rashidieh for shelter and protection. Over generations, Palestinian refugees in southern Lebanon have evolved a great degree of internal autonomy over camp political and social affairs, thus gaining a degree of sovereignty over these spaces. They exercised that sovereignty, making the decision to receive thousands of the displaced Lebanese citizens into their private homes and public buildings and care for their well-being. The study shows how the 2006 war inverted the normal relationship of host (Lebanon) and guest (Palestinian refugee). In addition, the war shifted Lebanese citizens' vision of the camps from dangerous, no-go zones to spaces of shelter and refuge. Finally, it demonstrated and reinforced Palestinian refugees' sovereignty over spaces originally conceived to be temporary and void of political status.

THE DIFFUSION OF POLITICAL VIOLENCE

Mass news media and political leaders often speak of the civil wars and internal armed conflicts that produce masses of refugees worldwide as chaotic and spotty in time and space. Unlike traditional interstate warfare with readily identifiable borders, civil wars are often diffuse with no discernable front. According to geographers Sebastian Schuttea and Nils B. Weidmann, that does not mean there are no regular spatial patterns to be found. In fact, they suggest, civil wars and conflicts exhibit common diffusion characteristics.

Many observers have recognized the relevance of **contagion diffusion** in the spread of armed conflict from one state to another. Schuttea and Weidman's study showed, however, that diffusion patterns are also relevant to analyzing the spread of a single conflict within state boundaries. Using data from four recent internal

contagious diffusion A type of expansion diffusion in which cultural innovation spreads by person-to-person contact, moving wavelike through an area and population without regard to social status.

expansion diffusion
The spread of innovations within an area in a snowballing process, so that the total number of knowers or users becomes greater and the area of occurrence grows.

escalation diffusion
The idea that civil wars may escalate through the diffusion of violence across ever-greater areas over time.

relocation diffusion
The spread of an innovation or other element of culture that occurs with the bodily relocation (migration) of the individual or group responsible for the innovation.

armed conflicts in Europe and Africa, they analyzed how violence spread in time and space. They found that in all cases violence spread in a pattern of **expansion diffusion.** They propose that such cases illustrate what they call **escalation diffusion** in civil wars and armed conflicts. By this, they mean that civil wars "intensify" or "escalate" through diffusion of violence across ever-greater areas over time. They also note that in one case in their study, **relocation diffusion**—the movement of violence to another location—took place as well, although expansion diffusion was the main pattern of spreading of violence. Such modeling of political violence may aid in predicting, and thus possibly preventing, the escalation of civil wars.

▚▚ GLOBALIZATION

6.3 LEARNING OBJECTIVE
Describe the effects of globalization on political geography.

The effects of globalization are complex. On the one hand, the idea of national self-determination continues to spread with the increase in global communications. Thus, new states continue to appear on the world map. On the other hand, the forces of globalization may be rendering some functions of the territorial state obsolete. As we will learn in this section, a great deal of speculation exists regarding the impact of globalization on political geography (see Subject to Debate on page 246).

THE NATION-STATE

As globalization has evolved, one of the most profound effects has been the division of the entire globe into nation-states. The link between global political and cultural patterns is epitomized by the nation-state, created when a nation—a people of common heritage, memories, myths, homeland, and culture; speaking the same language; and/or sharing a particular religious faith—achieves independence as a separate country.

Scholars generally trace the nation-state model to Europe and the European settler colonies in the Americas of the late eighteenth century. Political philosophers of the time argued that self-determination—the freedom to rule one's own country—was a fundamental right of all peoples. The ideal of self-determination spread, and over the course of the nineteenth century, a globalized system of nearly 100 nation-states emerged. Many of these nation-states, such as England, Portugal, Holland, and France, ruled overseas territories as colonies of the mother country. These territories were considered imperial holdings. Thus, in colonial territories, the important boundaries were those between European empires rather than nation-states. Under European Empire, self-determination did not apply to colonies.

However, when the United Nations was created in 1945, it incorporated the principle of national self-determination into its charter. This principle was adopted by independence movements in European colonies around the globe and led ultimately to decolonization. By the end of the 1960s, most colonial empires had dissolved into dozens of new nation-states. The Soviet Union, the last great empire to fall, disintegrated into 15 sovereign nation-states in 1991. The nation-state is now the primary political-geographic form for the entire globe (see Figure 6.1).

GLOBALIZATION AND SOVEREIGNTY

Many commentators have argued in recent years that globalization is eroding the sovereignty of the nation-state. Globalization, it is said, operates in networks and flows that are less and less affected by international borders. Best-selling journalist Thomas Friedman is a leading proponent of this position. He suggests that the power of states in the global order has been greatly diminished by free trade agreements, the Internet, satellite communications, and other such developments. In addition, many economists argue that "global markets" will increasingly dictate an individual state's social policies. That is, the global market will force states to reduce the cost of doing business so that they

SUBJECT TO *Debate*

WHITHER THE NATION-STATE?

Globalization presents powerful new challenges to the nation-state's dominant position in political geography. Undeniably, globalization has redefined the nation-state's role in regulating cultural, economic, and political life within its territorial boundaries. For instance, some scholars suggest that the increasing mobility of people on a global scale means that the ideal of a culturally homogeneous national identity is less and less attainable. In the economic sphere, production, trade, and finance are organized through transnational networks that operate as if national borders did not exist. Politically, the power of the nation-state is weakened by the rise of global institutions, such as the World Trade Organization, that have the power to penalize countries through trade sanctions.

There is, however, counterevidence suggesting that the nation-state is stronger than ever under globalization. In response to the greater transnational mobility of people, nationalist opposition has grown through so-called nativist movements that seek to stem the flow of immigrants at their countries' borders. Although finance and trade operate through global networks, transnational flows remain constrained and directed by the laws and policies of nation-states. Capital investment remains solidly grounded in the legal and institutional structures of the nation-state. The international pecking order, in place for more than a century, remains intact because globalization has done little to shift the structural inequalities among nation-states. Clearly, globalization's implications for the continued supremacy of the nation-state are complex and unresolved.

Continuing the Debate

Based on the discussion above, consider these questions:

- Is globalization a new phenomenon operating independently of the nation-state, or is it a set of processes structured by an underlying foundation of sovereign countries?

- Are the territorial barriers erected by nation-states being dissolved or reinforced by globalization?

- Under what circumstances and because of what cultural, economic, or political phenomena might globalization strengthen or weaken nation-states?

- Is a major reorganization of the world's political geography still unfolding, with the consequences of globalization yet to be fully evident?

- What new forms of political-geographic organization might emerge to supplant the territorial nation-state?

The Hungarian parliament building, built in 1904, is the seat of the Hungarian national government. The monumental size and historic architectural styles of many national capitols are meant to associate the nation-state with the qualities of stability, permanence, and antiquity. *(MIVA Stock/eStock Photo.)*

may remain competitive. The cumulative effect of globalizing processes, then, is a diminution of a nation-state's control over affairs within its boundaries.

According to geographer John Agnew, the idea that globalization is diminishing the role of states is based on a misunderstanding of the history and character of state sovereignty. Agnew argues that sovereignty has been (1) mistakenly equated to territory and (2) assumed to be equally distributed among states. That is, neither the assumption that a state has total control over affairs within its territorial boundaries nor the assumption that a state's sovereignty ends at its borders is historically accurate. These mistaken assumptions lead to an all-or-nothing view of sovereignty. Agnew suggests it is more accurate to think in terms of **effective sovereignty.** That, is we should give attention to the power of states to effectively enforce or ignore sovereignty claims irrespective of territorial boundaries.

> **effective sovereignty**
> The idea that states' power to effectively enforce or ignore sovereignty claims irrespective of territorial boundaries varies in time and from country to country.

An example will help to illustrate these points. The U.S. government began holding international prisoners in a detention center in a naval base at Guantánamo Bay, Cuba, in 2002 (**Figure 6.19**). The government argued that the Guantánamo center is outside of federal court jurisdiction because it is not part of U.S. territory. At the same time, the U.S. government ignored the Cuban state's claim of sovereignty over the area and enforced sole authority within the boundaries of the naval base. In short, the United States exercised *effective* sovereignty over a space that is contained within Cuba's territorial boundaries but completely outside of the U.S.'s. Sovereignty is not matched to nation-state territory in this case nor is it equally distributed between Cuba and the United States. We could present many similar examples related to international migration, Internet use, global transportation, and other such globalizing processes.

So how does Agnew's argument help us understand globalization's effects on state sovereignty? First, not all states will be affected by globalization in the same way and many stronger states will have a great deal of influence over the way globalization unfolds. For example, the U.S. government's new banking regulations implemented within the country's borders after the 2008–2009 credit crisis greatly altered the way global financial markets operate. Second, many of the

FIGURE 6.19 **A case of effective sovereignty.** Although outside of its territorial boundaries, the U.S. government exercises virtual sovereignty over Guantánamo Bay, Cuba. Far from a unique case, the United States Naval Base at Guantánamo illustrates the mistake of equating state sovereignty with state territory. *(AFP/Getty Images.)*

challenges to sovereignty presented by globalization have always existed in some form. Globalization did not create these challenges, but has made them vastly more complicated and wide ranging. As Agnew concludes, the problem is "simply too complex for the binary thinking—globalization versus states."

THE CONDITION OF TRANSNATIONALITY

Transnationalism and *transnationality* are terms that are increasingly being heard in the halls of academia as well as in popular media. Among other things, these terms suggest that the processes of globalization are raising new questions about cultural identities. Nationality and ethnicity may continue to be important in an era of globalization, but what should one make of the increasing number of people who live, work, and play in a way that is neither rooted in a tightly knit ethnic community nor territorially grounded in the traditional nation-state? What sorts of cultural identities do these people create and defend? Is a culture of transnationalism emerging?

Geographer Katharyne Mitchell suggests that we might think about some of the cultural implications of globalization in terms of a "condition of transnationality." She points out that the restructuring of the world economy caused by globalization has produced a great increase in the movement of people across borders. One of the key ways in which this movement differs from other historical migrations is that it tends not to be unidirectional and permanent. With the aid of new transportation and communications technologies, people are able to maintain social networks and physically move about in ways that transcend political boundaries. The experience of cultural "in-betweenness"—of living in and being linked to multiple places around the globe, while being rooted in no single place—has become fundamental to the identity of a growing group of people.

Mitchell's study of Hong Kong Chinese immigrants in Vancouver, British Columbia, Canada, illustrates the idea of transnationalism. Immigrants entering the country under a law designed to attract business investment had to establish businesses based in Canada. The main immigrants taking advantage of this law were Hong Kong Chinese businesspeople leaving the colony in anticipation of its handover to the People's Republic of China.

As it happened, these immigrants maintained business and social ties in both Hong Kong and Canada, moving freely and frequently between them. In the process, a whole set of cultural conflicts were ignited between longtime Vancouver residents and the transnational Chinese. Neighborhoods were transformed as the transnationals attempted to establish themselves economically and culturally. They rapidly bought up real estate, demolished old houses, built houses in uncharacteristic architectural styles, and redesigned residential landscaping. Their mobility—a culturally defining characteristic of transnationals—made them appear transient and rootless in the eyes of longer-term residents and weakened the legitimacy of their claims and practices. This case shows us that the cultural aspects of transnationalism are bound up with the economic and political processes of globalization.

ETHNIC SEPARATISM

Many independent countries—the large majority, in fact—are not nation-states but instead contain multiple national, ethnic, and religious groups within their boundaries. India, Spain, and South Africa provide examples of older multiethnic countries. **Figure 6.20** illustrates this point by dividing up South Africa linguistically. Many, though not all, multiethnic countries came into being in the second half of the twentieth century. These were former colonies in Africa, South Asia, and Southeast Asia whose boundaries were a product of colonialism. European colonial powers drew political boundaries without regard to the territories of indigenous ethnic or tribal groups. These boundaries remained in place when the countries gained the right of self-determination. Although these states are often culturally diverse, they are sometimes plagued by internal ethnic conflict. What's more, members of a single, territorially homogeneous ethnic group may find themselves divided among different states by culturally arbitrary international borders.

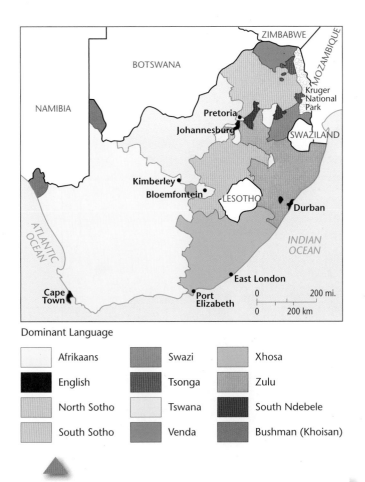

Dominant Language

☐ Afrikaans	☐ Swazi	☐ Xhosa
■ English	☐ Tsonga	☐ Zulu
☐ North Sotho	☐ Tswana	■ South Ndebele
☐ South Sotho	☐ Venda	☐ Bushman (Khoisan)

FIGURE 6.20 **Languages of the Republic of South Africa, a multinational state.** The mixture includes ten native tribal tongues and two languages introduced by settlers from Europe.

Many independent countries function as nation-states because the political power rests in the hands of a dominant, nationalistic cultural group, whereas sizable ethnic minority groups reside in the national territory as second-class citizens. This creates a centrifugal force, disrupting the country's unity. Many of the newest nation-states carved out of the former Soviet Union and Yugoslavia, such as Estonia, Armenia, and Serbia, are relatively linguistically and ethnically homogeneous. These states' cultural homogeneity represents a centripetal force that supports national unity.

Globalization, particularly the establishment of rapid global communications, has encouraged once-isolated and voiceless minority populations to appeal their rights of autonomy and self-determination to the global community beyond their nation-state borders. Ethnic groups and indigenous peoples are

cultural minorities living in a state dominated by a culturally distinct majority. Those who inhabit ethnic homelands (see Chapter 5) often seek greater autonomy or even full independence as nation-states. Even some old and traditionally stable multinational countries have felt the effects of separatist movements, including Canada and the United Kingdom. Certain other countries discarded the unitary form of government and adopted an ethnic-based federalism, in hopes of preserving the territorial boundaries of the state. The expression of ethnic nationalism ranges from public displays of cultural identity to organized protests and armed insurgencies. Occasionally, successful secessions occur, resulting in the birth of a new nation-state.

THE CLEAVAGE MODEL

Why do so many cultural minorities seek political autonomy or independence? The **cleavage model**, originally developed by Seymour Martin Lipset and Stein Rokkan to explain voting patterns in electoral geography, sheds light on this phenomenon. It proposes to explain persistent regional patterns in voting behavior (which, in extreme cases, can presage separatism) in terms of tensions pitting the national core area against peripheral districts, urban against rural, capitalists against workers, and the dominant culture against minority ethnic cultures. Frequently, these tensions coincide geographically: an urban core area monopolizes wealth and cultural and political power, while ethnic minorities, excluded from the power structure, reside in peripheral, largely rural, and less affluent areas.

The great majority of ethnic separatist movements, particularly those that have moved beyond unrest to violence or secession, are made up of peoples living in national peripheries, away from the core area of the country. Every republic that seceded from the defunct, Russian-dominated Soviet Union lay on the borders of that former country. Similarly, the Slovenes and Croats, who withdrew from the former Yugoslavia, occupied border territories peripheral to Serbia, which contained

> **■ cleavage model**
> A political-geographic model suggesting that persistent regional patterns in voting behavior, sometimes leading to separatism, can usually be explained in terms of tensions pitting urban against rural, core against periphery, capitalists against workers, and power group against minority culture.

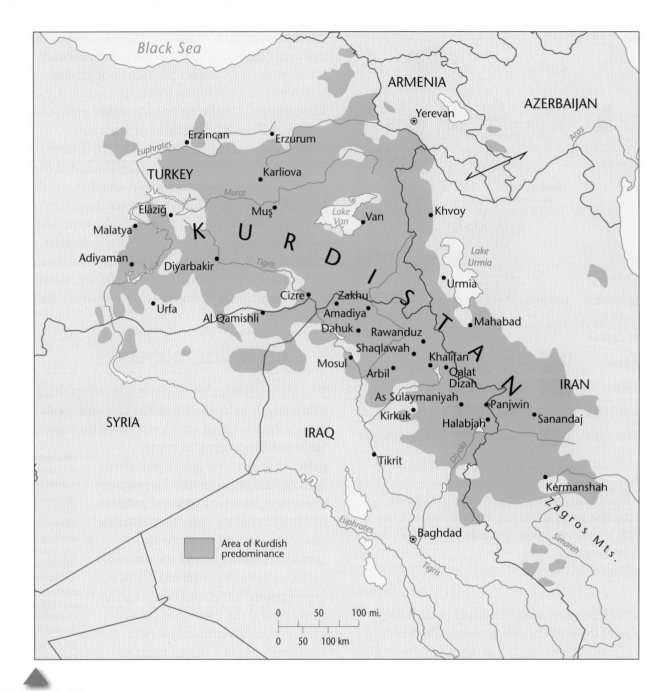

FIGURE 6.21 Kurdistan. This mountainous homeland of the Kurds now lies divided among several countries. The Kurds, numbering 25 million, have lived in this region for millennia. Many seek independence and have waged guerrilla war against Iran, Iraq, and Turkey, but, so far, independence has eluded them. ***What might cause so large and populous a nation to fail to achieve independence?*** (*Source: Office of the Geographer, U.S. Department of State.*)

the former national capital of Belgrade. Kurdistan is made up of the peripheral areas of Iraq, Iran, Syria, and Turkey—the countries that currently rule the Kurdish lands (**Figure 6.21**). Slovakia, long poorer and more rural than Czechia (the Czech Republic) and remote from the center of power at Prague, became another secessionist ethnic periphery. In a few cases, the secessionist peripheries were actually more prosperous than the political core area, and the separatists resented the confiscation of their taxes to support the less affluent core. Slovenia and Croatia both occupied such a position in the former Yugoslavia.

By distributing power, a federalist government reduces such core–periphery tensions and decreases the appeal of separatist movements. Switzerland, which epitomizes such a country, has been able to join Germans, French, Italians, and speakers of Raeto-Romansh into a single, stable independent country. Canada developed under Francophone pressure toward a Swiss-type system, extending considerable self-rule privileges even to the Inuit and Native American groups of the north. Russia, too, has adopted a more federalist structure to accommodate the demands of ethnic minorities, and 31 ethnic republics within Russia have achieved considerable autonomy. One of these, Chechnya (called Ichkeria by its inhabitants), has been fighting for independence.

Less common than the core–periphery pattern of ethnic separatism are cases where two competing cores exist within one state. Spain is a good example. Madrid, the political capital of Spain, is culturally and economically rivaled by Barcelona, one of Europe's largest and most vital industrial cities. Located on the northeast coast of Spain, Barcelona is the capital city of the Autonomous Community of Catalonia, home to Spain's large ethnic Catalan minority. Residents have a native tongue, Catalan, distinct from Spain's official language, Castilian (what English speakers call Spanish). Catalonian nationalists, active for nearly a century, have organized a major political movement for the secession of Catalonia from Spain. By turn of the twenty-first century, the movement had picked up a significant head of steam. A series of polls were conducted among Catalonian residents in 2011 and 2012, the results of which indicated small to overwhelming majorities in favor of secession and Catalonian independence (**Figure 6.22**). The central government in Madrid has made clear that independence for Catalan is not an option. As this edition of the textbook goes to press, the question of Catalan secession remains an open one.

On a more general level, one result of unrest and separatist desires is that the international political map reflects a strong linguistic-religious character. Nevertheless, the distribution of cultural groups is so confoundingly complicated and peoples are so thoroughly mixed in many regions that ethnographic political boundaries can rarely be drawn to everybody's satisfaction.

FIGURE 6.22 **Protest for an independent Catalonia.** Catalonian nationalists have been organizing to secede from Spain for decades. In recent years, the secessionist movement has gained public support in the region, raising the possibility of an independent Catalonia. *(Oscar Dominguez/Alamy.)*

NATURE–CULTURE

6.4 LEARNING OBJECTIVE
Recognize the importance of the physical environment to political geography.

How people use the land and natural resources is profoundly influenced by politics. Whether a particular habitat is conserved or degraded often has much to do with the structures of a country's land laws, tax codes, and agricultural policies. However, increasingly, politics is being defined by changing environmental conditions. National policies about environmental protection, guerrillas seeking a secure base for their operations, and the natural defense provided for an independent country by a surrounding sea—all reveal an intertwining of environment and politics. How governments respond to ecological crises like the loss of biodiversity, pollution, and climate change has become an important political issue. Let's examine this complex two-way interaction between politics and the environment.

CHAIN OF EXPLANATION

When geographers Piers Blaikie and Harold Brookfield used the term *political ecology,* they were interested in trying to understand how political and economic forces affect people's relationships to the land. They suggested that focusing on proximate or immediate causes—for example, the farmer dumping pesticides in a river or the poor peasant cutting a patch of tropical forest—provided an inadequate and misleading explanation of human–environment relations. As an alternative, they developed the idea of a "chain of explanation" as a method for identifying ultimate causes. The chain of explanation begins with the individual "land managers," the people with direct responsibility for land-use decisions—the farmers, timber cutters, firewood gatherers, or livestock keepers. The chain of explanation then moves up in spatial scale, tracing the land managers' economic, cultural, and political relationships from the local to the national and, ultimately, to the global scale.

One of the primary areas addressed by the chain-of-explanation approach is the character of the state, particularly the way in which national land laws, natural resource policies, tax codes, and credit policies influence land-use decision making. For example, if a state assesses high taxes on land improvements, such as terracing and channeling, its tax policies actually create disincentives for land managers to implement soil conservation measures. Conversely, if a state provides cheap loans to land managers to build such structures, its credit policies encourage soil conservation. There are many examples of state influences on individual land-use decisions, leading Blaikie and Brookfield to argue that one cannot fully explain the causes of environmental problems without analyzing the role of the state.

GEOPOLITICS

Spatial variations in politics and the spread of political phenomena are often linked to terrain, soils, climate, natural resources, and other aspects of the physical environment. The term **geopolitics** was originally coined to describe the influence of geography and the environment on political entities.

> **geopolitics** The influence of habitat on political entities.

Conversely, established political authority can be a powerful instrument of environmental modification, providing the framework for organized alteration of the landscape and for environmental protection.

Before modern air and missile warfare, a country's survival was often aided by some sort of natural protection, such as surrounding mountain ranges, deserts, or seas; bordering marshes or dense forests; or outward-facing escarpments. Political geographers named such natural strongholds **folk fortresses.** The folk fortress might shield an entire country or only its core area. In either case, it is a valuable asset. Surrounding seas have helped protect the British Isles from invasion for the past 900 years. In Egypt, desert wastelands to the east and west insulated the fertile, well-watered Nile Valley core. In the same way, Russia's core area was shielded by dense forests, expansive marshes, bitter winters, and vast expanses of sparsely inhabited lands. France—centered on the plains of the Paris Basin and flanked by mountains and hills such as the Alps, Pyrenees, Ardennes, and Jura along its borders—provides another good example (**Figure 6.23**).

> **folk fortress** A stronghold area with natural defensive qualities, useful in the defense of a country against invaders.

Expanding countries often regard coastlines as the logical limits to their territorial growth, even if those areas belong to other peoples, as the drive across the United States from the East Coast to the Pacific Ocean in the first half of the nineteenth century has made clear. U.S. expansion was justified by the doctrine of manifest destiny, which was based on the belief that the Pacific shoreline offered the logical and predestined western border for the country. Underlying the doctrine was a racist ideology of Anglo-Saxon superiority, which held that Native Americans were savages blocking the progress of civilization and the productive use of the western lands. A similar doctrine led Russia to expand in the directions of the Mediterranean and Baltic seas and the Pacific and Indian oceans.

THE HEARTLAND THEORY

Discussions of environmental influence, manifest destiny, and Russian expansionism lead naturally to the **heartland theory** of Halford Mackinder.

> **heartland theory** A 1904 proposal by Harold Mackinder that the key to world conquest lay in control of the interior of Eurasia.

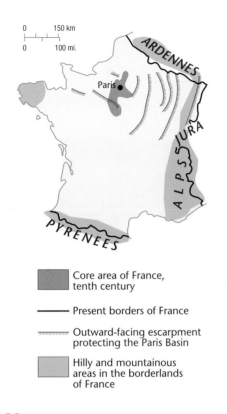

Core area of France, tenth century

Present borders of France

Outward-facing escarpment protecting the Paris Basin

Hilly and mountainous areas in the borderlands of France

FIGURE 6.23 **The distribution of landforms in France.** Terrain features such as ridges, hills, and mountains offer protection. Outward-facing escarpments form a folk fortress that protected the core area and capital of France until as recently as World War I. Hill districts and mountain ranges lend stability to French boundaries in the south and southeast.

Propounded in the early twentieth century and based on environmental determinism, the heartland theory addresses the balance of power in the world and, in particular, the possibility of world conquest based on natural habitat advantage. It held that the Eurasian continent was the most likely base from which to launch a successful campaign for world conquest.

In examining this huge landmass, Mackinder discerned two environmental regions: the **heartland,** which lies remote from the ice-free seas, and the **rimland,** the densely populated coastal fringes of Eurasia in the east,

heartland The interior of a sizable landmass, removed from maritime connections; in particular, the interior of the Eurasian continent.

rimland The maritime fringe of a country or continent; in particular, the western, southern, and eastern edges of the Eurasian continent.

south, and west (Figure 6.24, page 254). Far from the sea, the heartland was invulnerable to the naval power of rimland empires, but the cavalry and infantry of the heartland could spill out through diverse natural gateways and invade the rimland region. Mackinder thus reasoned that a unified heartland power could conquer the maritime countries with relative ease. He believed that the East European Plain would be the likely base for unification. As Russia had already unified this region at the time, Mackinder, in effect, predicted that the Russians would pursue world conquest.

Following Russia's communist revolution in 1917, the leaders of rimland empires and the United States employed a policy of containment. This policy, in no small measure, found its origin in Mackinder's theory and resulted in numerous wars to contain what was then considered a Russian-inspired conspiracy of communist expansion. Overlooked all the while were the fallacies of the heartland theory, particularly its reliance on the discredited doctrine of environmental determinism. In the end, Russia proved unable to hold together its own heartland empire, much less conquer the rimland and the world.

GEOPOLITICS TODAY

In the post–cold war period, geopolitics has reemerged as a dynamic field of political-geographic thought. As geographers Gerard Toal and John Agnew explain, the meaning of political geography is now reversed. Instead of focusing on the influence of geography and the environment on politics, "it now becomes the study of how geography is informed by politics." By this they mean the ways in which political goals and ideologies, based on preconceived notions of cultural identities, regional stereotypes, and regional development hierarchies, influence the ordering of the world. How does the geopolitical outlook of a state structure the world into places of crisis or stability, regions of opportunity or danger, and states of allies or enemies? Many geographers distinguish this new focus on culture by labeling their approach "critical geopolitics."

One of the important aspects of critical geopolitics is a concern with how geopolitics influences our understanding of human–environment

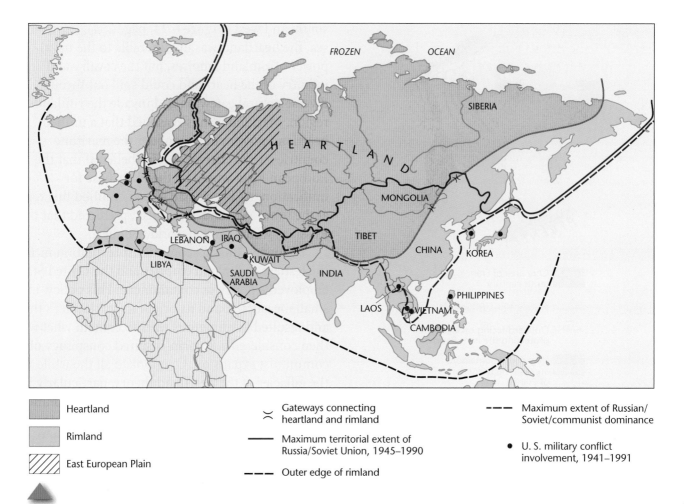

Heartland

Rimland

East European Plain

⌣ Gateways connecting heartland and rimland

— Maximum territorial extent of Russia/Soviet Union, 1945–1990

--- Outer edge of rimland

--- Maximum extent of Russian/Soviet/communist dominance

● U. S. military conflict involvement, 1941–1991

FIGURE 6.24   Heartland versus rimland in Eurasia. For most of the twentieth century, the heartland, epitomized by the Soviet Union and communism, was seen as a threat to America and the rest of the world, a notion based originally on the environmental deterministic theory of the political geographer Halford Mackinder. Control of the East European Plain would permit rule of the entire heartland, which, in turn, would be the territorial base for world conquest. During the cold war, 1945 to 1990, the United States and its rimland allies sought to counter this perceived menace by a policy of containment—resisting every expansionist attempt by the heartland powers. *(Sources: Mackinder, 1904; Spykman, 1944.)*

relations and affects the way we transform the environment. Consider, for example, the current scientific and policy debates over global climate change caused by greenhouse gas emissions. Worldwide debates must be placed in the historical geopolitical context of the global North's political and economic domination of the global South. From the South's perspective, according to Simon Dalby, "the North got rich by using fossil fuels for generations. Why should those in the South forgo the same possibilities just because they come on the development scene a little later?" According to advocates of the South, the North's ideas about restricting future emissions of greenhouse gases

globally will have a disadvantageous effect on the South's economic development. Thus, questions of global environmental management are not restricted to the realm of science and technology; they also fall squarely in the realm of geopolitics.

Another illustration of how geopolitics influences human–environment relations may be seen in the debates concerning the conservation of global biodiversity. Current global conservation policy suggests that the most effective way to save the world's biodiversity is by creating protected areas such as national parks and reserves. However, Roderick Neumann has demonstrated that these protected areas, whether in the North or the South, were created in the historical context of European conquest and colonization. In the British Empire, protected areas were part of a grand economic development strategy to spatially reorder colonies into separate spheres of nature and culture. In the United States, park creation was conducted in the context of manifest destiny and the removal of Native Americans from their homelands and their placement on reservations. In both cases, the environmental management strategies of native cultures were denigrated as immoral and irrational, providing the justification for discarding local land and resource claims and practices. Today, those who have lost their land to protected areas argue bitterly that they bear the main costs of conservation (**Figure 6.25**). Thus, as with the case of global climate change, the North–South debates over strategies for the conservation of global biodiversity fall under the domain of geopolitics.

STATE MILITARIES AND THE ENVIRONMENT

Nearly every state has a military to defend its flag and territory. The ecological impact of all of these armies and navies is complex and sometimes surprising. Interstate warfare delivers perhaps the most spectacular and devastating effects. "Scorched earth," the systematic destruction of resources, has been a favored practice of retreating armies for millennia. During the Vietnam War, the U.S. military sprayed defoliants on that country's tropical forests and croplands in an effort to deny the enemy of the cover of vegetation. Not only was vegetation destroyed, but the accompanying pollution is also blamed for 500,000 birth defects. In the 1991 Gulf War, the Iraqi military devised a scorched earth tactic of lighting over 600 oil wells on fire as it withdrew from Kuwait, thereby releasing tons of pollutants into the atmosphere and ground. People fleeing armies can have major environmental impacts. Wars create refugees and refugees tend to concentrate in encampments, sometimes comprising hundreds of thousands of people. For example, Kenya's Dadaab camp and its immediate surroundings have been largely deforested as the growing population scavenges for fuel, wood, and building materials.

Even military exercises and tests can be devastating. Certain islands in the Pacific were

FIGURE 6.25 Two visions of the landscape. Carter Camp speaks to a group of American Indians and others protesting a proposed tourist development near Bear Butte State Park in the Black Hills near Sturgis, South Dakota. The Oglala Sioux Tribe of the nearby Pine Ridge Indian Reservation consider the park area to be sacred ground. *What kinds of conflicts may arise between conservation goals and indigenous peoples' rights, and how might we resolve them?* (Doug Dreyer/AP Photo.)

rendered uninhabitable, perhaps forever, by American hydrogen bomb testing in the 1950s. The Hanford Nuclear Reservation in Washington State manufactured the plutonium for those bombs and many others (**Figure 6.26**, page 256). Now largely decommissioned, Hanford represents the single largest radioactive waste site in North America. Sited dangerously in the upper Colombia River drainage, the site's radioactive contamination poses a long-term threat to both terrestrial and marine ecosystems over a vast territory. About a third of the underground tanks containing radioactive liquids are leaking into the soil and groundwater. It is hoped that the U.S. military and other agencies can clean up the site before contaminants enter the Colombia River drainage.

Militaries and wars have less visible and dramatic effects on the environment, some of which may actually be ecological beneficial. During the civil war in Nicaragua, stretches of tropical forests became "no-person's lands," as people did not enter out of fear of both the army and insurgents who used the forests

FIGURE 6.26 **Nuclear waste storage at the Hanford Nuclear Reservation** The Hanford facility produced plutonium for nuclear weapons for the U.S. military. Located along the Columbia River, many fear that the site's stored nuclear waste may soon leak into the groundwater and spread radioactive pollution over a large region. ***Could modern war be waged without such pollution?*** *(Jeff T. Green/Getty Images.)*

FIGURE 6.27 **Wetlands of the former Mare Island Naval Shipyard.** Because military bases prohibit commercial development, they sometimes inadvertently protect vast areas of valued natural habitats. When the Mare Island Naval Shipyard in Vallejo, California, closed, some of its land holdings were added to the San Pablo Bay National Wildlife Refuge that protects scarce wetlands. *(Courtesy of Roderick Neumann.)*

as hideouts. As a result, wildlife populations recovered under the decreased hunting pressure. U.S. military bases have made vast areas off-limits to civilians and thereby inadvertently protected habitats for hundreds of endangered plant and animal species from commercial exploitation. When these bases are decommissioned (permanently closed), they are sometimes converted to conservation areas so that protection continues. For instance, part of the decommissioned Fort McClellen in Alabama became the 9,000-acre Mountain Longleaf Wildlife National Refuge in 2004, thereby continuing protection of one of the last remaining longleaf pine forests. Part of the Mare Island Naval Shipyard in San Francisco Bay was similarly passed to the San Pablo Bay National Wildlife Refuge when it was decommissioned (Figure 6.27).

Clearly, warfare—especially modern high-tech warfare—can be environmentally catastrophic. Yet ecological destruction is not always the only result of war or preparation for war. As with many nature–culture phenomena, the interactions are complex and have unpredictable and unintended outcomes.

CULTURAL LANDSCAPE

6.5 LEARNING OBJECTIVE
Analyze the role of cultural landscape in political geography.

The world over, national politics is literally written on the landscape. State-driven initiatives for frontier settlement, economic development, and territorial control have profound effects on the landscape. Conversely, political writers and politicians look to landscape as a source of imagery to support or discredit political ideologies.

IMPRINT OF THE LEGAL CODE

Many laws affect the cultural landscape. Among the most noticeable are those that regulate the land-survey system because they often require that land be divided

FIGURE 6.28 **Can you find the United States–Mexico border in this picture?** *Why does the cultural landscape so vividly reveal the political border?* (Courtesy of Terry G. Jordan-Bychkov.)

Legal imprints can also be seen in the cultural landscape of urban areas. In Rio de Janeiro, height restrictions on buildings have been enforced for a long time. The result is a waterfront lined with buildings of uniform height (Figure 6.29). By contrast, most American cities have no height restrictions, allowing skyscrapers to dominate the central city. The consequence is a jagged skyline for cities such as San Francisco or New York City. Many other cities around the world lack height restrictions, such as Malaysia's Kuala Lumpur, which has the world's tallest skyscrapers.

Perhaps the best example of how political philosophy and the legal code are written on the landscape is the so-called township and range system of the United States. The system is the brainchild of Thomas Jefferson—an early U.S. president and one of the authors of the U.S. Constitution—who chaired a national committee on land surveying that resulted in the U.S. Land Ordinance of 1785. Jefferson's ideas for surveying, distributing, and settling the western frontier lands as they were cleared of Native Americans were based on a political philosophy of "agrarian democracy." Jefferson believed that political democracy had to be founded on economic democracy, which in turn required a national pattern of equitable land-ownership by small-scale independent farmers. In order to achieve this agrarian democracy, the western lands would need to

into specific geometric patterns. As a result, political boundaries can become highly visible (Figure 6.28). In Canada, for example, the laws of the French-speaking province of Québec encourage land survey in long, narrow parcels, but most English-speaking provinces, such as Ontario, use a rectangular system. Thus, the political border between Québec and Ontario can be spotted easily from the air.

FIGURE 6.29 **Legal height restrictions, or their absence, can greatly influence urban landscapes.** Singapore (*left*), the city-state at the tip of the Malay Peninsula, lacks such restrictions, and its skyline is punctuated by spectacular skyscrapers. In Rio de Janeiro (*right*), by contrast, height restrictions allow the natural environment to provide the "high-rises." (Left: David Ball/Alamy; Right: SIME/eStock Photo.)

be surveyed into parcels that could then be sold at prices within reach of family farmers of modest means.

Jefferson's solution was the township and range system, which established a grid of square-shaped "townships" with 6-mile (9.6-kilometer) sides across the Midwest and West. Each of these was then divided into 36 sections of 1 square mile (2.6 square kilometers), which were in turn divided into quarter-sections, and so on. Sections were to be the basic landholding unit for a class of independent farmers. Townships were to provide the structure for self-governing communities responsible for public schools, policing, and tax collection. With the exception of the 13 original colonies and a few other states or portions of states, a gridlike landscape was imposed on the entire country as a result of Jefferson's political philosophy and accompanying land-survey system (**Figure 6.30**).

PHYSICAL PROPERTIES OF BOUNDARIES

Demarcated political boundaries can also be strikingly visible, forming border landscapes. Political borders are usually most visible where restrictions limit the movement of people between neighboring countries. Sometimes such boundaries are even lined with cleared strips, barriers, pillboxes, tank traps, and other obvious defensive installations. At the opposite end of the spectrum are international borders, such as that between Tanzania and Kenya in East Africa, that are unfortified, thinly policed, and in many places very nearly invisible. Even so, undefended borders of this type are usually marked by regularly spaced boundary pillars or cairns, customhouses, and guardhouses at crossing points (**Figure 6.31**).

Occasionally, present-day political boundaries reflect the durable landscape imprint of long-vanished cultures. Some of the best-known examples come from the ancient Roman Empire. In England, for example, many of today's parish and township boundaries surrounding the city of Bath follow the property markers of ancient Roman villas. Probably the best known of the Roman landscape's influence on present-day boundaries is Hadrian's Wall in northern England. Named after a Roman Emperor, the wall was constructed as a fortification to defend against the unconquered peoples to the north. Today,

FIGURE 6.30 **Imposing order on the land.** This aerial view of Canyon County, Idaho, farmland reveals a landscape grid pattern. This pattern was imposed on the landscape across much of the United States following the passage of the Land Ordinance of 1785. *(David Frazier/Image Works.)*

Hadrian's Wall parallels the modern border between England and Scotland (**Figure 6.32**).

THE IMPRESS OF CENTRAL AUTHORITY

Attempts to impose centralized government appear in many facets of the landscape. Railroad and highway patterns focused on the national core area, and radiating outward like the spokes of a wheel to reach the hinterlands of the country, provide good indicators of central authority. In Germany, the rail network developed largely before unification of the country in 1871. As a result, no focal point stands out. However, the superhighway system of autobahns, encouraged by Hitler as a symbol of national unity and power, tied the various parts of the Reich to such focal points as Berlin or the Ruhr industrial district.

The visibility of provincial borders within a country also reflects the central government's strength and stability. Stable, secure countries, such as the United States, often permit considerable display of provincial borders. Displays aside, such borders are easily crossed. Most state boundaries in the United

FIGURE 6.31 **Even peaceful, unpoliced international borders often appear vividly in the landscape.** The International Peace Garden park near Bottineau, North Dakota, and Boissevain, Manitoba, celebrates international cooperation between the United States and Canada—ironically, by emphasizing an otherwise discreet international border. *(Terrance Klassen/Alamy.)*

FIGURE 6.32 **Hadrian's Wall, U.K.** Now a World Heritage Site, this ancient Roman defensive fortification parallels the nearby modern boundary between England and Scotland. *Can you think of comparable modern structures? (Peter Mulligan/Getty Images.)*

States are marked with signboards or other features announcing the crossing. By contrast, unstable countries, where separatism threatens national unity, often suppress such visible signs of provincial borders. Also in contrast, such "invisible" borders may be exceedingly difficult to cross when a separatist effort involves armed conflict.

NATIONAL ICONOGRAPHY ON THE LANDSCAPE

The cultural landscape is rich in symbolism and visual metaphor, and political messages are often conveyed through such means. Statues of national heroes or heroines and of symbolic figures such as the goddess Liberty or Mother Russia form important parts of the political landscape, as do assorted monuments. The elaborate use of national colors can be visually very powerful as well. Landscape symbols such as the Rising Sun flag of Japan, the Statue of Liberty in New York Harbor, and the Latvian independence pillar in Riga (which stood untouched

throughout half a century of Russian-Soviet rule) evoke deep patriotic emotions (**Figures 6.33** and **6.34**). The sites of heroic (if often futile) resistance against invaders, as at Masada in Israel, prompt similar feelings of nationalism.

Some geographers theorize that the political iconography of landscape derives from an elite, dominant group in a country's population and that its purpose is to legitimize or justify its power and control over an area. The dominant group seeks both to rally emotional support and to arouse fear in potential or real enemies. As a result, the iconographic political landscape is often controversial or contested, representing only one side of an issue. Look again at Figure 6.34. The area in which Mount Rushmore stands, the Black Hills, is sacred to the Native Americans who controlled the land before whites seized it. How might these Native Americans, the Lakota Sioux, perceive this monument? Are any other political biases contained in it? Cultural landscapes are always complicated and subject to differing interpretations and meanings, and political landscapes are no exception.

World Heritage Site

Tiwanaku is an ancient example of how cultural landscapes can reflect and reinforce centralized political authority. Inscribed by UNESCO in 2000, Tiwanaku served as the political and religious center of an empire that dominated the Altiplano region of the Andes Mountains between 400 and 900 c.e. It retains political geographic significance today as a symbol of Bolivian national identity.

■ Tiwanaku was a sprawling planned city of 70,000 to 125,000 residents located near the southern shores of Lake Titicaca in Bolivia. Its ruins provide an exceptional example of pre-Incan civic architecture. The architecture and city layout symbolically reflect the site's former role as an imperial center.

■ Surrounding the imperial center were the towns and villages colonized by the Tiwanaku rulers. Even further out were the settlements that the center dominated, but did not colonize or conquer militarily. All of the areas were knitted together economically and politically by a network of alpaca caravan trade routes that stretched across the Altiplano and beyond. The central authority at Tiwanaku controlled the caravans.

■ The Tiwanaku rulers developed an ideological and religious iconography (visual images and symbols used in art) that was embedded in the center's architecture and then reproduced in the material life of the far-flung regions under its influence. For example, iconography originating from the center appeared on ceramic vessels and was reproduced on stone markers placed in peripheral settlements. Conversely, the rulers incorporated the iconography of the periphery into public architecture of the central city, thereby encouraging a variety of distant ethnic communities to identify themselves as part of a larger state. Thus, the symbolism embedded in civic architecture reinforces the idea of a periphery politically, culturally, and economically linked together through the center.

(AFP/Getty Images.)

Tiwanaku: Spiritual and Political Center of the Tiwanaku Culture

ANCIENT CIVIC ARCHITECTURE:

Much of the ancient city has been erased by modern development, but the ruins of the monumental stone buildings in the ceremonial center remain protected. The architecture of the center, which is oriented toward the cardinal points, reflects both the complexity of the Tiwanaku Empire's political structure as well as its religious nature.

(Lonely Planet Images/Getty Images.)

■ Tiwanaku culture perfected stone cutting, carving, and polishing and used those skills to create a monumental architecture as a projection of its power. A series of architectural structures—Kalasasaya's Temple, Akapana's Pyramid, and Pumapumku's Pyramid—define the space of the center.

■ Akapana's Pyramid, originally rising over 18 meters, is the most commanding of the structures. Though since destroyed, a temple stood at the top, as is also common in Mesoamerican architecture.

■ Kalasasaya's Temple is a rectangular structure that likely functioned as an observatory. In its interior are two carved monoliths as well as the Gate of the Sun, a monumental sculpture standing 3 meters tall, 4 meters wide, and carved from a single stone. Multiple human and anthropomorphic animal carvings decorate the Gate of the Sun, with what is likely a major deity appearing at the top of the monument, at its center. Some archeologists suggest that the carvings on the gate constituted an agricultural calendar.

■ Other architectural features in the ancient city include the Palace of Putuni and Kantatillita, which stands in tribute to the political and administrative authority of imperial rule.

TOURISM:
Located only 70 kilometers west of La Paz, Bolivia, and served by public transportation, Tiwanaku is relatively accessible.

(AFP/Getty Images.)

■ Popular literature in the United States and Europe speculated (wrongly) that the site provides evidence that an advanced race of extraterrestrials once visited Earth. This notoriety has been a major draw for international tourists, particularly during astronomical events.

■ Sometimes referred to as the "American Stonehenge," thousands of visitors gather during the southern hemisphere's winter solstice for a dawn ritual.

■ Visitation has sparked a local tourist industry that has become the most important source of income in surrounding communities.

■ Located near the southern shores of Lake Titicaca in Bolivia.

■ Tiwanaku craftspeople perfected stone cutting, carving, and polishing, creating a monumental architecture as a projection of the center's political power.

■ Kalasasaya's Temple, Akapana's Pyramid, and Pumapumku's Pyramid define the space of the center.

http://whc.unesco.org/en/list/567

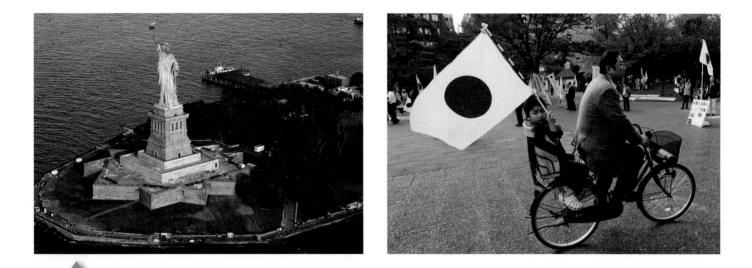

▲ **FIGURE 6.33** **The Statue of Liberty and the Rising Sun flag** are widely recognized national symbols of the United States and Japan, respectively. ***Can you think of other iconic landscapes that symbolize a nation?*** *(Left: Photodisc/Getty Images; Right: Kimimasa Mayama/EPA/Landov.)*

The landscapes of capital cities are important in shaping a sense of national identity and belonging. Geographer Diana Ter-Ghazaryan's study of

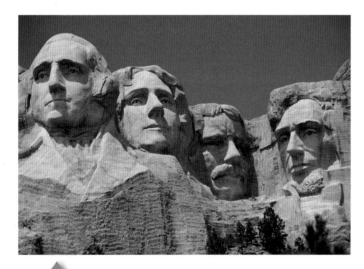

▲ **FIGURE 6.34** **Mount Rushmore,** in the Black Hills of South Dakota, presents a highly visible expression of American nationalism, an element of the political landscape. ***If political landscapes are created by an elite, in an effort to legitimize and justify their control over territory, then who might disapprove of this monument?*** *(Paul Damien/National Geographic Creative.)*

Armenia's national capital, Yerevan, shows how central and contentious urban landscape design can be to national identity. After gaining independence from the Soviet Union, the Armenian political elite of the country began redesigning the capital city as part of a process of symbolically constructing a national identity for a modern, democratic Armenia. Ter-Ghazaryan identified three key sites that symbolically anchored the Yervan landscape: Opera Square, Northern Avenue, and Republic Square (**Figure 6.35**). The post-Soviet development of each of these sparked street protests over what Yerevan will become and what it means to be Armenian. National identity, especially in times of major political transition such as Armenia has experienced, can be a very contentious issue that is expressed through struggles over the physical landscape of a nation's capital.

In a more contemporary study, geographer Gail Hollander has demonstrated the important symbolic role that the landscape and environment of the Florida Everglades have played in U.S. presidential elections (**Figure 6.36**). The symbolic role of the Everglades changed over time, from the 1928 presidential campaign to the present. In 1928 it was presented as worthless swampland. As such, it played a key role in the election of President Herbert Hoover,

FIGURE 6.35 Opera Square in Yerevan, Armenia's capital. Opera Square is an iconic place in the city's landscape and serves as a touchstone for Armenian national identity. As such, it has served as a locus for contentious debates over what it means to be Armenian in the post-Soviet era. *(Luis Dafos/Alamy.)*

FIGURE 6.36 Everglades National Park in Florida has long played a role in U.S. presidential politics. Visits to iconic national parks and protected landscapes are common in U.S. political campaigns and meant to symbolize a candidate's commitment to protecting the environment. Here, Vice President Joe Biden speaks during an Everglades tour while United States Senator Bill Nelson and Representative Alcee Hastings look on. *(Taimy Alvarez/Sun Sentinel/MCT via Getty Images.)*

who promised to drain it for agricultural development. By the 1970s, it was seen as an endangered wetland in need of protection and ecological restoration. Thus, in virtually every present-day presidential campaign, the candidates' positions on the Everglades are seen as indicators of their commitment to the environment. Photo opportunities in the park give both Democratic and Republican candidates alike a chance to symbolically link their political campaigns to the ecological restoration of what has come to be viewed as a national treasure.

CONCLUSION

Political spatial variations—from local voting patterns to the spatial arrangement of international power blocs—add yet another dimension to the complex human mosaic. In particular, nation-states operate as vital functional regions, which help shape many aspects of culture. Regions constantly change as political innovations ebb and flow across their surfaces. Political phenomena as varied as the nation-state, separatist movements, women's suffrage, and the territorial expansion of countries move along the paths of diffusion. Globalization interacts with political geography in complex and even contradictory ways. The forces of globalization have strengthened some features of the nation-state and weakened others.

Nature–culture relations are important to political geography, and the tools of political ecology help us understand the links between systems of power and the physical environment. Countries do not exist in an environmental vacuum. The spatial patterns of landforms often find reflection in boundaries, core areas, and geopolitical strategies. Likewise, political culture very much influences our ideas and judgments about landscape and environment. Finally, politics leaves diverse imprints on the cultural landscape, and landscapes often provide the symbolism and visual metaphors to support or refute political ideologies. Political geography is clearly important to understanding the human mosaic.

DOING GEOGRAPHY

The Complex Geography of Congressional Redistricting

Congressional redistricting normally happens every 10 years in the United States, following each national census. In some cases, such as Texas in 2002, redistricting occurs between censuses. Although soon proved wrong, newspaper accounts at the time predicted that Texas's midcensus redistricting would protect the Republican Party's majority in the U.S. Congress for the foreseeable future. Does this midcensus redistricting fall under the category of Republican gerrymandering, as some claim, or is it, as the Texas Republican Party argues, a case of necessary adjustments in response to population shifts? Either way, the case of Texas demonstrates how critically important the drawing of congressional district boundaries is to democratic governance.

This exercise requires you to identify cases of possible gerrymandering in your home state or an adjacent state. To do so, follow the steps below.

Steps to Identifying Gerrymandering

Step 1: Obtain a map of congressional district boundaries in your chosen state (http://nationalatlas.gov is one possible source). Once you have done so, see if you can visually identify districts that may have been gerrymandered. Figure 6.9 (page 236) and the discussion on pages 235–237 should be helpful to you in identifying such districts.

Step 2: Having identified your candidate(s) for gerrymandering, address the following questions:

■ What was it about the configuration of the boundaries that made you think the district(s) may have been gerrymandered?

■ When were the boundaries drawn?

■ Can you identify which of the major political parties was in power when the boundaries were drawn?

■ Which party do these boundaries favor and why? That is, what are the racial, economic, religious, and ethnic characteristics of the district(s) that may suggest a particular party affiliation?

Step 3 Look at the proportion of major party registration in nearby districts to see if you can determine whether the boundary lines were drawn in order to dilute or to concentrate opposition votes.

The complex, convoluted shapes of some voting district boundaries, such as U.S. Congressional District 4 in Illinois, are evidence of gerrymandering. *(Source: National Atlas.gov.)*

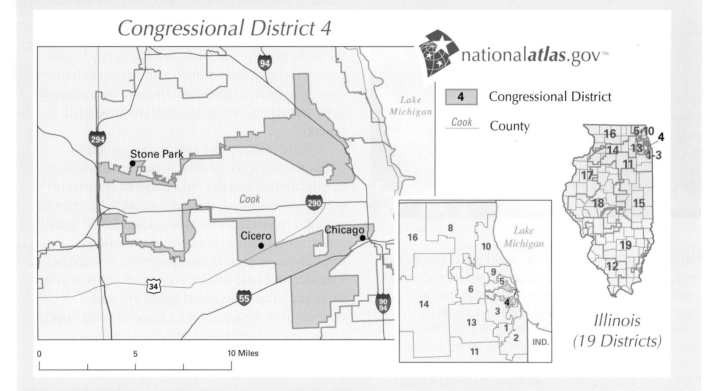

SEEING GEOGRAPHY Post-9/11 Security Fences

Are these border fences or walls?

Post-9/11 border security: the boundary between Israeli and Arab-Palestinian lands (*top*) and the U.S.–Mexico border (*bottom*).
(*Top: ImageBROKER/Alamy; Bottom: age fotostock/Getty Images.*)

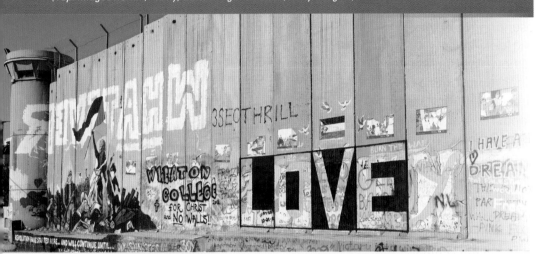

This question is more politically charged than it may seem at first. It is vehemently debated, and the side of the border from which one is observing greatly influences the answer. "Fence" suggests neighborliness; "wall" suggests isolation and exclusion. The two governments responsible for the structures, Israel and the United States, both argue that these are fences, which increase security for citizens on both sides. The U.S. Congress passed the Secure Fence Act following the 2001 terrorist attacks on the World Trade Center and the Pentagon. The act authorized the expenditure of more than $1 billion to build a 700-mile (1126-kilometer) fence on the U.S. border with Mexico to improve "homeland security." Critics argue that it is really a wall meant to stop the across-the-border flow of undocumented workers from Mexico and Central America, and that there is little hope of its stopping terrorists from entering the United States. They also note that the three coastlines and the border with Canada remain without similar barriers.

The case of the barrier between Israeli and Arab-Palestinian lands is even more hotly disputed. In 2004 the Israeli government began building a physical barrier between its territory and that controlled by the Palestinian Authority. The Israel Ministry of Foreign Affairs claims that the "antiterrorist fence" is an act of self-defense against terrorists entering from Palestinian-ruled lands, such as the Gaza Strip. They also note that only 3 percent of the barrier is a concrete wall, whereas the remainder is a chain-link fence. The Palestinian Authority counters that the "wall" violates civil and human rights by cutting off communities' access to schools, workplaces, and families. The International Court of Justice seems to support the Palestinian position, ruling that construction of "the wall" is contrary to international law. Whichever side of the fence you come down on, these barriers are powerful reminders of the continuing relevance of international borders.

THE VIDEO CONNECTION ▼

Border Patrol Body Slam
www.tinyurl.com/nm5az3p

This video documents a *lucha libre* match—a popular Mexican form of professional wrestling—in a California farming community. The match features two teams, one representing Mexican nationals with a Robin Hood–style professional wrestler known as Blue Demon Jr., and the other representing the U.S. Border Patrol with a menacing border agent known as El Patrón Oliver John. The match concludes with the "hero," Blue Demon Jr., defeating the "villain," El Patrón. Blue Demon Jr. explains that the performance provides something like a therapeutic release for the audience's frustration with their treatment at the hands of the Border Patrol and with their experiences as "illegals" in the United States.

Thinking Geographically

1. In this chapter, we learned about the concept of nationalism and some of the many ways it is expressed culturally. How are Mexican nationalism and national identity expressed in this video? Think about the performers, the makeup of the audience, and the audience's responses to different parts of the *lucha libre* performance.

2. Boundaries and borders are of key concern in political geography. For example, the Seeing Geography feature on page 265 discusses how different groups perceive the meaning of international border structures. What differences in the meaning of borders do you detect in this video? How do you think the *lucha libre* audience perceives the border between Mexico and the United States?

3. The theatrical aspects of professional wrestling rely on clear distinctions between good and evil, heroes and villains. Why are the two teams in this performance given their respective roles? How do these roles challenge or support prevailing political opinions regarding immigration in the United States?

Chapter 6
LEARNING OBJECTIVES REEXAMINED

6.1 Identify the importance of region to political geography.
See pages 223–239. Explain the terms *nationalism* and *nation-state*. How do these terms differ?

6.2 Identify the various ways that mobility and political geography interact.
See pages 239–245. Name two major political events that resulted in massive human migrations.

6.3 Describe the effects of globalization on political geography.
See pages 245–251. How do the forces of globalization both help and hinder nation-states?

6.4 Recognize the importance of the physical environment to political geography.
See pages 251–256. Name two ways that politics has affected the physical environment.

6.5 Analyze the role of cultural landscape in political geography.
See pages 256–263. Name two ways that politics influences cultural landscapes.

KEY TERMS

Political Geography on the Internet

You can learn more about political geography on the Internet at the following web sites:

European Union

http://europa.eu/index_en.htm

Here, you can find information about the 28-member supranational organization that is increasingly reshaping the internal political geography of Europe.

Global Geopolitics Net

http://globalgeopolitics.net/

Sponsored by the Eurasia Research Center, this site presents information, analysis, and opinion on global politics, problems of intelligence gathering and analysis, counterterrorism, human rights, globalization, and other world issues.

International Boundary News Archive

http://www.dur.ac.uk/ibru/resources/newsarchive/

This database contains more than 10,000 boundary-related reports from a wide range of news sources around the world dating from 1991 to March 2001, with additional reports from 2006 onward.

International Geographical Union (IGU): Commission on Political Geography

http://www.cas.muohio.edu/igu-cpg/

The objective of the commission is to study the main theoretical issues of political geography, including questions on the rise and fall of empires, the emergence of new geopolitical models, and contemporary challenges to the state. The site features the commission's newsletters.

Political Geography Specialty Group, Association of American Geographers

http://www.politicalgeography.org/

This site provides details about the activities and meetings of specialists in political geography and includes useful links to other sites featuring political geography and geopolitics.

United Nations

http://www.un.org

Search the worldwide organization with a membership that includes the large majority of independent countries. The site contains politically diverse information about such ventures as peacekeeping and conflict resolution.

Sources

Agnew, John. 1998. *Geopolitics: Re-Visioning World Politics.* London: Routledge.

Agnew, John. 2011. *Globalization and Sovereignty.* Lanham, Md.: Rowan & Littlefield.

Agnew, John, and Luca Muscarà. 2012. *Making Political Geography.* 2nd ed. Lanham, Md.: Rowan & Littlefield.

Berman, Ari. 2012. "How the GOP Is Resegregating the South." *The Nation,* February 20. Accessed July 4, 2013. http://www.thenation.com/article/165976/how-gop-resegregating-south#axzz2YaEuHCe4.

Blaikie, Piers, and Harold Brookfield. 1987. *Land Degradation and Society.* London: Methuen.

Blouet, Brian W. 1987. *Halford Mackinder: A Biography.* College Station: Texas A&M University Press.

Bradsher, Keith. 2012. "China Blocks Web Access To Times After Article." *New York Times,* October 26, A12.

Dalby, Simon. 2002. "Environmental Governance." In R. Johnston, P. Taylor, and M. Watts (eds.), *Geographies of Global Change: Remapping the World,* pp. 427–440. London: Routledge.

Draper, Robert. 2012. "The League of Dangerous Mapmakers." *The Atlantic,* October. Accessed July 1, 2013. http://www.theatlantic.com/magazine/archive/2012/10/the-league-of/309084/.

Elazar, Daniel J. 1994. *The American Mosaic: The Impact of Space, Time, and Culture on American Politics.* Boulder, Colo.: Westview Press.

Enghoff, Martin, Bente Hansen, Abdi Umar, Bjørn Gildestad, Matthew Owen, and Alex Obara. 2010. *In Search of Protection and Livelihoods: Socio-economic and Environmental Impacts of Dadaab Refugee Camps on Host Communities.* Nairobi: Government of Kenya.

Gastner, M., C. Shalizi, and M. Newman. 2004. "Maps and Cartograms of the 2004 U.S. Presidential Election Results." University of Michigan Department of Physics and Center for the Study of Complex Systems.

Hollander, Gail. 2005. "The Material and Symbolic Role of the Everglades in National Politics." *Political Geography* 24(4): 449–475.

Hyndman, Jennifer, and Wenona Giles. 2011. "Waiting for What? The Feminization of Asylum in Protracted Situations." *Gender, Place & Culture: A Journal of Feminist Geography* 18(3): 361–379.

Jones, Martin, and Rhys Jones. 2004. "Nation States, Ideological Power and Globalization: Can Geographers Catch the Boat? *Geoforum* 35: 409–424.

Mackinder, Halford J. 1904. "The Geographical Pivot of History." *Geographical Journal* 23: 421–437.

Mitchell, Katharyne. 1998. "Fast Capital, Race, and the Monster House." In R. George (ed.), *Burning Down the House: Recycling Domesticity,* pp. 187–212. Boulder, Colo.: Westview Press.

Mitchell, Katharyne. 2002. "Cultural Geographies of Transnationality." In K. Anderson, M. Domosh, S. Pile, and N. Thrift (eds.), *Handbook of Cultural Geography,* pp. 74–87. London: SAGE.

Neumann, Roderick P. 1995. "Local Challenges to Global Agendas: Conservation, Economic Liberalization, and the Pastoralists' Rights Movement in Tanzania." *Antipode* 27(4): 363–382.

Neumann, Roderick P. 2002. "The Postwar Conservation Boom in British Colonial Africa." *Environmental History* 7(1): 22–47.

Neumann, Roderick P. 2004. "Nature-State-Territory: Toward a Critical Theorization of Conservation Enclosures." In R. Peet and M. Watts (eds.), *Liberation Ecologies,* 2nd ed., pp. 195–217. London: Routledge.

O'Reilly, Kathleen, and Gerald R. Webster. 1998. "A Sociodemographic and Partisan Analysis of Voting in Three Anti-Gay Rights Referenda in Oregon." *Professional Geographer* 50: 498–515.

Ó Tuathail, Gearóid. 1996. *Critical Geopolitics.* Minneapolis: University of Minnesota Press.

Paulin, C., and John K. Wright. 1932. *Atlas of the Historical Geography of the United States.* New York: American Geographical Society and the Carnegie Institute.

Ramadan, Adam. 2008. "The Guests' Guests: Palestinian Refugees, Lebanese Civilians, and the War of 2006." *Antipode* 40(4): 658–677.

Rumley, Dennis, and Julian V. Minghi (eds.). 1991. *The Geography of Border Landscapes.* London: Routledge.

Sack, Robert D. 1986. *Human Territoriality: Its Theory and History. Studies in Historical Geography,* No. 7. Cambridge, U.K.: Cambridge University Press.

Schuttea, Sebastian, and Nils B. Weidmann. 2011. "Diffusion Patterns of Violence in Civil Wars." *Political Geography* 30: 143–152.

Smith, Dan, et al. 1997. *The State of War and Peace Atlas.* 3rd ed. New York: Penguin.

Smith, Graham. 1999. "Russia, Geopolitical Shifts and the New Eurasianism." *Transactions of the Institute of British Geographers* 24: 481–500.

Spykman, Nicholas J. 1944. *The Geography of the Peace.* New York: Harcourt Brace.

Ter-Ghazaryan, Diana K. 2013. "'Civilizing the City Center': Symbolic Spaces and Narratives of the Nation in Yerevan's Post-Soviet Landscape." *Nationalities Papers: The Journal of Nationalism and Ethnicity* 41(4): 570–589.

Toal, Gerard, and John Agnew. 2002. "Introduction: Political Geographies, Geopolitics and Culture." In K. Anderson, M. Domosh, S. Pile, and N. Thrift (eds.), *Handbook of Cultural Geography,* pp. 455–461. London: SAGE.

Ten Recommended Books on Political Geography

(For additional suggested readings, see the *Contemporary Human Geography* LaunchPad: http://www.macmillanhighered.com/launchpad/DomoshCHG1e.)

Agnew, John. 2011. *Globalization and Sovereignty.* Lanham, Md.: Rowan & Littlefield. A leading political geographer joins the debate over the effects of globalization on nation-state sovereignty. Very convincingly argued with helpful examples.

Agnew, John, and Luca Muscarà. 2012. *Making Political Geography.* 2nd ed. Lanham, Md.: Rowan & Littlefield. This book provides an excellent overview of the field of political geography, highlighting the contributions of key thinkers from the nineteenth century to the present.

Dalby, Simon, and Gearóid Ó Tuathail (eds.). 1998. *Rethinking Geopolitics.* London: Routledge. Fifteen contributors to this postmodernist collection address questions of political identity and popular culture, state violence and genocide, militarism, gender and resistance, cyberwar, and the mass media. They suggest that political geography needs to be reconceptualized for the twenty-first century.

Herb, Guntram H., and David H. Kaplan (eds.). 1999. *Nested Identities: Nationalism, Territory, and Scale.* Lanham, Md.: Rowman & Littlefield. This collection of essays by 14 leading political geographers focuses on the geographical issue of territoriality using case studies of troubled countries and regions at different scales.

Hooson, David (ed.). 1994. *Geography and National Identity.* Oxford: Blackwell. Essays examine the connection between identity and homeland in a wide variety of settings and argue that the globalization of culture has strengthened the bonds between place and identity.

Kearns, Gerry. 2009. *Geopolitics and Empire: The Legacy of Halford Mackinder.* New York: Oxford University Press. This book helps explain Mackinder's key ideas by placing his writings in their historical context. It presents convincing arguments on why his ideas are still salient to some political analysts. A definitive work on MacKinder's understanding of global geopolitics.

Olwig, Kenneth. 2002. *Landscape, Nature, and the Body Politic: From Britain's Renaissance to America's New World*. Madison: University of Wisconsin Press. This is an impressively researched historical study of the importance of landscape in shaping the ideas of nation and national identity in England and the United States.

Popescu, G. 2011. *Bordering and Ordering in the Twenty-first Century: Understanding Borders*. Lanham, Md.: Rowman & Littlefield. A comprehensive look at nation-state borders and their continuing importance in shaping daily lives. It includes discussions of the historical origins of modern state borders and contemporary questions such as the way mobility shapes and is shaped by borders.

Wallerstein, Immanuel. 1991. *Geopolitics and Geoculture: Essays on the Changing World-System*. Cambridge, U.K.: Cambridge University Press. A collection of Wallerstein's essays that link the collapse of the Soviet Union to the end of U.S. hegemony around the world.

Williams, Colin H. (ed.). 1993. *The Political Geography of the New World Order*. London: Belhaven. A collection of essays that explore the geopolitical consequences of the collapse of the Soviet Union and the rising importance of Europe and Japan.

Political-Geographical Journals

Geopolitics. This journal explores contemporary geopolitics and geopolitical change with particular reference to territorial problems and issues of state sovereignty. Published by Frank Cass. Volume 1 appeared in 1996. http://www.tandfonline.com/toc/fgeo20/.U3ZaBJhUOew

Political Geography. This is a journal devoted exclusively to political geography. Formerly titled Political Geography Quarterly, the journal changed its name in 1992. Published by Elsevier. Volume 1 appeared in 1982.

Space and Polity. This journal is dedicated to understanding the changing relationships between the state and regional/local forms of governance. It highlights the work of scholars whose research interests lie in studying the relationships among space, place, and politics. Published by Carfax Publishing. Volume 1 appeared in 1997. http://www.tandfonline.com/toc/cspp20/.U3ZanJhUOew

Answers

Figure 6.4 A, Armenia; B, Azerbaijan; C, Iran; a, Nagorno-Karabakh; b, the Nakhichevan Autonomous Republic; c, the Okhair Eskipara enclave; d, Sofulu enclave; e, Kyarki enclave; f, Bashkend enclave.

How can an ordinary landscape, such as a parking lot, become a sacred space?

Parking lot shrine to the Virgin of Guadalupe, Self-Help Graphics and Art, East Los Angeles. *(Courtesy of Patricia L. Price.)*

Go to "Seeing Geography" on page 313 to learn more about this image.

THE GEOGRAPHY OF RELIGION

Spaces and Places of Sacredness

Religion is a core component of culture. For many, **religion is the most profoundly felt dimension** of their identities. For this reason, it is important to clearly state what is meant by the term and provide a sense of the many ways in which religion can be manifest in people's lives. **Religion** can be

> **religion** A social system involving a set of beliefs and practices through which people seek harmony with the universe and attempt to influence the forces of nature, life, and death.

defined as a more or less structured set of beliefs and practices through which people seek mental and physical harmony with the powers of the universe. The rituals of religion mark the events in our lives—birth, puberty, marriage, having children, and death—that are observed and celebrated. Religions often attempt not only to accommodate but also to influence the awesome forces of nature, life, and death. Religions help people make sense of their place in the world. In literal terms, the word *religion*—derived from the Latin *religare*—means "to fasten loose parts into a coherent whole."

But religion goes beyond a merely pragmatic set of rules for dealing with life's joys and sorrows. Most religions incorporate a sense of the supernatural that can be manifest in the concept of a God or gods who

LEARNING OBJECTIVES

7.1 Define the differences among religions as well as the basic tenets of the world's principal religions, and locate their primary regional expressions on a world map.

7.2 Discuss how and why the world's religions became distributed as they are today.

7.3 Critically analyze how globalization has affected the practice of religion.

7.4 Discuss specific ways in which the natural environment shapes, and is shaped by, religious beliefs and practices.

7.5 Describe how religions have left their particular mark on the cultural landscape.

play a role in shaping human existence, in the notion of an afterlife that may involve a place of rest (or torment) for those who have died, or in ideas of a soul that exists apart from our physical bodies and that may be released, or even born again, once we have died. This sense of the otherworldly is often spatially demarcated through the designation of sacred spaces, such as cemeteries, religious buildings, and sites of encounters with the supernatural.

In addition, religion is often at the heart of how people with very different worldviews can come to understand one another. So, on the one hand, the conquest of the Americas by the Iberians (people of modern-day Spain and Portugal) was often a violent affair, whereby the religious conversion of indigenous peoples to Christianity accompanied the political takeover. Temples were destroyed and coercion was often used to convert the natives to Christianity. On the other hand, the Virgin of Guadalupe, said to have appeared to the indigenous Mexican convert Juan Diego in 1531—only 10 years after Cortes's conquest of Mexico—is believed to be one of the most powerfully healing figures in the Americas to this day. Her portrait is a mixture of European and indigenous American symbols (**Figure 7.1**). In her kind manner of speaking to Juan Diego in Nahuatl (an indigenous language spoken by central Mexicans) and her resemblance to the earth goddess Tonantzín, she made sense to native Mexicans. For the Spaniards, dark-skinned virgins had long been part of their religious symbolism. In the midst of the violence of conquest, then, the Virgin of Guadalupe provided a mother figure that was readily acceptable to native Mexicans, was familiar to Spaniards, and acted as a bridge by which people from these two very different cultures could understand one another.

FIGURE 7.1 The Virgin of Guadalupe. This is the image believed to have appeared to Juan Diego. As he opened his cape in the presence of Bishop Zumárraga of Mexico City, roses of Castile (a powerful symbol to Spaniards) fell onto the floor and this image was left behind on the fabric of the cape. The Virgin's downcast gaze and dark features spoke to Mexican Indians, as did the red belt about her waist, which indicates that she is pregnant. *(Mark Lennihan/AP Photo.)*

CLASSIFYING RELIGIONS

Each of the world's major religions is organized according to more or less standardized practices and beliefs, and each is practiced in a similar fashion by millions, even billions, of adherents worldwide. Yet many people also express their religious faith in individual ways. Rituals and prayers can be adapted to fit particular circumstances or performed at home alone, or over the Internet. Some religions, including the Taoic religions of East Asia, as well as Hinduism and Buddhism, are largely individual or family-oriented practices. Some people do not observe a widely recognized religion at all. They may be secular, holding no religious beliefs, or express skepticism—even hostility—toward organized religion. They may consider themselves to be faithful but not follow an organized expression of their beliefs. Or they may practice an unconventional belief system, or cult. The term *cult* is often used in a pejorative sense because it conjures images of mind control, mass suicide, and extreme veneration of a human leader. It is important to keep in mind that the practitioners of belief systems falling outside of the mainstream—for example, Mormonism, Scientology, and even Alcoholics Anonymous—strongly object to being labeled cult members.

Different types of religion exist in the world. One way to classify them is to distinguish between proselytic and ethnic faiths. **Proselytic religions,** such as Christianity and Islam, actively seek new members and aim to convert all humankind. For this reason, they are sometimes also referred to as **universalizing religions.** They instruct their faithful to spread the Word to all the Earth using persuasion and sometimes violence to convert the "heathen." The colonization of peoples and their lands is sometimes a result of the desire to convert them to the conqueror's religion. By contrast, each **ethnic religion** is identified with a particular ethnic or tribal group and does not seek converts. Judaism provides an example. In the most basic sense, a Jew is anyone born of a Jewish mother. Though a person can convert to Judaism, it is a complex process that has traditionally been discouraged. Proselytic religions sometimes grow out of ethnic religions—the evolution of Christianity from its parent Judaism is a good example.

Another distinction among religions is the number of gods worshipped. **Monotheistic religions**, such as Islam and Christianity, believe in only one God and may expressly forbid the worship of other gods or spirits. **Polytheistic religions** believe there are many gods. For example, Vodun (also spelled Voudou in Haiti or Voodoo in the southern United States) is a West African religious tradition with adaptations in the Americas wherever the enslavement of Africans was once practiced. Although, as with most major religions, there is one supreme God, it is the hundreds of spirits, or *iwa*, that Vodun adherents turn to in times of need.

Some of the better-known spirits in the Haitian Voudou tradition are Danbala Wedo, the peaceful snake-god who brings rain and fertility; Legba, the keeper of crossroads and doorways who is invoked at the beginning of all rituals; and Ezili Danto, the protective mother figure portrayed as a dark-skinned country woman.

Finally, the distinction between syncretism and orthodoxy is important. **Syncretic religions** combine elements of two or more different belief systems. Umbanda, a religion practiced in parts of Brazil, blends elements of Catholicism with a reverence for the souls of Indians, wise men, and historical Brazilian figures, along with a dash of nineteenth-century European spiritism, which is a set of beliefs about contacting spirits through mediums. Caribbean and Latin American religious practices often combine elements of European, African, and indigenous American religions. Sometimes, in order to continue practicing their religions, people in this region would hide statues of Afrocentric deities within images of Catholic saints. Or they would determine which Catholic figures were most like their own deities. Note in **Figure 7.2** the comparison between Danbala, the snake-god of Haitian Voudou, with Saint Patrick, who is also associated with snakes. **Orthodox religions,** by contrast, emphasize purity of faith and are generally not open to blending with elements of other belief systems. The word *orthodoxy* comes from Greek and literally means "right" (*ortho*) "teaching" (*doxy*). Many religions, including Christianity, Judaism, Hinduism, and Islam, have orthodox strains. So, for instance, although some orthodox Jews closely follow a strict interpretation of the Oral Torah (a specific version of the Jewish holy book), moderate but committed Jews may observe only some or perhaps none of the dietary, marriage, and worship proscriptions observed by orthodox Jews. Intolerance of other religions, or of those fellow

> ■ **proselytic** or **universalizing religions** Religions that actively seek new members and aim to convert all humankind.
>
> **ethnic religion** A religion identified with a particular ethnic or tribal group; does not seek converts.
>
> **monotheistic religion** The worship of only one God.
>
> **polytheistic religion** The worship of many gods.
>
> **syncretic religions** Religions, or strands within religions, that combine elements of two or more belief systems.
>
> **orthodox religions** Strands within most major religions that emphasize purity of faith and are not open to blending with other religions.

FIGURE 7.2 **Danbala and Saint Patrick.** Danbala in Haitian Voudou is parallel to the Catholic Saint Patrick; both are associated with snakes. *(Left: © 2003 Charles Walker/ Topham/The Image Works; Right: Mary Evans Picture Library.)*

> **fundamentalism** A movement to return to the founding principles of a religion, which can include literal interpretation of sacred texts, or the attempt to follow the ways of a religious founder as closely as possible.

believers not seeming to follow the "proper" ways, is associated with **fundamentalism** rather than orthodoxy. Many who consider themselves orthodox are, in fact, quite tolerant of other beliefs (see Subject to Debate on pages 280–281).

The importance of religion to the contemporary study of cultural geography will be explored through our five themes. First, religious beliefs differ from one place to another, producing spatial variations that can be mapped as culture regions. Second, a high degree of mobility is characteristic of religions as they have spread historically through conquest, trade, and missionary activity. Violent conquests, diasporas, and the geographic expansion of religions through conversion have all played a role in shaping the contemporary world map of religious belief. Pilgrimages, too, illustrate how people can literally be set in motion through religion. Third, the global spread of religious beliefs has led to the reach of some religions, such as Hinduism, beyond their traditional hearths, whereas other religions, such as Judaism, have become diluted in part through migration-related diffusion. Religions, like all elements of cultural geography, must either adapt, or not, to the globalizing world. But religions influence globalization as well; here, we will consider the faith-based dimensions of the Internet. Fourth, the natural environment frequently plays an important part in faith-based belief structures, either because the forces of nature are viewed as potentially negotiable or because features of the natural landscape, such as rivers and mountains, are thought to exert a powerful influence over human destinies. Thus, the nature–culture dimension of religion is a key aspect of its cultural geography. Fifth, and finally, religious beliefs are often visible on the cultural landscape. For example, religious architecture, such as mosques, temples, and shrines, literally marks the landscape with the imprint of particular religious beliefs. How the spiritual shaping of some spaces as sacred comes to be is an important statement about what, and who, matters, culturally speaking.

REGION

7.1 LEARNING OBJECTIVE
Define the differences among religions as well as the basic tenets of the world's principal religions, and locate their primary regional expressions on a world map.

Because religion, like all of culture, has a strong territorial association, religious culture regions abound. The most basic kind of formal religious culture region depicts the spatial distribution of organized religions (**Figure 7.3**). Some religions, such as Hinduism and Buddhism, are strongly associated with one world region, while others, such as Islam and Christianity, are dispersed across many regions (**Figure 7.4** on page 278). Some parts of the world exhibit an exceedingly complicated pattern of religious adherence, and the boundaries of formal religious culture regions, like most cultural borders, are rarely sharp. While roughly three-quarters of the world's inhabitants live in countries where their religion is in the majority, persons of different faiths often live in the same province or town.

JUDAISM

Judaism is a 4000-year-old religion and the first major monotheistic faith to arise in southwestern Asia. It is the parent religion of Christianity and is also closely related to Islam. Jews believe in one God who created humankind for the purpose of bestowing kindness on them. As with Islam and Christianity, people are rewarded for their faith, are punished for violating God's commandments, and can atone for their sins. The Jewish holy book, or Torah, comprises the first five books of the Hebrew Bible. In contrast to other monotheistic faiths, Judaism is not proselytic and has remained an ethnic religion through most of its existence.

Judaism has split into a variety of subgroups, partly as a result of the Diaspora, a term that refers to the forced dispersal of the Jews from Palestine in Roman times and the subsequent loss of contact among the various colonies. Jews, scattered to many parts of the Roman Empire, became a minority group wherever they went. In later times, they spread throughout much of Europe, North Africa, and Arabia. Those Jews who lived in Germany and France before migrating to central and eastern Europe are known as the Ashkenazim; those who never left the Middle East and North Africa are called Mizrachim; and those from Spain and Portugal are known as Sephardim. Spain expelled its Sephardic Jews in 1492, the same year that Christopher Columbus set sail for the New World. It was not until the quincentennial of both events, in 1992, that the Spanish government issued an official apology for the expulsion.

The late nineteenth and early twentieth centuries witnessed large-scale Ashkenazic migration from Europe to America. The Holocaust that befell European Judaism during the Nazi years involved the systematic murder of perhaps a third of the world's Jewish population, mainly Ashkenazim. Europe ceased to be the primary homeland of Judaism, and many of the survivors fled overseas, mainly to the Americas and later to the newly created state of Israel. Today, Judaism has about 15 million adherents throughout the world. At present, a roughly equal number—6 million—of the world's Jews reside in North America and Israel; Europe and Latin American are home to most of the rest.

CHRISTIANITY

Christianity, a proselytic faith, is the world's largest religion, both in area covered and in number of adherents, claiming about a third of the global population (**Figure 7.5** on page 278). Christians are monotheistic, believing that God is a Trinity consisting of three persons: the Father, the Son, and the Holy Spirit. Jesus Christ is believed to be the incarnate Son of God who was given to humankind for the sake of redemption some 2000 years ago. Through Jesus's death and resurrection, all of humanity is offered redemption from sin and provided eternal life in heaven and a relationship with God.

Christianity, Islam, and Judaism are the world's three great monotheisms and share a common culture

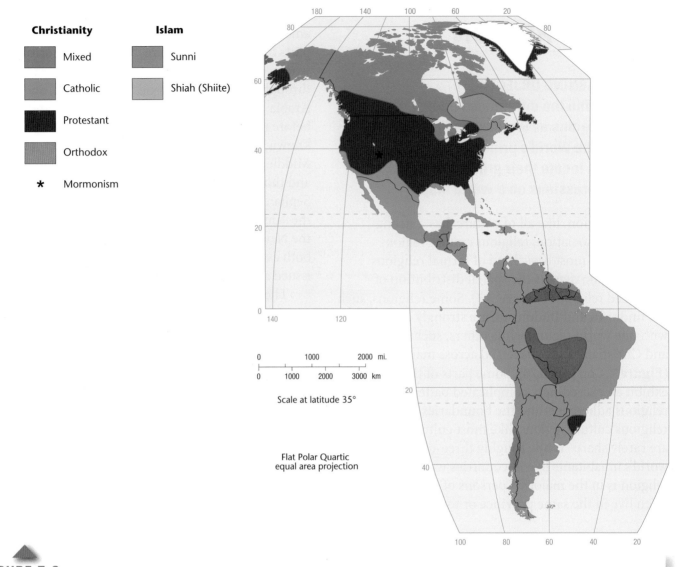

Christianity
- Mixed
- Catholic
- Protestant
- Orthodox
- ∗ Mormonism

Islam
- Sunni
- Shiah (Shiite)

0 1000 2000 mi.
0 1000 2000 3000 km
Scale at latitude 35°

Flat Polar Quartic
equal area projection

FIGURE 7.3 **The world distribution of major religions.** Much overlap exists that cannot be shown on a map of this scale. The attempt is to show which faith is dominant. "Animism" includes a wide array of diverse belief systems. Taoistic areas are cross-hatched to show the overlap with Buddhism in these areas. "Mixed" Christianity means that none of the three major branches of that faith has a majority.

hearth in southwestern Asia (see Figure 7.14). All three religions venerate the patriarch Abraham and thus are often called Abrahamic religions. Because Judaism is the parent religion of Christianity, the two faiths share many elements, including the Torah, the first five books of what Christians call the Old Testament (which forms part of the Christian Bible or holy book); prayer; and a clergy. This is the basis for the term *Judeo-Christian,* which is used to describe beliefs and practices shared by the two faiths. Because

Jesus was born into the Jewish faith, many Christians still accord Jews a special status as a chosen people and see Christianity as the natural continuation or fulfillment of Judaism.

Christianity has long been fragmented into separate branches (see Figure 7.3). The major division is threefold, made up of Roman Catholics, Protestants, and Eastern Christians. Western Christianity, which now includes Catholics and Protestants, was initially identified with Rome and

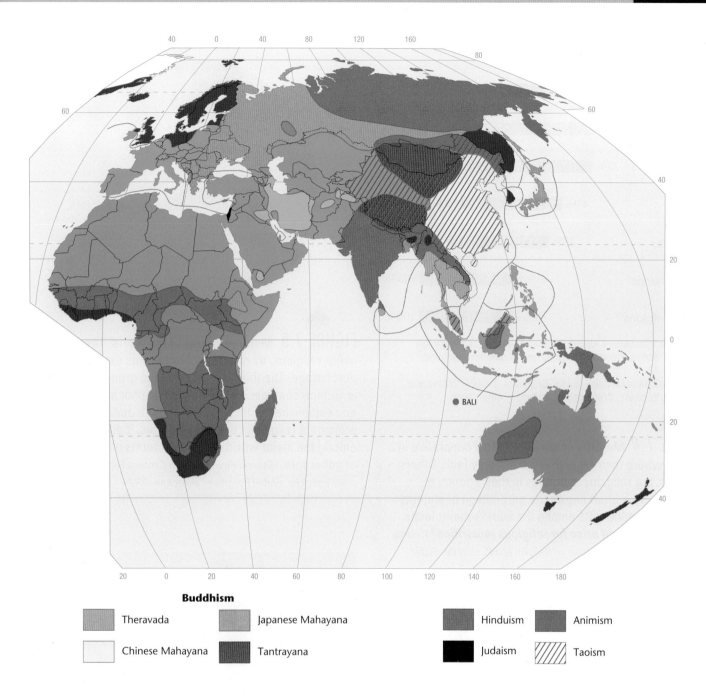

Buddhism

- Theravada
- Chinese Mahayana
- Japanese Mahayana
- Tantrayana
- Hinduism
- Judaism
- Animism
- Taoism

BALI

the Latin-speaking areas that are largely congruent with modern western Europe, whereas the Eastern Church dominated the Greek world from Constantinople (now the city of Istanbul, Turkey). Belonging to the Eastern group are the Armenian Church, reputedly the oldest in the Christian faith and today centered among a people of the Caucasus region; the Coptic Church, originally the religion of Christian Egyptians and still today a minority faith there, as well as being the dominant church among the highland people of Ethiopia; the Maronites, Semitic descendants of seventh-century heretics who disagreed with some of early Christianity's beliefs and retreated to a mountain refuge in Lebanon; the Nestorians, who live in the mountains of the Middle East and in India's Kerala state; and Eastern Orthodoxy, originally centered in Greek-speaking areas. Having converted many Slavic groups, Orthodox Christianity is today made up of a variety of national churches, such as Russian, Greek,

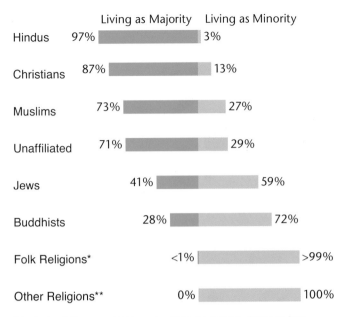

	Living as Majority	Living as Minority
Hindus	97%	3%
Christians	87%	13%
Muslims	73%	27%
Unaffiliated	71%	29%
Jews	41%	59%
Buddhists	28%	72%
Folk Religions*	<1%	>99%
Other Religions**	0%	100%

* Includes followers of African traditional religions, Chinese folk religions, Native American religions and Australian aboriginal religions.
** Includes Baha'i's, Jains, Sikhs, Shintoists, Taoists, followers of Tenrikyo, Wiccans, Zoroastrians, and many other faiths.

FIGURE 7.4 **Majority or minority?** Most people live in a country where their religion is the majority faith. Others, however, are religious minorities in their country of residence. This chart shows the percentage of majority and minority adherents among the world's major faiths. ***What issues might arise for religious minorities?*** *(Source: Pew Research Center's Forum on Religion and Public Life, Global Religious Landscape, 2012.)*

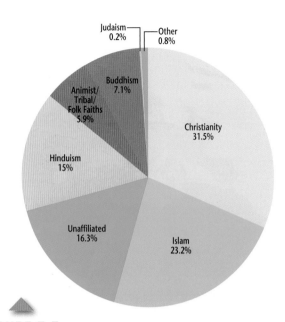

FIGURE 7.5 **Major religions of the world.** The world's major religions, by numbers of adherents expressed as a percentage. The third largest religious group in the world is, in fact, composed of those who are not affiliated with any established faith. Note that though Judaism is a prominent religion in the United States, it does not rank among the major world religions in terms of total number of adherents. *(Source: Pew Research Center's Forum on Religion and Public Life, Global Religious Landscape, 2012.)*

Ukrainian, and Serbian Orthodoxy, with a collective membership of some 260 million (**Figure 7.6**).

Western Christianity splintered, most notably with the emergence of Protestantism in the Reformation of the 1400s and 1500s. As with all religious reformation movements, Protestantism sought to overcome what were viewed to be the wrong practices of Roman Catholicism, such as the need for priestly or saintly mediation between humans and God, while still staying within the general framework of Christianity. Since then, the Roman Catholic Church, which alone includes 1.1 billion people, or more than one-sixth of humanity, has remained unified; but Protestantism, from its beginnings, tended to divide into a rich array of sects, which together and including Anglican and

FIGURE 7.6 **Greek Orthodox monastery.** The bells of St. John's Monastery in Patmos, Greece. *(Courtesy of Patricia L. Price.)*

independent Christians have a total membership today of about 801 million worldwide.

In the United States and Canada, the denominational map vividly reflects the fragmentation of Western Christianity and the resulting complex pattern of religious culture regions (**Figure 7.7**). Numerous denominations imported from Europe were later augmented by Christian denominations that originated in America. The American frontier was a breeding ground for new religious groups, as individualistic pioneer sentiment found expression in splinter Protestant denominations. Today, in many parts of the country, even a relatively small community may contain the churches of half a dozen religious

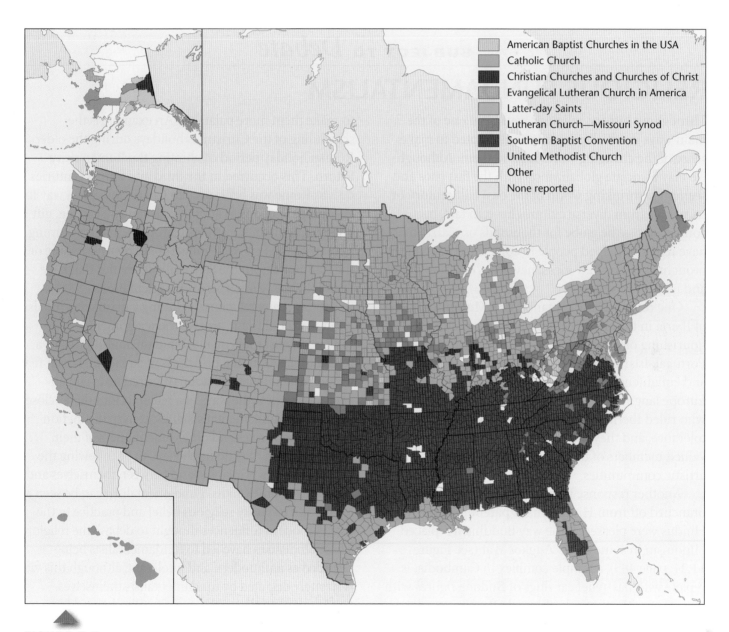

FIGURE 7.7 **Leading Christian denominations in the United States, shown by counties.** In the majority counties, the church or denomination indicated claim 50 percent or more of the total church membership. The most striking features of the map are Baptist dominance throughout the South, a Lutheran zone in the upper Midwest, Mormon (Latter-Day Saints) dominance in the interior West, and the zone of mixing in the American heartland. *Why do you think that so much of the western United States shows an adherence to Catholicism?*

groups. As a result, we can find about 2000 different religious denominations and cults in the United States alone. In a broad Bible Belt across the South, Baptist and other conservative fundamentalist denominations dominate, and Utah is at the core of the Mormon realm. A Lutheran belt stretches from Wisconsin westward through Minnesota and the Dakotas, and Roman Catholicism is dominant in southern Louisiana, the southwestern borderland, and heavily industrialized areas of the Northeast. The Midwest is a thoroughly mixed zone, although Methodism is the largest single denomination.

Today, Christianity is geographically widespread and highly diverse in its local interpretation. Although it is not as fast-growing as Islam, intense missionary efforts strive to increase the number of adherents.

SUBJECT TO *Debate*

RELIGIOUS FUNDAMENTALISM

Throughout history, religions have been one of the main ways in which people have attempted to make sense of the changing world around them. Although we may associate globalization with the fast pace and seemingly shrinking world of the past half century or so, people from diverse cultures have, in fact, been in contact across the globe for thousands of years. So how have religions helped people to cope with the changes brought about by new ideas, new ways of doing things, and new belief systems?

One response is acceptance. The Muslim conquest of Iberia in the eighth century brought a centuries-long flourishing of culture to modern-day Spain and Portugal. Islamic rule here was noted for its humane and enlightened nature, whereas the rest of western Europe languished in the Dark Ages. The Muslims who ruled Iberia were renowned for their religious tolerance, and they included Jews and Christians as valued members of their governmental, scientific, and artistic communities.

Another response is intolerance. When Buddhism branched off from Hinduism, its parent religion, not all Hindus were pleased at the way Buddhism replaced Hinduism in some areas. Angkor Wat (see Figure 11.24, page 483), a temple complex in Cambodia, is replete with bas-relief carvings of Buddha figures with their faces either entirely chipped off or recarved to resemble Hindu deities. Hindus intolerant of the Khmer King Jayavarman VII's Buddhist beliefs were responsible for defacing these sacred images on the king's death in 1220 C.E.

Yet another response is syncretism, where elements of preexisting faiths are blended with new ones in order to make them more palatable. An example is the scheduling of the Christmas holidays during the older Roman holiday period celebrating the Roman deity Saturn. This occurred in the third and fourth centuries C.E., as Rome was beginning to Christianize, as a way to encourage Romans to adopt Christianity. Feasting, gift giving, singing in the streets (caroling), and consuming human-shaped cookies (gingerbread men) are some of the traditions from the Saturnalia festivities that have found their way into Christmas celebrations.

Today many religions, including Christianity, Judaism, Hinduism, and Islam, are experiencing intense fundamentalist movements. Fundamentalism means a return to the founding principles of a religion, which may include a literal interpretation of sacred texts and an attempt to follow the ways of a religious founder as closely as possible. Fundamentalists draw a sharp distinction between themselves and other practitioners of their religion, whom they do not believe to be following the proper religious principles, and between themselves and adherents of other faiths. Fundamentalism can be seen as an attempt to purify religious belief and practice in the face of modern influences thought to debase the religion. These tendencies have led to fundamentalists being regarded as antimodern and intolerant, although this view is strongly disputed by fundamentalists themselves.

Fundamentalism is an emotionally charged term because it is often used to portray its followers derogatorily as radical extremists. The tendency of the U.S. media to use the term *Islamic fundamentalists* as a synonym for *terrorists* is an unfortunate example of this. Yet there are connections between politics and religious fundamentalism. The political agenda of the U.S.

Because of this proselytic work, Africa and Asia are the fastest-growing regions for Christianity.

ISLAM

Islam, another great proselytic faith, claims 1.6 billion followers, largely in the desert belt of Asia and northern Africa and in the humid tropics as far east as Indonesia and the southern Philippines (see Figures 7.3 and 7.5). Adherents of Islam, known as Muslims (literally, "those who submit to the will of God"), are monotheists and worship one absolute God known as Allah. Islam was founded by Muhammad, considered to be the last and most important in a long line of prophets. The word of

government on matters of abortion, adoption, marriage, foreign policy, domestic security, gay rights, and the curriculum in public schools has been notably influenced by politically conservative religious groups. Although not all of these groups are entirely fundamentalist in nature, many embrace fundamentalist Christian beliefs that espouse creationism and the sinfulness of homosexuality as well as question the separation of church and state. Islamism, a political ideology based in conservative Muslim fundamentalism, holds that Islam provides the political basis for running the state. Similar to the influence of Christian fundamentalism, Islamist influences in several Muslim-majority countries have set a conservative social agenda and have strongly questioned the separation of church and state.

Continuing the Debate

We often hear about religion in connection with violence. The next time you watch a news video or read a newspaper, keep these questions in mind:

- Do all religions have a dark, violent side to them, even those that profess a peaceful worldview?

- Is violence a necessary corollary to religious fundamentalism?

- What role does the media play in creating the perception—or inciting the practice—of religious antagonism?

Repent America, a fundamentalist Christian group, protests gay marriage outside San Francisco City Hall. *(Deborah Coleman/Getty Images.)*

FIGURE 7.8 **Muslims at prayer in Mecca, Saudi Arabia.** *(ayazad/Shutterstock.)*

Allah is believed to have been revealed to Muhammad by the angel Gabriel (Jibrail) beginning in 610 C.E. of the Christian calendar in the Arabian city of Mecca (**Figure 7.8**). The Qur'an, Islam's holy book, is the text of these revelations and also serves as the basis of Islamic law, or Sharia. Most Muslims consider both Jews and Christians to be, like themselves, "people of the Book," and all three religions share beliefs in heaven, hell, and the resurrection of the dead. Many biblical figures familiar to Jews and Christians, such as Moses, Abraham, Mary, and Jesus, are also venerated as prophets in Islam. Adherents to Islam are expected to profess belief in Allah, the one God whose prophet was Muhammad; pray five times daily at established times; give alms, or *zakat,* to the poor; fast from dawn to sunset during the holy month of Ramadan; and make at least one pilgrimage, if possible, to the sacred city of Mecca in Saudi Arabia. These duties are known as the Five Pillars of the faith.

Although not as severely fragmented as Christianity, Islam, too, has split into separate groups. Two major divisions prevail. Shiite Muslims, 16 percent of the Islamic total in diverse subgroups, form the majority in Iran and Iraq. Shiites believe that

Ali, who was Muhammad's son-in-law, should have succeeded Muhammad. Sunni Muslims, who represent the Islamic orthodoxy (the word *Sunni* comes from *sunnah,* meaning "tradition"), form the large majority worldwide (see Figure 7.3). Islam has not undergone a reformation parallel to that undergone by Christianity with Protestantism. However, reformation is the goal of liberal movements within Sunni and Shiite Islam.

Islam's strength is greatest in the Arabic-speaking lands in Southwest Asia and North Africa, although the world's largest Islamic population is found in Indonesia. Other large clusters live in Pakistan, India, Bangladesh, and western China. Because of successful conversion efforts in non-Muslim areas and high birthrates in predominantly Muslim areas, Islam is the fastest-growing world religion. In the United States, more people convert to Islam than to any other religion, and Islam will soon surpass Judaism to become America's second-largest faith.

HINDUISM

Hinduism, a religion closely tied to India and its ancient culture, claims about 1 billion adherents (see Figures 7.3 and 7.5). Hinduism is a decidedly polytheistic religion. Although Hindus recognize one supreme god, Brahman, it is his many manifestations that are worshipped directly. Some principal Hindu deities include Vishnu, Shiva, and the mother goddess Devi. Ganesha, the elephant-headed god depicted in **Figure 7.9**, is often revered by Hindu university students because Ganesha is the god of wisdom, intelligence, and education. Believing that no one faith has a monopoly on the truth, most Hindus are notably tolerant of other religions (but see Subject to Debate, pages 280–281).

Hindus strive to locate the harmonious and eternal truth, *dharma,* which is within each human being. Social divisions, or castes, separate Hindu society into four major categories, or *varna,* based on occupational categories: priests (Brahmins), warriors (Kshatriyas), merchants and artisans (Vaishyas), and workers (Shudras). Castes are related to dharma inasmuch as dharma implies a set of rules for each varna that regulate their behaviors with regard to eating, marriage, and use of space. All Hindus also share a belief in reincarnation, the idea that, although the

FIGURE 7.9 **Ganesha.** The elephant-headed god of wisdom, intelligence, and education is revered by many Hindu university students. *(Punit Puranjpe/Reuters/Landov.)*

share the Hindu belief in *ahimsa* and reincarnation. Jains adhere to a strict asceticism, a practice involving self-denial and austerity. For example, they practice veganism, a form of vegetarianism that prohibits the consumption of all animal-based products, including milk and eggs. Sikhism, by contrast, arose much later, in the 1500s, in an attempt to unify Hinduism and Islam. Centered in the Punjab state of northwestern India, where the Golden Temple at Amritsar serves as the principal shrine, Sikhism has about 23 million followers. Sikhs are monotheistic and have their own holy book, the Adi Granth.

No standard set of beliefs prevails in Hinduism, and the faith takes many local forms. Hinduism includes very diverse peoples. This is partly a result of its former status as a proselytic religion. A Hindu majority on the Indonesian island of Bali suggests the religion's former missionary activity (**Figure 7.10**). Today, most Hindus consider their religion an ethnic one, whereby one is acculturated into the Hindu community by birth; however, conversion to Hinduism is also allowed. Hindus form a majority in only three countries: Nepal, India, and Mauritius. Yet

physical body may die, the soul lives on and is reborn in another body. Related to this is the notion of *karma*. Karma can be viewed almost as a causal law, which holds that what an individual experiences in this life is a direct result of that individual's thoughts and deeds in a past life. Likewise, all thoughts and deeds, both good and bad, affect an individual's future lives. Ultimately, *moksha*, or liberation of the soul from the cycle of death and rebirth, will occur when the worldly bonds of the material self fall away and one's pure essence is freed. *Ahimsa*, or the principle of nonviolence, involves veneration of all forms of life. This implies a principle of noninjury to all sentient creatures, which is why many Hindus are vegetarians.

Hinduism has splintered into diverse groups, some of which are so distinctive as to be regarded as separate religions. Jainism, for instance, is an ancient outgrowth of Hinduism, claiming perhaps 4 million adherents, almost all of whom live in India, and traces its roots back more than 25 centuries. Although they reject Hindu scriptures, rituals, and priesthood, the Jains

FIGURE 7.10 **Hindu temple in Bali, Indonesia.** Besakih, known as the Mother Temple of Bali, is the largest Hindu temple complex on the island, occupying the slopes of Mount Agung. *(Courtesy of Patricia L. Price.)*

97 percent of the world's Hindus live in these three countries; thus, Hindus are more likely than any other faith to constitute a majority in the country where they reside (see Figure 7.4).

BUDDHISM

Hinduism is the parent religion of Buddhism, which began 25 centuries ago as a reform movement based on the teachings of Prince Siddhartha Gautama, "the awakened one" (**Figure 7.11**). He promoted the four "noble truths": life is full of suffering, desire is the cause of this suffering, cessation of suffering comes with the quelling of desire, and an Eightfold Path of proper personal conduct and meditation permits the individual to overcome desire. The resultant state of enlightenment is known as *nirvana*. Those few individuals who achieve nirvana are known as Buddhas. The status of Buddha is open to anyone regardless of social status, gender, or age. Because Buddhism derives from Hinduism, the two religions share many beliefs, such as dharma, reincarnation, and *ahimsa*. Buddhism and its parent religion, Hinduism, as well as the related faiths Jainism and Sikhism, are known as dharmic religions.

Today, Buddhism is the most widespread religion in South and East Asia, dominating a culture region stretching from Sri Lanka to Japan and from Mongolia to Vietnam. In the process of its proselytic spread, particularly in China and Japan, Buddhism fused with native ethnic religions such as Confucianism, Taoism, and Shintoism to form syncretic faiths that fall into the Mahayana division of Buddhism. Southern, or Theravada, Buddhism, dominant in Sri Lanka and mainland Southeast Asia, retains the greatest similarity to the religion's original form, whereas a variation known as Tantrayana, or Lamaism, prevails in Tibet and Mongolia (see Figure 7.3). Buddhism's tendency to merge with native religions, particularly in China, makes it difficult to determine the number of its adherents. Estimates range from 350 million to more than 500 million people (see Figure 7.5). Although Buddhism in China has become mingled with local faiths to become part of a composite ethnic religion, elsewhere it remains one of the three great proselytic religions in the world, along with Christianity and Islam.

FIGURE 7.11 **Buddhism is one of the religious faiths of South Korea.** Here, an image of the Buddha is carved from a rocky bluff to create sacred space and a local pilgrimage shrine. *(Courtesy of Terry G. Jordan-Bychkov.)*

TAOIC RELIGIONS

Confucianism, Shinto, and Taoism together make up the faiths that center on Tao, or the force that balances and orders the universe. Derived from the teachings of philosopher K'ung Fu-tzu (551–479 B.C.E.), Confucianism was later promoted by China's Han dynasty (206 B.C.E.–220 C.E.) as the official state philosophy. Thus, it has bureaucratic, ethical, and hierarchical overtones that have led some to question whether it is more a way of life than a religion in the proper sense. In China, Confucianism's formalism is balanced by Taoism's romanticism. Taoism, which is both an established religion and a philosophy, emphasizes the dynamic balance depicted by the Chinese yin-yang symbol. The "three jewels of Tao" are humility, compassion, and moderation. Shinto, once the state religion of Japan, has long been blended with Buddhism, as well as infused with a Confucian-derived legal system, but at its core Shinto is an animistic religion. Because Tao is found in nature, there is some overlap with the animistic faiths discussed next. In addition, because of their fluid nature, Taoic religions tend toward syncretism and have blended with other faiths, particularly Buddhism. People may simultaneously practice elements of all these faiths. For these reasons, it is difficult to provide an exact number of adherents, although estimates hold that there are about half a billion. Taoic religions center on East Asia (see Figure 7.3).

ANIMISM/SHAMANISM

Peoples in diverse parts of the world often retain indigenous religions and are usually referred to collectively as **animists** (Figure 7.12). For the most part, animists do not form organized or recognized religious groups but instead practice as ethnic religions common to clans or tribes. In addition, most animists follow oral, rather than written, traditions and thus do not have holy books. A tribal religious figure, called by some a shaman, usually serves as an intermediary between the people and the spirits.

animists Adherents of animism, the idea that souls or spirits exist in not only humans but also animals, plants, rocks, natural phenomena such as thunder, geographic features such as mountains or rivers, or other entities of the natural environment.

Currently numbering perhaps 240 million, animists believe that nonhuman beings and inanimate objects possess spirits or souls. These spirits are believed to live in rocks and rivers, on mountain peaks

FIGURE 7.12 **Druids from the Mistletoe Foundation bless mistletoe in Worcestershire, England.** Celtic Druids have their roots in the ancient, pre-Christian reverence for natural elements such as streams, hills, and plants. Mistletoe is sacred to the Druidic faith. *(Andrew Fox/Corbis.)*

FIGURE 7.13 **Pet grave marker.** Pumpkin the cat was apparently baptized into the Christian faith and now lies at rest in Pet Heaven Cemetery in Miami, Florida. *(Courtesy of Patricia L. Price.)*

and celestial bodies, in forests and swamps, and even in everyday objects. As such, objects are considered to be alive, and depending on the local beliefs, objects and animals may assume human forms or engage in human activities such as speech. For some animists, the objects in question do not actually possess spirits but rather are valued because they have a particular potency to serve as a link between a person and the omnipresent god. Followers of Wicca, a contemporary neopagan religion derived from pre-Christian European practices of reverence for the mother goddess and the horned god, worship the god and goddess who are thought to inhabit everything. Ritual tools used by Wiccans include the *athame* (dagger), *boline* (knife), and *besom* (broom), which are used in ceremonies to direct energy as practitioners communicate with the god and goddess.

Animistic elements can also pervade established religions. Japanese Shinto adherents worship *kami*, or spirits, that inhabit natural objects such as waterfalls and mountains. We should not classify such systems of belief as primitive or simple because they can be extraordinarily complex. Even in places such as the United States, where the majority of the inhabitants would not see themselves as animists, such beliefs are pervasive. For example, many people in the United States believe that their pets possess souls that ascend to heaven on death (Figure 7.13).

MOBILITY

7.2 LEARNING OBJECTIVE
Discuss how and why the world's religions became distributed as they are today.

Religions, according to religious studies scholar Thomas Tweed, can be thought of as "flows that intensify joy and confront suffering by drawing on human and suprahuman forces to make homes and cross boundaries." In other words, religions have always been on the move, crossing oceans and borders to establish new dwelling places. To a remarkable degree, the origin of the major religions was concentrated spatially in three principal culture hearth areas (**Figure 7.14**). A **culture hearth** is a focused geographic area where important innovations are born and from which they spread. Many religions mandate periodic return of the faithful to these culture hearths in order to confirm or renew their faith.

> **culture hearth** A focused geographic area where important innovations are born and from which they spread.

THE SEMITIC RELIGIOUS HEARTH

All three of the great monotheistic faiths—Judaism, Christianity, and Islam—arose among Semitic peoples who lived in or on the margins of the deserts of southwestern Asia, in the Middle East (see Figure 7.14). Judaism, the oldest of the three, originated some 4000 years ago. Only gradually did its followers acquire dominion over the lands between the Mediterranean and the Jordan River—the territorial core of modern Israel. Christianity, whose parent religion is Judaism, originated here about 2000 years ago. Seven centuries later, the Semitic culture hearth once again gave birth to a major faith when Islam arose in western Arabia, partly from Jewish and Christian roots.

Religions spread by both relocation and expansion diffusion. Recall from Figure 1.8 (page 15) that expansion diffusion can be divided into hierarchical and contagious subtypes. In hierarchical diffusion, ideas become implanted at the top of a society, leapfrogging across the map to take root in cities and bypassing smaller villages and rural areas. Because their main objective is to convert nonbelievers, proselytic faiths are more likely to diffuse than ethnic religions, and it is not surprising that the spread of monotheism was accomplished largely by Christianity and Islam, rather than Judaism. From Semitic southwestern Asia, both of the proselytic monotheistic faiths diffused widely, as shown in Figure 7.14.

Christians, observing the admonition of Jesus in the Gospel of Matthew—"Go ye therefore and teach all nations, baptizing them in the name of the Father, and of the Son, and of the Holy Ghost, teaching them to observe all things whatsoever I have commanded you"—initially spread through the Roman Empire, using the splendid system of imperial roads to extend the faith. In its early centuries of expansion, Christianity displayed a spatial distribution that clearly reflected hierarchical diffusion. The early congregations were established in cities and towns, temporarily producing a pattern of Christianized urban centers and pagan rural areas. Indeed, traces of this process remain in our language. The Latin word *pagus,* "countryside," is the root of both *pagan* and *peasant,* suggesting the ancient connection between non-Christians and the countryside.

The scattered urban clusters of early Christianity were created by such missionaries as the apostle Paul, one of Jesus's disciples who moved from town to town bearing the news of the emerging faith. In later centuries, Christian missionaries often used the strategy of converting kings or tribal leaders, setting in motion additional hierarchical diffusion. The Russians and Poles were converted in this manner. Some Christian expansion was militaristic, as in the reconquest of Iberia from the Muslims and the invasion of Latin America. Once implanted in this manner, Christianity spread farther by means of contagious diffusion. When applied to religion, this method of spread is called **contact conversion** and is the result of

> **contact conversion** The spread of religious beliefs by personal contact.

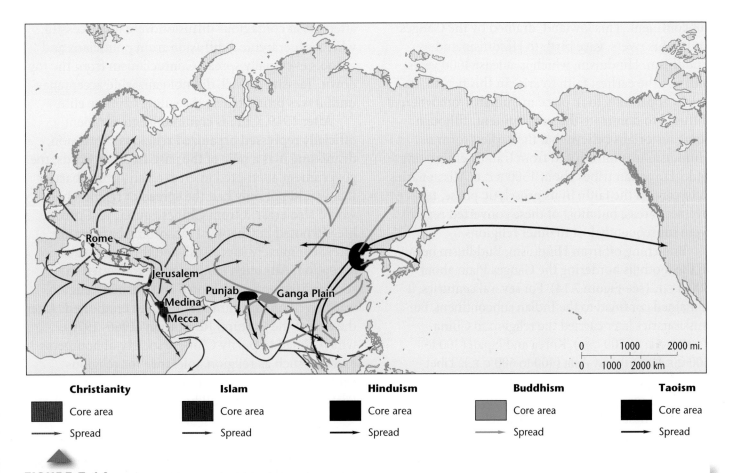

Christianity **Islam** **Hinduism** **Buddhism** **Taoism**

Core area | Core area | Core area | Core area | Core area

Spread | Spread | Spread | Spread | Spread

FIGURE 7.14 **The origin and diffusion of five major world religions.** Christianity and Islam, the two great proselytic monotheistic faiths, arose in Semitic southwestern Asia and spread widely through the Old World. Hinduism and Buddhism both originated in the northern reaches of the Indian subcontinent and spread throughout southeastern Eurasia. Taoic religions originated in East Asia, relocating with Chinese and Japanese migration regionally and, more recently, to North and South America.

everyday associations between believers and nonbelievers.

The Islamic faith spread from its Semitic hearth area in a predominately militaristic manner. Obeying the command in the Qur'an that they "do battle against them until there be no more seduction from the truth and the only worship be that of Allah," the Arabs expanded westward across North Africa in a wave of religious conquest. The Turks, once converted by the Arabs, carried out similar Islamic conquests. In a different sort of diffusion, Muslim missionaries followed trade routes eastward to implant Islam hierarchically in the Philippines, Indonesia, and the interior of China. Tropical Africa

is the current major region of Islamic expansion, an effort that has produced competition with Christians for the conversion of animists. As a result of missionary successes in sub-Saharan Africa and high birthrates in its older sphere of dominance, Islam has become the world's fastest-growing religion in terms of the number of new adherents.

THE INDUS-GANGES RELIGIOUS HEARTH

The second great religious hearth area lay in the plains fringing the northern edge of the Indian subcontinent. This lowland, drained by the Ganges and Indus rivers, gave birth to Hinduism and Buddhism. Hinduism, which is at least 4000 years old, was the earliest faith to arise in this hearth. Its origin apparently took place in Punjab, from where it diffused to dominate the subcontinent, although some historians believe that the earliest form of Hinduism was introduced from Iran by emigrating Indo-European tribes about 1500 B.C.E. Missionaries later carried the faith, in its proselytic phase, to overseas areas, but most of these converted regions were subsequently lost to other religions.

Branching off from Hinduism, Buddhism began in the foothills bordering the Ganges Plain about 500 B.C.E. (see Figure 7.14). For several centuries, it remained confined to the Indian subcontinent, but missionaries later carried the religion to China (100 B.C.E. to 200 C.E.), Korea and Japan (300 to 500 C.E.), Southeast Asia (400 to 600 C.E.), Tibet (700 C.E.), and Mongolia (1500 C.E.). Buddhism developed many regional forms throughout Asia, except in India, where it was ultimately accommodated within Hinduism.

The diffusion of Buddhism, like that of Christianity and Islam, continues to the present day. Some 2 million Buddhists live in the United States, about the same number as Episcopalians. Mostly, their presence is the result of relocation diffusion by Asian immigrants to the United States, where immigrant Buddhists outnumber Buddhist converts by 3 to 1.

THE EAST ASIAN RELIGIOUS HEARTH

K'ung Fu-tzu and Lao Tzu, the respective founders of Confucianism and Taoism, were contemporaries who reputedly once met with each other. Both religions were adopted widely throughout China only when the ruling elite promoted them. In the case of Confucianism, this was several centuries after the master's death. During his life, Kong Fu-tzu wandered about with a small band of disciples trying to convince rulers to put his ideas on good governance into practice. But he was shunned even by lowly peasants, who criticized him as "a man who knows he cannot succeed but keeps trying." Thus, early

attempts at contagious diffusion were unsuccessful, whereas hierarchical diffusion from politicians and schools eventually spread Confucianism from the top down. Taoism, as well, did not gain wide acceptance until it was promoted by the ruling Chinese elite.

After 1949, China's communist government officially repressed organized religious expression, dismissing it as a relic of the past. In other words, the government attempted to erect an absorbing barrier that would not only halt the spread of religion but would also erase it from Chinese public and private life. As noted in Chapter 1, absorbing barriers are rarely completely successful, and this example is no exception. Although temples were converted to secular uses, and even looted and burned, religion was driven underground rather than eradicated. After the end of the Cultural Revolution (1966–1976), which aimed to purify Chinese society of bourgeois excesses such as religion, tolerance for religious expression grew. However, the Chinese Communist Party's official stance still holds that religious and

FIGURE 7.15 **Shinto shrine in the United States.** With the migration of Japanese people in the nineteenth and twentieth centuries, particularly to the West Coast of the United States, Shinto traditions and religious structures followed. This image depicts the Tsubaki Shrine in Granite Falls, Washington State. *(© Alexander Marten Zhang.)*

party affiliations are incompatible; thus, some party officials are reluctant to divulge their religious status. This, combined with the fact that many Chinese practice elements of several Taoic faiths simultaneously, makes the precise enumeration of adherents difficult.

Taoism and Confucianism have spread with the Chinese people through trade and military conquest. Thus, people in Taiwan, Malaysia, Singapore, Korea, Japan, and Vietnam, along with mainland China, practice these beliefs or at least have been influenced by them. Today, Chinese and Japanese migrants alike have relocated their belief systems across the globe, as the image of the Tsubaki Shinto Shrine in Washington State shown in **Figure 7.15** confirms.

RELIGIOUS PILGRIMAGE

For many religious groups, journeys to sacred places, or **pilgrimages**, are an important aspect of faith-based mobility (**Figure 7.16**).

> **pilgrimages** Journeys to places of religious importance.

Pilgrimages are typical of both ethnic and proselytic religions. They are particularly significant for followers of Islam, Hinduism, Shintoism, and Roman Catholicism. Along with missionaries, pilgrims constitute one of the largest groups of people voluntarily on the move for religious reasons: religious pilgrimage is the number one driver of tourism. Pilgrims do not aim to convert people to their faith through their journeys. Rather, they enact in their travels a connection with the sacred spaces of their faith. Indeed, some religions mandate pilgrimage, as is the case with Islam, where pilgrimage to Mecca is one of the Five Pillars of the faith.

FIGURE 7.16 **Religious pilgrims.** Visitors number in the millions annually and can provide a holy site's main source of revenue. *Top:* the grotto where the Virgin Mary appeared to a girl in 1858 in Lourdes, France. *Middle:* the Temple of the Emerald Buddha in Bangkok, Thailand. *Bottom:* the Great Mosque in Touba, Senegal. *(Top: Franz-Peter Tschauner/dpa/Corbis; Middle: Kevin R. Morris/Corbis; Bottom: Richard List/Corbis.)*

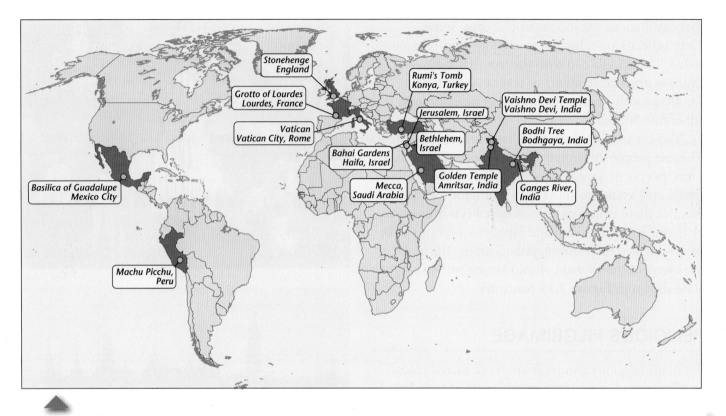

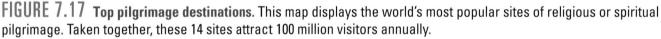

FIGURE 7.17 **Top pilgrimage destinations.** This map displays the world's most popular sites of religious or spiritual pilgrimage. Taken together, these 14 sites attract 100 million visitors annually.

The sacred places visited by pilgrims vary in character (**Figure 7.17**). Some have been the setting for miracles; a few are the regions where religions originated or areas where the founders of the faith lived and worked; others contain sacred physical features such as rivers, caves, springs, and mountain peaks; and still others are believed to house gods or are religious administrative centers where leaders of the religion reside. Examples include the Arabian cities of Mecca and Medina in Islam, which are cities where Muhammed resided and thus form Islam's culture hearth; Rome, the home of the Roman Catholic pope; the French town of Lourdes, where the Virgin Mary is said to have appeared to a girl in 1858; the Indian city of temples on the holy Ganges River, Varanasi, a destination for Hindu, Buddhist, and Jain pilgrims; and Ise, a shrine complex located in the culture hearth of Shintoism in Japan. Places of pilgrimage might be regarded as places of spatial convergence, or nodes, of functional culture regions.

Religion provides the stimulus for pilgrimage by offering those who participate in the purification of their souls or the attainment of some desired objective in their lives, by mandating pilgrimage as part of devotion, or by allowing the faithful to connect with important historic sites of their faith. For this reason, pilgrims often journey great distances to visit major shrines. Other sites of lesser significance draw pilgrims only from local districts or provinces. Pilgrimages can have tremendous economic impact because the influx of pilgrims is a form of tourism.

In some localities, the pilgrim trade provides the only significant source of revenue for the community. Lourdes, a town of about 16,000, attracts between 4 million and 5 million pilgrims each year, with many seeking miraculous cures at the famous grotto where the Virgin Mary reportedly appeared. Not surprisingly, among French cities, Lourdes ranks second only to Paris in its number of hotels, although

FIGURE 7.18 **Religious segregation.** Access to Mecca is restricted, allowing Muslims only. The Saudi Arabian government believes that permitting non-Muslim tourists to visit Mecca and Medina would disturb the sanctity of these sacred places. *(vario images GmbH & Co. KG./Alamy.)*

most of these are small. Mecca, a small city, annually attracts millions of Muslim pilgrims from every corner of the Islamic culture region as they perform the *hajj,* or Fifth Pillar of Islam, which all able-bodied Muslims who can afford the journey are encouraged to perform at least once in their lifetime. By land, by sea, and (mainly) by air, the faithful come to this hearth of Islam, a city closed to all non-Muslims (**Figure 7.18**). Such mass pilgrimages obviously have a major impact on the development of transportation routes and carriers, as well as other provisions such as inns, food, water, and sanitation facilities.

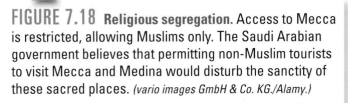

GLOBALIZATION

7.3 LEARNING OBJECTIVE
Critically analyze how globalization has affected the practice of religion.

Religion is seen by many people as providing a stable anchor, one that is particularly necessary amidst the flux and turmoil that characterizes globalization.

Religious customs, meetings, and obligations serve to form a strong basis of community. Typically, these communities are experienced locally; however, through long-distance pilgrimage, the faithful from across the globe may gather together in one place. Today's increased mobility and communication make possible religious communities at a truly global scale.

THE RISE OF EVANGELICAL PROTESTANTISM IN LATIN AMERICA

As economic, social, and political changes occur in places, cultural forces such as religion must adapt. For instance, in Chapter 4, we saw how languages change over time in response to immigration, urbanization, and trade-based exchanges. At other times, cultural forces drive changes in politics, economics, and societies. In Chapter 5, we saw how Hispanic immigration to the United States has resulted in major shifts in consumption, political behavior, and social attitudes.

A good example of how the cultural, economic, social, and political arenas work together is provided by the contemporary religious landscape of Latin America and the Caribbean. In this region, Roman Catholicism has dominated the religious landscape since the time of the Iberian conquerors in the late 1400s. More Catholics live in this region than in any other on Earth, with three out of every five Catholics residing here. Brazil is the largest Catholic country in the world in terms of population, with more than 120 million adherents. However, the Catholic Church has been on the decline in Brazil and throughout the region in recent decades. Only about 65 percent of Brazilians are Catholics today, whereas 50 years ago that figure was more than 95 percent. If Protestantism continues to grow as it has in recent years, by 2020 Brazil's Protestants will outnumber Catholics. What has happened to the region's Catholics?

Briefly, more and more Latin Americans believe that the Catholic Church has failed to keep in touch with the needs and concerns of contemporary urban societies. Indeed, the naming of Pope Francis of Argentina to the papacy in 2013 may, in part, have

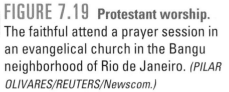

FIGURE 7.19 Protestant worship. The faithful attend a prayer session in an evangelical church in the Bangu neighborhood of Rio de Janeiro. *(PILAR OLIVARES/REUTERS/Newscom.)*

been the Church's attempt to underscore that humility and service to others are important to Catholic leadership. But birth control, divorce, and persistent poverty are issues that simply have not been addressed by Roman Catholicism to the satisfaction of many Latin Americans. The ongoing sexual abuse scandals that have rocked the Catholic Church have also led to diminished loyalty. As a result, more and more are turning to evangelical Protestant faiths, such as the Seventh-Day Adventist and Pentecostal churches (**Figure 7.19**). The focus of these churches on thrift, sobriety, and resolving problems directly

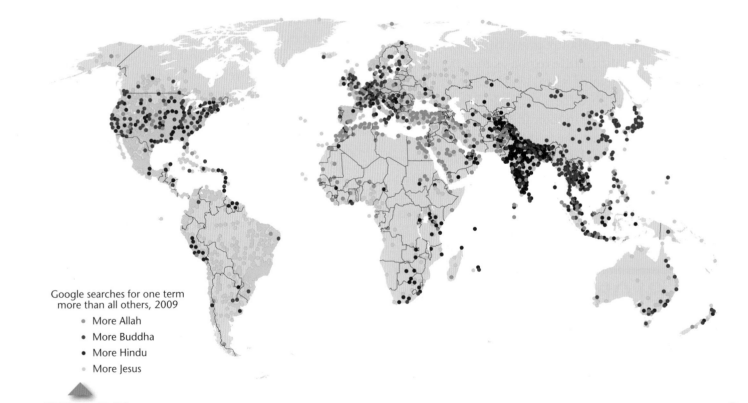

Google searches for one term more than all others, 2009

- More Allah
- More Buddha
- More Hindu
- More Jesus

FIGURE 7.20 The geography of religion, according to Google. This map of Google searches for one religious figure above all others depicts an online geography that mirrors regional religious patterns on the ground. *(Source: Google's Geographies of Religion—Allah, Buddha, Hindu, Jesus [blog post January 26, 2010] with permission from Matthew Zook.)*

rather than through the mediation of priests is appealing to many. Others find their needs are best addressed by an array of African-based spiritist religions, which include Umbanda, Candomblé, and Santería. Whether Catholicism will ultimately reform from within to become more current, or whether it will continue to lose out to rival denominations, is a key question in regard to the Latin American religious landscape.

RELIGION ON THE INTERNET

In the mid-fifteenth century, Gutenberg's printing press made the Bible available to a mass audience. The Internet, proponents argue, represents a similar technological revolution, one that will inevitably attract new adherents to religious faiths. Gathering together regularly in a holy place—a mosque, temple, church, or shrine—is at the heart of most organized religions. Traditionally, people have come together to receive sacraments, sing, pray, and celebrate major life events in houses of worship. With the tens of thousands of religious web sites now in existence, it is no longer necessary to physically go to a place of worship. Rather, you can sit in front of a computer, at any time of the day or night, and read a holy book, submit and read online prayers, engage in theological debate, or watch broadcasts of religious services (**Figure 7.20**).

For cultural geographers, the most interesting dimension of online worship concerns what it does to the role of place in a global society. Clearly, the Internet has made it possible to practice religion in a way that is not linked to a specific place of worship, but is this a good thing? Supporters of practicing religion online contend that it allows people to become members of virtual communities that would not otherwise have been available to them in times of spiritual need. Now illness, invalidism, or the pressures of a busy life need not present a barrier to worship. Detractors argue that solitary worship online erodes the place-based communities that are at the heart of many religions. Recall from Chapter 1 that debates center on whether globalization dilutes the role of place and of place-based differences. As people pick and choose elements from the different

faiths available online, will religions converge into a watered-down, homogeneous "McFaith"?

RELIGION'S RELEVANCE IN A GLOBAL WORLD

A 2012 Gallup poll revealed that 69 percent of Americans consider themselves very (40 percent) or moderately (29 percent) religious, making the United States one of the more religious nations on Earth. Religiosity will only increase in the United States over time, according to Gallup Editor Frank Newport. Yet the percentage of Americans who do not have a specific religious identification has also increased, and some believe that American culture is becoming more religiously mixed. Newer religious influences, too, are making an entrance onto the American religious stage. For example, if you have ever practiced yoga or meditation, you are engaging in Hindu and Buddhist practices (**Figure 7.21**). Many Americans do yoga or

FIGURE 7.21 Yoga class in session. Many people in the United States have incorporated Hindu and Buddhist spiritual practices into their lives. *(Ryan McVay/Getty Images.)*

FIGURE 7.22 **Britain's Atheist Bus Campaign.** This banner was part of an advertising campaign launched in 2009. It was intended to counter with a reassuring message the Christian bus advertising from the previous year, which had admonished non-Christians that they would "burn in Hell for all eternity." The so-called Atheist Bus Campaign was wildly successful, raising enough money to launch similar campaigns across the United Kingdom. *(David Wimsett/UPPA/Photoshot/Newscom.)*

meditate to enhance physical well-being, to promote spiritual growth, to relieve stress, or even to keep up with the latest trends, rather than as part of a religious practice. Yet they are among the growing number of Americans who have found a blend of Eastern and Western religious practices to be compatible with their beliefs and lifestyles.

The number of Americans who identify as atheists, agnostics, or adherents to "nothing in particular" stands at about one in five; whereas one in six worldwide, or 1.1 billion people, fall into this category of religious nonaffiliation known as "nones" (**Figure 7.22**). Across the United States, the importance of religion varies, as depicted in **Figure 7.23**. In some parts of the world, especially in much of Europe, religious affiliation has declined, giving way to *secularization* (**Figure 7.24**). In some instances, the retreat from organized religion has resulted from a government's active hostility toward a particular faith or toward religion in general. The

French government has long—since the French Revolution, in fact—discouraged public displays of religious faith. This extends not only to France's traditional Roman Catholicism but also to Jews and the rising Muslim population. In 2004 French law banned the wearing of skullcaps, headscarves, or large crosses in public schools, and in 2011 burqas—veils worn by some Muslim women—were outlawed in all public areas.

Yet nonbelief has always been a feature of the religious landscape. In Medieval Europe, according to historian Molly Worthen, "Ordinary people often skipped church and had a feeble grasp of basic Christian dogma. Many priests barely understood the Latin they chanted—and many parishes lacked any priest at all. Bishops complained about towns that used their cathedrals mainly as indoor markets or granaries." Such patterns once again reveal the inherent spatial variety of humankind and invite analysis by the cultural geographer.

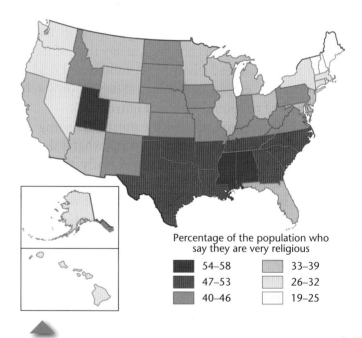

Percentage of the population who say they are very religious

■ 54–58	▨ 33–39
■ 47–53	▨ 26–32
▨ 40–46	□ 19–25

FIGURE 7.23 **Importance of religion.** This map of the United States shows the percent of people in each state who indicated that religion was very important in their lives. ***What factors might explain why this map illustrates these particular regional patterns?*** *(Source: www.pewforum.org.)*

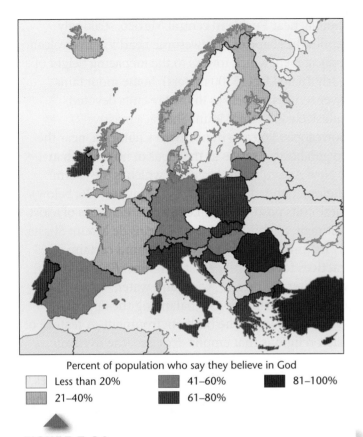

Percent of population who say they believe in God

Less than 20%	41–60%	81–100%
21–40%	61–80%	

FIGURE 7.24 **Secularized areas in Europe.** Places in which Christianity has ceased to be of much importance occur in a complicated pattern. ***What causal forces might have been at work?*** In all of Europe, some 190 million people report no religious faith, amounting to 27 percent of the population.

NATURE–CULTURE

7.4 **LEARNING OBJECTIVE**
Discuss specific ways in which the natural environment shapes, and is shaped by, religious beliefs and practices.

Our religious beliefs shape how we engage with the natural world. Some religions posit that humans transcend nature and are given license to dominate and exploit the natural world. Others see humans as merely co-dwellers in the natural world, alongside and interconnected with other inhabitants of nature. Thus, religions can inspire environmental stewardship when nature is viewed as a sacred part of creation, or encourage depletion and destruction of the natural world in the service of mankind.

APPEASING THE FORCES OF NATURE

One of the main functions of many religions is the maintenance of harmony between a people and their physical environment. Thus, religion is perceived by its adherents to be part of the **adaptive strategy** (one of the cultural tools needed to survive in a given environment); for that reason, physical environmental factors, particularly natural hazards and disasters, exert a powerful influence on the development of religions.

> **adaptive strategy** The unique way in which each culture uses its particular physical environment; those aspects of culture that serve to provide the necessities of life—food, clothing, shelter, and defense.

Environmental influence is most readily apparent in animistic faiths. In fact, an animistic religion's principal goal is to mediate between its people and the spirit-filled forces of nature. Animistic ceremonies are often intended to bring rain, quiet earthquakes, end plagues, or in some other way manipulate environmental forces by placating the spirits believed responsible for these events.

Rivers, mountains, trees, forests, and rocks often achieve the status of sacred space, even in the great religions. The Ganges River and certain lesser streams such as the Bagmati in Nepal are holy to the Hindus, and the Jordan River has special meaning for Christians, who often transport its waters in containers to other continents for use in baptism (**Figure 7.25**). Most holy rivers are believed to possess soul-cleansing abilities. Hindu geographer Rana Singh speaks of the "liquid divine energy" of the Ganges "nourishing the inhabitants and purifying them."

Mountains and other high places likewise often achieve sacred status among both animists and adherents of the great religions (**Figure 7.26**). Mount Fuji is sacred in Japanese Shintoism, and many high places are revered in Christianity, including Mount Sinai. The great pre-Columbian temple pyramid at

FIGURE 7.25 **Holy water from the Jordan River.** This water is taken from the Jordan River, at the site where Jesus is believed to have been baptized by John the Baptist. It can be purchased and used in rituals. *(Courtesy of Patricia L. Price.)*

Cholula, near Puebla in central Mexico, strikingly mimics the shape of the awesome nearby active volcano Popocatépetl, which towers to the menacing height of nearly 18,000 feet (5500 meters). Some mountains tower so impressively as to inspire cults devoted exclusively to them. Mount Shasta, a massive snowcapped volcano in northern California, near the Oregon border, serves as the focus of no fewer than 30 New Age cults, the largest of which is the "I Am" religion, founded in the 1930s (see Figure 7.26, below). These cults posit that the Lemurians, denizens of a lost continent, established a secret city inside Mount Shasta.

Animistic nature-spirits lie behind certain practices found in the great religions. *Feng shui,* which literally means "wind and water," refers to the practice of harmoniously balancing the opposing forces of nature in the built environment. A feature of Asian religions that emphasizes Tao, the dynamic balance found in nature, feng shui involves choosing environmentally auspicious sites for locating houses, villages, temples, and graves. The homes of the living and the resting places of the dead must be aligned with the cosmic forces of the world in order to assure good luck, health, and prosperity. Although the practice of feng shui dates back some 7000 years, contemporary people practice its principles. **Figure 7.27** depicts a high-rise condominium in

FIGURE 7.26 **Two high places that have evolved into sacred space.** *Left:* The reddish sandstone Uluru, or Ayers Rock, in central Australia is sacred in Aboriginal animism. *Right:* Snowy Mount Shasta in California is venerated by some 30 New Age cults. **Why do mountains so often inspire such worship?** *(Left: Hans-Peter Merten/Getty Images; Right: iStockphoto/Thinkstock.)*

FIGURE 7.27 Condominium in Hong Kong. The square opening in this building is supposed to allow for the passage of the dragon that resides in the hill behind. *(Courtesy of Ari Dorfsman.)*

Hong Kong's Repulse Bay neighborhood that incorporates feng shui principles in its design. The square opening in the building's center is said to provide passage to the bay for the dragon that dwells in the hill behind the building, allowing the dragon to drink from the waters of the bay and return to its abode unencumbered. Some Westerners have also adopted the principles of feng shui. Office spaces as well as homes are arranged according to its basic ideas. For example, artificial plants, broken articles, and paintings depicting war are thought to bring negative energy into living spaces and so should be avoided.

Although the physical environment's influence on the major Western religions is less pronounced, it is still evident. Some contemporary adherents to the Judeo-Christian tradition believe that God uses plagues to punish sinners, as in the biblical account of the 10 plagues inflicted on Egypt, which forced the Israelites into the desert. Modern-day droughts, earthquakes, and hurricanes are interpreted by some as God's punishment for wrongdoing, whereas others argue that this is not so. Environmental stress can, however, evoke a religious response not so different

from that of animistic faiths. Local ministers and priests often attempt to alter unfavorable weather conditions with special services, and there are few churchgoing people in the Great Plains of the United States who have not prayed for rain in dry years.

THE IMPACTS OF BELIEF SYSTEMS ON PLANTS AND ANIMALS

Every known religion expresses itself in food choices, to one degree or another. In some faiths, certain plants and livestock, as well as the products derived from them, are in great demand because of their roles in religious ceremonies and traditions. When this is the case, the plants or animals tend to spread or relocate with the faith.

For example, in some Christian denominations in Europe and the United States, celebrants drink from a cup of wine that they believe is the blood of Christ during the sacrament of Holy Communion. The demand for wine created by this ritual aided the diffusion of grape growing from the sunny lands of the Mediterranean to newly Christianized districts beyond the Alps in late Roman and early medieval times. The vineyards of the German Rhine were the creation of monks who arrived from the south between the sixth and ninth centuries. For the same reason, Catholic missionaries introduced the cultivated grape to California, in an example of relocation diffusion. In fact, wine was associated with religious worship even before Christianity arose. Vineyard-keeping and winemaking spread westward across the Mediterranean lands in ancient times in association with worship of the god Dionysus.

Religious taboos can even function as absorbing barriers, preventing diffusion of foods, drinks, and practices that violate the taboo. Mormons, who are encouraged to avoid caffeine, have not taken part in the American fascination with coffee. Sometimes these barriers are permeable. Certain Pennsylvania Dutch churches, for example, prohibit cigarette smoking but do not object to member farmers raising tobacco for sale in the commercial market.

Some religions set forth very specific guidelines for food preparation and consumption. Jews, for instance, are forbidden from consuming animals deemed "unclean," including shellfish, pork, and insects. Meat and milk must not be stored, prepared, or consumed together, which leads observant Jewish households to have two sets of dishes, two sinks, and even two ovens and refrigerators. *Kosher* food—food that is fit for consumption by observant Jews—allows only meat slaughtered by a trained individual using a sharp knife, drained of blood, and deemed disease-free. Muslims are commanded to consume only *halal,* or permissible, foods. Pork and alcohol are prohibited, while halal livestock should be fed with "clean" food that is free of animal by-products. Many non-Jews and non-Muslims view the kosher and halal certifications as an indication of wholesomeness, and seek such foods out for health rather than religious reasons. Indeed, the top three reasons given for purchasing kosher products were food quality, safety, and healthfulness; religious reasons ranked sixth (Rashtogi, 2010). But in the Netherlands, animal rights activists have deemed halal practices unacceptable because they do not stun an animal prior to slaughter, and have banned halal meat to the consternation of some who see this as a backlash against Muslim immigrants to this country.

The case of India's sacred cows provides an intriguing example of how religious beliefs shape the role of animals in society, in this instance avoiding the use of cows as food while emphasizing the animal's sacred as well as practical functions. Although only one out of every three of India's Hindus practices vegetarianism, almost none of India's Hindu population will eat beef, and many will not use leather. Why, in a populous country such as India where many people are poor and where food shortages have plagued regions of the country in previous decades, do people refuse to consume beef? There are several quite legitimate reasons. First, the dairy products provided by cow's milk and its by-products, such as yogurt and *ghee* (clarified butter), are central to many regional Indian cuisines. If the cow is slaughtered, it will no longer be able to provide milk. Second, in areas of the world that rely heavily on local agriculture for food production, such as India, cows provide valuable agricultural labor in tilling fields. Cows also provide free fertilizer in the form of dung. Finally, the value of the cow has been incorporated into Indian Hindu beliefs and practices over many centuries and has become a part of culture. Krishna, an incarnation of the major Hindu deity Vishnu, is said to be both the herder and the

protector of cows. Nandi, who is the deity Shiva's attendant, is represented as a bull.

Think for a moment about what you consider appropriate to eat. Perhaps beef is part of your diet, but would you eat horse meat? What is so different about a cow and a horse? Has the scare about mad cow disease or other beef contamination incidents affected your consumption of beef? Has the threat of so-called bird flu curtailed your poultry intake, or swine flu your pork consumption? How about dog or cat meat? What makes certain animals pets in some cultures and dinner in others? Through this sort of questioning, you may come to realize that practices that seem second nature—such as what you will and will not eat—are, in fact, the result of long histories that are very different from place to place.

ECOTHEOLOGY

Ecotheology is the name given to a rich and abundant body of literature studying the role of religion in habitat modification. More exactly, ecotheologians ask how the teachings and worldviews of religion are related to our attitudes about

> **ecotheology** The study of the influence of religious belief on habitat modification.

modifying the physical environment. In some faiths, human power over natural forces is assumed. In others, nature is a sacred creation that should be respected and protected by humans as an ethical component of the faith's practice.

The Judeo-Christian tradition also teaches that humans have dominion over nature, but it goes further, promoting the view that the Earth was created especially for human beings, who are separate from and superior to the natural world. This view is implicit in God's message to Noah after the Flood, promising that "every moving thing that lives shall be food for you, and as I gave you the green plants, I give you everything." The same theme is repeated in the Book of Psalms, where Jews and Christians are told that "the heavens are the Lord's heavens, but the Earth he has given to the sons of men." Humans are not part of nature but separate, forming one member of a God–nature–human trinity.

Believing that the Earth was given to humans for their use, early Christian thinkers adopted the view that humans were God's helpers in finishing the task of creation, that human modifications of the environment were therefore God's work. Small wonder, then, that the medieval period in Europe witnessed an unprecedented expansion of agricultural acreage, involving the large-scale destruction of woodlands and the drainage of marshes. Nor is it surprising that Christian monastic orders, such as the Cistercian and Benedictine monks, supervised many of these projects, directing the clearing of forests and the establishment of new agricultural colonies.

Subsequent scientific advances permitted the Judeo-Christian West to modify the environment at an unprecedented rate and on a massive scale. This marriage of technology and theology is one cause of our modern ecological crisis. The Judeo-Christian religious heritage, in short, has for millennia promoted an instrumentalist view of nature that is potentially far more damaging to the habitat than an organic view of nature in which humans and nature exist in balance. Yet there is considerable evidence to the contrary. For example, the Orthodox Church in Russia is working to create wildlife preserves on monastery lands. The Patriarch of Constantinople, leader of Eastern Orthodox Christianity, has made the fight against pollution a church policy, declaring that damaging the natural habitat constitutes a sin against God. The Church of England has declared that abuse of nature is blasphemous, and throughout the monotheistic religions, the green teachings of long-dead saints, such as Christianity's St. Francis of Assisi, who treasured birds and other wildlife, now receive heightened attention. In 1986 the Assisi Declarations united representatives of five of the world's major religions–Buddhism, Christianity, Hinduism, Islam, and Judaism–in Assisi, Italy, to discuss how faiths could work together to honor the sacredness of nature and stop its destruction.

Some fundamentalist Protestant sects herald ecological crisis and environmental deterioration as a sign of the coming Apocalypse, Christ's return, and the end of the present age. Thus, they welcome ecological collapse and, obviously, are unlikely to be of much help in solving the problem of environmental degradation. Other conservative Protestants, however, have adopted conservationist views, inspired by biblical admonitions such as the Flood story from the Old Testament, in which Noah saves animals by bringing them onto the ark. This story is viewed by some fundamentalists as a call to protect endangered species. An ecotheological focus underlies the multidenominational National Religious Partnership for the Environment, which includes many evangelical Protestant members. The hope is to mobilize the Christian Right against wanton environmental destruction in the same way in which they oppose abortion.

British religious studies scholar Emma Tomlin (2009) notes that "non-Abrahamic religions tend to be depicted as having more resonance with environmentalist thinking. For instance, eastern religions or Native American traditions are often seen as intrinsically oriented towards environmental protection." But is this correct? In fact, real-world practices do not always reflect the stated ideals of religions. Both Buddhism and Hinduism protect temple trees but demand huge quantities of wood for cremations (**Figure 7.28**). Traditional Hindu cremations, for example, place the corpse on a pile of

FIGURE 7.28 **Wood gathered for Hindu cremations at Pashupatinath, on the sacred river Bagmati in Nepal.** These cremations contribute significantly to the ongoing deforestation of Nepal and reveal the underlying internal contradiction in Hinduism between conservation as reflected in the doctrine of *ahimsa* and sanctioned ecologically destructive practices. *(Cormac McCreesh/Gallo Images/Getty Images.)*

wood, cover it with more wood, and burn it during an open-air rite that can last up to six hours. The construction of funeral pyres is estimated to strip some 50 million trees from India's countryside annually. In addition, the ashes are later swept into rivers, and the burning itself releases carbon dioxide into the atmosphere, contributing to the pollution of waterways and the atmosphere. The use of wood, and its placement on the ground, provide an important symbolic connection between the body and the earth. It does not, however, lead to efficient burning; on average, 880 pounds (400 kilograms) of wood are required to cremate a single corpse. A "green cremation system" is currently under development by a New Delhi–based nonprofit organization. By placing the first layer of wood on a grate, and placing a chimney over the pyre, wood use can be reduced by 75 percent. Although traditional Hindus balk at the notion of breaking with conventional practice, severe wood shortages and the escalating cost of wood will likely make green cremations an increasingly popular option in the future.

As this example illustrates, the idea of a link between godliness and greenness is a global one. In the years following a conference in Italy in the mid-1980s—which brought together environmentalists and religious leaders representing Buddhism, Christianity, Hinduism, Islam, and Judaism—some 130,000 projects have arisen linking the green teachings some see as inherent in these faiths to the ecology movement.

Ecofeminists have also entered this debate. They point out that the rise of the all-powerful male sky-deity of Semitic monotheism came at the expense of earth goddesses of fertility and sustainability. In their view, because the Judeo-Christian tradition elevated a sky-god remote from the Earth, the harmonious relationship between people and the habitat was disrupted. The ancient holiness of ecosystems perished, endangering huge ecoregions. The **Gaia hypothesis** possesses an ecofeminist spirit, in which the Earth is seen as a mother figure who reacts to humankind's environmental depredations through a variety of self-regulating mechanisms (**Figure 7.29**).

Gaia hypothesis The theory that there is one interacting planetary ecosystem, Gaia, that includes all living things and the land, waters, and atmosphere in which they live; further, that Gaia functions almost as a living organism, acting to control deviations in climate and to correct chemical imbalances, so as to preserve Earth as a living planet.

FIGURE 7.29 Earth-friendly license plate. *(UIG via Getty Images.)*

CULTURAL LANDSCAPE

7.5 LEARNING OBJECTIVE
Describe how religions have left their particular mark on the cultural landscape.

Religion is a vital aspect of culture; its visible presence can be quite striking, reflecting the role played by religious motives in the human transformation of the landscape. In some places, the religious aspect is the dominant visible evidence of culture, producing sacred landscapes. At the opposite extreme are areas almost purely secular in appearance. Religions differ greatly in visibility, but even those least apparent to the eye usually leave some subtle mark on the countryside.

RELIGIOUS STRUCTURES

The most obvious religious contributions to the landscape are the buildings erected to house divinities or to shelter worshippers. These structures vary greatly in size, function, architectural style, construction material, and degree of ornateness (**Figure 7.30**). To Roman Catholics, for example, the church building is literally the house of God, and the altar is the focus of key rituals. Partly for these reasons, Catholic churches

FIGURE 7.30 **Traditional religious architecture takes varied forms.** St. Basil's Cathedral on Red Square in Moscow (*top*) reflects a highly ornate Russian landscape presence, whereas the plain brown chapel in South Dakota (*middle*) demonstrates the opposite tendency of favoring visual simplicity by British-derived Protestants. The ornate Hindu temple in New Delhi, India (*bottom*), offers still another sacred landscape. *(Top: vladimir zakharov/Moment/Getty Images; Middle: Medioimages/Photodisc/Getty Images); Bottom: b n khazanchi/Moment Open/Getty Images.)*

are typically large, elaborately decorated, and visually imposing. In many towns and villages, the Catholic house of worship is the focal point of the settlement, exceeding all other structures in size and grandeur. In medieval European towns, Christian cathedrals were the tallest buildings, representing the supremacy of religion over all other aspects of life.

For many Protestants—particularly the traditional Calvinistic chapel-goers of British background, including Presbyterians and Baptists—the church building is, by contrast, simply a place to assemble for worship. The result is a small, simple structure that appeals less to the senses and more to personal faith. For this reason, traditional Protestant houses of worship typically are not designed for comfort, beauty, or high visibility but instead appear deliberately humble (see Figure 7.30, middle). Similarly, the religious landscape of the Amish and Mennonites in rural North America, the "plain folk," is very subdued because they reject ostentation in any form. Some of their adherents meet in houses or barns, and the churches that do exist are very modest in appearance, much like those of the southern Calvinists.

megachurch Large Protestant church structures, usually located in suburban areas of the United States, which have large congregations (2000–10,000 members) and utilize business models to tailor their spaces and services to their congregation's needs.

So-called **megachurches** present a recent twist on the Protestant house of worship, one that shuns humbleness and plainness for size and grandeur (**Figure 7.31**). Located in suburban areas, but also on the rise in Protestant areas worldwide, megachurches count from 2000 to 10,000 members and rely on a business-oriented approach to designing

FIGURE 7.31 **Megachurch.** Yoido Full Gospel Church, in Seoul, South Korea, is the largest megachurch in the world with approximately 830,000 members. *(Pascal Deloche/ GODONG/Godong/Corbis.)*

spaces and services in line with their congregation's wishes. These include plenty of parking and bathrooms, but also recreational facilities, high-tech audio visual sermon delivery, and casual dress codes. Megachurches also foster a sense of connectedness and community in an increasingly secular world. Geographers Barney Warf and Morton Winsberg (2010), who have studied megachurches, claim that they are "the new face of American religion, a threat that many traditional denominations have understandably viewed with alarm."

Islamic mosques are usually the most imposing structures in the landscape, whereas the visibility of Jewish synagogues varies greatly. Hinduism has produced large numbers of visually striking temples for its multiplicity of gods, but much worship is practiced in private households with their own personal shrines (Figure 7.32).

Nature religions such as animism generally place only a subtle mark on the landscape. Nature itself is sacred, and imposing features already present in the landscape—mountains, waterfalls, and grottos, for example—mean that few human-made shrines are needed.

Paralleling this contrast in church styles are attitudes toward roadside shrines and similar

manifestations of faith. Catholic culture regions typically abound with shrines, crucifixes, and other visual reminders of religion, as do some Eastern Orthodox Christian areas. Protestant areas, by contrast, are bare of such symbolism. Their landscapes instead display such features as signboards advising the traveler to "get right with God," a common sight in the southern United States (Figure 7.33).

FAITHFUL DETAILS

Not only buildings but also other architectural choices, such as color and landscaping elements, can convey spiritual meaning through the landscape. Consider the color green. The image of a Muslim mosque in northern Nigeria in Figure 7.34 depicts how green—sacred in Islam—appears often on mosque domes and minarets, the tall circular towers from which Muslims are called to prayer. The prophet Muhammad declared his favorite color to be green, and his cloak and turban were said to be green. In Christianity, green also has positive connotations, symbolizing fertility, freedom, hope, and renewal. It is

FIGURE 7.32 **Temples dedicated to ancestors in Bali, Indonesia.** Bali's population practices a blend of Hinduism, animism, and ancestor worship. These temples to ancestors are a feature of every Balinese home, and offerings of incense, food, and flowers are made three times per day to show reverence. *(Courtesy of Ari Dorfsman.)*

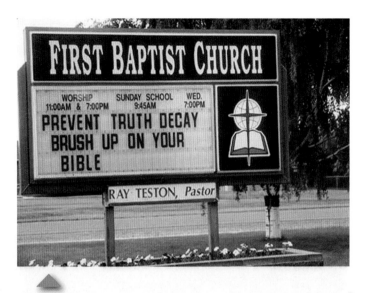

FIGURE 7.33 **Billboard in Three Forks, Montana.** Protestant billboards often use American-style advertising slogans to promote church attendance. *(Nancy H. Belcher.)*

LANDSCAPES OF THE DEAD

Religions differ greatly in the type of tribute each awards to its dead. These variations appear in the cultural landscape. The few remaining Zoroastrians, called Parsees, who preserve a once-widespread Middle Eastern faith now confined to parts of India, have traditionally left their dead exposed to be devoured by vultures. Thus, the Parsee dead leave no permanent mark on the landscape. Hindus cremate their dead. Having no cemeteries, their dead, as with the Parsees, leave no obvious mark on the land. Yet on the island of Bali in Indonesia, Hinduism has blended with animism such that temples to family ancestors occupy prominent places outside the houses

used in decorations, such as banners, and for clergy vestments, especially those associated with the Trinity and with the Epiphany season.

Water, often used as a decorative element in fountains, baths, and pools, has religious meaning. In fact, all three of the great desert monotheisms— Islam, Christianity, and Judaism—consider water to have a special status in religious rituals. In general, followers of all three religions believe water has the ability to purify and cleanse the body as well as to purify and cleanse the soul of sins. Muslims use water in ablutions, or washing, of the hands, face, or ears before daily prayers and other rituals. Footbaths, or ablution fountains, are important features located outside of mosques, and decorating with water, particularly pools and fountains, is a common practice in Islamic architecture. Central to Christianity is baptism, where the newborn infant or adult convert is sprinkled with or fully immersed in holy water, symbolizing the cleansing of original sin for infants and repentance or cleansing from sin for adults. Jews may engage in a ritual bath, known as a *mikveh*, before important events, such as weddings (for women), or on Fridays, the evening of the Jewish Sabbath (for men). The story of the Great Flood, found in the Book of Genesis (common to both Jews and Christians), depicts the washing away of the sins of the world so that it could be born anew.

FIGURE 7.34 **Muslim mosque in northern Nigeria.** Areas with Muslim populations enjoy a landscape where mosque minarets and domes are frequently green in color. *(Laurie Strachan/Alamy.)*

of the Balinese, a landscape feature that does not exist in India itself (see Figure 7.32).

In Egypt, spectacular pyramids and other tombs were built to house dead leaders. These monuments, as well as the modern graves and tombs of the rural Islamic folk of Egypt, lie on desert land not suitable for farming (Figure 7.35). Muslim cemeteries are usually modest in appearance, but spectacular tombs are sometimes erected for aristocratic persons, giving us such sacred structures as the Taj Mahal in India (Figure 7.36), one of the architectural wonders of the world.

Taoic Chinese typically bury their dead, setting aside land for that purpose and erecting monuments to their deceased kin. In parts of China—reflecting its pre-communist history—cemeteries and ancestral shrines take up as much as 10 percent of the land in some districts, greatly reducing the acreage available for agriculture.

Christians also typically bury their dead in sacred places set aside for that purpose. These sacred places vary significantly from one Christian denomination to another. Some graveyards, particularly those of Mennonites and southern Protestants, are very modest in appearance, reflecting the reluctance of these groups

FIGURE 7.36 **The Taj Mahal in Agra, India.** Built as a Muslim tomb, the Taj Mahal is perhaps the most impressive religious structure in the world. *(Pallava Bagla/Corbis.)*

to use any symbolism that might be construed as idolatrous. Among certain other Christian groups, such as Roman Catholics and members of the various Orthodox Churches, cemeteries are places of color and elaborate decoration (Figure 7.37).

Cemeteries often preserve truly ancient cultural traits because people as a rule are reluctant to change their practices relating to the dead. The traditional rural cemetery of the southern United States provides a case in point. Freshwater mussel shells are placed atop many of the elongated grave mounds, and rose bushes and cedars are planted throughout the cemetery. Recent research suggests that the use of roses may derive from the worship of an ancient, pre-Christian mother goddess of the Mediterranean lands. The rose was a symbol of this great goddess, who could restore life to the dead. Similarly, the cedar evergreen is an age-old pagan symbol of death and eternal life, and the use of shell decoration derives from an animistic custom in West Africa, the geographic origin of slaves in the American South. Although the present Christian population of the South is unaware of the origins of their cemetery symbolism, it seems likely that their landscape of the dead contains animistic elements thousands of years old, revealing truly ancient beliefs and cultural diffusions.

FIGURE 7.35 **Farmer with Great Pyramids in the background.** The foreground of this image, shot near Cairo along the fertile banks of the Nile River in Egypt, shows that this land is heavily utilized for agriculture. In the background are the Great Pyramids of Giza, originally built to house the dead, situated in the adjacent arid desert lands. *(Universal Images Group/DeAgostini/Alamy.)*

FIGURE 7.37 **Two different Christian landscapes of the dead.** In the Yucatán Peninsula of Mexico, the dead rest in colorful, aboveground crypts, whereas in Amana, Iowa, communalistic Germans prefer tidiness, order, and equality. *(Left: Macduff Everton/Corbis; Right: Alan Solomon/mct/ZUMAPRESS.com.)*

SACRED SPACE

sacred spaces Areas recognized by a religious group as worthy of devotion, loyalty, esteem, or fear to the extent that they become sought out, avoided, inaccessible to nonbelievers, or removed from economic use.

Sacred spaces consist of natural and/or human-made sites that possess special religious meaning that is recognized as worthy of devotion, loyalty, fear, or esteem. By virtue of their sacredness, these special places might be avoided by the faithful, sought out by pilgrims, or barred to members of other religions (see Figure 7.18). Often, sacred space includes the site of supposed supernatural events or is viewed as the abode of gods. Cemeteries, too, are also generally regarded as a type of sacred space. So is Mount Sinai, described by geographer Joseph Hobbs as endowed "with special grace," where God instructed the Hebrews to "mark out the limits of the mountain and declare it sacred." Conflict can result if two religions venerate the same space. In Jerusalem, for example, the Muslim Dome of the Rock, on the site where Muhammad is believed to have ascended to heaven, stands above the Western Wall, the remnant of the ancient Jewish temple (**Figure 7.38**; see Patricia's Notebook on page 306 and the World Historical Site on pages 308–309). These two religions literally claim the same space as their holy site, which has led to conflict.

Sacred space is receiving increased attention in the world. In the mid-1990s, the internationally funded Sacred Land Project began to identify and protect such sites, 5000 of which have been cataloged in the United Kingdom alone. Included are such places as ancient stone circles, pilgrim routes, holy springs, and sites that convey mystery or great natural beauty. This last type falls into the category of mystical places: locations unconnected with established religion where, for whatever reason, some people believe that extraordinary, supernatural things can happen. The Bermuda Triangle in the western Atlantic, where airplanes and ships supposedly disappear, is a mystical place and, in effect, a vernacular culture region. Some people find the expanses of the American Great Plains mystical, and writer Jonathan Raban spoke of them as "a landscape ideally suited to the staging of the

Patricia's Notebook

My Travels to Jerusalem

Patricia L. Price. *(American Council on Education.)*

One of the best things about being a geographer is that I always have a good excuse to travel the world! Among the most interesting places I have visited thus far is Israel. It's hard to think of any place on Earth where so much globally significant history, passion, and conflict are packed into such a small area. Because I'm a cultural geographer whose work emphasizes the humanistic theme of how it is that societies establish meaningful ties to places—in other words, how we dwell on Earth—I always have my eyes open for traces of meaning-making on the landscape.

In Jerusalem, for instance, you can walk the Via Dolorosa route, visiting all 14 Stations of the Cross walked by Jesus from his condemnation by the Romans to the site of his crucifixion and death. You can see the iconic golden Dome of the Rock on Jerusalem's Temple Mount, which shelters the rock (hence the name) from which Muhammad is said to have ascended to heaven. Jews also believe the site to be sacred because it is said to be the place where Abraham prepared to sacrifice his son Isaac and is the site of the Western Wall, the remnant of the ancient Jewish temple. Non-Muslims are forbidden from entering the Dome of the Rock for reasons of political conflict and control as well as pollution taboos (which are held by most religions, including Islam). Some Jews voluntarily decline to set foot on the Temple Mount as well, believing it too holy for mere mortal presence.

So much of what we hear and see about this place on our nightly news depicts only the political tensions existing between Israelis and Palestinians. So it is eye-opening to wander the streets of Jerusalem and witness Jews, Christians, and Muslims, orthodox and secular alike, living their daily lives in this city (see also the World Heritage site starting on page 308). Yes, there is tension. Mobility and access are sometimes blocked or at least slowed down, thanks mostly to security concerns. But, as with anyplace else on Earth, most people in Jerusalem are merely going about their business.

I always tell my students that, if given the chance to travel anywhere in the world, they should seriously consider visiting the Middle East. Directly experiencing peoples and places goes a long way toward dispelling the myths and stereotypes about those peoples and places, which are particularly prevalent in this part of the world.

Panoramic shot of Jerusalem, Israel, site of some of the world's most contested sacred spaces. In the foreground is the iconic Dome of the Rock, sacred to Muslims, which is located on the Temple Mount, site of the destroyed Second Temple and thus also sacred to Jews. *(Courtesy of Patricia L. Price.)*

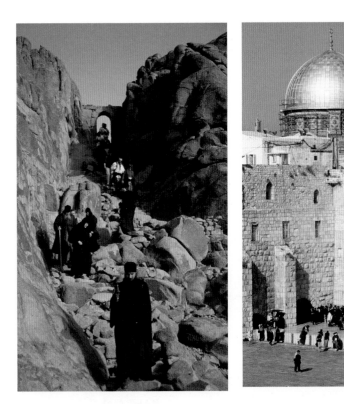

FIGURE 7.38 **Sacred spaces.** The three so-called Abrahamic faiths—Judaism, Christianity, and Islam—arose in the same place and thus their adherents revere some of the same holy sites. *Left:* Mount Sinai, in North Africa, is where Moses is said to have received the Ten Commandments as the Israelites continued their journey out of Egypt and is also mentioned several times in the Qur'an. *Right:* Jews pray at the Western Wall in Jerusalem, in Israel. Standing above the wall, on the site of the destroyed Jewish temple, is one of the holiest sites for Muslims: the golden-capped mosque called the Dome of the Rock. (See also the photograph in Patricia's Notebook, page 306.) *(Left: Konrad Wothe/ age fotostock; Right: Gary Cralle/Gettyone.)*

millennium, open to the gaze of the Almighty." Sometimes the sacred space of vanished ancient religions never loses—or later regains—the functional status of mystical place, which is what has happened at Stonehenge in England.

In his book *The Sacred and the Profane,* the renowned religious historian Mircea Eliade suggested that all societies, past and present, have sacred spaces. According to Eliade, sacred spaces are so important because they establish a geographic center on which society can be anchored. Through what he termed *hierophany,* a sacred space emerges from the profane, ordinary spaces surrounding it. A contemporary example of hierophany is provided by the appearance of an image of the Virgin of Guadalupe on a bank building in Clearwater, Florida, transforming the profane space of finance into a Catholic pilgrimage site (Figure 7.39).

FIGURE 7.39 **Our Lady of Clearwater.** This 60-foot (18-meter) image appeared in December 1996 in the window of a bank building in Clearwater, Florida. Because of its resemblance to Our Lady of Guadalupe, Catholics—including a sizable Mexican immigrant population as well as Catholic pilgrims from around the region—gather in the parking lot every day for the celebration of Mass. *(Courtesy of Patricia L. Price.)*

World Heritage Site

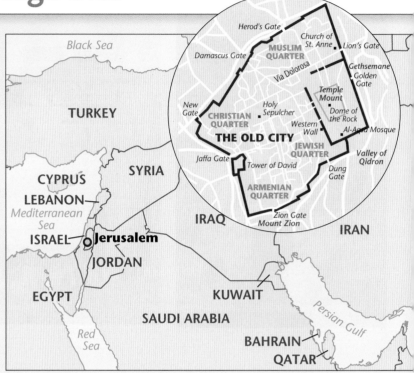

The Old City of Jerusalem and its walls were added to the list of World Heritage Sites in 1981. The area is less than one-half of a square mile, set within modern-day Jerusalem. The Old City has been occupied from the fourth millennium B.C.E., thus making Jerusalem one of the world's oldest urban settlements (see Chapter 10). Divided into four quarters—Armenian, Christian, Jewish, and Muslim—these different "neighborhoods" were formed based on major ethnic and religious divisions dating back centuries.

■ Greater Jerusalem holds tremendous significance to the major world religions as it is the place where Jews built and rebuilt the temple that served as a center of worship, Christ was crucified, and Mohammed ascended to heaven.

■ The Old City includes the city's holiest sites: the Temple Mount and the portion of the wall of the Jewish Second Temple known as the Western Wall, the Church of the Holy Sepulcher, the al-Aqsa Mosque, and the Dome of the Rock. Access to these sites, some of which are off-limits to Jews, is severely contested today and is discussed further in Patricia's Notebook (page 306).

■ The Dome of the Rock pictured here is considered to be the "most contested piece of real estate on Earth" due to its religious importance. For thousands of years, Jews, Christians, and Muslims have shared Jerusalem's city spaces—sometimes cooperatively and other times in conflict.

(Jon Arnold Images Ltd/Alamy.)

The Old City of Jerusalem and Its Walls

WAILING WALL: Pictured here is a portion of the Jewish Second Temple that was built in 515 B.C.E. after the first temple, constructed in 957 by King Solomon, was attacked by the Egyptian pharaoh and then destroyed by the Babylonians in 586. The Second Temple, itself threatened by Greek and Roman invaders, was ultimately destroyed by the latter in 70 C.E. A remnant of the Western Wall encloses the Second Temple courtyard and is one of Judaism's holiest sites.

(Lior Mizrahi/Getty Images.)

◼ It is often referred to as the "Wailing Wall" because of the Jewish custom of mourning the destruction of the Second Temple at this spot. Jews pray here and place slips of paper containing written prayers inside the crevices of the wall. More than 1 million notes are placed within the wall's crevices each year, and some organizations allow the faithful to e-mail notes that are later placed within the wall.

◼ As with holy sites from many faiths, the Wailing Wall is segregated by gender. Men visit one side, while women and children are restricted to a much smaller and crowded portion of the wall on the right-hand side.

◼ Some women argue that ultra-Orthodox restrictions against women and men sharing the same space in places of worship illegally restrict women's right to space, and that bans against women wearing prayer garments and reading aloud from the Torah at the Wall are unconstitutional under Israeli law.

TOURISM: Jews, Christians, and Muslims have for millennia shared Jerusalem's city spaces in ways that involve cooperation as well as conflict. Its streets, apartments, and public spaces bustle with nearly 1 million inhabitants of diverse faiths and ethnicities going about their daily lives. It is fascinating to experience the mixture of language, dress, and customs interacting in Jerusalem.

(MARCO LONGARI/AFP/Getty Images.)

◼ In addition, 3.5 million tourists visit every year—roughly 4 times the city's resident population. Given the number of religious sites in Jerusalem, pilgrimage alone accounts for many tourist visits. The country of Israel, where Jerusalem is located, is also home to many sites of historical, archaeological, cultural, and religious significance, as well as being near picturesque Mediterranean beaches and the greater Middle Eastern region.

◼ In 1982 Jerusalem's Old City and Walls were placed on the list of World Heritage Sites in danger, due in part to the deterioration and damage stemming from tourism and a lack of adequate conservation efforts. In other words, Jerusalem's Old City is literally in danger of being loved to death.

◼ Approximately 3 million tourists each year

◼ Divided into four quarters: Armenian, Christian, Jewish, and Muslim

◼ Home to key holy sites: The Temple Mount and Western Wall, Church of the Holy Sepulcher, the al-Aqsa Mosque, and the Dome of the Rock

◼ Sometimes a site of conflict due to shared religious spaces

www.whc.unesco.org/en/list/148

CONCLUSION

Religion is firmly interwoven in the fabric of culture. Religions vary greatly from one region to another, creating diversity so profound as to give special significance to James Griffith's admonition that we should all "learn, respect, and walk softly." Yet religions tend to share some common elements, the most basic being that they provide a structure of faith that anchors communities. This religious spatial variation leads us to ask how these distributions came to be, a question best answered by examining methods of cultural diffusion. Some religions actively encourage their own diffusion through missionary activity or conquest. Other religions erect barriers to expansion diffusion by restricting membership to one particular ethnic group. People, as well, are set in motion thanks to spiritual practices, most notably pilgrimages.

In today's globalizing world, it is worth exploring how—or even if—the world's great religions will become significantly modified in response to changing contexts. The common practice of Internet-based religious practice brings into focus the debates over the importance of place in a global world.

The theme of nature–culture reveals some fundamental ties between religion and the physical environment. One major function of many religious systems, particularly the animistic faiths, is to appease and placate the forces of nature in order to achieve and maintain harmony between humans and the physical environment. Plants and animals diffuse—or not—with the spread of religions. The followers of various religions differ in their outlooks on environmental modification by humans.

The cultural landscape abounds with expressions of religious belief. Places of worship—mosques, temples, churches, megachurches, and shrines—differ in appearance, distinctiveness, prominence, and frequency of occurrence from one religious culture region to another. These buildings provide a visual index to the various faiths, as do their related architectural elements. Cemeteries and sacred spaces also add a special effect to the landscape that tells us about the religious character of the population. In all these ways, and more, the five themes of cultural geography prove relevant to the study of the world's religions.

GEOGRAPHY @ WORK

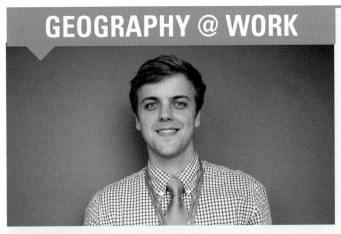

Robert Szypko. *(Courtesy of Robert Szypko.)*

Robert Szypko

First-Grade Teacher, Success Academy, Bedford-Stuyvesant, New York

Education:
BA Geography—Dartmouth College

Q. *Why did you major in geography and decide to pursue a career related to the field?*

A. I am fascinated by the ways that individuals negotiate cultural, social, and ethnic differences across space both in global and local (particularly urban) contexts. I initially chose geography because of the field's strong tradition in development studies, but grew to love the larger analytical frameworks and technological skills emphasized in the discipline. I appreciate nuance, and the cultural geography courses I took brought my understanding of the nuances of culture, landscapes, and other topics to a new level. The more geography courses I took, the more I realized the breath of geography's implications. Suddenly, my interests in music, education, graffiti, and journalism gained a greater richness as I saw them through a geographer's lens.

Q. *Please describe your job.*

A. I lead a classroom of 25 first-grade students who come from a variety of neighborhoods across Brooklyn. My responsibilities include planning and executing lessons, analyzing and responding to classroom achievement data, and maintaining a positive and hard-working classroom culture. I work closely with parents to leverage them in the education of their children, and work to better understand the backgrounds and home lives of each student in order to best serve them in the classroom.

Q. *How does your geographical background help you in your day-to-day work?*

A. My geographical background has helped me more fully understand and account for my students' varied backgrounds and the geographical diversity they represent (both in terms of where their family is from and where they live now, in different neighborhoods throughout Brooklyn). Geography's emphasis on complexity when dealing with cultures and communities at any scale has given me the critical eye to maintain an open mind about parents, students, the neighborhood of my school, and so on. It has also helped me as I try to lead students toward a better understanding of the world around them. Children's literature contains many different geographic representations, and I firmly believe that geographic nuance is important even at such a young age, from breaking down the homogenous identity of "Africa," to developing a deep understanding of the different landscapes and cultures of the arctic zone.

Q. *What advice do you have for students considering a career in the geography field?*

A. Keep an open mind about what geography can do for you in your career. The critical thinking skills and technological skills (e.g., GIS) you acquire as a geographer can serve you well in a variety of industries and jobs.

THE VIDEO CONNECTION

A Chinese Threat to Afghan Buddhas

http://www.tinyurl.com/mxxtkfs

Today, nearly everyone in Afghanistan practices Islam. But for around a millennium prior to the eighth century C.E., Buddhism was the prevalent religion. This video depicts the excavations of an ancient Buddhist settlement at a site called Mes Aynak, located southeast of the capital city of Kabul. Archaeologists' work has been accelerated by the threat of destruction of the site at the hands of a Chinese mining company. Directly under this site lies a rich deposit of copper, which the company—and the government of Afghanistan—would like to access. Previously, in 2001, the so-called Buddhas of Bamiyan—also located in Afghanistan but not at Mes Aynak—were blown up by the Taliban (a fundamentalist Islamic political group in Afghanistan) who declared them to be idols. The Buddhas of Bamiyan, which dated from the sixth century C.E and consisted of two large statues carved into a cliff, were a UNESCO World Heritage Site. The case study depicted in the video brings up the tensions often inherent at the crossroads of politics, religion, and economic development.

Thinking Geographically:

1. Find Afghanistan on a map. What is it about Afghanistan's location and physical geography that have, historically speaking, made it so susceptible to religious change?

2. Afghanistan's Buddhist cultural heritage seems to be under attack, first by the Taliban and now by a Chinese mining company. What arguments might the Taliban and Chinese mining company give to rationalize the destruction? What will be lost if the archaeological remains depicted in this video are destroyed? Is the loss Afghanistan's alone?

3. Describe some specific ways in which the case study in this video can also be understood in the context of Chapter 4 (geography of language), Chapter 6 (political geography), Chapter 9 (development geography), and Chapter 10 (urban geography). Why is it that religion has so many connections to the rest of our lives?

DOING GEOGRAPHY

The Making of Sacred Spaces

For this exercise, you will use Professor Kenneth Foote's ideas about how spaces become sacred. In his book *Shadowed Ground: America's Landscapes of Violence and Tragedy,* Foote argues that what a society chooses to forget about its past is at least as important as what it remembers in shaping an image of itself that can be presented to the world. Foote discusses how certain places in the United States have, or have not, become sacred sites. He notes, for example, that the battlefields of the Revolutionary and Civil wars are well marked and visited by thousands every year. Yet the 1692 execution site of the supposed witches of Salem, Massachusetts, is unmarked. How could the exact site of such an infamous event in U.S. history be impossible to locate? Foote realized that this scenario was not limited to the United States, writing, "Indeed, soon after my first trip to Salem, I found myself in Berlin before the reunification of Germany. There I came across similar places—Nazi sites such as the Gestapo headquarters and Reich chancellery—that have lain vacant since just after World War II and seem to be scarred permanently by shame."

Professor Foote suggests that sites of important events can be sanctified, obliterated, purposely ignored, or fall somewhere in between. Sanctified sites are set apart from other sites, are carefully maintained, are often publicly owned, and attract annual visitors and ceremonies. At the other end of the spectrum, the sites of particularly shameful, stigmatized, or violent events simply may be obliterated from the landscape in an attempt to forget about them.

Steps to Understanding Sacred Spaces

Step 1: Think about sites near where you now live and where a violent or tragic event occurred, for example, a murder, a freak accident, a natural disaster, an assassination, an act of terrorism, or a heinous crime. Virtually all places experience some such event. If nothing readily comes to mind, inquire, and inevitably you will discover an example for this exercise, even if you live in a rural area or a college town.

Step 2: Once you have identified a site, consider its current status. Is it sanctified, is it obliterated, or does it fall somewhere in between? Elements of the landscape, such as signs or monuments (or the lack of them), will indicate the status of the site.

Step 3: Take note of any debates surrounding what to do with this site. Has the status of the site changed over time? How?

When people are unsure of how an event fits into their history, the site where the event occurred may inhabit a sort of limbo, remaining unmarked until the society in question comes to terms with the event. This is the case with many sites of racially motivated, violent, or shameful acts in the United States, such as the Memphis, Tennessee, motel where Martin Luther King, Jr. was assassinated in 1968, or the Manzanar, California, internment camp where Japanese Americans were confined during World War II. Only recently have both sites become prominent on the landscape after decades of neglect. Consider the following questions:

- Are any newsworthy events happening now that may mark a place as a site of tragedy?

- What sites, whether nearby or far away, do you expect to become more sacred over time?

- Can you think of an event that is so horrific that the place where it occurred will be forever forgotten and erased from the map?

Cemetery at Manzanar National Historic Site. The Japanese inscription reads "Soul Consoling Tower." Fifteen of the 146 Japanese Americans who died at this Relocation Center are buried here. A total of 110,000 Japanese Americans were interned during World War II in camps like Manzanar. After the 1941 Japanese attack on Pearl Harbor, Hawaii, Japanese Americans were deemed "enemy aliens." It was not until 1992—nearly 50 years after the end of the war—that Manzanar was declared a National Historic Site. *(George Ostertag/age fotostock.)*

SEEING GEOGRAPHY Parking Lot Shrine

How can an ordinary landscape, such as a parking lot, become a sacred space?

Parking lot shrine to the Virgin of Guadalupe, Self-Help Graphics and Art, East Los Angeles.
(Courtesy of Patricia L. Price.)

This large statue of the Virgin of Guadalupe inhabits the corner of the parking lot of Self-Help Graphics and Art, an artists' cooperative in East Los Angeles. According to Michael Amezcua, a metalwork artist at Self-Help, neighborhood residents actively use this shrine, leaving offerings and petitions to the Virgin in the outdoor grotto. Every December 12th, neighborhood residents gather here to celebrate the feast day of the Virgin of Guadalupe.

We could interpret this unique parking lot shrine to be a variation on the yard shrine, a common feature of the residential landscape of the southwestern United States. Throughout the Mexican American homeland of the Southwest, it is common to see a small shrine to the Virgin of Guadalupe in people's front yards. This, in turn, is a variation on the Mexican custom of maintaining elaborate shrines and altars to important religious figures. These are sometimes located inside the house in a space reserved especially for altars (called a *nicho*). Or they can be located outside the house, near the front door, because Mexican homes usually do not have yard space in front of the house. On December 12th, altars all over Mexico and Mexican American areas in the United States are elaborately decorated with candles, ribbons, and flowers. Devotees of the Virgin of Guadalupe celebrate with family and neighbors throughout the night. Since turquoise blue is considered to be the favorite color of the Virgin of Guadalupe, this color abounds. Because the Virgin of Guadalupe is also the official patron saint of Mexico, the red, white, and green of the Mexican flag are prevalent in altar decorations.

In Miami, another Latino population center in the United States, yard shrines are built to La Virgen de la Caridad de Cobre. Like the Virgin of Guadalupe, La Virgen de la Caridad is an American manifestation of the Virgin Mary. This Virgin hails from the copper mining village of Cobre, in Cuba. She is believed to protect seafarers and has been adopted by Cuban rafters as their patron saint. Yemayá, who is an *orisha*, or spirit, in the Afrocentric Cuban religious tradition of Santería, is often equated with La Virgen de la Caridad de Cobre, as both are associated with the sea.

What does the fact that this particular shrine is built in a parking lot rather than a yard say about what, and how, spaces become sacred? Can you think of a more mundane place than a parking lot for such a hallowed figure as the Virgin of Guadalupe? Perhaps the drive-through architecture of strip malls, highways, and parking lots that is so closely associated with the urban landscape of Los Angeles has something to do with it. Where else, other than in Los Angeles, would you find a sacred parking lot?

Chapter 7
LEARNING OBJECTIVES REEXAMINED

7.1 Define the differences among religions as well as the basic tenets of the world's principal religions, and locate their primary regional locations on a world map.

See pages 275–285. Review Figure 7.3. Where are the largest concentrations of Muslim, Jewish, and Christians located in the world?

7.2 Discuss how and why the world's religions became distributed as they are today.

See pages 286–291. Explain what *contact conversion* means. What role does this term play in the distribution of the world's religions?

7.3 Critically analyze how globalization has affected the practice of religion.

See pages 291–295. What role has the Internet played in the globalization of religion?

7.4 Discuss specific ways in which the natural environment shapes, and is shaped by, religious beliefs and practices.

See pages 295–300. Explain the term *ecotheology*. What role does ecotheology play in Christianity?

7.5 Describe how religions have left their particular mark on the cultural landscape.

See pages 300–310. What role do sacred spaces play in the cultural landscape?

KEY TERMS

adaptive strategy	p. 295	monotheistic religion	p. 273
animists	p. 285	orthodox religions	p. 273
contact conversion	p. 286	pilgrimages	p. 289
culture hearth	p. 286	polytheistic religion	p. 273
ecotheology	p. 298	proselytic religions	p. 273
ethnic religion	p. 273	religion	p. 271
fundamentalism	p. 274	sacred spaces	p. 305
Gaia hypothesis	p. 300	syncretic religions	p. 273
megachurch	p. 301	universalizing religions	p. 273

Geography of Religion on the Internet

You can learn more about the geography of religion on the Internet at the following web sites:

Get Religion

http://www.getreligion.org/

Founded because (quoting journalist William Schneider) "the press . . . just doesn't get religion," this blog attempts to ferret out the ghosts lurking in news coverage of religion.

History of Religion

http://www.mapsofwar.com/ind/history-of-religion.html

"See 5000 years of religion in 90 seconds" by launching this dynamic map. The timeline shows major developments in the history of world religions.

National Religious Partnership for the Environment

http://www.nrpe.org/

An interfaith partnership encouraging environmental conservation in the name of religion.

Our Lady of Guadalupe

http://www.sancta.org/opro1.html

On the "ora pro nobis" ("pray for us") section of this web site, you can post your prayer or request to the Virgin of Guadalupe and read the heartfelt posts of other devotees.

Pew Forum on Religion & Public Life

http://pewforum.org

Seeking to promote a greater understanding of the intersection between religion and public affairs, the Pew Forum conducts polls, surveys, and demographic analysis in the United States and around the world.

Sources

BBC News Europe. 2011. *Halal and Kosher Hit by Dutch Ban.* 6 November video. http://www.bbc.co.uk/news/world-europe-15610142.

Eliade, Mircea. 1987 [1957]. *The Sacred and the Profane: The Nature of Religion.* Willard R. Trask (trans.). San Diego: Harcourt.

Foote, Kenneth E. 1997. *Shadowed Ground: America's Landscapes of Violence and Tragedy.* Austin: University of Texas Press.

Griffith, James S. 1992. *Beliefs and Holy Places: A Spiritual Geography of the Pimería Alta.* Tucson: University of Arizona Press.

Hobbs, Joseph J. 1995. *Mount Sinai.* Austin: University of Texas Press.

Huntsinger, Lynn, and María Fernández-Giménez. 2000. "Spiritual Pilgrims at Mount Shasta, California." *Geographical Review* 90: 536–558.

Newport, Frank. 2012. *God Is Alive and Well: The Future of Religion in America.* New York: Gallup Press.

Raban, Jonathan. 1996. *Bad Land: An American Romance.* New York: Pantheon.

Rashtogi, Nina Shen. 2010. "Is Kosher Better for the Planet?" *The Washington Post,* 2 February. http://articles.washingtonpost.com/2010-02-02/opinions/36839736_1_kosher-meat-halal-meat-kosher-products.

Singh, Rana P. B. 1994. "Water Symbolism and Sacred Landscape in Hinduism." *Erdkunde* 48: 210–227.

Warf, Barney and Morton Winsberg. 2010. "Geographies of Megachurches in the United States." *Journal of Cultural Geography* 27(1): 33–51.

Worthen, Molly. 2012. "One Nation Under God?" *The New York Times Sunday Review,* 22 December.http://www.nytimes.com/2012/12/23/opinion/sunday/american-christianity-and-secularism-at-a-crossroads.html?pagewanted=all&_r=0.

Ten Recommended Books on the Geography of Religion

(For additional suggested readings, see the *Contemporary Human Geography* LaunchPad: http://www.macmillanhighered.com/launchpad/DomoshCHG1e.)

Esposito, John, Susan Tyler Hitchcock, Desmond Tutu, and Mpho Tutu. 2004. *Geography of Religion: Where God Lives, Where Pilgrims Walk.* Washington, D.C.: National Geographic. In the tradition of National Geographic magazine, this comprehensive reference book is beautifully illustrated.

Falah, Ghazi-Walid, and Caroline Nagel (eds.). 2005. *Geographies of Muslim Women: Gender, Religion, and Space.* New York: Guilford Press. A collection of chapters by geographers that highlights the diversity of women's experiences across the Islamic world.

Hart, John. 2006. *Sacramental Commons: A Christian Ecological Ethics.* Lanham, Md.: Rowman & Littlefield. The author suggests that local places and communities provide "sacraments," or signs of divine presence, which give rise to an awareness of human interconnectedness with all living things and our moral duty to care for the planet.

Murray, Stuart A. P. 2009. *Hammond Atlas of World Religions: A Visual History of Our Great Faiths.* Irvington, N.Y.: Hylas. Takes a historical look at the development of the world's religions. This atlas contains beautifully rendered historical and topical maps, along with many photographs.

O'Brien, Joanne, and Martin Palmer. 2007. *The Atlas of Religion: Mapping Contemporary Challenges and Beliefs.* Berkeley and Los Angeles: University of California Press. Maps the nature, extent, and influence of the world's major religions. Depicts how religions spread across space, their role in global issues such as hunger as well as conflict, and the locations of sacred places.

Orsi, Robert A. 2005. *Between Heaven and Earth: The Religious Worlds People Make and the Scholars who Study Them.* Princeton, N.J.: Princeton University Press. Examines twentieth-century American Catholicism and how people forge ties to heaven and Earth through their relationships to the Virgin Mary and the saints.

Stump, Roger W. 2008. *The Geography of Religion: Faith, Place, and Space.* Lanham, Md.: Rowman & Littlefield. An overview of a geographer's approach to studying religion, encompassing their growth and spread, the diverse geographical scales that religion engages, and sacred spaces.

Timothy, Dallen J., and Daniel H. Olsen (eds.). 2006. *Tourism, Religion and Spiritual Journeys.* London: Routledge. Acknowledging that faith is still the primary reason people travel, this collection of essays provides case studies of religious tourism and pilgrimage.

Tomalin, Emma. 2009. *Biodivinity and Biodiversity: The Limits of Religious Environmentalism.* Aldershot, U.K.: Ashgate. Argues that environmentalism may be a romantic Western notion which poorer nations cannot easily subscribe to. Case studies from India and Britain trace environmental concerns in divergent religious and material contexts.

Tweed, Thomas. 2008. *Crossing and Dwelling: A Theory of Religion.* Cambridge, Mass.: Harvard University Press. This book's premise is that religion is about movement, connection, and life as lived, rather than a static entity pinned to a certain place.

CHAPTER 8

What differences can you "read" in these
landscapes? Can you determine their locations?

Two types of contemporary agricultural landscapes. *(Left: Jim Wark/AirPhoto; Right: Michael Busselle/Corbis.)*

Go to "Seeing Geography" on page 362 to learn more about these images.

AGRICULTURE
The Geography of the Global Food System

Every one of us depends, either directly or indirectly, on agriculture for our survival. It is easy to forget that urban-industrial society relies on the food surplus generated by farmers and herders and that, without agriculture, there would be no cities, universities, factories, or offices.

Agriculture, the tilling of crops and rearing of domesticated animals to produce food, feed, drink, and fiber, has been the principal enterprise of humankind throughout recorded history. Even today, agriculture remains by far the most important economic activity in the world, using more land than any other activity and employing about 40 percent of the working population. In some parts of Asia and Africa, more than 75 percent of the labor force is devoted to agriculture. North Americans, on the other hand, live in an urban society in which less than 2 percent of the population work as agriculturists. Likewise, Europe's labor force is as thoroughly nonagricultural as North America's. Nearly half of the world's population, however, continues to live in farm villages.

> **agriculture** The cultivation of domesticated crops and the raising of domesticated animals.

REGION

8.1 LEARNING OBJECTIVE
Explain the connection between region and agriculture.

For thousands of years, farmers have found ways to cope with a range of environmental conditions, creating in the process an array of different types of food-producing systems. Collectively, these farming

LEARNING OBJECTIVES

8.1 Explain the connection between region and agriculture.

8.2 Analyze the role mobility plays in the spatial and cultural patterns of agricultural production and food consumption.

8.3 Describe how the processes of globalization alter the geography of agriculture.

8.4 Understand how nature–culture relations are expressed through the production and consumption of food.

8.5 Relate how agriculture is expressed within the cultural landscape.

agricultural region
A geographic region defined by a distinctive combination of physical and environmental conditions; crop type; settlement patterns; and labor, cultivation, and harvesting practices.

extensive agriculture
The practices of farming and livestock raising using low levels of labor and capital relative to the areal extent of land under production, relying chiefly on natural soil fertility and prevailing climate.

intensive agriculture
The practices of farming and livestock raising using high levels of labor and capital relative to the size of the land holding, thus overcoming environmental constraints through irrigation, land reclamation, synthetic fertilizers, chemical pesticides and herbicides, and genetic engineering.

practices have constructed formal **agricultural regions.** During the past 500 years, colonialism, industrialization, and globalization have greatly altered existing practices and created new types of agricultural regions, such as plantations. One prominent trend over the past five centuries is the progressive, worldwide movement from extensive to intensive cultivation and husbandry. **Extensive agriculture** refers to the practices of farming and livestock raising that require low levels of labor and capital relative to the areal extent of land under production. Thus, extensive agriculture relies mostly on natural soil fertility and prevailing climate conditions.

In contrast, **intensive agriculture** uses high levels of labor and capital relative to the size of the land holding. Increasingly, intensive agriculture overcomes local environmental constraints through irrigation; land reclamation; and the use of synthetic fertilizers, chemical pesticides and herbicides, and genetic engineering. Consequently, in order to understand agricultural regions, we need to consider the role of the environment, cultural factors, and the political and economic forces that influence a region.

SWIDDEN CULTIVATION

Many of the peoples of tropical lowlands and hills in the Americas, Africa, and Southeast Asia practice a land-rotation agricultural system known as **swidden cultivation.** The term *swidden* is derived from an old English term meaning "burned clearing." Using machetes, axes, and chainsaws, swidden

swidden cultivation A type of agriculture characterized by land rotation in which temporary clearings are used for several years and then abandoned to be replaced by new clearings; also known as slash-and-burn agriculture.

FIGURE 8.1 **A farmer in Tojo, central Sulawesi, Indonesia, surveys the results of clearing and burning a patch of forest for planting.** What looks like destruction is actually part of a cycle that begins and ends in forest when swidden cultivation is practiced sustainably. *(Reuters/Yusuf Ahmad/Landov.)*

cultivators chop away the undergrowth from small patches of land and kill the trees by removing a strip of bark completely around the trunk. After the dead vegetation dries out, the farmers set it on fire to clear the land. Because of these clearing techniques, swidden cultivation is also called slash-and-burn agriculture. Working with digging sticks or hoes, the farmers then plant a variety of crops in the ash-covered clearings, varying from the maize (corn), beans, bananas, and manioc of the Americas to the yams and non-irrigated rice grown in Southeast Asia (Figure 8.1). Different crops typically share the same clearing, a practice called **intercropping.** This technique allows taller, stronger crops to shelter lower, more fragile ones; reduces the chance of total crop losses from disease or pests; and provides the farmer with a varied diet. The complexity of many intercropping systems reveals the depth of knowledge acquired by swidden cultivators over many centuries. Relatively little tending of the plants is necessary until

intercropping The practice of growing two or more different types of crops in the same field at the same time.

harvest time, and no fertilizer is applied to the fields because the ashes from the fire are a sufficient source of nutrients.

The planting and harvesting cycle is repeated in the same clearings for perhaps three to five years, until soil fertility begins to decline as nutrients are taken up by crops and not replaced. Subsequently, crop yields decline. These fields then are temporarily left fallow, and new clearings are prepared to replace them. Because the farmers periodically shift their cultivation plots, another commonly used term for the system is *shifting cultivation*. The abandoned cropland lies fallow for 10 to 20 years before farmers return to clear it and start the cycle again. Swidden cultivation represents one form of **subsistence agriculture:** food production mainly for the family and local community rather than for market.

> **subsistence agriculture** Farming to supply the minimum food and materials necessary to survive.

Although the technology of swidden may be simple, it has proved to be an efficient and adaptive strategy. Swidden farming, unlike some modern systems, is ecologically sustainable and has endured for millennia. Indeed, contemporary studies suggest that some swidden systems have actually enhanced biodiversity. Furthermore, swidden returns more calories of food for the calories spent on cultivation than does modern mechanized agriculture. In many tropical forest regions, swidden cultivation, unlike Western plantations, has left most of the forest intact over centuries of continuous use.

Nonetheless, swidden cultivation can be environmentally destructive under certain conditions. In poor countries with large landless populations, one often finds a front of pioneer swidden farmers advancing on the forests. In such situations, a range of institutional, economic, political, and demographic factors restrict poor farmers' abilities to employ the methods of swidden agriculture in a sustainable way. In many tropical countries, for example, a small proportion of the population owns most of the best agricultural land, forcing the majority of farmers to clear forests to gain access to land. Another condition that may diminish the sustainability of swidden cultivation occurs when a population experiences a sudden increase in its rate of growth and political or social conditions restrict its mobility. Swidden cultivation, still widely practiced throughout the tropics, is thus a highly variable system, occurring in both sustainable and unsustainable forms.

PADDY RICE FARMING

Peasant farmers in the humid tropical and subtropical parts of Asia practice a highly distinctive type of agriculture called **paddy rice farming.** Rice, the dominant paddy crop, forms the basis of civilizations in which almost all the caloric intake is of plant origin. From the monsoon coasts of India through the hills of southeastern China and on to the warmer parts of Korea and Japan stretches a broad region of diked, flooded rice fields, or paddies, many of which are perched on terraced hillsides (**Figure 8.2**). The terraced paddy fields form a striking cultural landscape (see Figure 1.15, page 30).

> **paddy rice farming** The cultivation of rice on a paddy, or small flooded field enclosed by mud dikes, practiced in the humid areas of the Far East.

A paddy rice farm of only 3 acres (1 hectare = 2.47 acres) is usually adequate to support a family because irrigated rice provides a very large output of food per unit of land. Still, paddy farmers must till their small patches intensively to harvest enough food. A system of irrigation that can deliver water when and where it is needed is key to success. In addition, large

FIGURE 8.2 **Cultivation of rice on the island of Bali, Indonesia.** Paddy rice farming traditionally entails enormous amounts of human labor and yields very high productivity per unit of land. ***What are the disadvantages of such a system?*** *(Denis Waugh/Stone/Getty Images.)*

amounts of fertilizer must be applied to the land. Paddy farmers often plant and harvest the same parcel of land twice per year, a practice known as **double-cropping.** These systems are extremely productive, yielding more food per acre than many forms of industrialized agriculture in the United States.

The modern era has witnessed a restructuring of paddy rice farming in more developed countries, such as Japan, Korea, and Taiwan. In some cases, the terrace structure has been reengineered to produce larger fields that can be worked with machines. In addition, dams, electric pumps, and reservoirs now provide a more reliable water supply, and high-yielding seeds, pesticides, and synthetic fertilizers boost production further. Most paddy rice farmers now produce mainly for urban markets.

PEASANT GRAIN, ROOT, AND LIVESTOCK FARMING

In colder, drier Asian farming regions that are climatically unsuited to paddy rice farming—as well as in the river valleys of the Middle East, in parts of Europe, in Africa, and in the mountain highlands of

FIGURE 8.3 **Peasant grain, root, and livestock agriculture in Africa.** Peasant framers in the Democratic Republic of the Congo prepare their field for planting. The photo illustrates the important role of women's labor in African food production. *(Brent Stirton/Getty Images for WWF-Canon.)*

FIGURE 8.4 **Peasant producers in global trade.** Miguel, a peasant farmer in Mexico, joined a fair trade cooperative in an effort to get better prices for his coffee crop in the global market. *(Erick Bonnier/Sipa Press/Newscom.)*

Latin America and New Guinea—farmers practice a diverse system of agriculture based on bread grains, root crops, and herd livestock (**Figures 8.3**). Many geographers refer to these farmers as **peasants,** recognizing that they often represent a distinctive **folk culture** and socioeconomic class strongly rooted in the land. Peasants are small-scale farmers who own their fields, rely chiefly on family labor, and produce both for their own subsistence and for sale in the market. The dominant grain crops in these regions are wheat, barley, sorghum, millet, oats, and maize. Common cash crops— some of them raised for export— include cotton, flax, hemp, coffee, and tobacco (**Figure 8.4**).

These farmers also raise herds of cattle, pigs, sheep, and, in South America, llamas and alpacas. The livestock pull the plow; provide milk, meat, and wool; serve as beasts of burden; and produce manure for the fields. They also consume a portion of the grain harvest. In some areas, such as the Middle Eastern river valleys, the use of irrigation helps support this peasant system. In general, however, most modern agricultural technologies are beyond the financial reach of most peasants.

PLANTATION AGRICULTURE

plantation agriculture A system of monoculture for producing export crops requiring relatively large amounts of land and capital; originally dependent on slave labor.

plantation A large landholding devoted to specialized production of a tropical cash crop.

In certain tropical and subtropical areas, Europeans and Americans introduced a commercial agricultural system called **plantation agriculture.** A **plantation** is a landholding devoted to capital-intensive, large-scale, specialized production of one tropical or subtropical crop for the global market. Each plantation district in the tropical and subtropical zones tends to specialize in one crop. Plantation agriculture has long relied on large amounts of manual labor, initially in the form of slave labor and later as wage labor. The plantation system originated in the 1400s on Portuguese-owned sugarcane-producing islands off the coast of tropical West Africa—São Tomé and Principe—but the greatest concentrations now exist in the American tropics, Southeast Asia, and tropical South Asia (**Figure 8.5**). Historically, most plantations have been located near the seacoast, close to the shipping lanes that carry their produce to nontropical lands such as Europe, the United States, and Japan.

Workers usually live right on the plantation, where a rigid social and economic segregation of labor and management produces a two-class society of the wealthy and the poor. As a result of the concentration of ownership and production, a handful of multinational corporations, such as

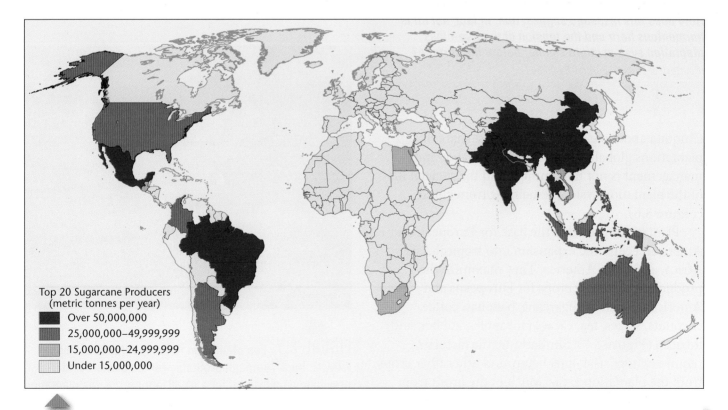

Top 20 Sugarcane Producers
(metric tonnes per year)

- Over 50,000,000
- 25,000,000–49,999,999
- 15,000,000–24,999,999
- Under 15,000,000

FIGURE 8.5 **Sugar producers of the twenty-first century.** Sugarcane plantations have expanded into nearly every tropical and subtropical region in the world. Brazil's production now dwarfs that of the historically prominent producers of the Caribbean. *(Source: UNFAO.)*

FIGURE 8.6 **Plantation agriculture.** This sign was erected by management at the entrance to a banana plantation in Costa Rica. "Welcome to Freehold Plantation, a workplace where labor harmony reigns; in mutual respect and understanding, we united workers produce and export quality goods in peace and harmony." *How does this message suggest that, in fact, not all is harmonious here and the tension of the two-class plantation system might simmer below the surface?* (Courtesy of Terry G. Jordan-Bychkov.)

Chiquita and Dole, control the largest share of plantations globally. Tension between labor and management is not uncommon, and the societal ills of the plantation system remain far from cured (**Figure 8.6**).

Plantations provided the base for European and American economic expansion into tropical Asia, Africa, and Latin America. They maximize the production of luxury crops for Europeans and Americans, such as sugarcane, bananas, coffee, coconuts, spices, tea, cacao, pineapples, rubber, and tobacco (**Figure 8.7**). Similarly, textile factories require cotton, sisal, jute, hemp, and other fiber crops from the plantation areas. Much of the profit from these plantations is exported, along with the crops themselves, to Europe and North America, another source of political friction between countries of the global North and South.

MARKET GARDENING

The growth of urban markets in the last few centuries also gave rise to other commercial forms of agriculture, including **market gardening,** also known as truck farming. Unlike plantations, truck farms are located in developed countries and specialize in intensively cultivated nontropical fruits, vegetables, and vines. They raise no livestock. Many districts concentrate on a single product, such as wine, table grapes, raisins, olives, oranges, apples, lettuce, or potatoes, and the entire farm output is raised for sale rather than for consumption on the farm (**Figure 8.8**). Many truck farmers participate in cooperative marketing arrangements and depend on migratory seasonal farm laborers to harvest their crops. Market garden districts appear in most industrialized countries. In the United States, a broken belt of market gardens extends from California eastward through the

> **market gardening** Farming devoted to specialized fruit, vegetable, or vine crops for sale rather than consumption.

FIGURE 8.7 **Tea plantation in the highlands of Papua New Guinea.** Although profitable for the owners and providing employment for a small labor force, the plantation recently displaced a much larger population of peasant grain, root, and livestock farmers. This is one result of globalization. *Should the government have prevented such a displacement?* (Courtesy of Terry G. Jordan-Bychkov.)

FIGURE 8.8 **Market gardening in the United States' Pacific Northwest.** Marketing gardening (or truck farming) often specializes in one kind of crop. A common market garden crop in the Pacific Northwest is hops, grown mostly for beer brewing. *(Gary J. Weathers/Tetra images RF/Getty Images.)*

Gulf and Atlantic coast states, with scattered districts in other parts of the country. The lands around the Mediterranean Sea are dominated by market gardens. In regions with mild climates, such as California, Florida, and the Mediterranean region, winter vegetables are a common market crop raised for sale in colder regions of the higher latitudes.

LIVESTOCK FATTENING

In **livestock fattening,** farmers raise and fatten cattle and hogs for slaughter. One of the most highly developed fattening areas is the famous Corn Belt of the U.S. Midwest, where farmers raise corn and soybeans to feed cattle and hogs. Typically, slaughterhouses are located close to feedlots, creating a new meat-producing region, which is often dependent on mobile populations of cheap immigrant labor. A similar system prevails over much of western and central Europe, though the feed crops there more

> **livestock fattening** A commercial type of agriculture that produces fattened cattle and hogs for meat.

commonly are oats and potatoes. Other zones of commercial livestock fattening appear in overseas European settlement zones such as southern Brazil and South Africa.

One of the central traditional characteristics of livestock fattening is the combination of crops and animal husbandry. Farmers breed many of the animals they fatten, especially hogs. In the last half of the twentieth century, livestock fatteners began to specialize their activities; some concentrated on breeding animals, others on preparing them for market. In the factorylike **feedlot,** farmers raise imported cattle and hogs on purchased feed (**Figure 8.9**). Increases in the amount of land and crop harvest dedicated to beef production have accompanied the growth in feedlot size and number. In the United States, across Europe, and in European settlement zones around the globe, 51 to 75 percent of all grain raised goes to livestock fattening.

The livestock fattening and slaughtering industry has become increasingly concentrated, on both the

> **feedlot** A factorylike farm devoted to either livestock fattening or dairying; all feed is imported and no crops are grown on the farm.

FIGURE 8.9 **Cattle feedlot for beef production.** This feedlot, in Colorado, is reputedly the world's largest. ***What ecological problems might such an enterprise cause?*** *(Glowimages/Getty Images.)*

national and global scales. In 1980 in the United States, the top four companies accounted for 41 percent of all slaughtered cattle. By 2000 the top four companies were slaughtering 81 percent of all feedlot cattle. One company alone accounted for 35 percent. Corporate conglomerates such as ConAgra and Cargill control much of the beef supply through their domination of the grain market and ownership of feedlots and slaughterhouses. The concentration of the industry has extended north and south across the border, primarily spurred by the 1994 North American Free Trade Agreement (NAFTA). Cargill owns beef operations in more than 60 countries. One industry observer concludes that three multinational companies control the entire global beef industry.

GRAIN FARMING

Grain farming is a type of specialized agriculture in which farmers grow primarily wheat, rice, or corn for commercial markets. The United States is the world's leading wheat and corn exporter. The United States, Canada, Australia, the European Union (EU), and Argentina together account for more than 85 percent of all wheat exports, while the United States alone accounts for about 70 percent of world corn exports (**Figure 8.10**). Wheat belts stretch through Australia, the Great Plains of interior North America, the steppes of Russia and Ukraine, and the pampas of Argentina. Farms in these areas generally are very large, ranging from family-run wheat farms to giant corporate operations (**Figure 8.11**).

Widespread use of machinery, synthetic fertilizers, pesticides, and genetically engineered seed varieties enables grain farmers to operate on this large scale. The planting and harvesting of grain are more completely mechanized than any other form of agriculture. Commercial rice farmers employ such techniques as sowing grain from airplanes. Harvesting is usually done by hired migratory crews operating corporation-owned machines (**Figure 8.12**). Perhaps

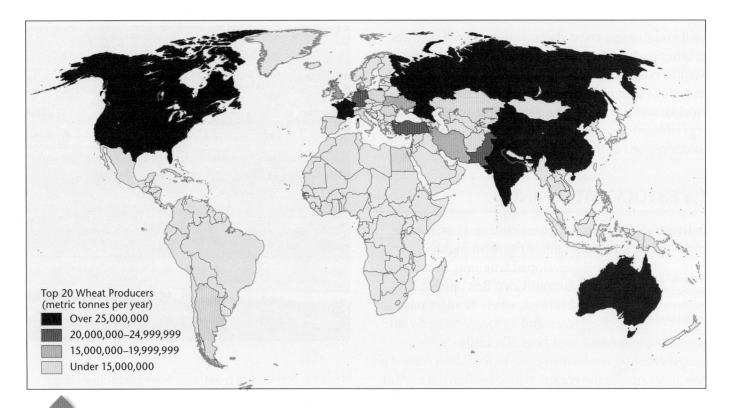

Top 20 Wheat Producers (metric tonnes per year)

- Over 25,000,000
- 20,000,000–24,999,999
- 15,000,000–19,999,999
- Under 15,000,000

FIGURE 8.10 **World's top wheat producers.** Mechanized grain farming has spread rapidly over the past few decades, introducing new leading players into the global spotlight. *(Source: UNFAO.)*

FIGURE 8.11 A "wheat landscape" in the Palouse, a grain farming region on the borders of Washington and Oregon. Grain elevators are a typical part of such agricultural landscapes. The raising of one crop such as wheat over entire regions is called monoculture. *What problems might be linked to monoculture? (Courtesy of Terry G. Jordan-Bychkov.)*

FIGURE 8.12 Mechanized wheat harvest on the Great Plains of the United States. An aerial view of combines harvesting wheat near Colorado. North American grain farmers operate in a capital-intensive manner, investing in machines, chemical fertilizers, and pesticides. *What long-term problems might such methods cause? What benefits are realized in such a system? (Glowimages/Getty Images.)*

> **suitcase farm** In American commercial grain agriculture, a farm on which no one lives; planting and harvesting are done by hired migratory crews.

grain farming's ultimate development is the **suitcase farm,** found in the Wheat Belt of the northern Great Plains of the United States. The people who own and operate these farms do not live on the land. Most of them own several suitcase farms, lined up in a south-to-north row through the Plains states. They keep fleets of farm machinery, which they send north with crews of laborers along the string of suitcase farms to plant, fertilize, and harvest the wheat. The progressively later ripening of the grain as one moves north allows these farmers to maintain and harvest crops on all their farms with the same crew and the same machinery. Except for visits by migratory crews, the suitcase farms are uninhabited.

Such highly mechanized, absentee-owned, large-scale operations, or *agribusinesses,* have mostly replaced the small, husband- and wife-operated American family farm, an important part of the U.S. rural heritage. Geographer Ingolf Vogeler documented the decline of small family farms in the American countryside and argued that U.S. governmental policies, prompted by the forces of globalization, have consistently favored the interests of agribusiness, thereby hastening this decline. Family-owned farms continue to play an important role, but they now operate mostly as large agribusinesses that own or lease many far-flung grain fields.

DAIRYING

In many ways, the specialized production of dairy goods closely resembles livestock fattening. In the large dairy belts of the northern United States from New England to the upper Midwest, western and northern Europe, southeastern Australia, and northern New Zealand, the keeping of dairy cows depends on the large-scale use of pastures. In colder areas, some acreage must be devoted to winter feed crops, especially hay. Dairy products vary from region to region, depending, in part, on how close the farmers are to their markets. Dairy belts near large urban centers usually produce milk, which is more perishable, while those farther away specialize in butter, cheese, or processed milk. An extreme case is New Zealand, which, because of its remote location from world markets, produces much butter.

As with livestock fattening, in recent decades a rapidly increasing number of dairy farmers have adopted the feedlot system and now raise their cattle on feed purchased from other sources. Often situated on the suburban fringes of large cities for quick access to market, the dairy feedlots operate like factories. Like industrial factory owners, feedlot dairy owners rely on hired laborers to help maintain their herds. Dairy feedlots are another indicator of the rise of globalization-induced agribusiness and the decline of the family farm. By easing trade barriers, globalization compels U.S. dairy farmers to compete with producers in other parts of the world. Huge feedlots, a factory-style organization of production, automation, the concentration of ownership, and the increasing size of dairy farms are responses to this intense competition.

NOMADIC HERDING

In the dry or cold lands of the Eastern Hemisphere, particularly in the deserts, prairies, and savannas of Africa, the Arabian Peninsula, and the interior of Eurasia, **nomadic livestock herders** graze cattle,

> **nomadic livestock herder** A member of a group that continually moves with its livestock in search of forage for its animals.

sheep, goats, and camels. North of the tree line in Eurasia, the cold tundra forms a zone of nomadic herders who raise reindeer. The common characteristic of all nomadic herding is mobility. Herders move with their livestock in search of forage for the animals as seasons and range conditions change. Some nomads migrate from lowlands in winter to mountains in summer; others shift from desert areas during the rainy season to adjacent semiarid plains in the dry season or from tundra in summer to nearby forests in winter. Some nomads herd while mounted on horses, such as the Mongols of East Asia, or on camels, such as the

Bedouin of the Arabian Peninsula. Others, such as the Rendile of East Africa, herd cattle, goats, and sheep on foot.

Their need for mobility dictates that the nomads' few material possessions be portable, including the tents used for housing (**Figure 8.13**). Their mobile lifestyle also affects how wealth is measured. Typically, in nomadic cultures, wealth is based on the size of livestock holdings rather than on the accumulation of property and personal possessions. Usually, the nomads obtain nearly all of life's necessities from livestock products or by bartering with the farmers of adjacent river valleys and oases.

For a number of different reasons, nomadic herding has been in decline since the early twentieth century. Some national governments established policies encouraging nomads to practice **sedentary cultivation** of the land. This practice

> **sedentary cultivation** Farming in fixed and permanent fields.

FIGURE 8.13 **Nomadic pastoralists in West Africa.** This household in Chad is moving their livestock to new pastures. Mobility is key to their successful use of the variable and unpredictable environment of this region of Africa. They are taking with them all of their possessions, including their shelter—a hut, which they have packed on top of their cows. *(Michael Nichols/National Geographic/Getty Images.)*

was begun in the nineteenth century by British and French colonial administrators in North Africa because it allowed greater control of the people by the central governments. Today, many nomads are voluntarily abandoning their traditional life to seek jobs in urban areas or in the Middle Eastern oil fields. Recent severe drought in sub-Saharan Africa's Sahel region, which decimated nomadic livestock herds, was a further impetus to abandon nomadic life.

In recent decades, research conducted by geographers and anthropologists in Africa's semiarid environments has revealed the sound logic of nomadic herding practices. These studies demonstrate that nomadic cultures' pasture and livestock management strategies are rational responses to an erratic and unpredictable environment. Rainfall is highly irregular in time and space, and herding practices must adjust. The most important nomadic strategy is mobility, which allows herders to take fullest advantage of the resulting variations in range productivity. These findings have led to a new appreciation of nomadic herding cultures and may cause governments to reconsider sedentarization programs, thus postponing the demise of herding cultures.

LIVESTOCK RANCHING

Superficially, **ranching** might seem similar to nomadic herding. It is, however, a fundamentally different livestock-raising system. Although both nomadic herders and livestock ranchers specialize in animal husbandry to the exclusion of crop raising and both live in arid or semiarid regions, livestock ranchers have fixed places of residence and operate as individuals rather than within a communal or tribal organization. In addition, ranchers raise livestock on a large scale for market, not for their own subsistence.

ranching The commercial raising of herd livestock on a large landholding.

Livestock ranchers are found worldwide in areas with environmental conditions that are too harsh for crop production. They raise only two kinds of animals in large numbers: cattle and sheep. Ranchers in the United States, Canada, tropical and subtropical Latin America, and the warmer parts of Australia specialize in cattle raising. Midlatitude ranchers in cooler and wetter climates specialize in sheep. Sheep production is geographically concentrated to such a degree that only three countries—Australia, China, and New Zealand—account for 56 percent of the world's export wool.

URBAN AGRICULTURE

The United Nations (UN) calculated that in 2008 the human species passed a milestone. For the first time in history, more people live in cities than in the countryside. As this global-scale rural-to-urban migration gained momentum, a distinct form of agriculture rose in significance. We might best call this **urban agriculture.** Millions of city dwellers, especially in Third World countries, now produce enough vegetables, fruit, meat, and milk from tiny urban or suburban plots to provide most of their food, often with a surplus to sell (**Figure 8.14**). In China, urban agriculture now provides 90 percent or more of all the vegetables consumed in the cities. In the African metropolises of Kampala and Dar es Salaam, 70 percent of the poultry and 90 percent of the leafy vegetables consumed in the cities, respectively, come from urban lands. In 2010 the UN Food and Agriculture Organization convened the first symposium on urban agriculture in Africa in recognition of the important role it will play in feeding the continent's twenty-first-century urban residents.

urban agriculture The raising of food, including fruit, vegetables, meat, and milk, inside cities, especially common in the Third World.

Geographer Susanne Freidberg has conducted research demonstrating the importance of urban agriculture to family income and food security in West Africa. Focusing on the city of Bobo-Dioulasso in Burkina Faso, Freidberg showed that though plots were small, urban agriculture offered residents "a culturally meaningful way to fulfill their roles as food producers and family providers." In its heyday in the 1970s and 1980s, urban farming provided substantial incomes from vegetable sales in both the domestic and export markets. Since then, collapsing demand and the deterioration of environmental conditions have threatened the enterprise and undermined cooperation and trust within Bobo-Dioulasso's urban agricultural communities.

FIGURE 8.14 **Urban agriculture.** Farmers work a rice field near Antananarivo, Madagascar. Urban agriculture is a critical food source for many African city dwellers. *(Martin Harvey/Gallo Images/Getty Images.)*

In the developed world, urban agriculture is becoming more prevalent despite, or perhaps in reaction to, the fact that food sourcing is increasingly globalized. The reasons for the boom in urban agriculture in cities such as Vancouver, Canada, and London are complex. However, almost everywhere it is seen as a partial solution to the problems of food insecurity and lack of access to nutritious foods in cities. The work of geographer Nik Heynen has shown that solving food insecurity is integral to improving urban social justice in the United States. It tends to be poorer residents, often members of marginalized minority groups, who lack access to a healthy diet. Similarly in Europe, the World Health Organization

(WHO) has identified the poor availability of and inequitable access to nutritious vegetables and fruits in cities, especially for vulnerable groups, as a major twenty-first-century problem for the region. According to the WHO, the promotion of local food production in European cities will aid in reducing urban poverty and inequalities.

FARMING THE WATERS

Most of us don't think of the ocean when discussing agriculture. In fact, however, every year more and more of our animal protein is produced through **aquaculture:** the cultivation and harvesting of aquatic organisms under controlled conditions. Aquaculture includes **mariculture,** shrimp farming, oyster farming, fish farming, pearl cultivation, and more. In the heart of Brooklyn, New York, urban aquaculture thrives in a laboratory, where fish bound for New York City restaurants are harvested from indoor tanks. Aquaculture has its own distinctive cultural landscapes of containment ponds, rafts, nets, tanks, and buoys (**Figure 8.15**). Like terrestrial agriculture, there is a marked regional character to aquacultural production.

> **aquaculture** The cultivation, under controlled conditions, of aquatic organisms, primarily for food but also for scientific and aquarium uses.
>
> **mariculture** A branch of aquaculture specific to the cultivation of marine organisms, often involving the transformation of coastal environments and the production of distinctive new landscapes.

Aquaculture is an ancient practice, dating back at least 4500 years. Older local practices, sometimes referred to as traditional aquaculture, involve simple techniques such as constructing retention ponds to trap fish or "seeding" flooded rice fields with shrimp. Commercial aquaculture, the industrialized, large-scale protein factories driving today's production growth, is a contemporary phenomenon. The greatest leaps in technology and production have occurred mostly since the 1970s.

At the turn of the twenty-first century, aquaculture was experiencing phenomenal growth worldwide, growing nearly four times faster than all terrestrial animal food-producing sectors combined. Aquaculture is a key reason that per-capita protein availability has more than kept pace with population growth in recent

FIGURE 8.15 Emerging mariculture landscapes. As mariculture expands, coastal landscapes and ecosystems are transformed, as in the case of marine fish farming off Langkawi Island, Malaysia. *(© age fotostock/SuperStock.)*

decades. For example, China's population grew 63 percent from 1970 to 2010, while its per-capita protein supply derived from aquaculture went from 1 kilogram to over 36 kilograms. By 2011 aquaculture accounted for nearly two-thirds of all fish and seafood produced worldwide, compared with only 4 percent in 1970. Aquaculture's share of production is projected to continue growing as demand for seafood increases and wild fish stocks decline.

The extraordinary expansion of food production by aquaculture has come with high costs to the environment and human health. As with industrialized agriculture, most commercial aquaculture relies on large energy and chemical inputs, including antibiotics and artificial feeds made from the wastes of poultry and hog processing. Such production practices tend to concentrate toxins in farmed fish, creating a potential health threat to consumers. The discharge from fish farms, which is equivalent to the sewage from a small city, can pollute nearby natural aquatic ecosystems. Around the tropics, especially tropical Asia, the expansion of commercial shrimp farms is contributing to the loss of highly biodiverse coastal mangrove forests.

Although aquaculture can take place just about anywhere that water is found, strong regional patterns do exist. Mariculture is prevalent along tropical coasts, particularly in the mangrove forest zone. Marine coastal zones in general, especially in protected gulfs and estuaries, have high concentrations of mariculture. On a global scale, China dwarfs all other regions, producing over two-thirds of the world's farmed seafood (**Figure 8.16**). Asia and the Pacific regions combined produce over 90 percent of the world's total.

If you order shrimp, trout, or salmon for dinner at any of the myriad restaurant chains, you will almost certainly be eating farmed protein. As the populations of wild species decline and the price of captured fish increases, farmed seafood will increasingly move into the breach. If trends continue, we may be among the last generations to enjoy affordable fresh-caught wild fish.

NONAGRICULTURAL AREAS

Areas of extreme climate, particularly deserts and subarctic forests, do not support any form of agriculture. Such lands are found predominantly in much of Canada and Siberia. Often these areas are inhabited by **hunting-and-gathering** groups of native peoples, such as the Inuit, who gain a livelihood by hunting game, fishing where possible, and gathering edible and medicinal wild plants. Twelve thousand years ago, all humans lived as hunter-gatherers. Today, fewer than 1 percent of humans do. Given the various inroads of the modern world, even these people rarely depend entirely on hunting and gathering. In most hunting-and-gathering societies, a division of labor by gender occurs. Males perform most of the hunting and fishing, whereas females carry out the equally important task of gathering harvests from wild plants. Hunter-gatherers generally rely on a great variety of animals and plants for their food.

> **hunting and gathering** The killing of wild game and the harvesting of wild plants to provide food in traditional cultures.

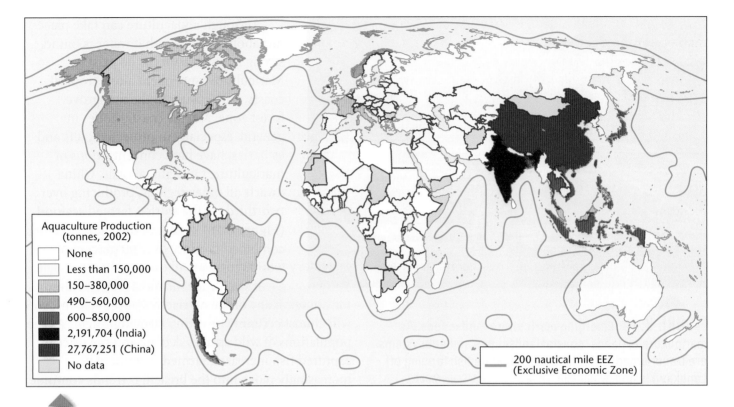

FIGURE 8.16 **Global aquaculture and fisheries.** Aquaculture is expanding rapidly around the globe, led by China, the world's top producer. Virtually all mariculture takes place within countries' 200-mile coastal territory known as the EEZ (exclusive economic zone). The EEZ is also the site of most commercial fishing, which has greatly depleted wild fish stocks and raised the need for increased mariculture production. *(Source: Global Education Project, Fishing and Aquaculture.)*

▦ MOBILITY

8.2 **LEARNING OBJECTIVE**
Analyze the role mobility plays in the spatial and cultural patterns of agricultural production and food consumption.

> **cultural diffusion** The spread of elements of culture from the point of origin over an area.

Some of the variation among the agriculture regions we've discussed results from **cultural diffusion.** Agriculture and its many components are inventions; they arose as innovations in certain source areas and diffused to other parts of the world. Mobility, as we shall see, is key in the expansion and functioning of the modern global food system.

ORIGINS AND DIFFUSION OF PLANT DOMESTICATION

Agriculture probably began with the domestication of plants. A **domesticated plant** is one that is deliberately planted, protected, cared for, and used by humans. Such plants are genetically distinct from their wild ancestors because they result from selective breeding by agriculturists. Accordingly, they tend to be bigger than wild species, bearing larger and more abundant

> **domesticated plant** A plant deliberately planted and tended by humans that is genetically distinct from its wild ancestors as a result of selective breeding.

fruit or grain. For example, the original wild "Indian maize" grew on a cob only one-tenth the size of the cobs of domesticated maize.

Plant domestication and improvement constituted a process, not an event. It began as the gradual culmination of hundreds, or even thousands, of years of close association between humans and the natural vegetation. The first step in domestication was perceiving that a certain plant was useful, which led initially to its protection and eventually to deliberate planting.

Cultural geographer Carl Johannessen suggests that the domestication process can still be observed. He believes that by studying current techniques used by native subsistence farmers in places such as Central America, we can gain insight into the methods of the first farmers of prehistoric antiquity. Johannessen's study of the present-day cultivation of the *pejibaye* palm tree in Costa Rica revealed that native cultivators actively engage in seed selection. All choose the seed of fresh fruit from superior trees, those that bear particularly desirable fruit, as determined by size, flavor, texture, and color. Superior seed stocks are built up gradually over the years, with the result that elderly farmers generally have the best selections. Seeds are shared freely within family and clan groups, allowing rapid diffusion of desirable traits.

The widespread association of female deities with agriculture suggests that it was women who first worked the land. Recall the almost universal division of labor in hunting-gathering-fishing societies. Because women had day-to-day contact with wild plants and their mobility was constrained by childbearing, they probably played the larger role in early plant domestication.

LOCATING CENTERS OF DOMESTICATION

When, where, and how did these processes of plant domestication develop? Most experts now believe that the process of domestication was independently invented at many different times and locations. Geographer Carl Sauer, who conducted pioneering

research on the origins and dispersal of plant and animal domestication, was one of the first to propose this explanation.

Sauer believed that domestication did not develop in response to hunger. He maintained that necessity was not the mother of agricultural invention, because starving people must spend every waking hour searching for food and have no time to devote to the leisurely experimentation required to domesticate plants. Instead, he suggested this invention was accomplished by peoples who had enough food to remain settled in one place and devote considerable time to plant care. The first farmers were probably sedentary folk rather than migratory hunter-gatherers. He reasoned that domestication did not occur in grasslands or large river floodplains, because primitive cultures would have had difficulty coping with the thick sod and periodic floodwaters. Sauer also believed that the hearth areas of domestication must have been in regions of great biodiversity where many different kinds of wild plants grew, thus providing abundant vegetative raw material for experimentation and crossbreeding. Such areas typically occur in hilly districts, where climates change with differing sun exposure and elevation above sea level.

Geographers, archaeologists, and, increasingly, genetic scientists continue to investigate the geographic origins of domestication. Because the conditions conducive to domestication are relatively rare, most agree that agriculture arose independently in at most nine regions (Figure 8.17). All of these have made significant contributions to the modern global food system. For example, the Fertile Crescent in the Middle East is the origin of the great bread grains of wheat, barley, rye, and oats that are so key to our modern diets. This region is also home to the first domesticated grapes, apples, and olives. China and New Guinea provided rice, bananas, and sugarcane, while the African centers gave us peanuts, yams, and coffee. Native Americans in Mesoamerica created another important center of domestication, from which came crops such as maize (corn), tomatoes, and beans. Farmers in the Andes domesticated the potato. While crop diffusions out of these nine regions have occurred over the millennia, the forces of globalization have now

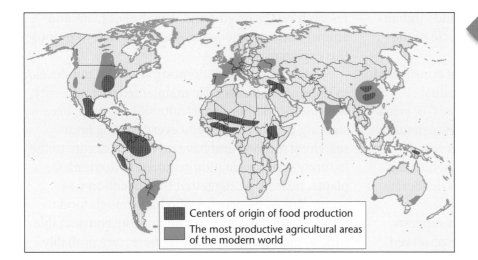

FIGURE 8.17 **Ancient centers of plant domestication.** New archaeological discoveries and new technologies such as genetic science are changing our understanding of the geography and history of domestication. This map represents a synthesis of the latest findings. *(Source: Diamond, 2002.)*

Legend:
- Centers of origin of food production
- The most productive agricultural areas of the modern world

made even the rarest of local domesticates available around the world.

The dates of earliest domestication are continually being updated by new research findings. Until recently, archaeological evidence suggested that the oldest center is the Fertile Crescent, where crops were first domesticated roughly 10,000 years ago. However, domestication dates for other regions are constantly being pushed back by new discoveries. Most dramatically, in the Peruvian Andes archaeologists recently excavated domesticated seeds of squash and other crops that they dated to 9240 years before the present. These seeds were associated with permanent dwellings, irrigation canals, and storage structures, suggesting that farming societies were established in the Americas 10,000 years ago, similar to the dating of the Fertile Crescent.

PETS OR MEAT? TRACING ANIMAL DOMESTICATION

A **domesticated animal** is one that depends on people for food and shelter and that differs from wild species in physical appearance and behavior as a result of controlled breeding and frequent contact with humans. Animal domestication apparently occurred later in prehistory than did the first planting of crops—with the probable exception of the dog, whose companionship with humans appears

> **domesticated animal**
> An animal kept for some utilitarian purpose whose breeding is controlled by humans and whose survival is dependent on humans; domesticated animals differ genetically and behaviorally from wild animals.

to be much more ancient. Typically, people value domesticated animals and take care of them for some utilitarian purpose. Certain domesticated animals, such as the pig and the dog, probably attached themselves voluntarily to human settlements to feast on garbage. At first, perhaps, humans merely tolerated these animals, later adopting them as pets or as sources of meat.

The early farmers in the Fertile Crescent deserve credit for the first great animal domestications, most notably that of herd animals. The wild ancestors of major herd animals—such as cattle, pigs, horses, sheep, and goats—lived primarily in a belt running from Syria and southeastern Turkey eastward across Iraq and Iran to central Asia. Farmers in the Middle East were also the first to combine domesticated plants and animals in an integrated system, the antecedent of the peasant grain, root, and livestock farming described earlier. These people began using cattle to pull the plow, a revolutionary invention that greatly increased the acreage under cultivation. In other regions, such as southern Asia and the Americas, far fewer domestications took place, in part because suitable wild animals were less numerous. The llama, alpaca, guinea pig, Muscovy duck, and turkey were among the few American domesticates.

MODERN MOBILITIES

Over the past 500 years, European exploration and colonialism were instrumental in redistributing a wide variety of crops on a global scale: maize and potatoes from North America to Eurasia and Africa, wheat and grapes from the Fertile Crescent to the Americas, and West African rice to the Carolinas and Brazil.

The diffusion of specific crops continues, extending the process begun many millennia ago. The introduction of the lemon, orange, grape, and date palm by Spanish missionaries in eighteenth-century California, where no agriculture existed in the Native American era, is a recent example of **relocation diffusion.** This was part of a larger process of multidirectional diffusion. Eastern Hemisphere crops were introduced to the Americas, Australia, New Zealand, and South Africa through the mass emigrations from Europe over the past 500 years. Crops from the Americas diffused in the opposite direction. For example, chili peppers and maize, carried by the Portuguese to their colonies in South Asia, became staples of diets all across that region (**Figure 8.18**).

> **relocation diffusion** The spread of an innovation or other element of culture that occurs with the bodily relocation (migration) of the individual or group responsible for the innovation.

In cultural geography, our understanding of agricultural diffusion focuses on more than just the crops; it also includes an analysis of the cultures and **indigenous technical knowledge** systems in which they are embedded. For example, geographer Judith Carney's study of the diffusion of African rice (*Oryza glaberrima*), which was domesticated independently in the inland delta area of West Africa's Niger River, shows the importance of indigenous knowledge. European planters and slave owners carried more than seeds across the Atlantic from Africa to cultivate in the Americas. The Africans taken into slavery, particularly women from the Gambia River region, had the knowledge and skill to cultivate rice. Slave owners actively sought slaves from specific ethnic groups and geographic locations in the West African rice-producing zone, suggesting that they knew about and needed Africans' skills and knowledge. Carney argues that the "association of agricultural skills with certain African ethnicities within a specific

> **indigenous technical knowledge** Highly localized knowledge about environmental conditions and sustainable land-use practices.

FIGURE 8.18 **Chili peppers in Nepal and Korea.** A Tharu tribal woman of lowland Nepal prepares a condiment made of chili peppers from her garden, and in South Korea chili peppers dry under a plastic-roofed shed. This crop comes from the Indians of Mexico. ***How might it have diffused so far and become so important in Asia?*** For the answer, see Andrews, 1993. (*Courtesy of Terry G. Jordan-Bychkov.*)

geographic region" means that research on agricultural diffusion must address the relation of culture to technology and the environment.

Not all innovations involve expansion diffusion and spread wavelike across the land; less orderly patterns are more typical. The **green revolution** in Asia provides an example. The green revolution is a product of modern agricultural science that involves the development of high-yield hybrid varieties of crops, increasingly genetically engineered, coupled with extensive use of chemical fertilizers. The high-yield crops of the green revolution tend to be less resistant to insects and diseases, necessitating the widespread use of pesticides. The green revolution, then, promises larger harvests but ties the farmer to greatly increased expenditures for seed, fertilizer, and pesticides. It enmeshes the farmer in the global corporate economy. In some countries, most notably India, the green revolution diffused rapidly in the latter half of the twentieth century. By contrast, countries such as Myanmar resisted the revolution, favoring traditional methods. An uneven pattern of acceptance still characterizes the paddy rice areas today.

> **green revolution** The recent introduction of high-yield hybrid crops and chemical fertilizers and pesticides into traditional Asian agricultural systems, most notably paddy rice farming, with attendant increases in production and ecological damage.

The green revolution illustrates how cultural and economic factors influence patterns of diffusion. In India, for example, new hybrid rice and wheat seeds first appeared in 1966. These crops required chemical fertilizers and protection by pesticides, but with the new hybrids, India's 1970 grain production output was double its 1950 level. However, poorer farmers—the great majority of India's agriculturists—could not afford the capital expenditures for chemical fertilizer and pesticides, and the gap between rich and poor farmers widened. Many of the poor became displaced from the land and flocked to the overcrowded cities of India, aggravating urban problems. To make matters worse, the use of chemicals and poisons on the land heightened environmental damage.

The widespread adoption of hybrid seeds has created another problem: the loss of plant diversity or genetic variety. Before hybrid seeds diffused around the world, each farm developed its own distinctive seed types through the annual harvest-time practice of saving seeds from the better plants for the next season's sowing. Enormous genetic diversity vanished almost instantly when farmers began purchasing hybrids rather than saving seed from the last harvest. "Gene banks" have belatedly been set up to preserve what remains of domesticated plant variety, not just in the areas affected by the green revolution but also in the American Corn Belt and many other agricultural regions where hybrids are now dominant. In sum, the green revolution has been a mixed blessing.

LABOR MOBILITY

Agriculture, more than any other modern economic endeavor, is constrained by the rhythms of nature. Biological cycles associated with planting and harvesting are reflected in cycles of labor demand. Seeds must be planted at the right moment to take advantage of seasonal conditions. When crops ripen, they must be harvested, often in a matter of days or weeks. In between, labor demand is minimal, so farmers face a dilemma. They need to mobilize a large labor force for harvesting crops, but keeping a year-round work force raises farm production costs. One solution has been to rely on migratory labor.

In the United States, the use of **migrant workers** has been central to the growth and profitability of farming. In his study of labor and landscape in California, geographer Don Mitchell found that the mechanization and intensification of farming created ever greater extremes in seasonal labor demand, lessening the demand for labor during nonharvest periods while still requiring manual labor for the increased harvests. Farmers thus developed an agricultural industry based on migratory labor, employing cultural and racial stereotypes to depress farm wages and tighten employers' control over farmworkers. Mitchell noted that through much of the twentieth century, California growers surmised that "hispanic, black, or Asian workers . . . were 'naturally' better suited to agricultural tasks," and prevailing racist attitudes allowed them to "pay nonwhite workers a lower wage

> **migrant workers** Most broadly, this term refers to people working outside of their home country. Migrant workers are particularly critical to large-scale commercial agriculture.

FIGURE 8.19 **Migrant farmworkers: a global phenomenon.** International migrant farmworkers, such as these African immigrants harvesting lettuce in southern Spain, are critical to production in the global food system. *(Mark Eveleigh/Alamy.)*

than white workers." Ultimately, the federal government provided the legal mechanism, called the *Bracero Program,* by which contract workers were brought from Mexico to California during periods of peak labor demand. Migrant workers lived in substandard housing, were paid less than a living wage, and were deported back to Mexico if they complained.

The case of California's migrant workers is not unusual. Geographer Gail Hollander has documented the use of migratory labor in the development of Florida's sugarcane region. Florida produces 20 percent of the U.S. sugar supply, which until the 1990s was harvested entirely by hand by migrant workers imported seasonally from "former slave plantation economies of the Caribbean." Like their

California counterparts, growers relied on racial stereotypes to argue that only blacks were suitable for cutting cane in Florida. The federal government also established a special federal immigration program like Bracero to import Caribbean migrant workers for the Florida harvest and repatriate them afterward. Migrant farmworkers remain ubiquitous today in other regions, too, such as the European Union. A recent agricultural boom in Mediterranean coastal Spain relies heavily on North African migrant workers, many of them undocumented immigrants, who are subject to persistent racial prejudices (Figure 8.19).

GLOBALIZATION

8.3 LEARNING OBJECTIVE
Describe how the processes of globalization alter the geography of agriculture.

For most of human history, people obtained their provisions locally and maintained their own distinct dietary cultures. The development of global markets over the past 500 years has shifted cultural food preferences and altered the ecology of vast areas of the planet.

LOCAL–GLOBAL FOOD PROVISIONING

As European maritime explorers brought far-flung cultures into contact, a multitude of crops were diffused around the globe. The processes of exploration, colonization, and globalization created new regional cuisines (imagine Italian cuisine without tomatoes from the Americas) and at the same time simplified the global diet to a disproportionate reliance on only three grains: wheat, rice, and maize.

The expansion of European empires in the seventeenth and eighteenth centuries was inseparable from the expansion of tropical plantation agriculture. Plantations in warm climates produced what were then luxury foods for markets in the global North, which had developed seemingly insatiable appetites for sugar, tea, coffee, and other tropical crops. The

FIGURE 8.20 **An oil palm plantation in Malaysia.** Plantation agriculture continues to expand in many Third World countries. While plantation-grown export commodities can be important to national economies, the accompanying destruction of tropical rain forests is a high ecological price to pay. *(Stuart Franklin/Magnum.)*

expansion of plantation agriculture had profound effects on local ecology, all but obliterating, for example, the forests of the Caribbean and the tropical coasts of the Americas (**Figure 8.20**).

In short, we have witnessed the development of a global food system that, for better or worse, has freed consumers in the affluent regions of the world from the constraints of local ecologies. Fresh strawberries, bananas, pears, avocados, pineapples, and many, many other types of temperate, subtropical, and tropical produce are available in our urban supermarkets any day, any time of the year. On the other hand, the emphasis on a relatively small number of staple crops desired by northern consumers can mean the abandonment of local crop varieties and a decline in the associated biological diversity. Imported refined wheat from the global North enters poor tropical countries by the shipload, altering local dietary cultures and undercutting the ability of local farmers to sell their crops at a profitable price.

These are the general patterns, but the globalization of food and agriculture has complex effects on culture and ecology that vary by location and spatial scale. These complexities are best illustrated by a case study from the Peruvian Andes. Geographer Karl Zimmerer (see Rod's Notebook, page 337, for a different look at the globalization of food and agriculture) conducted extensive fieldwork among Quichua peasant farmers in the Paucartambo Andes to determine the effects of economic change on indigenous agricultural practices and the genetic diversity of local crops. He wanted to test the general hypothesis that as globalization and national economic policies integrate indigenous farmers into market production, the diversity of crops declines, ultimately resulting in genetic erosion (i.e., a decline in the genetic diversity of cultivars).

Zimmerer's study produced surprising findings on the complex relationships among culture, economy, and the environment. On the question of whether farmers must abandon crop diversity in order to adopt new, commercially oriented high-yielding varieties, he found there was no simple answer. In fact, it was the more well-off peasants, heavily involved in commercial farming, who had the resources and land to cultivate diverse crops and "enjoy their agronomic, culinary, cultural, and ritual values." Among these values was the use of diverse, noncommercial potato varieties in local bartering. The ability to use noncommercial varieties in this way is valued in the local culture because it is a traditional way to cement interpersonal bonds. Such uses emphasize the cultural importance of crop diversity. Zimmerer discovered that the cultural relevance of crops was a strong motivation for planting by well-off farmers. At least 90 percent of the genetically diverse crops had been conserved, even as the Quichua were further integrated into commercial production for the market.

A number of lessons can be drawn from Zimmerer's study, chief among them the need to carefully examine the effects of globalization on culture and environment, rather than simply assuming that local agricultural practices will give way to the demands of the marketplace. The Quichua farmers who benefited most from their participation in the market were those best able to cultivate traditional varieties, which functioned as an expression of cultural identity and their sense of place. Cultural values, not merely a strict economic or ecological calculus, critically influenced farming decisions.

Rod's Notebook

The Importance of Place in the Global Food System

Rod Neumann. *(Courtesy of Roderick Neumann.)*

December is olive harvest season in southern Spain, when farmers large and small bring their crops to nearby mills for processing. On a recent research trip to Jaén Province in southern Spain, my colleague Gail Hollander and I observed the process at the mill of the Sierra de Cazorla Olive Growers' Cooperative. We watched as families with 5-foot tow trailers lined up next to huge corporate-owned dumptrucks to unload their harvests onto 150-foot conveyer belts leading to the oil presses. We were witnessing a centuries-old harvest ritual refitted to agroindustrial mass production.

Olives and olive oil are staple foods in the Mediterranean diet. Thus, people in the region are sensitized to varying tastes, colors, and consistencies of olive oils to which most North Americans are oblivious. Approach most North Americans about olive oil and they are likely to respond with something like, "Italian food." In one sense, this is correct, since Italy is indeed the largest exporter of processed olive oil. But Spain is the world's largest producer of olives, the greatest share coming from Jaén Province. Most of Spain's harvest, however, is sent to Italy for repackaging and export under Italian brand names that are more familiar to North American shoppers.

This system of mass production of olive oil for Italian export firms worked well for Spanish farmers until recent years. The forces of globalization are now challenging Spain's dominance of the market. North African countries, such as Tunisia, are expanding olive production and competing with Spain for access to the global mass market of undifferentiated oil. In interviews with farmers in Jaén, we discovered that they are looking for ways to differentiate their product from mass-produced oils. Similar to the way fine wines are marketed, they seek to emphasize the unique qualities of oils that are produced in specific places with particular climate and soil characteristics. The idea is to find a niche market of high-end consumers who are appreciative of the distinct qualities and characteristics of local olive varieties.

Jaén farmers have responded to the challenges of globalization by creating new place-based products. Olive oil is being bottled with *denominación de origen* labeling, a third-party certification system that guarantees consumers that the product originates from a specific place. We discovered that the number of marketing cooperatives using the *denominación de origen* designation has exploded across Jaén. Each co-op offers an array of olive-based products. We are still enjoying, for example, our bottle of extra-virgin Royal Aniversario 10 olive oil from the Sierre de Cazorla *denominación de origen,* a product that many Spaniards consider to be of the highest quality anywhere. Currently, *denominación de origen* olive oil is sold mostly to an appreciative domestic Spanish market. Farmers are eyeing North America and East Asia as future export markets as affluent consumers learn of the complexities of olive oil quality and seek new culinary experiences. We were reminded that place matters in the global food system.

Olive growers large and small deliver their harvests for milling into oil at the Sierra de Cazorla Olive Growers' Cooperative in Andalusia, Spain. Such local co-ops are critical in the place-based marketing of global agricultural commodities such as olive oil. *(Courtesy of Roderick Neumann.)*

THE VON THÜNEN MODEL

Geographers and others have long tried to understand the distribution and intensity of agriculture based on transportation costs to market. Long before globalization took hold, the nineteenth-century German scholar-farmer Johann Heinrich von Thünen developed a **core–periphery** model to address the problem. In his model, von Thünen proposed an "isolated state" that had no trade connections with the outside world; possessed only one market, located centrally in the state; had uniform soil and climate; and had level terrain throughout. He further assumed that all farmers located the same distance from the market had equal access to it and that all farmers sought to maximize their profits and produced solely for market. Von Thünen created this model to study the influence of distance from market and the concurrent transport costs on the type and intensity of agriculture.

core–periphery A concept based on the tendency of both formal and functional culture regions to consist of a core or node, in which defining traits are purest or functions are headquartered, and a periphery that is tributary and displays fewer of the defining traits.

Figure 8.21 presents a modified version of von Thünen's isolated-state model, which reflects the effects of improvements in transportation since the 1820s, when von Thünen proposed his theory. The model's fundamental feature is a series of concentric zones, each occupied by a different type of agriculture, located at progressively greater distances from the central market.

For any given crop, the intensity of cultivation declines with increasing distance from the market. Farmers near the market have minimal transportation costs and can invest most of their resources in labor, equipment, and supplies to augment production. Indeed, because their land is more valuable and subject to higher taxes, they have to farm intensively to make a bigger profit. With increasing distance from the

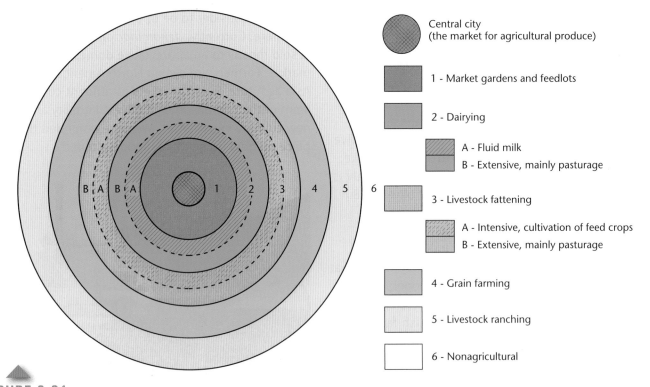

Central city
(the market for agricultural produce)

1 - Market gardens and feedlots

2 - Dairying

A - Fluid milk
B - Extensive, mainly pasturage

3 - Livestock fattening

A - Intensive, cultivation of feed crops
B - Extensive, mainly pasturage

4 - Grain farming

5 - Livestock ranching

6 - Nonagricultural

FIGURE 8.21 **Von Thünen's isolated-state model.** The model is modified to fit the modern world better, showing the hypothetical distribution of types of commercial agriculture. Other causal factors are held constant to illustrate the effect of transportation costs and differing distances from the market. The more intensive forms of agriculture, such as market gardening, are located nearest the market, whereas the least intensive form (livestock ranching) is most remote. Compare this model to the real-world pattern of agricultural types in Uruguay, South America, shown in Figure 8.22. *Why does the model have the configuration of concentric circles?*

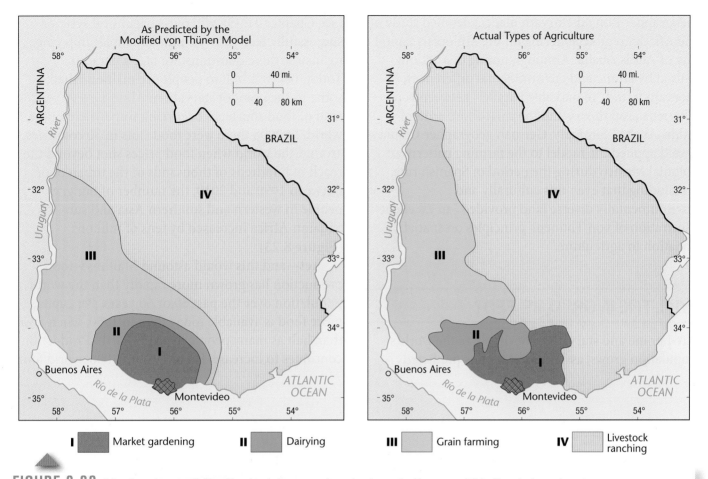

FIGURE 8.22   **Ideal and actual distribution of types of agriculture in Uruguay.** This South American country possesses some attributes of von Thünen's isolated state, in that it is largely a plains area dominated by one city. *In what ways does the spatial pattern of Uruguayan agriculture conform to von Thünen's model? How is it different? What might cause the anomalies?* For the answers, see Griffin, 1973.

market, farmers invest progressively less in production per unit of land because they have to spend progressively more on transporting produce to market. The effect of distance means that highly perishable products such as milk, fresh fruit, and garden vegetables need to be produced near the market, whereas peripheral farmers have to produce nonperishable products or convert perishable items into a more durable form, such as cheese or dried fruit.

The concentric-zone model describes a situation in which highly capital-intensive forms of commercial agriculture, such as market gardening and feedlots, lie nearest to market. The increasingly distant, successive concentric belts are occupied by progressively less intensive types of agriculture, represented by dairying, livestock fattening, grain farming, and ranching.

How well does this modified model describe reality? As we would expect, the real world is far more complicated. For example, the emergence of **cool chains** for agricultural commodities—the refrigeration and transport technologies that bring fresh produce from fields around the globe to our dinner tables—have collapsed distance. Still, on a world scale, we can see that intensive commercial types of agriculture tend to occur most commonly near the huge urban markets of northwestern Europe and the eastern United States. An even closer match can be observed in smaller areas, such as in the South American nation of Uruguay (**Figure 8.22**).

The value of von Thünen's model can also be seen in the underdeveloped countries of the world.

> **cool chain** The refrigeration and transport technologies that allow for the distribution of perishables.

Geographer Ronald Horvath made a detailed study of the African region centering on the Ethiopian capital city of Addis Ababa. Although noting disruptions caused by ethnic and environmental differences, Horvath found "remarkable parallels between von Thünen's crop theory and the agriculture around Addis Ababa." Similarly, German geographer Ursula Ewald applied the model to the farming patterns of colonial Mexico during the period of Spanish rule, concluding that even this culturally and environmentally diverse land provided "an excellent illustration of von Thünen's principles on spatial zonation in agriculture."

WILL THE WORLD BE FED?

Are famine and starvation inevitable as the world's population grows, as Thomas Malthus predicted (see Chapter 3)? Or, will our agricultural systems successfully feed nearly 7 billion people? In trying to answer these questions, we face a paradox. Today, nearly 1 billion people are malnourished, some to the point of starvation. Almost every year, we read of food shortages occurring somewhere in the world. In 2008 there were food riots in 30 countries around the world when food prices shot beyond the reach of hundreds of thousands of the urban poor. Between 1990 and 2010, the number of hungry people in western and southern Asia and sub-Saharan Africa increased by tens of millions (Figure 8.23).

Yet—and this would astound Malthus—food production has grown more rapidly than the world population over the past 50 or 60 years. Per capita, more food is available today than in 1950, when fewer than half as many people lived on Earth. Production continues to increase. From 1996 to 2006, world food

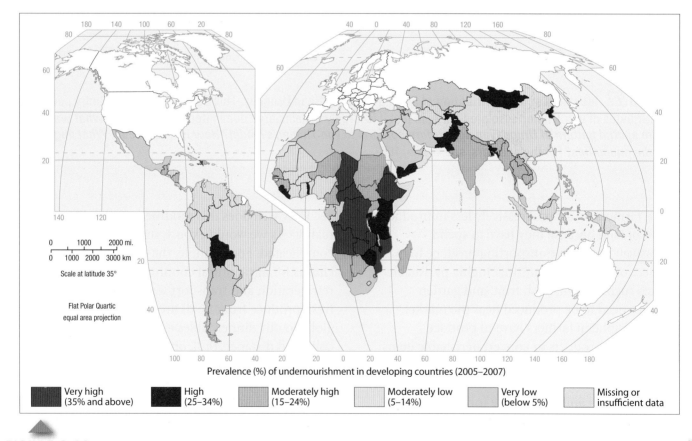

Prevalence (%) of undernourishment in developing countries (2005–2007)

Very high (35% and above)	High (25–34%)	Moderately high (15–24%)	Moderately low (5–14%)	Very low (below 5%)	Missing or insufficient data

FIGURE 8.23 **Mapping hunger worldwide.** While food supply has outpaced population growth on a global scale, many world regions continue to suffer from malnutrition. ***What geographic patterns does this map reveal? How might we explain them?*** *(Source: UNFAO, 2013.)*

production increased at an annual rate of 2.2 percent, and hunger was reduced by 30 percent in more than 30 countries. Global projections foresee continuing annual production increases of 1.5 percent through 2030. Thus, paradoxically, on a global scale there is enough food produced to feed everyone, while famines and malnutrition prevail. If the world food supply is sufficient to feed everyone and yet hunger afflicts one of every six or seven persons, then Malthus was wrong about the limits on population growth but right about the persistence of deprivation.

What explains this paradox of dearth amidst plenty? As geographer Thomas Bassett and economist Alex Winter-Nelson show in their *Atlas of World Hunger,* one must ask both where and why people are hungry. We see in Figure 8.23 where people are going hungry on a global scale. Why they are hungry is complex and varies geographically and historically. For example, Bassett and Winter-Nelson explain that the crisis of HIV/AIDS in southern Africa affects mostly 15- to 49-year-olds, the most productive segment of the population. The epidemic thus negatively affects food production and raises the level of food insecurity, especially in rural areas.

To a great extent, international political economics, not global food shortages, causes hunger and starvation. International trade favors the farmers of wealthier countries through systems of government subsidies that keep the prices of their agricultural exports artificially low. Third World farmers find it difficult to compete. Many Third World countries do not grow enough food to feed their populations, and they cannot afford to purchase enough imported food to make up the difference. As a result, famines can occur even when plenty of food is available. Millions of Irish people starved in the 1840s while the adjacent nation of Great Britain possessed enough surplus food to have prevented this catastrophe. Bangladesh suffered a major famine in 1974, a year of record agricultural surpluses in the world.

Internal government policies are also an important cause of famine. The roots of the largest famine of the twentieth century may be traced to the agricultural policies of the Chinese government's 1958–1961 Great Leap Forward. The Chinese government required peasants to abandon their individual fields and work collectively on large, state-run farms. This policy of collectivization succeeded in boosting food production in some cases but failed in most. Thirty million rural Chinese died of starvation during the Great Leap Forward. Misguided government policies triggered one of the first famines of the twenty-first century as well. In the early 2000s, Zimbabwe's President Robert Mugabe clung to power by demonizing white commercial farmers. In 2002 he threatened Zimbabwe's commercial farmers with imprisonment if they continued to farm. Other government policies discouraged planting and cultivation, thus producing another human-caused famine.

Even when major efforts are made to send food from wealthy countries to famine-stricken areas, the poor transportation infrastructure of Third World countries often prevents effective distribution. Political instability can disrupt food shipments, and the donated food often falls into the hands of corrupt local officials. Such was the case in Somalia during the famine of 2011–2012, where warring factions prevented food aid from reaching starving populations. Over one-quarter of a million people died needlessly during that period of drought. So while the trigger for famine may be environmental, there are frequently deep-seated political and economic problems that conspire to block famine relief.

THE GROWTH OF AGRIBUSINESS

Globalization and its impact on agriculture have been referred to throughout this chapter. Globalization, you will recall, involves the restructuring of the world economy by multinational corporations thriving in an era of free-trade capitalism, rapid communications, improved transport, and computer-based information systems. When applied to agriculture, globalization tends to produce **agribusiness:** a modern farming system that is totally commercial, large-scale, mechanized, and dependent upon chemicals, hybrid seeds, genetic engineering, and the practice of

> **globalization** The binding together of all the lands and peoples of the world into an integrated system driven by capitalistic free markets, in which cultural diffusion is rapid, independent states are weakened, and cultural homogenization is encouraged.

> **agribusiness** Highly mechanized, large-scale farming, usually under corporate ownership.

> **monoculture** The raising of only one crop on a huge tract of land in agribusiness.

monoculture (raising a single specialty crop on vast tracts of land). Furthermore, agribusinesses are often vertically integrated; that is, corporations own the land as well as the processing and marketing facilities. Vertical integration takes a variety of forms depending on the nature of production processes and markets. The case of the "global chicken" provides a useful illustration of vertically integrated agribusiness as well as the interconnections of changing cultural values and global food production.

The origins of the "global chicken" can be found in the shift from beef to poulty as the preferred protein source in American dietary culture, which underwent fundamental changes during the post–World War II period. From 1945 to 1995, per-capita consumption of chicken in the United States rose from 5 to 70 pounds (2.25 to 31.5 kilograms), and by 1990 it had surpassed that of beef. This was a startling development in American culture, where the myth of the cowboy herding cattle on the open range has been so central to an imagined national identity. Since the 1990s, the per-capita consumption of chicken has continued to grow as that of beef continues to decline, especially following the publicity about mad cow disease.

Advances in U.S. agrotechnologies for the breeding, nutrition, housing, and processing of chickens largely account for increases in production efficiency (**Figure 8.24**). This has allowed the U.S. poultry industry, largely centered in the South, to become the world's single largest supplier of broilers. As the taste for chicken spread worldwide, U.S. producers gained an increasing share of an expanding global market. In the last decades of the twentieth century, world trade in broilers grew nearly 500 percent, while the U.S. share of that trade doubled. China has been the hottest import market, because rising affluence there has led to increasing per-capita consumption. At the same time, China, along with Brazil, Turkey, and Japan, is increasing its production and its exports of poultry. Thus, U.S. share of the global market has shrunk in the twenty-first century even as per-capita poultry consumption continues to rise worldwide.

Farmers who produce a single commodity, such as poultry, must produce far more than the local

FIGURE 8.24   The "global chicken." Poultry consumption has skyrocketed worldwide, propelled by new industrialized systems of meat production and changing food preferences. *How might cultural differences influence the structure of international trade in poultry products?* (Robert Nickelsburg/Time Life Pictures/Getty Images.)

market can consume in order to be profitable. They must therefore sell in both national and global markets, access to which requires a dependence on multinational agribusinesses. So pervasive is the reach of agribusiness that many poultry farmers no longer own the chickens they produce; multinational corporations do. Farmers contract with multinationals to receive farm inputs such as chicks and feed and to cover other expenses related to marketing and transport. When the chickens mature, they are trucked to the contracting corporation's processing plant where they are weighed and corporate expenditures are deducted from the farmer's shares. The farmers take their earnings to pay the mortgages on their lands and buildings, and the chickens are processed for the global food system.

THE ONGOING GREEN REVOLUTION

The green revolution generally refers to the transfer of agroindustrial technological packages to Third World countries. It can also been seen as one component of agricultural globalization, along with countless "rural development" projects in Third World countries, usually funded by the World Bank or the International Monetary Fund. These projects often displace small-scale peasant farmers to make way for larger enterprises and even multinational agribusinesses. The green revolution is thus a continuation and geographic expansion of the industrialization of agriculture. Key to the industrialization process are seeds, which are the foundation of cultivation. Whoever controls seeds controls access to the next crop harvest.

Control of seeds has been consolidated in fewer and fewer companies. The five biggest hybrid vegetable seed suppliers control 75 percent of the global market, and the ten largest agrochemical manufacturers command 85 percent of the world supply. Four corporations supply more than two-thirds of the U.S. consumption of hybrid seed maize. Sometimes single companies—Monsanto, for example—both supply the seeds and manufacture the pesticides. What's more, the genetic engineering of seeds is often done in-house. This arrangement allows Monsanto to genetically engineer "Roundup Ready" seed varieties. Roundup is an herbicide manufactured by Monsanto, and its Roundup Ready gene builds in greater tolerance to higher doses. The seeds essentially became vehicles to sell more herbicide.

Genetically modified (GM) crops, the products of biotechnology, are seen by many as another aspect of globalization. Genetic engineering produces new organisms through gene splicing. Pieces of DNA can be recombined with the DNA of other organisms to produce new properties, such as pesticide tolerance or disease resistance. DNA can be transferred between not only species but also plants and animals, which makes this technology truly revolutionary and unlike any other development since the beginning of domestication. Agribusinesses are often able to patent the processes and resulting genetically engineered organisms and, thus, claim legal ownership of new life-forms.

genetically modified (GM) crops Plants whose genetic characteristics have been altered through recombinant DNA technology.

Political, economic, and environmental problems resulting from the concentration of ownership of seeds and the production of pesticide-resistant GM

seeds are beginning to emerge in the United States. The Department of Justice began an antitrust investigation of Monsanto's activities in the seed market. One reason for the investigation was the sharp rise in corn and soybean seed prices, which have more than doubled since 2001. The use of Roundup Ready seeds has resulted in the emergence and spread of so-called superweeds, weeds that have evolved resistance to increasing doses of Roundup herbicide. Thus, the gains in yield initially produced by Roundup Ready seeds are starting to decline, suggesting the environmental limitations of GM seeds that promote the use of pesticides.

Commercial production of GM crops began in the United States in 1996. The technology has now spread around the globe, but the United States still dominates, accounting for two-thirds of the world's acreage (**Figure 8.25**). Two crops, soybeans and corn, account for the rapid growth of GM food production in the United States. By 2010, 93 percent of all soybeans and 86 percent of all corn produced in the United States were genetically modified. Their genetically engineered resistance to disease and drought are an important reason for the spread of GM crops, but that's only part of the story. For example, all of the GM soybean seeds in the United States are engineered to tolerate greater doses of synthetic herbicides produced by agrochemical companies.

If you provision your household from a U.S. supermarket, you have undoubtedly ingested GM foods. Whether or not one finds this troubling is closely related to the strength of certain cultural norms and values that vary from region to region and country to country. In England and western Europe, where national identities are strongly linked to the countryside, agrarian culture, and regional cuisines, there has been a lot of opposition to biotechnology in agriculture. In response to public pressure, major supermarket chains, such as Sainsbury's in England, have refused since 1998 to sell GM foods. In the United States, the response has been far more muted, so much so that the expansion of GM crop planting has proceeded virtually without public debate. These cultural differences are now coming to the fore of globalization debates as the European Union challenges the United States over international trade in GM seeds and foods.

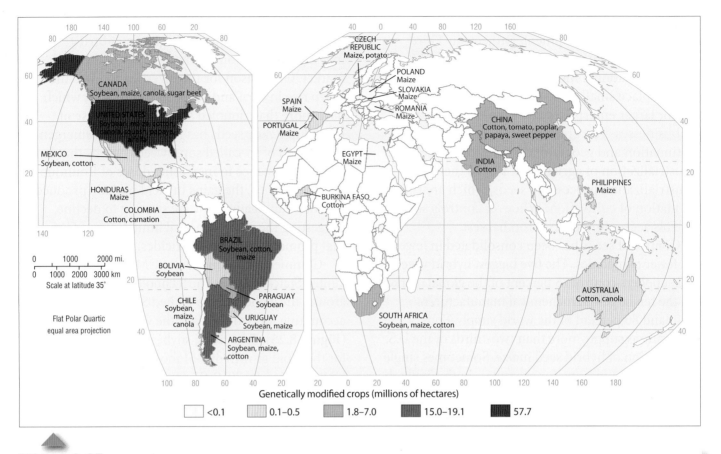

FIGURE 8.25 **Worldwide use of genetically altered crop plants, especially maize and soybeans.** This diffusion has occurred despite the concerns of health scientists, environmentalists, and consumers. ***What problems might arise?*** *(Source: ISAAA, 2007.)*

FOOD FEARS

A globalized food system is vulnerable to events that threaten food safety. Globally significant events, such as the 1990s outbreaks of mad cow disease in Great Britain that shut down the country's export of live cattle and devastated consumer demand for beef, make international headlines. Likewise, national-level recalls of contaminated foods in the United States seem to be in the news almost daily. News of disease outbreaks, some life-threatening, has raised anxiety levels among consumers about eating food transported from distant places. Responses to such crises in the global food system vary culturally and individually and may be based more on the perception of risk than on the probability of illness or death. The way in which national governments and consumers respond to a particular food safety problem can fundamentally reshape geographic patterns of agricultural production on a global scale. Conversely, the idea that a culturally symbolic food may be tainted and life-threatening can shake the strongest of cultural identities. For example, the case of mad cow disease in Great Britain—where beef and dairy production and consumption have long been associated with British cultural identity—challenged culturally defined notions of a proper British meal.

There have been many more cases of foodborne disease outbreaks around the world since mad cow disease made headlines. The U.S. Centers for Disease Control (CDC) has responded by taking on a larger role in monitoring and investigating **foodborne outbreaks**—illnesses that

■ **foodborne outbreaks**
When two or more people acquire the same illness from the same contaminated food or drink.

occur when two or more people consume the same food or drink. CDC records show the most frequent contaminants are various genetic strains and species of common viruses and bacteria, principally the bacteria *Salmonella* spp. and *Escherichia coli* (abbreviated as *E. coli*) and the viruses hepatitis A and norovirus. Infections can result in hospitalization and even death, mostly in the cases of children, the elderly, and the infirm. *Salmonella* infections are the single largest source of fatalities.

Ironically, many of the most common foodborne outbreaks occur in what nutritionists consider the basics of a healthy diet—fresh fruits and vegetables, nuts, seeds, and lean protein sources such as chicken and fish. In particular, pre-packaged leafy greens have repeatedly been implicated in outbreaks of *E. coli* and *Salmonella*. In 2012 mixed greens from a Massachusetts grower infected 33 people in 5 states with *E. coli*. In 2013 Taylor Farms, a giant grower with multiple processing facilities in the United States and Mexico, recalled from 39 states its baby green spinach, which was potentially contaminated with *E. coli* (**Figure 8.26**). Such preemptive recalls are becoming commonplace. Taylor Farms conducted three recalls in 2011 and three again in 2012. The growth in consumer demand for fresh produce combined with the consolidation and centralization of food production leads to increased incidents, outbreaks, and recalls.

The centralization and globalization of food production systems have made foodborne outbreaks both more common and difficult to control. Take the case of a 2013 outbreak of hepatitis A in 10 U.S. states that afflicted 161 people coast to coast, of whom 70 were hospitalized. The product common to all cases was "Townsend Farms Organic Antioxidant Blend." The strain of hepatitis A was identified as common in the Middle East and North Africa. Using this genetic clue and information about the geographic origins of the product's component ingredients, investigators narrowed the likely source to pomegranate seeds from a global exporter in Turkey. The location of the seeds' production has not been identified, but the CDC noted that a similar outbreak in Europe was traced to seeds grown in Egypt. This case illustrates how the growing complexity of the global food system both increases consumer choice while

FIGURE 8.26 Vulnerabilities in the global food system. While leafy greens like this lettuce are considered key to a healthy diet, large-scale production and processing has created dangerous bacterial contamination problems. *(David Gomez, Getty Images.)*

simultaneously exposing consumers to risks that transcend national borders.

A final case illustrates how under globalization the dangers of foodborne outbreaks can alter agricultural practices in far-flung parts of the world. Disproportionately high levels of *Salmonella* contamination have been discovered in common spices found in kitchens worldwide. A 2010 outbreak that sickened 250 in the United States, for example, was traced to red and black pepper. India is the world's largest spice producer and has been for centuries, so it is vulnerable to huge economic losses when such outbreaks occur. Responding to *Salmonella* outbreaks, the Indian government has mandated changes in the way pepper is harvested and processed. The costs of the changes are out of reach for many farmers, prompting the government to issue subsidies and organize grower cooperatives.

The cases of foodborne disease outbreaks demonstrate the global interconnectivity of food-producing and food-consuming regions, the vulnerability of the global food system, and the role of cultural norms and values surrounding questions of food safety.

NATURE–CULTURE

8.4 LEARNING OBJECTIVE
Understand how nature–culture relations are expressed through the production and consumption of food.

Agriculture has been the fundamental encounter between nature and culture for more than 10,000 years, as human labor is mixed with nature's bounty to produce our sustenance. What we eat and how we eat it are a basic source of cultural identity. At the same time, thousands of years of agricultural use of the land have led to massive alterations in our natural environment.

TECHNOLOGY OVER NATURE?

Historically, climate and the physical environment have exerted the greatest influence on shaping agriculture. People have had to adjust their subsistence strategies and techniques to the prevailing regional climate conditions. In addition, soils have played an influential role in both agricultural practices and food provisioning. Swidden cultivation, in part, reflects an adaptation to poor tropical soils, which rapidly lose their fertility when farmed. Peasant agriculture, by contrast, often owes its high productivity to the long-lived fertility of local volcanic soils. Terrain has also influenced agriculture, as farmers tended to cultivate relatively level areas (Figure 8.27). In sum, the constraints of climate, soil, and terrain have historically limited the types of crops that could be grown and the cultivation methods that could be practiced.

In recent centuries, markets, technology, and capital investment have greatly altered the spatial patterns of agriculture that climate and soils had historically shaped. Expanding global-scale markets for agricultural commodities such as sugar, coffee, and edible oils have reduced millions of acres of biologically diverse tropical forests to monocrop plantations. Synthetic fertilizers and petroleum-based insecticides and herbicides, widely available in developed countries after World War II, helped boost agricultural productivity to unimagined levels. Massive dams and large-scale irrigation systems have

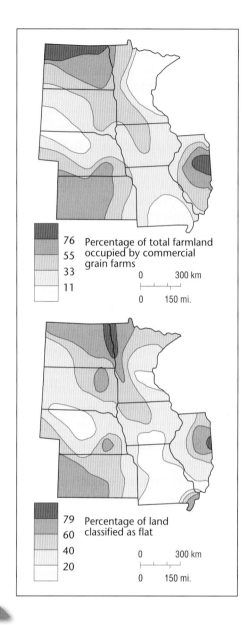

76 / 55 / 33 / 11 Percentage of total farmland occupied by commercial grain farms

0   300 km
0   150 mi.

79 / 60 / 40 / 20 Percentage of land classified as flat

0   300 km
0   150 mi.

FIGURE 8.27 **The influence of terrain on agriculture.** The spatial relationship of commercial grain farming and flat terrain appears in the American Midwest, about 1960. "Flat" terrain is defined as any land with a slope of 3 degrees or less. Commercial grain farming is completely mechanized, and flat land permits more efficient machine operation. The result is this striking correlation between a type of agriculture and a type of terrain. *What other factors might attract mechanized grain farming to level land?* (Source: After Hidore, 1963, pp. 86, 87.)

caused the desert to bloom from central Asia to the Americas, converting, for example, the semiarid Central Valley of California into the world's most productive agricultural region.

FIGURE 8.28 **Reaching the limits of irrigated agriculture.** In the mid-twentieth century, irrigation water from Lake Mead, the reservoir behind the Hoover Dam, enabled the conversion of millions of acres of arid lands to agriculture in the Southwest. The reservoir has not reached capacity for three decades, however (drought marks are clearly shown in this photo), and likely never will again, putting the future of some irrigated lands into question. *(iStock/360/Getty Images.)*

The ecological price of such technological miracles is high. Drainage and land reclamation destroy wetlands and associated biodiversity. The application of synthetic fertilizers results in nutrient-rich runoff from farms that enters freshwater systems as pollution, lowering water quality, destroying aquatic habitat, and reducing biodiversity. Agrochemicals also enter the environment as runoff, as residue on food crops, and in the tissues of livestock. These chemicals ultimately reduce biodiversity, pollute water systems, and cause increases in the rates of cancer and birth defects in humans and animals.

In the context of global climate change, great concern exists about the long-term sustainability of modern agricultural practices that rely on large-scale irrigation. In many parts of the world, groundwater is being pumped to the surface faster than it can be replenished and reservoirs are approaching the end of their life spans. Well-and-pump irrigation has drastically lowered the water table in parts of the American Great Plains, particularly Texas, causing ancient springs to go dry. Climate change data from the distant past and computer models of future climate conditions suggest that the U.S. Southwest is likely entering a drier climate regime.

The Colorado River Basin provides a disturbing example of the challenges climate change poses for agriculture in the Southwest. Its dams and reservoirs irrigate 5.5 million acres of farmland. Lake Mead, the giant reservoir in Arizona and Nevada that helps supply water to California's farms, has not reached full capacity since 1983 and is unlikely ever to be full again (Figure 8.28). By 2010 its capacity stood at only 39 percent capacity and the future looks drier than ever. The year 2013 marked 14 years of drought in the region, the worst in 100 years. Most climate change models predict that this region's climate is shifting toward a hotter and drier norm with more frequent droughts. The U.S. Southwest, the most agriculturally productive in the world, will face increasingly difficult questions regarding the viability of current farming and land-use practices.

Another area where arid land irrigation has had severe ecological consequences lies on the borderland between Kazakhstan and Uzbekistan in central Asia. The once-huge Aral Sea has become so diminished by the diversion of irrigation water from the rivers flowing into it that large areas of dry lakebed now lie exposed (Figure 8.29). Not only was the local fishing industry destroyed, but also noxious, chemical-laden dust storms now blow from the barren lakebed onto nearby settlements, causing assorted health problems. Irrigation water diverted to huge cotton fields, then, destroyed an ecosystem and produced another desert.

SUSTAINABLE AGRICULTURE

As cultural geography studies by Zimmerer and many others have shown, local and indigenous knowledge about ecological conditions can be a foundation for sustainable agriculture. **Sustainability**—the survival of a land-use system for centuries or millennia without destruction of the environmental base—is the central ecological issue confronting agriculture

> **sustainability** The survival of a land-use system for centuries or millennia without destruction of the environmental base, allowing generation after generation to continue to live there.

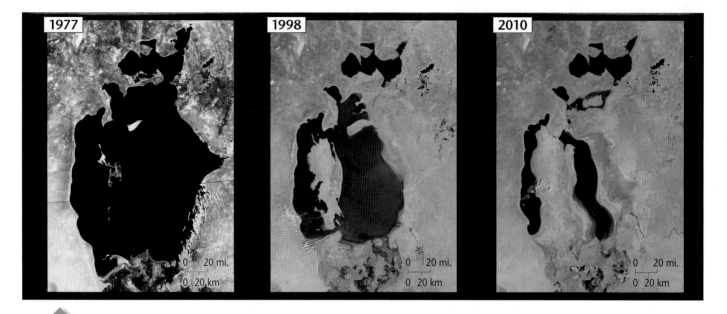

FIGURE 8.29 **The incredible shrinking sea.** The Aral Sea in Central Asia has nearly disappeared over the course of four decades due to the diversion of water for irrigated agriculture. While agricultural productivity increased, the price paid was environmental devastation. *(U.S. Geological Survey Department of the Interior/USGS U.S. Geological Survey/photo by Jane Doe.)*

today. The case of the Quichua peasants offers an optimistic assessment of indigenous knowledge as the basis for long-term sustainability. Their response to contemporary market pressures suggests that development and conservation can be compatible. Their knowledge of complex and variable ecological conditions in the Andes has allowed them to farm highly diverse crop varieties, a practice that in some cases has been strengthened by economic development.

Another example of sustainable indigenous agriculture is the paddy rice farming that occurs near the margins of the Asian wet-rice region, where unreliable rainfall causes harvests to vary greatly from one year to the next. Farmers have developed complex cultivation strategies to avert periodic famine, including growing many varieties of rice. These farmers, including those in parts of Thailand, almost universally rejected the green revolution. The simplistic advice given to them by agricultural experts working for the Thai government was inappropriate for their marginal lands. Based on generations of experimentation, the local farmers knew that their traditional adaptive strategy of diversification was superior. In West Africa, peasant grain, root, and livestock farmers have also developed adaptations to

local environmental influences. They raise a multiplicity of crops on the more humid lands near the coast. Moving inland toward the drier interior, farmers plant fewer kinds of crops but grow more drought-resistant varieties. Having observed many cases like these in which local practices have proved effective and sustainable, most geographers now agree that agricultural experts need to consider indigenous knowledge when devising development plans.

INTENSITY OF LAND USE

Great spatial variation exists in the intensity of rural land use. In **intensive agriculture,** a large amount of human labor or investment capital, or both, is put into each acre or hectare of land, with the goal of obtaining the greatest output. Intensity can be calculated by measuring either energy input or level of productivity. In much of the world, especially the paddy rice areas of Asia, high intensity is achieved through the prodigious use of human labor, which results in a rice output per unit of land that is the

> **intensive agriculture** The expenditure of much labor and capital on a piece of land to increase its productivity. In contrast, *extensive agriculture* involves less labor and capital.

highest in the world. In Western countries, high intensity is achieved through the use of massive amounts of investment capital for machines, fertilizers, and pesticides, resulting in the highest agricultural productivity per capita found anywhere.

Many geographers support the theory that increased land-use intensity is a common response to population growth. As demographic pressure mounts, farmers systematically discard the more geographically extensive adaptive strategies to focus on those that provide greater yield per unit of land. In this manner, the population increase is accommodated. The resultant farming system may be riskier, because it offers fewer options and possesses greater potential for environmental modification, but it does yield more food—at least in the short run. Other geographers reject this theory, arguing instead that increases in population density follow innovations, such as the introduction of new high-calorie crops, which lead to greater land-use intensity.

GLOBAL WARMING AND THE FUTURE OF AGRICULTURE

For centuries, increased food production has been achieved by converting forests and grasslands to pastures and plowed fields. In many parts of China, India, and the Mediterranean lands, forests virtually vanished. In transalpine Europe (an area north of the Alps), the United States, and some other areas, they were greatly reduced (**Figure 8.30**). Such large-scale land-use changes led geographers to conclude that agriculture can actually alter regional environmental conditions. In extreme cases, land-use change over an extensive area can produce **desertification,** a process of desert expansion triggered by human actions such as overgrazing and deforestation. For example, geographer Rhoads Murphey suggested over 60 years ago that farmers caused substantial parts of North Africa to be added to the margins of the Sahara Desert. More recent studies indicate that the Sahara margins expand and contract over time, depending on temporary shifts in rainfall rather than as a result of farmers' mismanagement.

While desertification's relation to farming is still a concern, currently it is the interaction among agricultural land use, food supply, and global warming that is drawing geographers' attention.

> ■ **desertification** A process whereby human actions unintentionally turn productive lands into deserts through agricultural and pastoral misuse, destroying vegetation and soil to the point where they cannot regenerate.

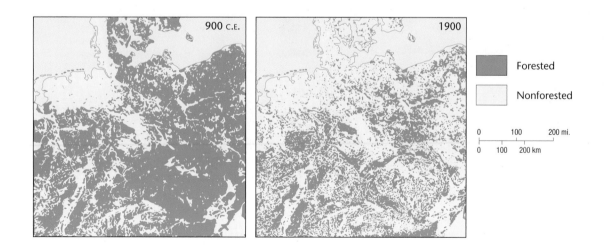

FIGURE 8.30 **The agricultural impact on the forest cover of central Europe from 900 to 1900** c.e. Extensive clearing of the forests, mostly before 1350, was tied largely to expansion of farmland. The distribution of forests in 1900 closely resembles that of hills and mountain ranges. **Why might this pattern have developed?** (Source: Redrawn from Darby, 1956, pp. 202–203.)

World Heritage Site

The Cultural Landscape of Honghe Hani Rice Terraces in Yunnan, China, is one of the most recent additions to the World Heritage Site list. Inscribed in 2013, the site offers visually striking examples of humans' creative response to environmental limits and of long-term sustainable agricultural production.

■ Thousands of terraced rice paddies, built over the course of 1300 years, reach down the slopes of the Ailao Mountains to the valley of the Hong River. The entire site covers 1000 square kilometers and three distinct terraced areas: Bada, which is for the most part gently sloped; Duoyishu, which is steeper; and Laohuzui, which is very steep. Eighty-two villages of 50 to 100 households each are located among the terraced fields.

■ The Hani people, whose ancestors migrated to the region some 2000 years ago, are the predominant ethnic group resident in it. As far back as China's Tang dynasty (seventh to tenth century C.E.) outside commentators remarked on the Hani's skill and ingenuity in modifying the mountainsides for agriculture. The Hani migrants initially encountered a forested ridge cut by numerous surface springs fed by underground seepage. The Hani left forest pockets intact to protect the spring sources, built rock terraces to create level ground, and then channeled the water from the springs to irrigate the fields.

■ Hani religion is rooted in a belief in the sacredness of nature. This spirituality supports ecologically beneficial land-use practices, such as the maintenance of sacred forests and trees in and around Hani villages. The Hani's efforts over the centuries have produced an integrated nature–culture system that unites forest, dwelling, water, and land to sustainably produce both abundant food and environmental benefits.

(Chen Haining/Xinhua Press/Corbis.)

Cultural Landscape of Honghe Hani Rice Terraces

ON THE TERRACES:
Each household has access to one or two segments of the irrigated terrace (as well as to a house site, fuel and building material from the forest, and pasture). The terraced fields are maintained and cultivated through a complex system of communal obligations and individual entitlements. A fundamental task of this system is the regulation of the movement of water onto and off of tens of thousands of individual fields

(Chen Haining/Xinhua Press/Corbis.)

■ Roughly 445 kilometers of water channels, including 4 trunk canals and 392 branch ditches, crisscross the mountainside. Residents employ bars of wood or stone to direct water through these channels. Each bar is carved with a different-sized outlet so that sufficient water flows to the lower ditches on the slope and each ditch receives the correct amount of water. Ditch leaders are appointed from the villages to maintain the ditches, allocate water, and settle disputes. The villagers, in turn, pay the ditch leader in rice for the service of managing the ditches.

■ The water itself is nutrient rich, having passed through the forest litter and picked up key minerals. In late winter, this nutrient-laden water is diverted to the fields to fertilize the soil. In addition, residents have established manure ponds where livestock manure is collected during the winter to create a source of organic fertilizer. At the start of spring planting, nutrient-rich water from the ponds is diverted onto the terraces to fertilize the crop.

■ Along with multiple varieties of rice, residents harvest from the terraces many aquatic herbs and vegetables. Eels, bivalves, snails, fish, shrimp, and other aquatic animals are raised in the water of the flooded fields. Ducks are also part of the system and feed on aquatic animals and vegetation. Soybeans and livestock, including pigs, cattle, and buffalo, are raised on the land between flooded paddies.

TOURISM:
The region has no history of mass tourism, though its scenic beauty and dramatic irrigation structures have long attracted photographers from all over the world.

■ Recent developments, however, including its WHS designation and local transportation infrastructure upgrades, are likely to initiate a wave of touristic activity. The Chinese government has formulated a plan for an ecotourism industry, founded on the region's reputation as an example of sustainable agriculture within an intact traditional culture.

■ Future projects include the designation of scenic tourist routes, an information center focused on local culture, the maintenance of traditional architecture, and the promotion of place-based agricultural products.

(Yang Zongyou/Xinhua Press/Corbis.)

■ Beginning 1300 years ago, the Hani people of China's Ailao Mountains constructed an ingenious system of terraced agriculture.

■ By protecting forests, cycling organic nutrients, and carefully controlling the distribution of water, Hani farming is both highly productive and environmentally sustainable.

■ Long appreciated by photographers for its scenic beauty, the landscape of the Honghe Hani rice terraces will soon be developed for tourism.

http://whc.unesco.org/en/list/1111

FIGURE 8.31 Regions at risk in a warming future. People in Africa's Sahel region, which marks the transition from the Saharan Desert to moister conditions to the south, will likely suffer disproportionately from global climate change. This region is characterized by low income levels, high birthrates, and low agricultural productivity, which means the predicted future of drier conditions in the Sahel will probably result in regional food shortages. ***Can you think of other regions in the world that might face similar challenges if global warming leads to less rainfall?*** (Ariadne Van Zandbergen/Alamy.)

The Intergovernmental Panel on Climate Change (IPCC), the scientific body that the United Nations designated to track global climate change, released its *Fifth Assessment Report* in 2013. Evaluating the research of hundreds of scientists from around the world, the report concludes that the evidence for unprecedented global warming is "unequivocal." Without reductions in greenhouse gasses, such as CO_2, climate conditions will soon be significantly different from those under which humans invented agriculture. That is, the climate conditions that prevailed for millennia and to which our food production systems are adapted are beginning to shift, perhaps radically and irreversibly.

The interactions are complex. On the one hand, modern agriculture contributes to global warming in several ways. For example, industrial agriculture is powered by fossil fuels, which contribute greenhouse gasses to the atmosphere. Also, the clearing of forests for cultivation or pasture reduces the amount of CO_2 removed from the atmosphere by trees. On the other hand, global warming can have negative impacts on food production. For example, one of the IPCC's predictions for global warming's effects is an increase in extreme weather, such as droughts and floods, that can reduce regional crop yields. Also, regional climates are expected to shift geographically, which means that some areas currently cultivated will no longer be suitable for agriculture.

In regard to the last point, global warming will first be felt in those regions where poorer farmers operate without significant capital investment at the margins of suitable agricultural environments. Sub-Saharan Africa is a case in point (**Figure 8.31**). As we have seen, sub-Saharan Africa faces relatively greater challenges than other regions, including a rapidly growing population, high poverty rates, and high rates of malnutrition. In short, the region needs to increase food productivity quickly and significantly. Unfortunately, recent studies indicate declining productivity under most climate change models. In particular, under global warming scenarios, productivity projections for the main African grain crops of sorghum, maize, and millet are negative throughout the region with the exception of a few high-elevation countries.

ENVIRONMENTAL PERCEPTION BY AGRICULTURISTS

People perceive the physical environment through the lenses of their culture. Each person's agricultural heritage can be influential in shaping these perceptions. This is not surprising, because human survival depends on how successfully people can adjust their ways of making a living to environmental conditions.

The Great Plains of the United States provides the setting for how an agricultural experience in one environment influenced farmers' environmental perceptions and subsequent behavior when they immigrated to the Great Plains. The Plains farmers who came from the humid eastern United States

Milk of Human Kindness

http://www.tinyurl.com/p4bsva5

This video features dairy farmer Bob Bansen of Yamhill, Oregon, and his philosophy of sustainable farming. Bansen's method of raising dairy cattle is contrasted with that of corporate-owned industrialized dairy farms. Industrialized dairy farms are run as food factories where cows are kept confined indoors, sustained with concentrated feeds, and given antibiotics to reduce the spread of disease. Bansen's family-owned farm produces dairy products by allowing cows to graze outdoors on natural grass, which reduces disease exposure. As a result, the use of antibiotics can be avoided. The video concludes that modern dairy farms can be environmentally sustainable and still profitable, and provide for the humane treatment of animals.

Thinking Geographically:

1. Chapter 8 featured a discussion of sustainable agriculture. How does this case study of an Oregon dairy farm compare with that discussion? Can you identify any underlying principles or philosophy that would link this video with the sustainability discussion in the textbook?

2. In the video, the interviewer and farmer emphasize the fact that the farmer names and recognizes each individual cow in his herd. What does this practice have to do with his farming philosophy? How does naming his cows help the farmer turn a profit from his dairy farm?

3. Chapter 8 introduced some of the chemical-intensive practices used in both industrialized cattle feedlots and dairy farms. In the video, the interviewer states, "One of the big worries about industrial agriculture is the overuse of antibiotics." How has Bansen addressed this concern on his farm? How did his practices change when he converted to organic production? From this one example, what general insights might we draw regarding chemical inputs in industrial farming?

---

consistently underestimated the problem of drought in their new home. In the 1960s, geographer Thomas Saarinen pointed out that almost every Great Plains farmer, even the oldest and most experienced of the farmers, underestimated the frequency of the dry periods. By contrast, **culturally preadapted** German-speaking immigrants from the steppes of Russia and Ukraine, an area very much like the American Great Plains, accurately perceived the new land and experienced fewer problems due to drought.

Farmers rely on climatic stability. A sudden spell of unusual weather events can change agriculturists' environmental perceptions. Geographer John Cross studied Wisconsin agriculture following a series of floods, droughts, and other anomalies. He found that two-thirds of all Wisconsin dairy farmers now believe the climate is changing for the worse, and fully one-third told him that continued climatic variability threatened their operations. Perhaps they perceive the environmental hazard to be greater than it really is, but they make decisions based on their perceptions.

A study of more recent immigration from Mexico into the Midwest by geographer Eric Carter (and associates) shows that interactions with the receiving culture influence the environmental perception of immigrants in complex ways. Most of the immigrants in the study came from agricultural communities in rural Mexico, drawn to the Midwest to work in the livestock and meat-packing industry. On the one hand, immigrants readily adopted some local environmental values, such as the importance of clean, well-maintained public spaces. On the other hand, immigrants criticized the consumerism and materialism of the host culture, which contradicted

culturally preadapted A complex of adaptive traits and skills possessed in advance of migration by a group, giving it survival ability and competitive advantage in occupying the new environment.

the environmental values they carried with them from rural Mexico. Thus, it is important to recognize environmental perception as evolving through interactions with nature and among culture groups rather than as frozen in time.

DON'T PANIC, IT'S ORGANIC

Alarmed by the ecological and health hazards of chemical-dependent, industrialized agriculture, a small counterculture movement emerged in the United States and Europe in the 1960s and 1970s. Geographer Julie Guthman labels this the organic farming movement. For Guthman, the movement in the United States saw in **organic agriculture** a solution to a range of social, cultural, political, and environmental ills. These included the loss of small family-owned and -operated farms, environmental pollution from industrial agriculture, corporate control of the food system, and the nutritional deficiencies of highly processed foods.

When the organic food movement was in its infancy, there was no way to differentiate organically produced animals and crops from the product of what has come to be called **conventional agriculture.** Movement advocates, many of them based in California, invented new certification systems in the 1970s that focused on the technical aspects of organic agriculture, particularly the absence of artificial fertilizers and petrochemical-based pesticides and herbicides. Certification systems created uniform standards and definitions for organic production but lacked enforcement powers to control fraud. In 1979 California passed the first organic law in the United States, but it took 11 more years before the state added enforcement powers. At the federal level, opposition from corporate agribusiness interests delayed regulatory legislation on organic farming until 1990 and full implementation for another decade.

> **organic agriculture** A form of farming that relies on manuring, mulching, and biological pest control and rejects the use of synthetic fertilizers, insecticides, herbicides, and genetically modified crops.
>
> **conventional agriculture** The widely adopted commercial, industrialized form of farming that uses a range of synthetic fertilizers, insecticides, and herbicides to control pests and maximize productivity; a term that emerged following the creation of alternative forms, such as organic farming.

By legislating regulatory standards for organic agriculture, the state and federal governments provided organic farmers the basis for differentiating their product in the marketplace. This, in turn, allowed producers and retailers to charge consumers a premium. Premium pricing has encouraged many producers to switch to organic agriculture. The organic food market is now the most rapidly growing (and most profitable) agricultural sector. Affluent consumer demand for organics in the United States, while only 4 percent of national food and beverage sales, is growing at the phenomenal pace of 20 to 25 percent annually. Land-use practices reflect this growing demand. In the United States, organic crop acreage increased at an annual average rate of 15 percent between 2002 and 2008, with even larger increases in vegetables and fruits. Despite the dramatic growth in acreage, by 2011 organic farms accounted for only .6 percent of total land area in agricultural use.

Organic production in many parts of the world matches or exceeds the United States' trends. Europe's expansion of land in organic agriculture is happening at an even greater rate, increasing 61 percent between 2005 and 2011. In several European countries, including Austria, Estonia, and Sweden, at least 15 percent of all agricultural land is used in organic production. In terms of area of land in organic production, Australia has roughly six times the amount as the United States and Argentina nearly double. On the other hand, the United States is by far the world's largest consumer of organics, accounting for 44 percent of global sales, more than twice the world's second largest, Germany.

In Guthman's assessment, the United States organic farming movement, because it focused solely on the technical aspects of organic production, fell short of addressing many of the social and political ills it sought to correct. Her study of California organic agriculture demonstrated that organic farming is readily incorporated into large-scale agribusiness enterprises. Indeed, some existing agribusinesses merely purchased existing organic farms as a means of diversifying their operations and tapping the profits from premium pricing. While organic agriculture has produced environmental and health benefits (for farm laborers and consumers), its

FIGURE 8.32 Corn Belt ethanol plant. Plants such as this one in Colorado have been springing up in cornfields across the Midwestern United States as demand for alternative fuel sources increases. Corn is the main source of ethanol as a gasoline additive in the United States. *(Rick Wilking/Reuters/Corbis.)*

effect on the historic decline of family farms and their rural communities has been minimal.

GREEN FUELS FROM AGRICULTURE

Henry Ford fueled his first car with alcohol, and Rudolf Diesel ran his engine on fuel made from peanut oil. They soon abandoned these **biofuels,** however, for nonrenewable fossil fuels derived from "rock oil," and the rest is history. The modern global economy is dependent on fossil fuels to produce everything that we eat, wear, listen to, read, and live in. Now fossil fuels, particularly oil, have become scarce, causing prices and political uncertainties to increase. In addition, fossil-fuel combustion is a source of atmospheric greenhouse gases, which are partly responsible for global warming. Thus, an urgent search is under way to find alternative, renewable fuel supplies, and agriculture has become one of the main sources.

How can agriculture be a source of energy for industry? By tapping the simple process of

> **biofuel** Broadly, this term refers to any form of energy derived from biological matter, increasingly used in reference to replacements for fossil fuels in internal combustion engines, industrial processes, and the heating and cooling of buildings.

fermentation, many plant materials can be converted to a combustible alcohol, ethanol. In the United States, corn is the main source of ethanol, and new ethanol plants dot the landscape of the Midwestern Corn Belt (Figure 8.32). Responding to federal legislation, U.S. ethanol production expanded more than 600 percent during the first decade of the twenty-first century. By 2013, however, ethanol production had saturated the fuels market to the point where supply exceeded demand. Subsequently, the U.S. government recommended that efforts to expand the share of ethanol in the country's energy mix be slowed as Midwestern ethanol plants began boarding up.

This shift in corn production from food to fuel affects nearly every aspect of the field crop sector, as well as livestock production, habitat protection, food retailing, and the global grain trade. For example, total planted corn acreage in 2013 was the highest in 75 years. Corn acreage was gained by replacing crops such as cotton and by the environmentally damaging practices of reducing fallow periods and taking over pastures and land set aside for conservation. Also, ethanol for fuel creates a new source of consumer demand that competes with existing demand for food. As a result, the price of corn on the global market rose, with anticipated negative consequences for the poor and undernourished. The situation is made worse when weather, such as the extended droughts of 2012, reduces yields and drive corn prices even further out of reach of the world's most vulnerable populations.

Brazil, which is the closest international rival of the United States in ethanol production, grows and ferments sugarcane, which yields twice as many gallons of ethanol per acre as does corn. The policies of the Brazilian government have encouraged the creation of a delivery infrastructure (plants, tanks, and pumps) and the manufacture and sale of "flex" cars that can run on either gasoline or ethanol. By 2006 the country had freed itself from a dependence on imported oil, an achievement that many countries only dream of emulating.

Will biofuels from agriculture produce a sustainable, environmentally friendly alternative to fossil fuels? Biofuels appear to offer tremendous promise, yet not without drawbacks. For example, switching farms from food to fuel production raises the specter of food shortages. The environmental benefits of biofuels also may be reduced by some of the associated costs of increasing ethanol and biodiesel production. Biodiesel, fuel made from vegetable oils, takes less energy to produce than crop-based ethanol, but it is typically more expensive than petroleum-derived diesel. These issues are explored further in Subject to Debate (page 357).

CULTURAL LANDSCAPE

8.5 LEARNING OBJECTIVE
Relate how agriculture is expressed within the cultural landscape.

According to the UN Food and Agriculture Organization, 37.3 percent of the world's land area is cultivated or pastured. In this huge area, the visible imprint of humankind might best be called the **agricultural landscape.** The agricultural landscape often varies even over short distances, telling us much about local cultures and subcultures. Moreover, it remains in many respects a window on the past, and archaic features abound. For this reason, the traditional rural landscape can teach us a great deal about the cultural heritage of its occupants.

In Chapter 3, we discussed some aspects of the agricultural landscape, in particular the rural settlement forms. We saw the different ways in which farming people situate their dwellings in various cultures. In Chapter 2, we considered traditional rural architecture, another element in the agricultural landscape. In this chapter, we attend to a third aspect of the rural landscape: the patterns of fields and property ownership created as people occupy land for the purpose of farming.

agricultural landscape The cultural landscape of agricultural areas.

SURVEY, CADASTRAL, AND FIELD PATTERNS

cadastral pattern The shapes formed by property borders; the pattern of land ownership.

A **cadastral pattern** is one that describes property ownership lines, whereas a field pattern reflects the way that a farmer subdivides land for agricultural use. Both can be greatly influenced by **survey patterns,** the lines laid out by surveyors prior to the settlement of an area. Major regional contrasts exist in survey, cadastral, and field patterns, for example, unit-block versus fragmented landholding and regular, geometric survey lines versus irregular or unsurveyed property lines.

Fragmented farms are the rule in the Eastern Hemisphere. Under this system, farmers live in farm villages or smaller **hamlets.** Their landholdings lie splintered into many separate fields situated at varying distances and lying in various directions from the settlement. One farm can consist of 100 or more separate, tiny parcels of land (**Figure 8.33**). The individual plots may be roughly rectangular in shape, as in Asia and southern Europe, or they may lie in narrow strips. The latter pattern is most common in Europe, where farmers traditionally worked with a bulky plow that was difficult to turn. The origins of the fragmented farm system date back to an early

survey pattern A pattern of original land survey in an area.

hamlet A small rural settlement, smaller than a village.

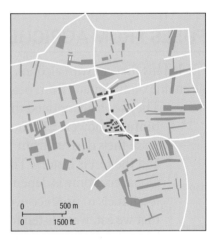

Buildings of the village

Holdings of one farmer

FIGURE 8.33 **Fragmented landholdings surround a French farm village.** The numerous fields and plots belonging to one individual farmer are shaded. Such fragmented farms remain common in many parts of Europe and Asia. ***What are the advantages and disadvantages of this system?*** (Source: After Demangeon, 1946.)

SUBJECT TO *Debate*

CAN BIOFUELS SAVE THE PLANET?

In 2005, in response to diminishing oil reserves and global warming, the U.S. government mandated that biofuels be added to gasoline. Governments around the world have implemented similar initiatives to increase renewable fuel use. Globally, the most promising environmental outcome of increased biofuel use is a decrease in greenhouse gasses. Growing plants consume atmospheric carbon dioxide. Using them for fuel thus recycles an important greenhouse gas, in contrast with fossil fuels, which release stored carbon into the atmosphere when combusted.

Biofuel demand is transforming agriculture around the world, but the energy and environmental benefits are uncertain. Because U.S. corn cultivation is so thoroughly industrialized, ethanol production consumes as much fossil fuel as it replaces. By some estimates, corn ethanol production *uses more* energy than it supplies. Brazil's sugarcane ethanol industry has a far better energy balance of 1 unit of fossil fuel input to 8 units of biofuel output. These energy gains may be offset by other environmental costs. Sugarcane cultivation has created a monocultural desert that is expected to double in acreage by 2014. Many fear this

will contribute to deforestation. Likewise, in the United States, portions of 35 million acres of land set aside for soil and wildlife conservation have been plowed to grow corn for ethanol.

Continuing the Debate

Contemplate the future role biofuels will play in addressing the linked crises of energy supply and global warming and consider these questions:

- What do you think can be done to make biofuels more promising environmentally?

- How can food security for the poor be assured as biofuel use increases?

- Who do you think will benefit the most from the expansion of biofuel use? Small farmers or agribusiness? High-income countries or low-income countries? Consumers or corporations?

The biofuel controversy. Many U.S. politicians and farm lobbyists promote ethanol from corn as an alternative fuel source for cars. Many scientists and environmentalists argue otherwise. *(Pat LaCroix/Photographer's Choice/Getty Images.)*

period of peasant communalism. One of its initial justifications was a desire for peasant equality. Each farmer in the village needed land of varying soil composition and terrain. Travel distance from the village was to be equalized. From the rice paddies of

Japan and India to the fields of western Europe, the fragmented holding remains a prominent feature of the cultural landscape.

Unit-block farms, by contrast, are those in which all of the farmer's property is contained in a single, contiguous piece of land. Such forms are found mainly in the overseas area of European settlement, particularly the Americas, Australia, New Zealand, and South Africa. Most often, they reveal a regular, geometric land survey. The checkerboard of farm fields in the rectangular survey areas of the United States provides a good example of this cadastral pattern (Figure 8.34).

The American township and range system, discussed in Chapter 6, first appeared after the Revolutionary War as an orderly method for parceling out federally owned land for sale to pioneers. It imposed a rigid, square, graphpaper pattern on much of the American countryside; geometry triumphed over physical geography (Figure 8.35). Similarly, roads follow section and township lines, adding to the checkerboard character of the American agricultural landscape. Canada adopted an almost identical survey system, which is particularly evident in the Prairie Provinces.

Equally striking in appearance are long-lot farms, where the landholding consists of a long, narrow unit

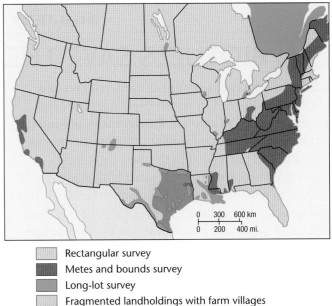

- Rectangular survey
- Metes and bounds survey
- Long-lot survey
- Fragmented landholdings with farm villages
- Irregular rectangular survey (mixture of rectangular and metes and bounds)

FIGURE 8.35 Original land-survey patterns in the United States and southern Canada. The cadastral patterns still retain the imprint of the various original survey types. The map is necessarily generalized, and many local exceptions exist. *What impact on rural life might the different patterns have?*

FIGURE 8.34 American township and range survey creates a checkerboard, illustrated well by irrigated agriculture in the desert of California's Imperial Valley. *(Glowimages/Getty Images.)*

block stretching back from a road, river, or canal (Figure 8.36). Long lots lie grouped in rows, allowing this cadastral survey pattern to dominate entire districts. Long lots occur widely in the hills and marshes of central and western Europe, in parts of Brazil and Argentina, along the rivers of French-settled Québec and southern Louisiana, and in parts of Texas and northern New Mexico. These unit-block farms are elongated because such a layout provides each farmer with fertile valley land, water, and access to transportation facilities, either roads or rivers. In French America, long lots appear in rows along streams, because waterways provided the chief means of transport in colonial times. In the hill lands of central Europe, a road along the valley floor provides the focus, and long lots reach back from the road to the adjacent ridge crests.

Some unit-block farms have irregular shapes rather than the rectangular or long-lot patterns. Most of these result from metes and bounds surveying,

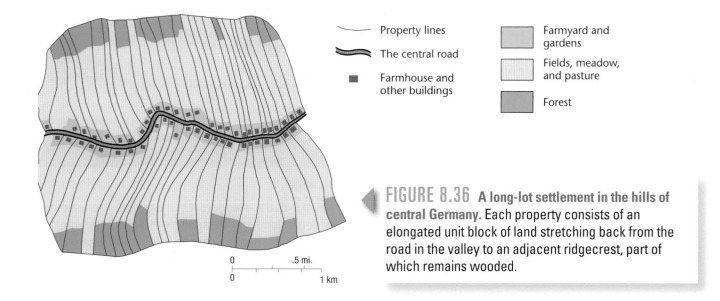

Property lines

The central road

Farmhouse and other buildings

Farmyard and gardens

Fields, meadow, and pasture

Forest

0   .5 mi.

0   1 km

> FIGURE 8.36 **A long-lot settlement in the hills of central Germany.** Each property consists of an elongated unit block of land stretching back from the road in the valley to an adjacent ridgecrest, part of which remains wooded.

which makes much use of natural features such as trees, boulders, and streams. Parts of the eastern United States were surveyed under the metes and bounds system, with the result that farms there are much less regular in outline than those where rectangular surveying was used (**Figure 8.37**).

Original Survey Lines

Property Lines, About 1955
(Those that follow original survey lines are shown by thicker lines.)

Field and Woodlot Borders, About 1955

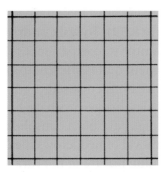

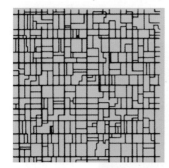

U.S. RECTANGULAR SURVEY, HANCOCK AND HARDIN COUNTIES, OHIO

METES AND BOUNDS SURVEY, UNION AND MADISON COUNTIES, OHIO

FIGURE 8.37 **Two contrasting land-survey patterns, rectangular and metes and bounds.** Both types were used in an area of west-central Ohio. Note the impact these survey patterns had on modern cadastral and field patterns. ***What other features of the cultural landscape might be influenced by these patterns?*** (*Source: After Thrower, 1966, pp. 40, 63, 84.*)

FENCING AND HEDGING

Property and field borders are often marked by fences or hedges, heightening the visibility of these lines in the agricultural landscape. Open-field areas, where the dominance of crop raising and the careful tending of livestock make fences unnecessary, still prevail in much of western Europe, India, Japan, and some other parts of the Eastern Hemisphere, but much of the remainder of the world's agricultural land is enclosed.

Fences and hedges add a distinctive touch to the cultural landscape (Figure 8.38). Because different cultures have their own methods and ways of enclosing land, types of fences and hedges can be linked to particular groups. Fences in different parts of the world are made of substances as diverse as steel wire, logs, poles, split rails, brush, rock, and earth. Those who visit rural New England, western Ireland, or the Yucatán Peninsula will see mile upon mile of stone fence that typifies those landscapes. Barbed-wire fences swept across the American countryside a century ago, but remnants of older styles can still be seen. In Appalachia, the traditional split-rail zigzag

FIGURE 8.38 **Traditional fence in the mountains of Papua New Guinea.** The fence is designed to keep pigs out of sweet potato gardens. The modern age has had an impact, as revealed in the use of tin cans to decorate and stabilize the fence. Each culture has its own fence types, adding another distinctive element to the agricultural landscape. *(Courtesy of Terry G. Jordan-Bychkov.)*

FIGURE 8.39 **Hedgerows in England.** Archetypical hedgerow landscapes, such as this one in Devon, England, historically common throughout western Europe, are now disappearing as land ownership has become increasingly concentrated. ***How might agricultural landscapes with hedgerows support greater biodiversity than those without?*** *(Tony Eveling/Alamy.)*

fence of pioneer times survives here and there. As do most visible features of culture, fence types can serve as indicators of cultural diffusion.

The hedge is a living fence. In Brittany and Normandy in France and in large areas of Great Britain and Ireland, hedgerows are a major aspect of the rural landscape (Figure 8.39). To walk or drive the roads of hedgerow country is to experience a unique feeling of confinement quite different from the openness of barbed wire or unenclosed landscapes. In recent decades, hedgerows have been disappearing as landholdings have been consolidated and grown larger. The removal of hedgerows means not only the loss of a defining feature of the rural landscape but also a decline in habitat for many rare plants, mammals, and birds. In response, the U.K. government passed regulations in 1997 to protect hedgerows in England and Wales; these regulations appear to have slowed their removal.

CONCLUSION

We have seen that the ancient human endeavor called agriculture varies markedly from region to region and is reflected in formal agricultural regions; we have also seen that we can better understand this complicated pattern through the themes of mobility, the global food system, nature–culture interactions, and agricultural landscape. Once again, we have seen the interwoven character of the five themes of cultural geography. In many fundamental ways, the agricultural revolution changed humankind. In equally dramatic fashion, the Industrial Revolution sparked further changes. We will use the five themes to guide an exploration of the industrial world in the next chapter.

DOING GEOGRAPHY

The Global Geography of Food

Until recently, people throughout history obtained the food they needed either by growing it themselves or procuring it directly from nearby farmers. The choice and availability of food were limited and changed seasonally. About 100 years ago, this situation began to change dramatically as the pace of urbanization and industrialization accelerated. Today, very few people in developed countries grow their own food or even know where the food they eat was produced or who produced it. Where does our food come from? Who produces and sells it? Your task for this exercise is to find out.

This exercise can be organized as a group or individual project. As a group project, students can be assigned to research particular categories of food, such as meat and poultry, cereals and grains, fruits and vegetables, and dairy products. As an individual project, you should begin with a typical day's meals and identify all their ingredients (don't forget the seasonings and cooking oils used in preparing them).

Steps to Tracing the Global Geography of Food

Step 1: The project starts at the food markets where you usually shop. For much of the information you will need, you can refer to the labels on the food items. For some items, such as fish, poultry, and meat, you may need to speak to the butcher or store manager. Find out, as specifically as possible, where the food item was produced. Find out the name of the company that marketed the product and, if available, the name of the parent company.

Step 2: After collecting this basic information, you will need to log on to the Internet to do further research (the web sites listed in this chapter should be helpful). Organize a list of companies and the food products they market; then locate the geographic origins of each food product.

Step 3: Now look for patterns.

■ Which and how many companies are involved and what proportion of the food supply does each control?

■ What proportion and which kinds of food are produced in other countries? Do certain kinds of foods tend to be produced closer to the market than others?

■ Do certain kinds of food more so than others tend to be marketed by large corporations?

■ Can you think of explanations for the patterns you identify?

You may wish to take this investigation to a greater depth.

■ Can you determine from your research what conditions exist where the food is produced? For example, what landscape changes occur when regions begin producing for the global food system?

■ How is production structured? Is it organized into large corporate plantations or small peasant farm plots?

■ What are the ethical and social justice dimensions of food production?

■ What are conditions like for workers?

■ Have concerns over the treatment of animals been raised?

■ Have environmental or human health concerns been raised?

A bountiful produce display, common in U.S. supermarkets. Where does the bounty come from, who grows it, and how is it made available for our tables no matter what the season? *(Photodisc/Getty Images.)*

SEEING GEOGRAPHY   Reading Agricultural Landscapes

What differences can you "read" in these landscapes?
Can you determine their locations?

Two types of contemporary agricultural landscapes.
(Left: Jim Wark/AirPhoto/Corbis; Right: Michael Busselle.)

Take a careful look at the two photos shown here and systematically identify the differences in each, beginning with the one on the left. The most striking aspect of this aerial landscape shot is the abrupt division between the cultivated land at the top and the noncultivated land at the bottom. Looking closely, we see that an irrigation channel forms the boundary between the two. A second prominent feature of the landscape is the checkerboard pattern of the fields and the straight roads forming their boundaries. Other details emerge as you look more closely. For example, the settlement pattern consists of isolated, sparsely arranged farmsteads separated by large expanses of cultivated fields. You might also note that the uncultivated land is brown and treeless and that trees in the cultivated portion are found only along the watercourses.

The landscape features in the photo on the right are nearly the opposite of those in the image on the left. Settlement is clustered in a densely populated village centered on a church and town square. The fields are of irregular sizes and shapes and form a band of cultivated land around the concentrations of houses, some of which are built of stone. There is no clear evidence of irrigation. Trees and shrubs are concentrated in the outermost band but also occur throughout the landscape, which overall appears verdant.

Putting all these visual clues together leads us to conclude that the landscape on the left must be somewhere in the western United States. We know this region was surveyed and settled under the township and range system, which explains the isolated farmsteads and checkerboard pattern. We also know that much of the western United States is arid or semiarid, which explains the need for irrigation and the general lack of trees and green vegetation in the bottom half of the photo. In fact, this is a photo of Mack, Colorado, where irrigation meets the desert. The landscape on the right is probably located in Europe. The large church in the center and dense cluster of houses suggest the settlement pattern of a historical market town. The irregular fields and their close proximity to the town are explained by deep historical patterns of land ownership and the reliance on foot travel in preindustrial agriculture. The verdant landscape and absence of irrigation suggest the temperate climate characteristic of western Europe. In fact, you are looking at the vineyard region of Saône-et-Loire, France.

Chapter 8
LEARNING OBJECTIVES REEXAMINED

8.1 Explain the connection between region and agriculture.

See pages 317–329. How is the theme of region relevant to agriculture?

8.2 Analyze the role mobility plays in the spatial and cultural patterns of agricultural production and food consumption.

See pages 330–335. What role does cultural diffusion play in the variation of agriculture regions?

8.3 Describe how the processes of globalization alter the geography of agriculture.

See pages 335–345. How does globalization affect the availability and variety of food in specific places?

8.4 Understand how nature–culture relations are expressed through the production and consumption of food.

See pages 346–356. What role does sustainability play in nature–culture relations?

8.5 Relate how agriculture is expressed within the cultural landscape.

See pages 356–360. What is meant by the agricultural landscape and what does it tell us about cultures?

KEY TERMS

Agricultural Geography on the Internet

You can learn more about agricultural geography on the Internet at the following web sites:

Agriculture, Food, and Human Values (AFHVS)

http://afhvs.org

Founded in 1987, AFHVS promotes interdisciplinary research and scholarship in the broad areas of agriculture and rural studies. The organization sponsors an annual meeting and publishes a journal by the same name.

Food First

http://foodfirst.org

Founded in 1975 by author-activist Francis Moore Lappé, Food First is a nonprofit, "people's" think tank and clearinghouse for information and political action. The organization highlights root causes and value-based solutions to hunger and poverty around the world, with a commitment to establishing food as a fundamental human right.

International Food Policy Research Institute

http://www.ifpri.org

Learn about strategies for more efficient planning for world food supplies and enhanced food production from a group concerned with hunger and malnutrition. Part of this site deals with domesticated plant biodiversity.

Resources for the Future

http://www.rff.org

This well-respected center for independent social science research was the first U.S. think tank on the environment and natural resources. This site contains a great deal of information related to the environmental aspects of global food and agriculture.

United Nations Food and Agriculture Organization (FAO)

http://www.fao.org

Discover an agency that focuses on expanding world food production and spreading new techniques for improving agriculture as it strives to predict, avert, or minimize famines.

United States Department of Agriculture

http://www.usda.gov

Look up a wealth of statistics about American farming from the principal federal regulatory and planning agency dealing with agriculture.

Urban Agriculture Notes

http://www.cityfarmer.org

This is the site of Canada's Office of Urban Agriculture. It concerns itself with all manner of subjects, from rooftop gardens to composting toilets to air pollution and community development. It encompasses mental and physical health, entertainment, building codes, rats, fruit trees, herbs, recipes, and much more.

World Bank Group

http://www.worldbank.org

Read about an agency that provides development funds to countries, particularly in economically distressed regions. It is a driving force behind globalization and agribusiness.

Worldwatch Institute

http://www.worldwatch.org

Learn about a privately financed organization focused on long-range trends, particularly food supply, population growth, and ecological deterioration.

Your Food Environment Atlas

http://maps.ers.usda.gov/FoodAtlas/

A powerful, interactive mapping site that currently includes 90 indicators of the food environment ranging from store/restaurant proximity to income and poverty measures.

Sources

Andrews, Jean. 1993. "Diffusion of Mesoamerican Food Complex to Southeastern Europe." *Geographical Review* 83: 194–204.

Bassett, Thomas, and Alex Winter-Nelson. 2010. T*he Atlas of World Hunger.* Chicago: University of Chicago Press.

Bourne, Joel, Jr. 2007. "Biofuels: Boon or Boondoggle." *National Geographic Magazine* 212(4): 38–59.

Boyd, William, and Michael Watts. 1997. "Agroindustrial Just-In-Time: The Chicken Industry and Postwar American Capitalism." In M. Watts and D. Goodman (eds.), *Globalizing Food: Agrarian Questions and Global Restructuring,* pp. 192–225. London: Routledge.

Carney, Judith. 2001. *Black Rice: The African Origins of Rice Cultivation in the Americas.* Cambridge, Mass.: Harvard University Press.

Carter, Eric, Bianca Silva, and Graciela Guzman. 2013. "Migration, Acculturation, and Environmental Values: The Case of Mexican Immigrants in Central Iowa." *Annals of the Association of American Geographers* 130(1): 129–147.

Centers for Disease Control. *Tracking and Reporting Foodborne Disease Outbreaks.* http://www.cdc.gov/Features/dsFoodborneOutbreaks/.

Chakravarti, A. K. 1973. "Green Revolution in India." Annals of the Association of American Geographers 63: 319–330.

Cowan, C. Wesley, and Patty J. Watson (eds.). 1992. *The Origins of Agriculture: An International Perspective.* Washington, D.C.: Smithsonian Institution Press.

Cross, John A. 1994. "Agroclimatic Hazards and Fishing in Wisconsin." *Geographical Review* 84: 277–289.

Darby, H. Clifford. 1956. "The Clearing of the Woodland in Europe." In William L. Thomas, Jr. (ed.), *Man's Role in Changing the Face of the Earth,* pp. 183–216. Chicago: University of Chicago Press.

Demangeon, Albert. 1946. *La France.* Paris: Armand Colin.

Diamond, Jared. 2002. "Evolution, Consequences and Future of Plant and Animal Domestication." *Nature* 418: 700–707.

Dillehay, T., J. Rossen, T. Andres, and D. Williams. 2007. "Preceramic Adoption of Peanut, Squash, and Cotton in Northern Peru." *Science* 316(5833): 1890–1893.

Global Education Project. 2014. *Earth: A Graphic Look at the State of the World.* "Fisheries and Agriculture." http://www.theglobaleducation-project.org/earth/fisheries-and-aquaculture.php.

Ewald, Ursula. 1977. "The von Thünen Principle and Agricultural Zonation in Colonial Mexico." *Journal of Historical Geography* 3: 123–133.

Freidberg, Susanne. 2001. "Gardening on the Edge: The Conditions of Unsustainability on an African Urban Periphery." *Annals of the Association of American Geographers* 91(2): 349–369.

Griffin, Ernst. 1973. "Testing the von Thünen Theory in Uruguay." *Geographical Review* 63: 500–516.

Griliches, Zvi. 1960. "Hybrid Corn and the Economics of Innovation." *Science* 132 (26 July): 275–280.

Guthman, Julie. 2004. *Agrarian Dreams: The Paradox of Organic Farming in California.* Berkeley: University of California Press.

Harris, Gardiner. 2013. "Farmers Change over Spices' Link to Food Ills." *New York Times,* 27 August, A1.

Hewes, Lewlie. 1973. *The Suitcase Farming Frontier: A Study in the Historical Geography of the Central Great Plains.* Lincoln: University of Nebraska Press.

Heynen, Nik. 2010. "Cooking Up Non-Violent Civil Disobedient Direct Action for the Hungry: Food Not Bombs and the Resurgence of Radical Democracy." *Urban Studies* 47(6): 1225–1240.

Hidore, John J. 1963. "Relationship Between Cash Grain Farming and Landforms." *Economic Geography* 39: 84–89.

Hollander, Gail. 2008. *Raising Cane in the 'Glades: The Global Sugar Trade and the Transformation of Florida.* Chicago: University of Chicago Press.

Horvath, Ronald J. 1969. "Von Thünen's Isolated State and the Area Around Addis Ababa, Ethiopia." A*nnals of the Association of American Geographers* 59: 308–323.

ISAAA. 2007. International Service for the Acquisition of Agri-biotech Applications web site. http://www.isaaa.org.

Johannessen, Carl L. 1966. "The Domestication Processes in Trees Reproduced by Seed: The Pejibaye Palm in Costa Rica." *Geographical Review* 56: 363–376.

Mathews, Kenneth, Jason Bernstein, and Jean Buzby. 2003. "International Trade of Meat and Poultry Products and Food Safety Issues." In Jean Buzby (ed.), *International Trade and Food Safety: Economic Theory and Case Studies.* Washington, D.C.: U.S. Department of Agriculture Economic Research Service.

Millstone, Erik, and Tim Lang. 2003. *The Penguin Atlas of Food.* New York: Penguin.

Mitchell, Don. 1996. *The Lie of the Land: Migrant Workers and the California Landscape.* Minneapolis: University of Minnesota Press.

Murphey, Rhoads. 1951. "The Decline of North Africa Since the Roman Occupation: Climatic or Human?" *Annals of the Association of American Geographers* 41: 116–131.

Norberg-Hodge, Helena, Todd Merrifield, and Steven Gorelick. 2002. *Bringing the Food Economy Home: Local Alternatives to Global Agribusiness.* London: Zed.

Popper, Deborah E., and Frank Popper. 1987. "The Great Plains: From Dust to Dust." *Planning* 53(12): 12–18.

Saarinen, Thomas F. 1966. *Perception of Drought Hazard on the Great Plains.* University of Chicago, Department of Geography, Research Paper No. 106. Chicago: University of Chicago Press.

Sauer, Carl O. 1952. *Agricultural Origins and Dispersals.* New York: American Geographical Society.

Sauer, Jonathan D. 1993. *Historical Geography of Crop Plants.* Boca Raton, Fla.: CRC Press.

Schlenker, Wolfram, and David B. Lobell. 2010. "Robust Negative Impacts of Climate Change on African Agriculture." *Environmental Research Letters* 5(1): 1–8.

Strom, Stephanie. 2013. "Food Supplier Grapples with Frequent Recalls." *New York Times,* 29 August, B1.

Thrower, Norman J. W. 1966. *Original Survey and Land Subdivision.* Chicago: Rand McNally.

United Nations Food and Agricultural Organization. 2013. *Food Outlook.* Rome: UNFAO.

U.S. Department of Agriculture. 2004. Economic Research Service web site. http://www.ers.usda.gov.

Vogeler, Ingolf. 1981. *The Myth of the Family Farm: Agribusiness Dominance of United States Agriculture.* Boulder, Colo.: Westview Press.

von Thünen, Johann Heinrich. 1966. *Von Thünen's Isolated State: An English Edition of Der Isolierte Staat.* Carla M. Wartenberg (trans.). Elmsford, N.Y.: Pergamon.

Wallander, Steven, Roger Claassen, and Cynthia Nickerson. 2011. *The Ethanol Decade: An Expansion of U.S. Corn Production, 2000–09.* Economic Information Bulletin Number 79. Washington, D.C.: U.S. Department of Agriculture Economic Research Service.

Westcott, Paul. 2007. *Ethanol Expansion in the United States: How Will the Agricultural Sector Adjust?* Washington, D.C.: U.S. Department of Agriculture Economic Research Service.

Wilken, Gene C. 1987. *Good Farmers: Traditional Agricultural and Resource Management in Mexico and Central America.* Berkeley: University of California Press.

Zimmerer, Karl. 1996. *Changing Fortunes: Biodiversity and Peasant Livelihood in the Peruvian Andes.* Berkeley: University of California Press.

Ten Recommended Books on Agricultural Geography

(For additional suggested readings, see the *Contemporary Human Geography* LaunchPad: macmillanhighered.com/launchpad/DomoshCHG1e.)

Bassett, Thomas, and Alex Winter-Nelson. 2010. *The Atlas of World Hunger.* Chicago: University of Chicago Press. The authors map out the geography and causes of world hunger from a critical social science perspective.

Clay, Jason. 2004. *World Agriculture and the Environment: A Commodity-by-Commodity Guide to Impacts and Practices.* Washington, D.C., and Covelo, Calif.: Island Press. This book describes the environmental effects resulting from the production of 22 major crops; it is global in scope and encyclopedic in detail.

Denham, Tim, Jose Iriarte, and Luc Vrydaghs (eds.). 2007. *Rethinking Agriculture: Archeological and Ethnoarcheological Perspectives.* Walnut Creek, Calif.: Left Coast Press. An edited volume bringing together geographers, anthropologists, and archaeologists to present the latest research findings on the origins of early agriculture in non-Eurasian regions.

Freidberg, Susanne. 2010. *Fresh: A Perishable History.* Cambridge, Mass.: Harvard University Press. A wonderfully written cultural and historical geography study of the pursuit of freshness in the global food system. The range of topics is sweeping and includes every link in the commodity chain from farm to table.

Gertel, Jorg, and Richard Le Heron (eds.). 2011. *Economic Spaces of Pastoral Production and Commodity Systems.* Williston, Vt.: Ashgate Press. A deeply ethnographic collection of studies on pastoralism in the twenty-first century. It provides an enlightening set of comparative cases from around the world, including countries typically underrepresented in such collections.

Millstone, Erik, and Tim Lang. 2008. *The Atlas of Food: Who Eats What, Where, and Why.* 2nd ed. Berkeley: University of California Press. A book packed with information on global agriculture on a wide range of topics, including genetically modified crops, fast food, organic farming, and more, all presented in brilliantly detailed maps.

Morgan, Kevin, Terry Marsden, and Jonathan Murdoch. 2006. *Worlds of Food: Place, Power and Provenance in the Food Chain.* Oxford: Oxford University Press. A valuable contribution to twenty-first-century studies of the global food system. Its authors link recent cultural shifts in food preference to the emergence of an alternative geography of agriculture that emphasizes the importance of place and region.

Sauer, Carl O. 1969. *Seeds, Spades, Hearths, and Herds.* Cambridge, Mass.: MIT Press. The renowned American cultural geographer presents his theories on the origins of plant and animal domestication—the beginnings of agriculture.

Watts, M., and D. Goodman (eds.). 1997. *Globalizing Food: Agrarian Questions and Global Restructuring.* London: Routledge. This edited volume, containing primarily the work of geographers, analyzes globalization and the biotechnological revolution in agriculture.

Woods, Michael. *Rural.* 2011. New York: Routledge. Part of the "Key Ideas in Geography" series from the publisher. This is wide-ranging synopsis of the latest geographic research on the rural, including but not limited to issues of food and agriculture. It is particularly strong in addressing the themes of globalization and landscape in contemporary rural spaces.

Journals in Agricultural Geography

Agriculture and Human Values. An interdisciplinary journal dedicated to the study of ethical questions surrounding agricultural practices and food. Published by Kluwer. Volume 1 appeared in 1984. Visit the homepage of the journal at http://link.springer.com/journal/10460.

Journal of Agrarian Change. A journal focusing on agrarian political economy, featuring both historical and contemporary studies of the dynamics of production, property, and power. Published by John Wiley & Sons. Volume 1 appeared in 2000. Visit the homepage of the journal at http://onlinelibrary.wiley.com/journal/10.1111/(ISSN)1471-0366.

Journal of Peasant Studies. One of the leading journals of rural development, especially focused on marginalized agricultural communities and social groups in Third World regions. Published by Taylor & Francis. Volume 1 appeared in 1973. Visit the homepage of the journal at http://www.tandfonline.com/loi/fjps20#.UokN85hZ6ew.

Journal of Rural Studies. An international interdisciplinary journal ranked as the best of its kind. Published by Pergamon, an imprint of Elsevier Science, Amsterdam, the Netherlands. Volume 1 appeared in 1985. Visit the homepage of the journal at http://www.journals.elsevier.com/journal-of-rural-studies/.

Have you ever wondered where your smartphone started its life, and where it will go when it dies?

Clockwise from top left: China's Mongolia mining region; Foxconn workers assembling smartphones; using the latest iPhone; children sifting through electronic waste in Manila, Philippines. *(Ren Junchuan/ Xinhua/Sipa Press/Newscom; YM YIK/EPA/Newscom; Sergio Azenha/Alamy; REUTERS/Cheryl Ravelo/ Newscom.)*

Go to "Seeing Geography" on page 408 to learn more about this image.

DEVELOPMENT GEOGRAPHY

Transforming Landscapes of Well-Being

Being **"developed" as opposed to "underdeveloped"** is typically understood as primarily an economic issue. Wealthy places are considered to be developed, whereas poor places are labeled as underdeveloped. Indeed, the key terms relating to development geography have traditionally been economically based. The most common measure of development, for instance, is **Gross Domestic Product,** or **GDP**. Gross Domestic Product totals the dollar value of all goods and services produced in a country over a specific period, usually a year. The map shown in **Figure 9.1** shows the vast disparities in GDP from country to country.

Gross Domestic Product (GDP) The total dollar value of all goods and services produced in a country over a specific period, usually a year.

What do those differences really mean? In fact, not much can be inferred from GDP numbers except that some countries have economies that are far larger— and thus produce more total goods and services—than the economies of other countries. For instance, Brazil's economy, as measured by the total value of goods and services produced annually, is the seventh largest in the world. It is approximately 10 times the size of Chile's economy. Is the average Brazilian 10 times richer than

LEARNING OBJECTIVES

9.1 Discuss the history of development and how it has shaped, and reshaped, regions.

9.2 Specify the ways in which mobilities of various sorts—of individuals as well as things and ideas—are important for development.

9.3 Identify how formerly poor places are today significant agents of development.

9.4 Recognize the role that nature has played over time as concepts and practices of development have changed.

9.5 Observe how development projects have shaped the landscapes around us.

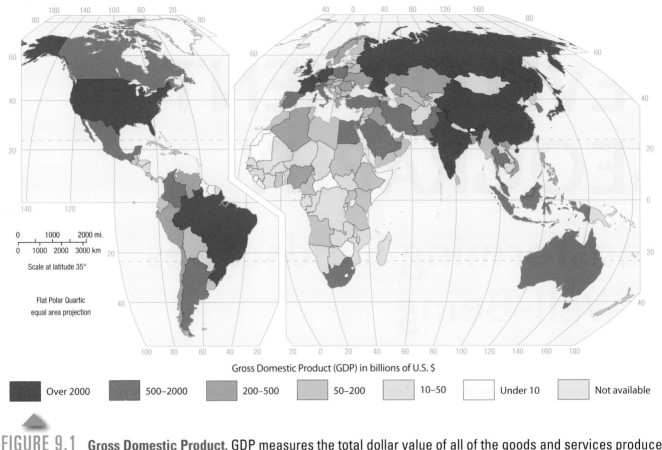

Gross Domestic Product (GDP) in billions of U.S. $

Over 2000	500–2000	200–500	50–200	10–50	Under 10	Not available

FIGURE 9.1 **Gross Domestic Product.** GDP measures the total dollar value of all of the goods and services produced in a country annually. This map illustrates the regional patterns of economic output.

the average Chilean? Ten times more productive? Not so fast. Comparing Brazil's GDP to Chile's is hardly fair. Chile's population is only about 8 percent of Brazil's. So, it is understandable that far more populous Brazil can produce more total goods and services in a year than Chile.

How then can we compare apples to apples—in this case, the share of GDP that corresponds to each Brazilian or Chilean resident? To do that, we need to divide GDP by the national population, which gives the **GDP per capita.** GDP per capita provides a crude way to compare how wealthy or poor the "average" Brazilian or Chilean may be. In our example, Chile's GDP per capita—which in 2012 was measured at $15,410—was actually larger than Brazil's at $12,079.

> ■ **GDP per capita** The mathematical result of dividing a country's GDP by its national population. (The Latin *per capita* means "per head.")

Figure 9.2 shows GDP per capita figures globally. Now you can begin to assess how different nations stack up against one another in terms of their per capita economic production, bearing in mind that GDP per capita figures cannot demonstrate how incomes are actually distributed across national populations. Brazil, for instance, is notoriously unequal in its income distribution. (For a measure of income inequality, see Figure 9.3.)

It is certainly true that without sufficient economic resources, societies are unable to provide for their citizens. But—to put it crudely—is money all there is to development? Or, to turn this question around, does having monetary resources necessarily mean that societies will choose to spend those resources to improve the well-being of their populations?

Development geographers and others who study such questions largely agree that development is indeed

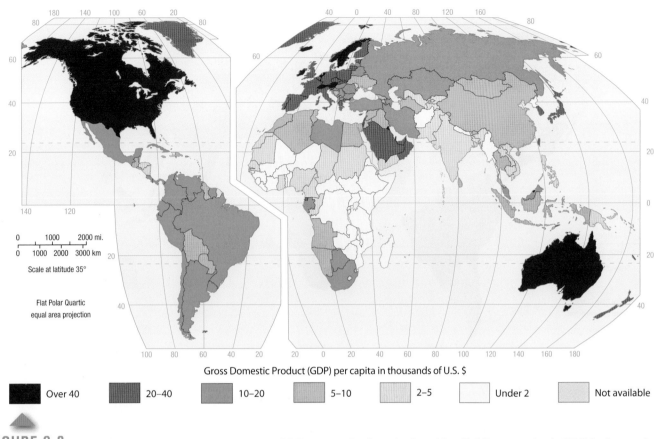

Gross Domestic Product (GDP) per capita in thousands of U.S. $

| Over 40 | 20–40 | 10–20 | 5–10 | 2–5 | Under 2 | Not available |

FIGURE 9.2 **Gross Domestic Product per capita.** GDP per capita is calculated by dividing a nation's GDP by its total population. On this map you can begin to discern areas of poverty and wealth. This statistic cannot capture how equally (or unequally) income is actually distributed among people on the ground, however.

about more than just money. Development is, at its heart, a much more encompassing idea: it is about enhancing individual and societal quality of life. To paraphrase the great Indian leader Mahatma Gandhi, development is about each human being's ability to reach his or her full positive potential. In a material sense, we can understand development as the fulfillment of basic needs for food, clean water, adequate shelter, and clothing. Some understandings of basic needs add education, health care, and sanitation to this list. **Figure 9.4** shows a world map of the **Human Development Index,** or **HDI,** which is a traditional measure of how well people's basic needs are being met, while **Figure 9.5** takes a closer

Human Development Index (HDI) A numerical value devised by the United Nations Development Programme that is used to measure how well basic needs are being met. It is a composite index, derived from three areas: life expectancy, education, and income.

look at well-being in the United States. (See also the discussion of HDI in Chapter 1.)

Although access to a certain baseline amount of economic resources is necessary in order to fulfill these basic needs, merely having the economic means to provide them is no guarantee that basic needs will, in fact, be met for all members of society. The principal allocators of such resources—in most nations, that is governments, corporations, and nongovernmental entities such as churches and foundations—may choose to direct economic resources to other areas. For instance, purchasing military equipment and funding armies may be a higher priority than making sure basic human needs are met. Or, some form of prejudice—for instance, discrimination by gender, religion, age, race, sexuality, or ethnicity—may exist, such that only some members of society have their basic needs met while others

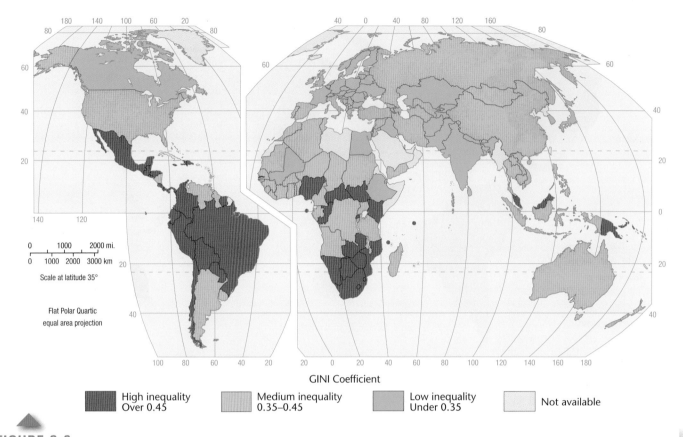

180   140   100   60   20
80   80

60   60

40   40

20   20

140   120

0   1000   2000 mi.
0   1000 2000 3000 km   20

Scale at latitude 35°

Flat Polar Quartic
equal area projection   40

100   80   60   40   20

40   0   40   80   120   160
                              80

60   60

40   40

20   20

0

20   20

40   40

20   0   20   40   60   80   100   120   140   160   180

GINI Coefficient

High inequality   Medium inequality   Low inequality   Not available
Over 0.45   0.35–0.45   Under 0.35

FIGURE 9.3 **Income inequality.** The Gini coefficient (named after the Italian sociologist who developed it) provides a numerical measure of income inequality. Zero represents perfect income equality, thus those nations that score lowest are the most equal with respect to income distribution, while the high scorers are more unequal. Not all nations regularly gather the data necessary to calculate this statistic.

systematically do not. For their part, individuals simply may not have sufficient economic resources to meet their own basic needs or those of their dependent family members. And, as with larger entities such as governments or corporations, individuals may choose to direct their economic resources in ways that do not fulfill their basic needs; for instance, forgoing medical care in order to gamble.

In 2000 the United Nations alongside other international development agencies established eight Millennium Development Goals. As you can see in **Figure 9.6**, some of these—eradication of extreme poverty and hunger (Goal 1), for example—are focused on basic needs. Other goals, such as promoting gender equality and empowering women,

are broader in scope. In retrospect, these goals were probably too ambitious and broad for all participating nations to achieve: in practice, the results have been uneven. Yet, the Millennium Development Goals did offer a bold and sweeping statement on the universal elements of a dignified, comfortable, and productive human existence.

Thinking beyond the material, social, and infrastructural dimensions of development for a moment, what other conditions might be important for a place to be considered truly developed? Development economist Amartya Sen has proposed that freedom lies at the heart of development. Sen intends for freedom to be interpreted as encompassing political and civic freedoms, as well as freedom from the vulnerabilities (such as illiteracy, or

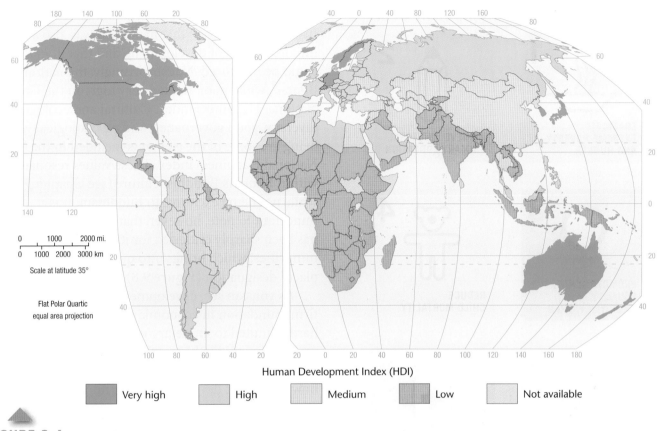

FIGURE 9.4 Human Development Index. The HDI was developed by the United Nations to categorize nations into tiers according to the well-being of their citizens. The HDI is a composite index derived from three separate statistics for life expectancy, education, and income.

early death) associated with poverty. Geographers might add to this notion of development the freedom to move about without restrictions.

In the 1970s, the king of Bhutan coined the term *Gross National Happiness index* to counter what he viewed to be an overly economy-centered approach to

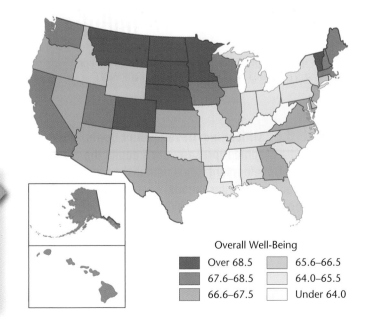

FIGURE 9.5 Well-Being Index. This map displays a state-by-state assessment of well-being. Here, well-being is understood to be composed of life evaluation, emotional health, physical health, healthy behavior, work environment, and basic access. Notice the regional patterns of those places that score highest, and lowest, on this index.

2015 Millennium Development Goals

1 ERADICATE EXTREME POVERTY AND HUNGER

2 ACHIEVE UNIVERSAL PRIMARY EDUCATION

3 PROMOTE GENDER EQUALITY AND EMPOWER WOMEN

4 REDUCE CHILD MORTALITY

5 IMPROVE MATERNAL HEALTH

6 COMBAT HIV/AIDS, MALARIA AND OTHER DISEASES

7 ENSURE ENVIRONMENTAL SUSTAINABILITY

8 A GLOBAL PARTNERSHIP FOR DEVELOPMENT

FIGURE 9.6 **Millennium Development Goals.** These eight simple graphic representations depict the eight Millennium Development Goals established by the UN in 2000. The ambitious international development targets were to be achieved by 2015. While some countries have reached all eight goals, other countries will not achieve any of them. *(UNDP Brazil.)*

development taken by Western nations (**Figure 9.7**). Instead, he proposed that happiness was also an important part of development, and that it, in fact, could be measured. Not surprisingly, the nation of Bhutan, while poor, ranked very high on the Gross National Happiness index. Cultural and ecological diversity, good governance, and psychological well-being are some of the elements included in the Gross National Happiness index. These values resonate within Bhutan's Buddhist culture (see Chapter 7), while not being exclusive to Buddhism. It is important to note, however, that Bhutan's government has been sharply criticized for its oppression of ethnic minorities, which brings into question its "happiest place" designation (**Figure 9.8**).

As you can see, development geography has a firm foundation in economic well-being, but it is hardly limited to monetary measures. In this chapter, we will explore the history and contemporary profile of development geography, in order to gain a sense of how important development is to human well-being across the globe.

REGION

9.1 **LEARNING OBJECTIVE**
Discuss the history of development and how it has shaped, and reshaped, regions.

You may have had the experience of traveling to countries or regions where the everyday conditions of life seem so much more difficult than the conditions you are accustomed to at home. If you haven't traveled to such places, you certainly have seen enough television shows, movies, and Internet content to understand that the everyday lives of people in different parts of the world, including their access to housing, food, and health care, may differ greatly from yours. You may also notice differences within your own neighborhood, town, or city, in the quality of life of the people who live around you.

We've already touched on many development-related topics throughout the book, which tells you

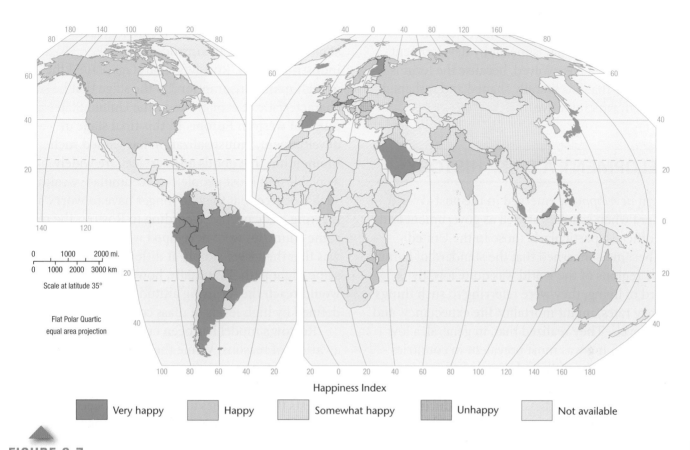

Happiness Index

| Very happy | Happy | Somewhat happy | Unhappy | Not available |

FIGURE 9.7 **Net happiness by country.** On New Year's Eve, 2012, individuals in 54 countries worldwide were asked, "As far as you are concerned, do you personally feel happy, unhappy, or neither happy nor unhappy about your life?" Countries were then rank-ordered by their net happiness score, calculated as the percentage of respondents who were happy minus the percentage of respondents who were unhappy. The resulting map depicts very happy countries that scored between 75% and 49% net happiness; happy countries that scored between 48% and 30% net happiness; somewhat happy countries that scored between 29% and 5% net happiness; and unhappy countries where the unhappy outnumbered the happy. The global net happiness average is 40%.

that development is a concept that reaches into many areas of human geography. In the chapter introduction that you just read, we also noted that development includes but exceeds mere wealth as

FIGURE 9.8 **Bhutan: The happiest place on Earth?** From this photo you might guess this to be the case. However, Bhutan has been criticized for some unhappy conditions, including the oppression of ethnic minority populations, pervasive poverty, and unemployment. Some claim Bhutan's Gross National Happiness Index is merely a clever marketing ploy intended to lure tourists to visit the tiny mountain nation. *(Keren Su/China Span/Alamy.)*

defined in economic terms. It is important to understand exactly what is meant by development and how this idea has evolved, because the term is a powerful one that has literally shaped and reshaped the world in which we live.

DEVELOPMENT: A BRIEF HISTORY

The term *development* emerged in the post–World War II era when scholars as well as government and policy officials, particularly those in the United States, became concerned that the standard of living in many regions of the world was quite low. By standard of living, they were referring to such things as literacy rates, infant mortality, life expectancy, and poverty levels. They realized that, in general, a low standard of living was most prevalent in countries whose economies were based primarily on subsistence agriculture (see Chapter 8). These officials and scholars believed that introducing new technologies and skills would enable these countries to develop more productive forms of agriculture, and that *industrialization*—the transformation of raw materials into commodities—would follow. These new economic initiatives would then lead to a higher standard of living. Transforming the economic structure of a region in order to raise the standard of living is what at that time was known as development. The term *development* referred to a process of both economic and social transformation, and provided a way of categorizing different regions of the world.

One of the earliest and most influential models of the process of economic development was suggested by the economist W. W. Rostow in 1962. Rostow posited that economic development was a process that all regions of the world would go through and experience in similar ways. His model assumed that regions would progress in a linear fashion through defined economic stages as they developed, much like an airplane pulling out of the gate, taxiing along the runway, taking off, and ultimately flying high. The first stage, which he called a "traditional" economy, is one that is based on agriculture and has limited access to or knowledge of advanced technologies: in other words, the economy is undeveloped. This first stage would be followed by a series of other stages characterized by increasing technological sophistication and the introduction of industrialization. Rostow's final stage, called the "age of high mass consumption," is the stage that he saw as characteristic of economies in the United States and much of western Europe. In this final stage of development, industrialization has gained such momentum that goods and services are widely available and most people can accumulate wealth to such a degree that they no longer have to worry about meeting basic needs. According to Rostow, the reason some countries were developed while others were not was that they were simply at different stages along the same flight path. Eventually, however, all nations would reach the "cruising altitude" of development as defined by high levels of mass consumption.

Rostow's model has been criticized by scholars for a variety of reasons. While the simplicity of the model is appealing, its assumptions about the world often don't match up to reality. For example, the model suggests that countries and regions proceed along the path to development in isolation from one another. But as we will see in this chapter, different countries and their economies are interlinked in complex ways; for example, if one country's economy is based primarily on producing goods and services, it needs other regions and economies to supply its food and raw materials. In addition, the model assumes that all economies will develop without obstacles from other countries, and we know that this is rarely true, since, for example, some countries may deliberately try to block economic development in other countries that they think will become their competitors. Indeed, some critics of Rostow's approach claimed that developed countries—those enjoying high levels of mass consumption—depended on the exploitation and impoverishment of other regions in order to achieve their success. The cheap natural and human resources of less developed regions were exploited in order to permit more developed regions to enrich themselves. In other words, underdevelopment of some places is an active process that goes hand-in-hand with the development of others. Finally, the end goal of Rostow's development model—consuming lots of material goods—may not be considered as desirable as other goals, such as political freedom or decent working conditions.

In this chapter, we use the terms *developing* and *developed* to refer to different regions of the world, but we don't assume, like the Rostow model does, that there is a normal and linear progression from one stage of development to another. We use the term *developing* to refer to regions whose economies include not only a good deal of subsistence activities but also some manufacturing and service activities. In these regions, people are often unable to accumulate wealth, since they produce just enough, or at times not enough, food and other resources for their own immediate needs. These countries therefore have a lower GDP and are considered relatively poor. We use the term *developed* to refer to those regions of the world whose economies are based more on manufacturing, services, and information. These regions have a higher GDP and are considered wealthier because people living there are better able to accumulate resources. The landscape of development is a complex one, as individual cities, regions, and countries can contain a mixture of developed as well as underdeveloped places.

There have been other ways of using notions of development to categorize the world into regions. For instance, during the mid-twentieth century, the world was seen as being divided into the First, Second, and Third Worlds (**Figure 9.9**). The **First World** represented the world's wealthy, democratic, capitalist nations, while the **Third World** was made up of poor countries. Because the so-called Cold War was the principal frame for power from the end of World War II in 1945 to the fall of the Soviet Union in 1991, these labels also had political meanings. The **Second World** consisted of those countries that were ruled by communist or socialist governments and had central or command economies, rather than capitalist ones. The Third World nations were not aligned with either the First or Second World. After 1991 the terms *First, Second,* and *Third World* had primarily economic connotations, along

> **First World** A term used to describe the group of wealthy, democratic, capitalist nations across the world.
>
> **Third World** A term used to describe the world's poor nations as a group. During the Cold War era, these were also the nations that were not communist or socialist; in other words, they were not aligned with either the First or Second World.
>
> **Second World** A term used during the Cold War era to describe the group of countries that were communist or socialist in their governmental philosophy, and had centrally planned rather than free-market economic systems.

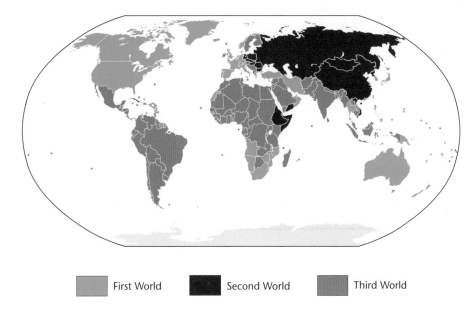

First World   Second World   Third World

FIGURE 9.9 **The three worlds of the Cold War era.** During the middle decades of the twentieth century, when the world order was shaped by the geopolitical tensions among the liberal democracies and communist nations, the world was divided into those nations aligned with these two positions and those non-aligned nations known as the third world. As the Cold War progressed, the terms First and Third World became synonymous with wealthy and poor nations, respectively.

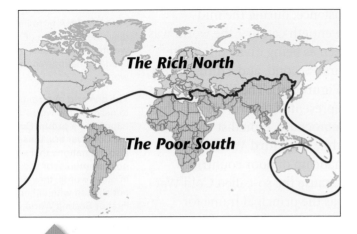

The Rich North

The Poor South

FIGURE 9.10 **The Brandt Line.** This line divides the wealthy nations of the Global North from the poor nations of the Global South. It provided another way of conceptualizing world regions from a development perspective.

Fourth World A term used to describe the world's poorest nations.

Brandt Line Division of the countries of the world into two regions: wealthy and poor. The Brandt Line is named after former West German Chancellor Willy Brandt, and was a concept commonly applied in the 1970s through the 1980s.

with the **Fourth World**, those regions considered to be so poor that they existed beyond the reach of contemporary society.

In the 1970s, West German Chancellor Willy Brandt proposed what came to be known as the Brandt Line as a way of understanding the world. The **Brandt Line** divided the countries of the world into two regions: wealthy and poor (Figure 9.10). Notice that wealthy countries tend to be in the Northern Hemisphere, while poorer ones are in the Southern Hemisphere.

By the 1980s, however, many places in the so-called Third World had made significant economic advances, and the Cold War had lost some of its traction as an organizing framework with which to understand power in the world. Thus, the division of the globe into different regions, by the political and economic criteria represented by the four worlds, or the "have" versus "have-not" distinction demarcated by the Brandt Line, fell out of fashion. Today, you can find divisions of the globe that focus on its more contemporary development aspects—for instance,

highlighting the world's fast-emerging economies, or those with access to resources such as technology, which distinguish the current haves from have-nots (Figure 9.11).

CATEGORIZING TYPES OF ECONOMIC ACTIVITY BY SECTOR

In order to fully understand the differences and similarities between regions, it is helpful to distinguish the types of economic activities that characterize them. In general, scholars divide types of economic activities into four broad categories, or sectors. **Primary sector** activities involve extracting natural resources from the Earth. Fishing, farming, hunting, lumbering, oil extraction, and mining are examples of primary industries. **Secondary sector** activities process the raw materials extracted by primary industries, transforming them into more usable forms. Ore is converted into steel; logs are milled into lumber; and fish are processed and canned. *Manufacturing* is a more common way of referring to secondary industries. Manufacturing activities are found throughout the world, but, as we will discuss later in this chapter, they tend to cluster in particular areas because of favorable circumstances such as cheap power sources or available labor.

In parts of the world where people import the bulk of their manufactured products, economic activities are dominated by services. Services comprise the **tertiary sector**, which refers to all the different types of work necessary to move goods and resources around and deliver them to people. So wide is the range of services that some geographers find it useful to distinguish three different types: transportation/communication services, producer services, and consumer services. Finally, the fastest-emerging set of economic activity is occurring in the

primary sector The set of economic activities that involves extracting natural resources from the Earth.

secondary sector The set of economic activities that involves processing the raw materials extracted by primary sector activities into usable goods. Commonly referred to as manufacturing.

tertiary sector The set of economic activities that refers to all the different types of work necessary to move goods and resources around and deliver them to people; in other words, service activities.

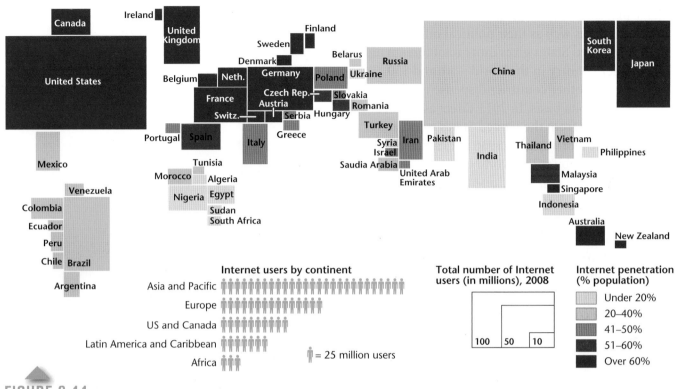

FIGURE 9.11   **The digital divide.** This map is a cartogram, whereby the size of the country is scaled to the size of its Internet user base. Large countries with relatively low Internet penetration—India, for example—appear much smaller than they really are. By the same token, countries that are geographically small but have high Internet penetration—such as Japan—appear large on this map. *(Source: Graham, M., S. Hale, and M. Stephens. 2012. Digital Divide: The Geography of Internet Access. Environment and Planning A, 44(5). 1009–1010 © Pion, Ltd. www.Pionco.uk and www.envplan.com.)*

---

quaternary sector
Intellectual and informational economic activities.

quaternary sector. The quaternary sector includes intellectual and informational activities.

DEVELOPMENT ACTORS

Now that you have some idea of the various definitions and approaches to development, it is logical to ask who "does" development and what kinds of activities count as development?

Traditionally, there have been several major actors in the field of development. International bodies specifically charged with some aspect of economic development, many of them emerging from the aftermath of World War Two, include the World Bank, the International Monetary Fund, and the United Nations Development Programme. Others are more focused on basic needs fulfillment, such as the

World Health Organization that targets disease eradication or Oxfam that provides hunger relief. Such bodies have played longstanding and visible roles in international development by setting agendas for development, providing expertise, and coordinating policies and people. Some churches and religiously affiliated nongovernmental organizations (NGOs) also provide development assistance in cash or in kind. Sometimes their assistance is offered internationally, but it can be locally targeted at specific neighborhoods.

National governments, both individually and working together, are also significant development players. Groups of countries, particularly wealthy ones such as member nations of the Organisation for Economic Cooperation and Development (OECD) or the Group of 8 (G-8), have been very influential in shaping the course of economic development through promoting international trade. Individual nations also have played decisive roles thanks to their economic might as well as their political influence. Though the United States gives more total foreign aid than any other country in the world—around $50 billion per year—other countries dispense much more on a per capita basis. In 2011 direct assistance to other countries from the United States stood at $97 per capita, while Norway gave $936 per capita and Canada $150 per capita. The world average was $134 per capita (**Figure 9.12**). The United States has been criticized for channeling much of its international development assistance toward promoting its foreign policy agenda and supporting military and political allies around the world (**Figure 9.13**).

Corporations can be significant development agents as well. *Business Insider* recently noted that if Walmart were a country, its revenues would be equivalent to the GDP of the 25th largest national economy in the world. In other words, Walmart

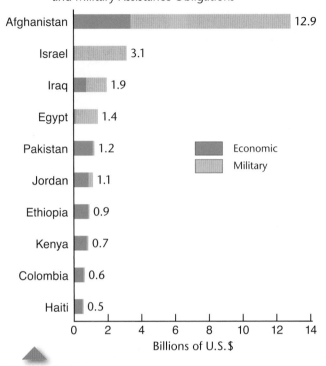

Top 10 Recipients of U.S. Economic and Military Assistance Obligations

FIGURE 9.13 **The world's top 10 recipients of U.S. foreign aid.** This graph displays where U.S. foreign aid goes and how it is distributed between military and economic purposes. *(Source: USAID Foreign Assistance Database.)*

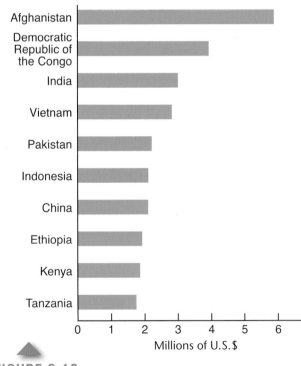

Top 10 Recipients of OECD Official Direct Assistance, 2011-2012

FIGURE 9.12 **The world's top 10 recipients of foreign aid.** The 34 wealthy nations that make up OECD provide foreign aid (called Official Development Assistance by the OECD) to developing nations. Some receive more than others.

would be bigger than Norway, Exxon Mobil would be larger than Thailand, General Electric would outrank New Zealand, and Apple would come in ahead of Ecuador. Of course, corporations are not countries and in some important ways the comparison is an unfair one. Yet, the analogy does a good job of highlighting just how economically powerful corporations can be. Some corporations are noted for their development-related investments, which are framed as an aspect of corporate social responsibility. Walmart, for instance, gave cash and in-kind (primarily food) donations nationally and internationally in 2012 amounting to over a billion dollars, or 4.5 percent of its pretax profits for 2011. While corporate leaders argue that national governments and NGOs have until now done little to alleviate poverty and provide basic needs, their critics accuse corporations of being interested in donating solely in order to grow their market share and brand loyalty in impoverished places.

Individuals associated with corporations can also become quite active in the arena of international development. Perhaps the most recognizable illustration of this is Bill Gates, who amassed a multibillion dollar fortune as co-founder of Microsoft. Now the head of the Bill and Melinda Gates Foundation, he directs his energy and wealth toward supporting development efforts aimed at alleviating poverty and disease and promoting education globally. In 2008 investment guru Warren Buffett pledged multiple billions to the foundation. He was at the time the world's wealthiest individual. Buffett and Gates are the first and second most generous philanthropists in the United States, respectively (Figure 9.14). Here, criticisms focus on the individuals' motivations for giving; namely, tax breaks.

Despite the criticisms, all of these development actors are "on the radar." In other words, they are visible, recognized, and legitimate participants in the development enterprise. But, it is important to know that there are development actors and activities that are not sanctioned or recognized. For instance, Mexico's drug lords consider themselves to be the primary source of development aid to regions of the country that have been neglected by the Mexican government for many years. In some states, such as Sinaloa, drug cartels have long based their activities both as producers of locally grown marijuana and heroin, and traffickers for these drugs and for cocaine from South America. Drug lords are often the only investors concerned with building local schools, parks, and infrastructure in rural areas. In 2012 one Sinaloa cartel kingpin, Joaquin "El Chapo" Guzmán, was ranked at #63 on Forbes's list of the world's most powerful people.

Indeed, the story that such cartels tell the local people who grow and traffic narcotics on their behalf is that they are merely taking from the wealthy, drug-addicted gringos and giving back to Mexico's forgotten poor. In other words, drug lords paint themselves as Robin Hood figures enacting social justice on behalf of the downtrodden. It is a powerful narrative, told in quasi-religious terms through the patron saint of narcotraffickers, Jesus Malverde, and in the lyrics of so-called narco-corridos: traditional Mexican ballads (*corridos*) singing the praises of drug lords.

It is important to emphasize that acknowledging how drug cartels and kingpins have channeled some of their wealth into local development activities in the places where they operate is not the same as condoning these activities. Indeed, Mexico's drug cartels are engaged in some of the most violent criminal behavior ever seen, seriously threatening the rule of law. Yet it is worth asking, are the sorts of criticisms leveled at the drug lords in some ways similar to those directed at legitimate development actors? More broadly, is development just about helping people, or is it also about controlling them?

FIGURE 9.14 **Warren Buffet and Bill Gates at a press conference.** Buffett and Gates launched the Giving Pledge in 2010, challenging billionaires to donate at least half of their wealth to charitable causes. They illustrate the power that high-profile individuals can exert in encouraging philanthropic behavior. *(FREDERIC J. BROWN/ AFP/Getty Images.)*

Finally, and importantly for you as students of geography, what is the future role of powerful and wealthy countries as the world's primary development actors? The emergence of development alliances among countries that were formerly the recipients of development aid, discussed under Globalization on page 390, provides yet another example of how the contemporary regional geographies of development are changing.

MOBILITY

9.2 LEARNING OBJECTIVE
Specify the ways in which mobilities of various sorts—of individuals as well as things and ideas—are important for development.

The movement of people, goods, and ideas from place to place has rarely been symmetrical. Instead, some places are net exporters of people, goods, and ideas, while others import them. This has led to the enrichment of some places at the expense or the impoverishment of others. In this section, we consider aspects of how the mobility of people, goods, and ideas has shifted over time and from place to place, and how this affects people's lives. We start by considering the importance of transportation and transportation-related infrastructure.

TRANSPORTATION AND DEVELOPMENT

Development, industrialization, and urbanization are profoundly connected. Development as we know it—both of the narrow economic sort as well as the broader attainment of basic needs and loftier human aspirations—could not have occurred without industrialization (i.e., the shift from primary to secondary sector activities) or the existence of cities (see Chapter 10). And, it is the movement of goods, people, and ideas from one place to another—transportation, in other words—that provides perhaps the most important element linking development, industrialization, and urbanization.

The **Industrial Revolution** saw a radical transformation in economic activities and processes. The Industrial Revolution began in England in the early 1700s. Machines, new power sources, and novel chemical processes replaced human hands in the making of products, initially in the cotton textile industry. In addition, many aspects of how people lived their lives in places were altered, including concepts of time, money, work, home, and gender roles.

The Industrial Revolution both encouraged and was furthered by the development of new forms of transportation. Traditional wooden sailing ships gave way to steel vessels driven by steam engines, and later railroads became more prevalent. The need to move raw materials and finished products from one place to another cheaply and quickly was the main stimulus that led to these transportation breakthroughs. Without them, the impact of the Industrial Revolution would have been minimized.

Once in place, the railroads and other innovative modes of transport associated with the Industrial Revolution fostered additional cultural diffusion. Ideas and, in fact, the Industrial Revolution itself spread more rapidly and easily because of this efficient transportation network.

Great Britain maintained a virtual monopoly on the Industrial Revolution well into the 1800s. Indeed, the British government actively tried to prevent other countries from acquiring the various inventions and innovations that distinguished the Industrial Revolution. After all, they gave Britain an enormous economic advantage and contributed greatly to the growth and strength of the British Empire. Nevertheless, this technology finally spread beyond the bounds of the British Isles, with continental Europe feeling the impact first. In the last half of the nineteenth century, the Industrial Revolution took firm root in the coalfields of Belgium, Germany, and other nations of northwestern and central Europe. The growth of railroads in Europe provides a good index of the spread of the Industrial Revolution there. The United States began rapid adoption of this new

Industrial Revolution Beginning in England in the early 1700s, the Industrial Revolution saw the rapid transformation of the economy through the introduction of machines, new power sources, and novel chemical processes. These replaced human hands in the making of products, initially in the cotton textile industry.

technology around 1850, followed half a century later by Japan, the first major non-Western nation to undergo full industrialization. In the first third of the twentieth century, industry and modern transport spilled over into Russia and Ukraine.

TRANSPORTATION AND THE COLONIAL LEGACY

As you have already seen in previous chapters, Europe's colonization of large parts of the world from the fifteenth century through the early decades of the twentieth century has shaped much of today's cultural geographies. That colonization was, in great measure, what allowed Europe to become wealthy and industrialized—developed—and also led to the progressive impoverishment and underdevelopment of the colonized areas.

The physical transportation of raw materials from the site of their extraction from the Earth, to places where they were processed into manufactured goods, was an essential aspect of industrialization. Before the advent of the steam engine in the late 1700s, for instance, the running water of streams and rivers drove the mechanical textile looms of northwestern England. However, once the steam engine came into use, textile factories were no longer geographically tied to locations with running water. They could be moved to cities where labor was cheaper and where transportation of their finished product to market—in ships and locomotives also powered by steam—was more centralized and efficient.

Thanks then to the steam engine and other transportation innovations, colonizer and colonized became drawn together in a geographically ever-wider and increasingly complex orbit. Raw materials such as cotton could be brought into England's mill towns from as far away as India, to be spun, woven into fabric, dyed, and made into clothing. These finished goods were then re-exported to the colonies, to be sold at far higher prices than the original raw cotton had commanded.

All these activities involved transportation from one point to another, with many stops along the journey from raw material to final consumer. Transportation technologies and infrastructure (such as roads, railways, and ports) played a central role in the relationship between colonizer and colonized. Thus, it should not come as a surprise that struggles for independence were often waged on and around the means of transportation. In many colonial places, **nationalization**—the government takeover of the transportation infrastructure and fuel sources from private foreign ownership—was a crucial element. For example, Japan nationalized its railways in 1906, Mexico nationalized its petroleum industry in 1938, and Egypt nationalized the Suez Canal in 1956. Airlines, banks, and a wide variety of natural resources have also been subject to nationalization as a part of decolonization and independence movements.

> ■ **nationalization**
> Government takeover of the transportation infrastructure and fuel sources from private foreign ownership, occurring in many previously colonized nations in the twentieth century.

Today, many formerly colonized countries have inherited the transportation infrastructure that was built during the colonial era. Because colonies were seen primarily as sources of raw materials, the primary focus of colonial-era infrastructure was moving the products extracted from the Earth—lumber, ores, minerals, and agricultural products—to ports and transporting them to factories in the cities of England, France, Germany, Belgium, Spain, and other colonial power centers, to be processed into finished goods. Latin America and Africa, in particular, have transportation infrastructures oriented toward the coasts and port cities. They do not, however, necessarily move people and goods about efficiently within the country.

Countries formerly colonized by the same European power are also not likely to be connected to each other. Long-distance air travel to African nations tends to make at least one stop in a European capital city en route, rather than flying direct. And, you may find it cheaper and even quicker to fly from one African city to a European capital first in order to get to a second African city, rather than flying directly within Africa. In other words, today even modern transportation infrastructure such as airports and flight routes may still be oriented toward transporting people to and from former colonial powers than within formerly colonized regions.

This is not to say that precolonial societies did not develop transportation networks and infrastructure: indeed, many did. For instance, the roadways of the

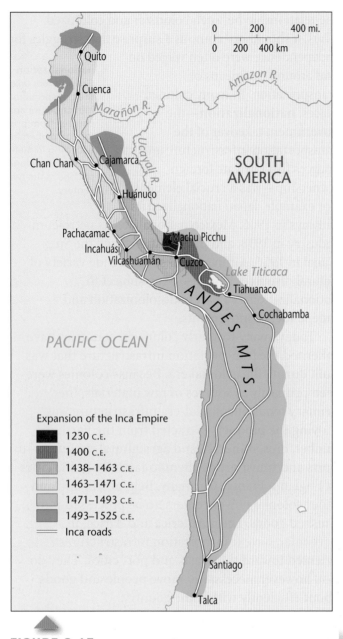

FIGURE 9.15 **Incan road system.** Much as with the Roman Empire, the expansion of the Incan Empire in South America in the 15th Century C.E. depended on an extensive system of roads, bridges, tunnels, and causeways. The Incas did not use wheeled vehicles, so the roads were intended for human and animal foot traffic.

Incan and Mayan societies in the Americas were noted for their complexity and high quality. The sixteenth-century Spanish conquerors of Peru observed the Incan road system to be better than the Roman roads at the time (**Figure 9.15**). Other remarkable roadways detailed in *Highways, Byways,*

and Road Systems in the Pre-Modern World (see Ten Recommended Books on Development Geography, page 410) include those built by imperial China, in the indigenous U.S. Southwest, and trans-regional roadways across the Sahara Desert and the Silk Road through Asia. The major difference between precolonial and colonial-era transportation infrastructure was their orientation. Precolonial transportation routes focused inward, on moving people and goods across interiors and connecting places to each other; whereas colonial-era infrastructure for the most part was concerned with getting people and goods to sea and river ports as expediently as possible.

Major regional differences exist in the relative importance of the various modes of transport. In Russia and Ukraine, for example, highways are not very important to industrial development; instead, railroads—and, to a lesser extent, waterways—carry much of the transport load. Indeed, Russia still lacks a paved transcontinental highway. In the United States, by contrast, highways reign supreme, while the railroad system has declined in importance. Western European nations have a greater balance among rail, highway, and waterway transport.

AUSTRALIA'S COLONIAL RAILWAY WOES

Even some countries that we would today consider to be "developed," for instance, Australia, Canada, and Russia, developed rail systems and to a lesser extent roads that are primarily focused on moving raw materials to port and then on to European colonial powers. **Figure 9.16** shows a map of the present-day Australian passenger rail system. Notice how the geographic focus of the routes is on the coastal areas, and that interior and transcontinental railways are few and far between.

Australia is coping with an additional colonial transportation legacy as well. The gauges (or widths) of railway tracks differ from state to state. Before the unified nation of Australia came into being in 1901, the continent consisted of six British colonies. Each developed its own railway system and, for a variety of personal and political reasons, chose different railway gauges. In 1848 the Secretary of State for the Colonies

FIGURE 9.16 **Australia's railway network.** Notice that the country's railways are concentrated along the length of the eastern coastline, and that just one line bisects the nation north–south and one east–west. This is a wonderful illustration of a colonial era extraction-oriented transportation infrastructure. **Why do you think the focus is primarily on Australia's eastern coast?**

in London, Earl Grey (yes, the man associated with the tea that bears his name), advised the colonial governors to adopt a uniform railway gauge, the English Standard Gauge. Three of the colonies did so, while the others later adopted different gauges or a mixture of gauges.

Before unification, this system of differing gauges was not foreseen to be a problem because the colonial governors could not envision the need to connect their railways across the continent to one another. Rather, their focus was moving people and raw materials to the port cities and back and forth from Europe. Transcontinental freight or passenger traffic had to transfer lines as it crossed from one colony to another. Thus, the Australian railway network became a mixture of narrow, standard, and broad gauge tracks. Significant personnel and resources had to be devoted to so-called break-of-gauge activities in places where routes of different gauges became connected to one another. Several initiatives to convert to a standard railway gauge nationwide have

been undertaken, but the work is slow, politically complicated, and expensive (**Figure 9.17**).

TRANSPORTATION HAVES AND HAVE-NOTS

As the Australian railroad gauge mismatch problem illustrates, modernizing transportation infrastructure can be a costly and difficult undertaking. Yet, having a modern transportation infrastructure in place is vital to a place's ability to develop. For example, the port of San Francisco declined in importance relative to the port of Oakland just across the bay for this very reason. The local authorities were unwilling or unable to undertake the difficulties and expense of modernizing San Francisco's port facilities to accommodate containerized shipping, whereas Oakland's decision-makers took

FIGURE 9.17 **Dual gauge railroad tracks near Melbourne, Australia.** Multiple gauges, or widths between railroad tracks, are a colonial legacy in many nations. While it is costly and politically difficult to adopt a standard gauge, it is also costly as well as inefficient to operate a system of different gauges. *(Paul A. Souders/CORBIS.)*

the plunge and modernized that port's infrastructure (see Chapter 10).

Because of the colonial legacy, many of the world's underdeveloped places have an inadequate transportation infrastructure. Old and dilapidated, failing to connect places to one another in an efficient fashion, mismatched, and often running on outdated technologies, these ports, roadways, and rail systems prevent a place from becoming developed. Much of Africa, Latin America, and former communist countries in Europe are facing this very predicament. Rightly or wrongly, their decision makers have determined that it is simply too expensive an undertaking to demolish existing facilities and build new, modern ones.

It is typically the poor that have the least access to everything related to development: income, health care, clean water and adequate food, housing, and so on. Where access to transportation infrastructure and mobility are concerned, this axiom generally holds true. For example, in Africa it is a fact that 90 percent of the continent's land area and 80 percent of the population are 100 kilometers (62.1 miles) or more away from the sea coast or a navigable river. Given Africa's colonial legacy of an extraction-oriented transportation infrastructure, this means that the majority of Africans do not have easy access to the infrastructure of mobility. As we have seen already, this makes moving about from place to place within the continent difficult and costly, and inhibits the sort of regional cooperation and integration that would allow African nations to work together to establish a stronger global economic presence. In other words, Africa's colonial mobility legacy impedes the region's development potential, making it all that much more difficult for an already poor place to progress.

Yet, we have also seen that places which are today thought of as developed—Russia, Canada, the United States, and Australia, in particular—have similarly extraction-focused transportation infrastructures. This is true because these countries (or regions within them) were themselves former colonies. At least in the United States, a persistent unwillingness to modernize the nation's transportation infrastructure has and will continue to hurt the country economically. On the other hand, formerly underdeveloped areas that have emphasized the modernization and expansion of their transportation infrastructure, including many East Asian places, have become global hubs of economic mobility for this very reason.

We can also drill down deeper into the structure of mobility haves and have-nots within individual countries or cities.

Singapore's Transportation Infrastructure Reboot

Some formerly underdeveloped places have prioritized building a state-of-the-art transportation infrastructure. Many Asian nations fall into this category. Fifty years ago, for example, Singapore was an overcrowded, housing- and transportation-poor city-state situated at the southern the tip of the Malay Peninsula. Since the 1970s, the Singaporean government has prioritized transportation infrastructure expansion and modernization as a development strategy, to position Singapore as a regional Asian as well as global hub of economic exchange. Singapore expanded and modernized its deep-water port facilities. Today, one-fifth of the entire world's container traffic and half of global crude oil shipments pass through the port of

FIGURE 9.18 **Singapore's subway station.** Singapore's MRT system efficiently moves half the population of the city-state about on a daily basis. *(ROSLAN RAHMAN/AFP/ Getty Images.)*

FIGURE 9.19 **Japan's Maglev train.** This Maglev (short for "magnetic levitation") train is a test version that Japan and other nations hope to adopt for long-distance transportation in the near future. Maglev transportation uses electromagnets to levitate and float the train forward along a guideway. The result is a fast, smooth, quiet, efficient, and clean ride when compared to conventional wheeled trains. Traveling at speeds of over 310 miles (499 kilometers) per hour, this train is still in the prototype and testing stage, and is scheduled to connect Tokyo and Nagoya in 2027. *(Tomohiro Ohsumi/Bloomberg via Getty Images.)*

investments in transportation infrastructure, on a much larger scale. Japan's transportation infrastructure leads the world in high-tech innovation (Figure 9.19). South Korea as well is rapidly building a modern national transportation infrastructure. East Asia is "on track," literally and figuratively speaking, to lead the world in transportation.

The United States: A Transportation Infrastructure Left Behind The United States provides an interesting example of a wealthy country whose aging transportation infrastructure is fast becoming a serious deterrent to the well-being of its citizens and a drag on economic growth. Think about recent "infrastructure incidents" in your local news: a collapsed bridge, airport delays arising from congestion, railway accidents due to damaged track or misrouted signals, or unsafe roadways closed to traffic (Figure 9.20). Why does the United States have so many problems with its transportation infrastructure?

Singapore. Singapore's airport facilities were moved from a residential area to reclaimed land away from the urban population. The new Changai Airport was expanded with the addition of terminals and runways, and is today an ultra-modern facility handling, on average, over a million passengers per week. Finally, Singapore's Mass Rapid Transit (MRT) rail system, built in 1987, is a model of cleanliness, safety, and efficiency (Figure 9.18). The MRT claims to transport 2.5 million people per day around a city-state of roughly 5 million inhabitants. That's half of the entire population moving about the city on mass transit.

The investment has clearly paid off, though Singapore's small size arguably makes such undertakings easier. In addition, Singapore's government is criticized for being highly restrictive with respect to the civil liberties allowed their population: when a course of national investment is set, there is not much room for meaningful protest. The Chinese government is making similar

FIGURE 9.20 **The U.S.'s aging transportation infrastructure.** This image shows a collapsed bridge section of Interstate 5 near Mount Vernon, in Washington State. This dramatic infrastructure failure occurred in 2013 and is one of many such incidents across the United States of late. Unfortunately such dramatic infrastructure failures are becoming more common in the United States. *(Stephen Brashear/Getty Images.)*

FIGURE 9.21 **Deteriorated roadways.** The image on the left is from Los Angeles, California, and the one on the right was taken in South Africa. The similarity in road conditions underscores that differences between developed and underdeveloped places aren't always hard and fast, especially with respect to the quality of infrastructure. *(Jonathan Alcorn/ Bloomberg via Getty Images; EggImages/Alamy.)*

As in many developing regions, the transportation infrastructure of the United States is generally old, outdated, and not adequately maintained (**Figure 9.21**). Most of the country's railways, airports, seaports, bridges, and roadways were built decades, if not centuries, ago. The American Society of Civil Engineers publishes a Report Card for America's Infrastructure. The 2013 report awarded an overall grade of D+ to the transportation infrastructure of the United States, estimating that some $3.6 trillion would need to be invested by the year 2020 to update it.

The World Economic Forum ranked America's infrastructure 23rd worldwide, between Spain and Chile. Compared to northern Europe, the infrastructure for U.S. roads, railways, ports, and air transport is ranked "mediocre." Traffic congestion, delays, and fatalities are significantly higher when compared to western Europe. By comparison to those in Europe, U.S. railways are slow and delays common. When it comes to air travel, a 1950s-era ground-tracking system forces flights to take inefficient routes to maintain contact with controllers, airports are overbooked, and cumbersome security measures cause passenger delays. The United States' inland waterways and rivers, what the American Society of Civil Engineers calls "the hidden backbone of our freight network . . . inland

marine highways," received a grade of "D–," which puts it on the line between poor and failing. Aging, unrepaired, and inadequate facilities cause an average of 52 service interruptions per day throughout this system.

The deplorable state of the U.S. transportation infrastructure poses economic costs and real dangers to its users. The United States is today the world's largest economy, but as **Figure 9.22** shows, it spends far less as a percentage of GDP than other nations on infrastructure. This stands in direct contrast to a sustained history of large-scale transportation infrastructure construction in the United States, with the nation's canal system constructed in the late 1700s, the transcontinental railway linking the East and West coasts created in the late 1800s, and the interstate highways built in the mid-1900s. What happened?

For one thing, Americans have insisted on low automobile and fuel taxes, yet in most countries these are a principal source of funding for building and maintaining roadways. Poor regional planning and political squabbling often lead to reluctance to fund construction and repairs. Budget crises at all levels— federal, state, and local—mean less public investment in transportation infrastructure, particularly in a deficit-averse environment. And, solutions that would privatize these costs, such as tolls, taxes, and fees

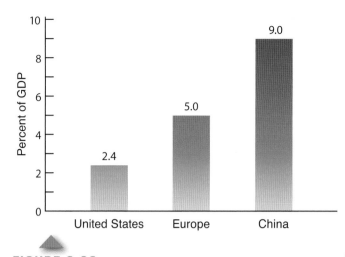

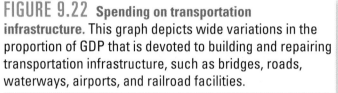

FIGURE 9.22 **Spending on transportation infrastructure.** This graph depicts wide variations in the proportion of GDP that is devoted to building and repairing transportation infrastructure, such as bridges, roads, waterways, airports, and railroad facilities.

levied on users, are unpopular. Only one thing is clear: the political gridlock preventing the prioritization of investment to repair and improve the U.S. transportation infrastructure is resulting in literal gridlock on the nation's roads, railways, waterways, and airports.

Nicaragua's Disembedded Elite Often it is the case that the elite residents of a place—whether the place in question is a city, a country, or a world region—have access to different and better (safer, quicker, more comfortable) transportation options than the less-well-off residents. This discrepancy is an example of how development geographies, in general, tend to be uneven, with the benefits accruing disproportionately to those who are already relatively well-off. Uneven development has spatial implications as well, with some places becoming progressively better-connected, wealthier, and more modern while others grow isolated, impoverished, and dilapidated.

If we consider the scale of most cities, the way that uneven development typically plays out is in the creation of enclaves or areas of relative wealth that exist side by side with areas of relative poverty. You might recall driving through a city where luxurious perhaps gated communities are located right next to

impoverished ghettos or *barrios.* (See also Chapter 5, pages 187–189.) In this case, spatial differences in development are very clear on the ground. Travel through such cities is seldom a straightforward exercise of moving from Point A to Point B, because roadways frequently bypass the poorer neighborhoods, while walls and fences make it impossible to transit freely through wealthier ones. As we will discuss in Chapter 10, this compromises the notion of public space, which so many have regarded as central to the exercise of democracy.

Research from Managua, Nicaragua, demonstrates how uneven development in some metropolises has become verticalized. Dennis Rodgers found that "rather than fragmenting into an archipelago of self-sustaining and isolated islands of wealth within a wider sea of poverty, urban space has undergone a process whereby a whole layer of the metropolis has been disconnected from the general fabric of the city." Wealthy Managuans attempt to escape from rising crime rates by going up, rather than further out into the suburbs. Managua's urban elites inhabit what Rodgers calls a "fortified network," moving back and forth from high-rise apartments, office towers, and shopping and leisure spaces located above the poverty, insecurity, and pollution found in the street-level areas of the city. Roadways connecting elite urban spaces to each other were modernized so that elite car traffic could move at higher speeds across urban space. Roundabouts were constructed to reduce carjackings at intersections. Use of these roads presupposes ownership of a private vehicle and the ability to pay tolls. Pedestrians face significant dangers simply trying to cross multilane high-speed roadways: Rodgers reports that pedestrian fatalities constitute fully 40 percent of all traffic fatalities in Managua.

The next time you move through a large city, look around and see if you can find evidence of the verticalization of privilege. More and more cities around the globe look like Managua, where the urban elites have literally detached their lives from the social and spatial fabric of the city, hovering above it. Indeed, globalization in many areas—economic, technological, and even of celebrity—has reshaped places and their relationships to other places in ways that bring into question some of the fundamental assumptions of development.

Paraíso
http://www.tinyurl.com/kfy8ajy

Three Mexican immigrants who work as window washers on Chicago's downtown high-rise buildings are profiled in this video. The process of window washing is depicted, and the men speculate about the dangers—but also the beauty and craft—of the work they do. In addition, the men discuss more abstract notions: death, heaven, and the future that awaits them. Their conversations also bring up comparisons to the lives they might have lived had they stayed in Mexico. A great deal of nuance characterizes this video, and the three men's understandings of central contrasts—between danger and beauty, poverty and wealth, heaven and earth—are not what you might expect.

Thinking Geographically:

1. One of the topics covered in this chapter's Mobility section is the verticalization of privilege. How does this look in Chicago? In what specific ways do these men see and experience the verticalization of privilege from their vantage point as window washers?

2. Window washing is a dangerous occupation. How do these three men rationalize that danger? What do they view as the positive aspects of their job?

3. The three men talk a lot about wealth and poverty. Give some examples from the video of their understandings of wealth and poverty. Does anything strike you as unexpected?

GLOBALIZATION

9.3 LEARNING OBJECTIVE
Identify how formerly poor places are today significant agents of development.

As we have already seen in this chapter, development today looks a lot different than it did right after World War II. In some very important ways, economic, technological, cultural, and even media-related transformations have brought different actors and sites into prominence. These transformations have fundamentally reoriented the landscape of global development geographies.

POSTDEVELOPMENT AND THE RISE OF THE GLOBAL SOUTH

Were today's less developed places always poor, isolated, and largely engaged in primary sector activities? Not necessarily. As you have learned already, and will see again in the next chapter on urban geographies, many civilizations now considered "underdeveloped" formerly were hubs of economic activity, technological innovation, cultural diversity, and political dominance.

Consider the arc of Islamic influence, stretching from Rabat in northern Africa to Manila in Southeast Asia. As former university president William Frawley describes, "There is a fifteen-century-old line of faith and business that runs straight from Morocco to the Philippines and that has drawn Muslims and workers of need to countries of religion and economic prosperity." This 7,600 mile (12,300 kilometer) sweep of Islamic influence constituted a vital corridor of economic, cultural, and political activity: an arc of development. It arose in a world where Europe had languished for centuries in the Dark Ages, and the so-called New World was populated by indigenous societies that were themselves centers of innovation and prosperity (see also Chapter 10, page 421).

The stark division between "developed" and "underdeveloped" places discussed earlier in the Region section has itself been largely redrawn. In reality, very few of the world's places are impoverished, isolated, and predominantly agricultural anymore. Individual countries, like China, or entire regions, for instance, Sub-Saharan Africa or Southeast Asia, that did fit this profile as recently as 30 years ago have witnessed the transformation of their economic, political, technological, and cultural systems. Even within countries that may still be less developed as a whole, there are pockets of accelerated urbanization, innovation, and connection to global networks that make these locales look, feel, and act wealthy, contemporary, and plugged-in (Figure 9.23).

By the same token, places that several decades ago unquestionably occupied the ranks of the wealthy, connected, and modern—in other words, developed countries and regions—have slipped. Earlier, we discussed the United States' crumbling transportation infrastructure and noted the high toll placed on economic productivity by this situation. Europe, too, has seen its seemingly secure First World status called into question by debt, financial crisis, an aging workforce, fears of terrorism, conflicts over immigration, and racist violence. In addition, places

FIGURE 9.24 **Detroit's third-world cityscape**. This abandoned house is one of many in Detroit. The city estimates one-fifth of its housing stock is abandoned and blighted. *(Getty Images/WIN-Initiative RM.)*

FIGURE 9.23 **Songdo International Business City**. This new development is located on reclaimed land along South Korea's Incheon Waterway, near the capital city of Seoul. It includes ultra-modern residential, commercial, education, leisure, and healthcare facilities. Forty percent of the area is devoted to green spaces such as the park you see here. Songdo is a "smart city": sustainable, green, and wired. Though South Korea is overall not a wealthy country, Songdo represents the cutting edge of urban design.*(Ann Hermes/The Christian Science Monitor via Getty Images.)*

within these countries and regions, once centers of economic, technological, political, and cultural dynamism, have fallen on hard times.

The city of Detroit is a perfect example. Once a world leader in the auto industry and the United States' fourth largest city, Detroit was home to secure jobs and seen as a cultural embodiment of quintessentially Midwestern American values and lifestyle. In 2013 Detroit declared bankruptcy, the largest U.S. city ever to do so. It has lost over half of its population since its heyday in 1950, and with it the tax base needed to pay for city services and maintenance. Financial mismanagement and the burden of health care and pensions for city workers have further impoverished Detroit. In many ways, Detroit's landscape today is reminiscent of a Third World place (Figure 9.24).

Some scholars see these changes as evidence of **postdevelopment**: the idea that "development" was simply a way to categorize most of the world as needing the assistance of a few, wealthy nations. However, postdevelopment scholars argue, development as we knew it—in the form of monetary aid, investments,

postdevelopment
A theoretical approach that is critical of standard development practices, asserting that traditional development efforts do not work because they are ultimately about controlling, not empowering, poor nations and people.

expertise, and projects—was seldom successful. Instead, development in reality constituted a modern form of **imperialism**, whereby wealthy nations ruled poor nations by controlling their economic, political, and cultural systems. The age of development, they argue, is over.

Whether or not one agrees with this statement, it is clear that places formerly dismissed as underdeveloped have emerged as major players in a global world. There are several countries that are particularly important in what is commonly called the **Global South**: a term that has largely replaced Third World when referring to the countries of Latin America, Africa, and most of Asia. The term also represents an attempt to move beyond the negative connotations of "Third World," recognizing that much of the world's growth—economic, population, and urban—is happening in these countries, and that powerful contributions and alliances are emerging among them without the involvement of traditionally wealthy nations.

South–South cooperation is a major dimension of the dynamism occurring within the Global South. Several countries traditionally not part of the developed nations—Brazil, China, India, South Africa, Venezuela, Malaysia, and Indonesia among them—have large populations and growing economies. In many respects, they are the new centers for technical and economic development vis-à-vis their poorer neighbors. These countries are emerging as major global lenders, investors, researchers, educators, and employers (**Figure 9.25**). A report by the World Bank finds that South–South trade, for instance, has grown on average by 13 percent every year since 1995. They are also standing shoulder-to-shoulder with the world's former political powerhouses, demanding a voice in issues like global climate change, energy use, and the world's political organization.

> ■ **imperialism** A relationship whereby wealthy nations dominate poor ones by controlling their economic, political, and cultural systems.
>
> **Global South** A term that has largely replaced "Third World" when referring to regions of Latin America, Africa, and most of Asia, in recognition that much of the world's dynamism, growth, and power resides in these places.
>
> **South–South cooperation** Trade, technological innovation, and other forms of exchange and assistance that occur between the larger and wealthier nations of the Global South, and poorer nations.

FIGURE 9.25 **India's Tata Nano.** Billed as "the world's cheapest car," the Nano, designed and manufactured by Tata Motors, is a good example of technology developed by and for the Global South. Retailing for around $1,700, the Nano is lightweight, compact, simply-designed, and fuel-efficient. *(Burhaan Kinu/Hindustan Times via Getty Images.)*

TECHNOLOGY'S ROLE IN CLOSING THE DIGITAL DIVIDE

Technology has long played a central role in economic development. Earlier in this chapter, we discussed Europe's transformation from an agricultural to an industrial society, beginning in the eighteenth century. Recall that technological developments, particularly in powering machinery for manufacturing goods and transporting them, were the key to this transformation. Throughout the centuries that followed and today as well, revolutions in transportation and communications technologies continue to play a huge role in economic development. Small- to medium-sized enterprises in countries of the Global South are the catalysts for providing low-cost solutions to common problems there. Tata Motors, one of whose best-selling products is shown in Figure 9.25, is just one example. Other innovations are occurring in the fields of energy, environment, public health, education, and transportation.

India's Aakash tablet computer, for instance, is priced around $35 U.S. with a government subsidy (it retails for around $60 U.S.). It is targeted for use in the educational sector, to expose Indian students to interactive online learning, testing, and computer programming. In another example, the One Laptop per Child initiative aims to facilitate children's engagement with education by distributing low-cost, rugged, low-power, and connected laptops to students in 46 countries of the Global South, with the stated goal of "collaborative, joyful, and self-empowered learning."

Mobile phones provide another field of South–South technological innovation and exchange. In 2012 the World Bank reported that three out of every four people on the planet have access to a mobile phone, and usage in the Global South has surpassed that in wealthy Northern countries. "Mobile phones are arguably the most ubiquitous modern technology: in some developing countries, more people have access to a mobile phone than to a bank account, electricity, or even clean water." Mobile phones and applications (*apps*) are used by farmers, fishers, and traders to monitor weather conditions, track schools of fish, and follow market prices. Health care providers use mobile phones and apps to remotely diagnose ailments and to text patients medication reminders. Mobile phones are used by migrants to send money home, by citizens to monitor elections, and by workers to apply for jobs.

All these projects attempt to close the digital divide. The **digital divide** is the rift between those

> **digital divide** The rift between those who benefit from easy access to the Internet and, more importantly, the vast storehouse of knowledge available online; and those who do not have easy access.

who benefit from easy access to the Internet and, more importantly, the vast storehouse of knowledge available online; and those who do not have easy access. The digital divide is a contemporary aspect of the longstanding gap between development's haves and have-nots. Those who do not have easy access to the Internet are disadvantaged to the extent that they cannot readily participate in today's globally connected world. Though Internet access is growing rapidly, 65 percent of the world's inhabitants do not use the Internet (**Figure 9.26**). Even within wealthy countries, some segments of the population—the elderly, disabled people, and the poor—are far less likely to go online than are the young, healthy, and wealthy.

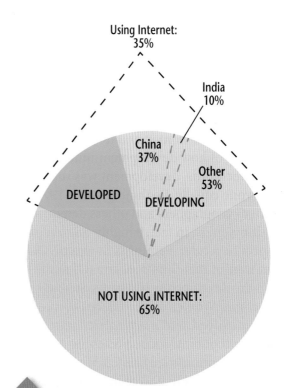

FIGURE 9.26 **The global digital divide.** Only 35 percent of the world's 7.1 billion inhabitants uses the Internet. The majority of Internet users reside in the developing world. While non-users stood at 65 percent of the world's population in 2011, that was down from the 2006 figure of 82 percent non-users. The number of Internet users in the developing world has now surpassed those in the developed world. We can only expect these trends to continue.

Part of the problem may be that the creators of the Internet and its content represent a narrow sector of the world's population. According to Chris Csikszentmihályi, director of the MIT Center for Future Civic Media, "What you end up with is an Internet that assumes a particular kind of user, one that resembles the authors. So in a sense, almost everyone who uses the Internet has to sort of pass as a white, 20-something, urban-dwelling kind of person." Consider, for instance, that since 2000 the Middle East has been the most active world region for growth in online use. Yet, less than 1 percent of Internet content is in Arabic. It is interesting to think about the ways that the growing Global South presence online might significantly reshape the Internet's content and delivery.

GENDER, GLOBALIZATION, AND DEVELOPMENT

The traditional development models formulated after World War II were gender-blind. They did not distinguish between women and men in their understanding of how development unfolds in a place. In other words, development was supposed to affect everyone equally. Beginning in the 1970s, development geographers and others demonstrated that development is, in fact, a highly uneven process with respect to gender. Feminists in wealthy Northern countries were at that time concerned that the labor market in the developed world had systematically left women behind. Women were underpaid, unemployed and underemployed, and marginalized into certain sectors of the economy, such as teaching and nursing. The situation, feminists speculated, was likely to be the same in countries that were just then undergoing the process of development.

Evidence shows that women in the developing as well as the developed world earn less than their male counterparts, have fewer employment choices, and are far more likely to lose their jobs during economic downturns. In some societies in the developing world, there are cultural prohibitions against women working for wages outside of the home. These women are entirely economically dependent on their fathers, brothers, husbands, and sons for economic support. Research conducted in many areas of the developing world has demonstrated that men are more likely to spend their wages on alcohol and leisure activities, whereas women tend to use money for food, education, and health care (see also Subject to Debate on page 393).

Women's economic vulnerability often translates into vulnerabilities in the other key arenas of development: among them, health, housing, safety, and education. The logical solution, according to this understanding, is to more fully include women in the development process, particularly in paid employment outside of the home. Women's inclusion in the labor market would reduce their economic dependence on men; in addition, it would raise overall economic productivity. **Figure 9.27** shows how the GDPs of several countries would rise if women were incorporated in the labor force at the same levels as men in those countries. Yet, such a solution overlooks several important factors.

First, many women are already involved in the development process, even if they do not participate in paid employment outside of the home. All of the labor performed by women in the home for free, including child care, meal preparation, in-home health care, and cleaning, subsidizes the world economy. This is true everywhere, not just in the Global South. Says New Zealand economist Marilyn Waring, "Unpaid work makes all the rest of work possible. The market wouldn't survive if it wasn't able to survive on the backbone of unpaid work." Waring calculates unpaid labor, the vast majority of it performed by women, to be the largest sector of any national economy. If this so-called nonmarket household production was incorporated into the economic output of the United States, for instance, it would raise the national GDP by 26 percent.

Second, women—even those who are actively participating in paid employment outside of the

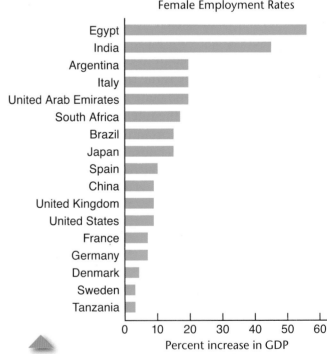

FIGURE 9.27 **The economic impact of women.** Nearly one billion women are poised to enter the global labor market over the next decade. This graph depicts their impact on the GDPs of several nations, if women were employed at the same levels as men in those nations.

SUBJECT TO *Debate*

MICROFINANCE AND DEVELOPMENT

In the 1970s, an economics professor by the name of Mohammed Yunnus was troubled. On the one hand, he was teaching textbook economic theory; while on the other hand, he was surrounded by the poverty and hunger of so many inhabitants of his native Bangladesh. One thing that stood out to Professor Yunnas was the difficulty faced by poor people, particularly women, in securing credit. Banks didn't want to lend to them because they had no assets or credit history. The loan sharks who did lend to women demanded exorbitant interest rates in return. Yunnus himself initially loaned a total of $27, which was enough to erase the debts of 42 individuals.

In 1983 Yunnus founded the Grameen ("Village") Bank. The Grameen Bank lends only to those without assets or credit, and requires loan recipients to educate their children and establish savings. Most of the microloan recipients are women or groups of women, who—far from being credit risks—have very high loan repayment rates. The loans, which average $100, are used to establish small businesses and to lift families out of poverty. In 2006 Yunnus was awarded the Nobel Peace Prize; at that time, over 7 million individuals had received loans through the Grameen Bank.

Why do microfinance efforts like the Grameen Bank focus on women? Research and experience have shown that poor women in the Global South by and large prioritize expenditures on family well-being, including health care, nutrition, and education. Poor men, on the other hand, often spend money on short-term pursuits that do not enhance family well-being, such as leisure activities and alcohol. In addition, women are far more likely than men to repay loans on time and in full and less likely to default.

Microfinance efforts throughout the developing world challenge the traditional focus of economic development, which emphasizes lending significant sums of money to organizations and governments to finance large-scale projects. Women are typically not well represented in the ranks of formal development organizations and national governments. In some ways, you might think of microfinance as a postdevelopment undertaking.

Continuing the Debate

Consider the debates surrounding microfinance, and ask yourself these questions:

- Do you believe that women are truly better credit risks than men? Why or why not?

- Microloans are for such small amounts of money—can they really make a difference?

- Are microfinance initiatives under way where you live?

Female entrepreneurs benefit from micro-loans by Bangladesh's Grameen bank. *(Robert NICKELSBERG/ Gamma-Rapho via Getty Images.)*

home—face systematic barriers that men do not. For instance, women often confront discrimination, harassment, and violence in the labor market and specific workplaces, but also in other public spaces as well as inside their homes. Prevalent cultural associations of women with motherhood and the spaces of domesticity (see also Chapter 3, pages 106–108) mean that women who work outside the home are seen as transgressing into male-dominated spaces. In order to force these women back into their homes and mothering roles, men—and other women as well as public policy and the media—subtly (and not so subtly) retaliate in a variety of ways.

Because such scenarios occur throughout the world, you are probably familiar with some of these retaliatory tactics. For instance, women moving through public spaces such as city streets and on public urban transit are subjected to catcalls, groping, and sexual assault. Working Indian women have been raped on public transit and in city spaces, reports of which have only recently become international news. Since the 1990s, hundreds of female factory workers in Northern Mexico have been murdered as they traveled to and from their jobs.

Women's mobility is restricted in other ways as well. In Saudi Arabia, for instance, women are still legally forbidden to drive automobiles. Many societies expect women to dress in restrictive clothing that limits their mobility and comfort in the name of modesty. You might think of high-heeled shoes as an example of this tactic. Even the persistently lower pay for women the world over who hold the same jobs as men can be seen as a tactic to discourage women who work. In the United States, young college-graduate women working full time, on average, make 82 percent of what their male counterparts are paid; in South Korea and Japan, women earn less than 70 cents for every dollar earned by a man. In many ways, women are still second-class citizens, a reality that hinders progress in development by holding back fully one-half of humanity.

CELEBRITIES AND DEVELOPMENT

Technology is not just good for solving logistical problems in poor areas, such as bringing electricity to remote villages or tracking the height and weight of children. In the development field, technology has also been used to create communities of concern and action that are truly global in scope. Many times, celebrity figures from the world of music, film, and sports are the faces of such communities. This is not a new development. In 1940, for instance, silent movie star Charlie Chaplin's first talking film *The Great Dictator* condemned Hitler, Nazi Germany, and fascism. Film star Jane Fonda was a very vocal participant in the antiwar movement of the 1960s. Singer Johnny Cash performed at California's Folsom State Prison, in part to protest conditions there in 1968.

The Ethiopian famine of 1984, however, heralded the advent of contemporary high-profile celebrity-led development efforts televised worldwide. The 1985 Live Aid charity concert, assembled by the Boomtown Rats' front man Bob Geldof, was the most visible of these early efforts (**Figure 9.28**). Geldof had been deeply affected by coverage of the famine, which claimed more than 1 million lives. Geldof assembled a superstar musical lineup, including Bono, Mick Jagger, Elton John, Madonna, Paul McCartney, and B. B. King, performing in one of two shows broadcast via satellite technology. Phil Collins actually performed in both events: first in London and then, after flying across the Atlantic to perform again with the reunited superband Led Zeppelin, in Philadelphia. Each performer had no more than 17 minutes on stage, and the performances were interspersed with film clips depicting the famine and calls by Geldof to donate money to the famine relief efforts.

MTV estimated that 1.4 billion of the world's then 5 billion inhabitants watched, and that at one point 95 percent of the world's television sets were tuned in to the concert. More than $200 million was raised. Keep in mind that this huge participation occurred in the mid-1980s, when cell phones, the Internet, and Twitter did not exist. Indeed, the 1981 birth of MTV—and the music video—can itself be seen as a technological innovation facilitating massive charity rallies such as Live Aid. Broadcasts like this one allowed virtual participation, leading to a widening of the concerned community away from just the face-to-face audience at live concerts. MTV thrived on the repeated playing of music video clips, and those shot at Live Aid were soon iconic, such as U2's performance of "Sunday Bloody Sunday." These performances resonated with the career development of this era's pop

FIGURE 9.28 **Live Aid Concert.** Organized by Bob Geldof in 1985, Live Aid was at the time the biggest and most ambitious concerts ever staged. *(Michael Ochs Archives/Getty Images.)*

musicians, which would become increasingly video-based rather than linked solely to live performances.

Since then, there have been any number of celebrities associated with causes in the Global South. Actors, musicians, and television personalities make up a large portion of such celebrity ranks: Angelina Jolie, Madonna, George Clooney, Oprah Winfrey, and Leonardo DiCaprio have all engaged in social justice charity work in Africa, for instance. To be sure, the wealth, mobility, and cultural visibility of celebrities can go a long way toward bringing worldwide attention to events that might otherwise remain under the digital radar. These individuals leverage their fame to raise millions of dollars for development-related causes, sometimes far more effectively than the traditional economic development agencies officially charged with funding development efforts in the Global South.

Nevertheless, there is also widespread criticism of celebrity development efforts. Most media celebrities are not trained economists, climate change scientists, medical doctors, educators, or social workers. In some ways, they are playing a role—that of development expert—they are not really qualified to play. A few celebrities acknowledge this and have teamed up with international development organizations or experts. Figure 9.29 depicts Angelina Jolie's professional association with renowned development economist Jeffrey Sachs, for example. Others, however, seem

oblivious to their limitations, as seen in the example of Madonna's failed \$3.8 million Malawian school project, apparently a victim of financial mismanagement.

Critics have noted that celebrity development efforts tend to portray the Global South and its residents as passive victims of calamity who are waiting for wealthy Westerners to rescue them. Writes Andrew M. Mwenda, "So some [celebrities] come to save orphans, others to defend human rights, feed the hungry, treat the sick, educate our children, protect the environment, end civil wars, negotiate aid and promote family planning, lest we overproduce ourselves. It is almost as if Africans cannot do anything by themselves and need a combination of Mother Teresa and Santa Claus to survive." Critics of celebrity development figures suggest that their efforts are more about attempting to enhance the individual celebrity's brand, rather than reflecting any true concern with living people and places.

In 2010 Twitter technology spawned the Digital Death campaign on December 1, World AIDS day. Music, sports, and entertainment personalities such as Usher, Swizz Beats, Kim Kardashian, Ryan Seacrest, Justin Timberlake, and Alicia Keys posed in caskets. They threatened to leave their collective 30 million Twitter followers "in the dark" until \$1 million was raised. Videos narrating each celebrity's "last tweet and

FIGURE 9.29 **Celebrity and international development.** Actress Angelina Jolie and development economist Jeffrey Sachs joined forces to film a video diary of their travels to various African development projects addressing the U.N. Millennium Goals discussed on pp. 370–372. *(Andrew Kent/Getty Images.)*

testament" implored viewers to buy back their favorite star's digital life; until then, there would be no Facebook or Twitter, no beats and no music (see Development Geography on the Internet, page 409, for a link to the last tweets and testament videos of participating stars). Justin Timberlake, for instance, claims to have "sacrificed my digital life in order to give real life to millions of people who are affected by HIV and AIDS in Africa and India." The Digital Death campaign raised $1 million within a week, but there is speculation that the stars kicked in cash of their own to reach the goal and buy back their digital lives. In all likelihood, celebrity development efforts are not entirely naïve and self-serving undertakings. In some ways, they underscore a longstanding and problematic relationship between the world's haves and have-nots.

As we have seen, technological, cultural, and political transformations at the global scale will certainly continue to shape and reshape how development is practiced and who benefits.

⠿ NATURE–CULTURE

9.4 LEARNING OBJECTIVE
Recognize the role that nature has played over time as concepts and practices of development have changed.

As we have seen already in this chapter, the extraction of natural resources from the Earth, and their subsequent transformation through manufacturing into products that people need and want, have for centuries been the engine of economic development. Nature, in other words, is transformed into (material) culture. Nature is also transformed symbolically into culture. Specifically, "pristine" natural environments (see Figure 9.8, for example) are cultural images often positively associated with poorer places on the planet. It is as if a loss of purity, both of the natural world and of man's place in nature, is the primary trade-off of economic development. Some scholars would go so far as to say that we have tacitly agreed to the ongoing destruction of the planet in exchange for living comfortably today, for most economic activity creates serious ecological problems. Thus, nature–culture is a very important theme to explore with respect to development geography.

RENEWABLE RESOURCE CRISES

Human beings have a long history of depleting the natural world around them, in the name of progress. Deforestation, for example, has been a problem since the beginning of human civilization. Ancient humans used fire to burn down trees and drive woodland animals out of the forest in order to more easily hunt them. Neolithic-era agriculture involved clearing forest land for cultivation (see World Heritage Site, pages 398–399). The ancient Greeks faced problems with the silting of their harbors as the lands cleared for domestic crop cultivation eroded, a scenario also experienced by the ancient Romans and medieval Europeans. Cities were often abandoned when their ports became unusable. Europe's heavy involvement in the trans-Atlantic slave trade, warfare, and voyages of exploration to distant lands resulted in massive deforestation of Northern forests in order to build wooden ships for these activities.

It might seem that only finite resources are affected by industrialization, but even renewable resources such as forests and fisheries are endangered. The term **renewable resources** refers to those primary sector products that can be replenished naturally at a rate sufficient to balance their depletion by human use. But, that balance is often critically disrupted by the effects of industrialization. So, while deforestation is a longstanding process, the Industrial Revolution drastically increased the magnitude of the problem. Today, the Earth Policy Institute has identified Latin America as the region experiencing the greatest loss of forest cover, thanks primarily to tree clearance in the Amazon to make way for cattle ranches. Indeed, we are witnessing the rapid destruction of one of the last surviving great woodland ecosystems, the tropical rain forest. The most intensive rain forest clearing is occurring in Indonesia and Brazil. Although trees represent a renewable resource when properly managed, too many countries are, in effect, mining their forests at a rate faster than they can be replaced. Even when forests are replaced by

> **renewable resources** Those primary sector products that can be replenished naturally at a rate sufficient to balance their depletion by human use.

scientifically managed "tree farms," as is true in most of the developed world, ecosystems are often destroyed. Natural ecosystems engender plant and animal diversity that cannot be sustained under the monoculture of commercial forestry.

Similarly, overfishing has brought a crisis to many ocean fisheries, a problem compounded by the pollution of many of the world's seas. On average, people consume four times as much fish as they did in 1950, reaching an all-time high of 37 pounds (17 kilos) per person annually in 2011. More than 85 percent of the world's fish stocks are compromised by overfishing. A BBC report suggests that "large areas of seabed in the Mediterranean and North Sea now resemble a desert—the seas having been expunged of fish. . . ." Commercial fishing methods are now depleting the world's tropical waters as well, with a 40 percent drop in tropical fish catch expected by 2050. As with tree farms, fish farming has significant drawbacks, including pollution of the surrounding waters from the antibiotic and fecal waste of the concentrated fish farm populations, along with a lack of nutrients in the meat when farmed fish are fed vegetarian diets. Some good news was reported in the North Sea, though: since 2009 the size of the spawning stock of cod reached a level that scientists agreed was sustainable. Experts credited an array of conservation techniques for the turnaround.

GLOBAL CLIMATE CHANGE

Virtually all climate change scientists now agree that we have entered a phase of **global climate change** (commonly called **global warming**) caused by human activity and, most particularly, by the greatly increased amount of carbon dioxide (CO_2) produced by burning fossil fuels. (However, see Subject to Debate in Chapter 1, page 31, to examine more fully why some disagree about the extent to which industrial activities contribute to the increase in CO_2. See also Chapter 10, pages 440–446.) Fossil fuels—coal, petroleum, and natural gas—are burned to create the energy that powers the world's factories, and as that burning has increased, so has the amount of CO_2 released into the atmosphere. The 10 hottest years on record have all occurred since 1998. Rising sea surface temperatures melt Arctic ice. The extent of the Arctic Sea ice measured by NASA in 2012 was the smallest ever recorded. Melting ice raises ocean levels, threatening to obliterate coastal areas and even entire countries, such as the island nation of Maldives. Thanks to global climate change, some areas of Earth are hotter, wetter, drier, or colder than normal. As displayed in **Figure 9.30**, these changes have led to disruptions in agriculture, weather, tourism and leisure activities, and human health.

At issue is the so-called **greenhouse effect**. Every year automobiles and industry produce billions of tons of CO_2 worldwide by burning fossil fuel. By some estimates, the atmospheric concentration of CO_2 has climbed to the highest level in 180,000 years (Figure 9.30).

> **global climate change** Also known as global warming, global climate change involves alterations in climate caused by human activity and, most particularly, by the greatly increased amount of carbon dioxide (CO_2) produced by burning fossil fuels.

> **greenhouse effect** The effect of increased atmospheric carbon dioxide and other absorbing gases, which permits solar short-wave heat radiation to reach the Earth's surface but acts to block or trap long-wave outgoing radiation, causing a thermal imbalance and global heating.

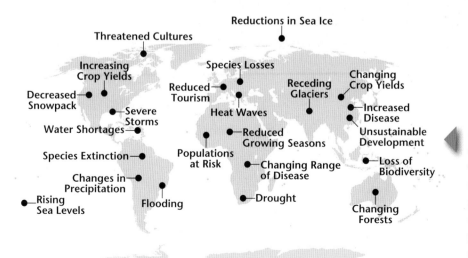

FIGURE 9.30 **Effects of global climate change.** The effects of climate change vary from place to place. This map is rendered from an interactive map produced by *National Geographic,* which can be found at http://environment. nationalgeographic.com/environment/ global-warming/gw-impacts-interactive/.

World Heritage Site

Two hundred and ninety seven of the UNESCO World Heritage Sites pre-date the modern era, and are considered ancient places having outstanding universal value to humankind. The Neolithic flint mines at Spiennes is one such place. Located near the city of Mons, modern-day Spiennes is a small village in Wallonia, a French-speaking region in southern Belgium. Spiennes dates back to the middle Neolithic period (4400–3500 B.C.E.) in central Europe. Associated with the ancient settlement above ground is a vast complex of mine shafts from the same period.

■ The mines span an area of over 100 hectares and represent the largest concentration of early primary sector activity in Europe.

■ The mine shafts are among the deepest ever sunk anywhere to retrieve flint.

■ The area was actively mined from 4400 to 2000 B.C.E., though there is some evidence of human activity prior to that, as early as 6000 B.C.E.

■ The Neolithic era was a turning point in the development of technology for human progress, which is why the flint mines at Spiennes are designated as an Industrial Heritage Site.

■ The mine complex is the earliest and largest example of primary sector activity in Europe.

■ The Neolithic flint mines at Spiennes were designated a World Heritage Site in 2000.

www.whc.unesco.org/en/list/1006

(iStockphoto/Getty Images.)

Neolithic Flint Mines at Spiennes

TECHNOLOGY AND DEVELOPMENT:

The Neolithic era, also known as the New Stone Age, was a pivotal one in human history. It saw the domestication of crops and animals, and coincided with the development of early cities (see Chapter 10). The tools developed during this period were made of stone—the Metal Ages would come later. Flint proved to be an exceptionally useful material as it can be worked into a very sharp edge through a process known as "knapping" and polishing. The first flint tools were hand-held; later, wooden handles were added to make axes.

(Charlie Waite/English Heritage/Arcaid/Corbis.)

■ As discussed in the nature–culture theme for this chapter, the Neolithic period saw the rise of agriculture and the related clearance of vast areas of forested land across Europe and the Mediterranean. The felled timber was used in the construction of housing and boats. The axes—known as the Swiss army knives of their time—were used to fell trees as well as for hunting and mining. Many were constructed of the flint mined at Spiennes.

■ As you know from reading this chapter, industrial activities are an important component of development. Mining is a primary sector activity because it involves the extraction of raw material from the Earth. Large flint nodules were extracted from underground at Spiennes using flint picks, deer antler, and cow bone.

TOURISM:

Some ancient World Heritage Sites, such as Stonehenge in England and the Pyramids at Giza, Egypt, are visited by thousands of tourists annually. Others, such as the Neolithic flint mines at Spiennes, are by contrast off the beaten path. Until quite recently, the mines had hardly been marked and were difficult for tourists to find. The mine shafts remain almost entirely untouched, except for some that have been fortified to accommodate the visits of tourists.

(Heritage Images/Getty Images.)

■ Only a few hundred tourists visit the Neolithic flint mines at Spiennes annually, often on a day trip that is part of a broader tourist itinerary in the area.

■ A modern visitors center is under construction and scheduled to open in 2015; it is hoped that this will help to attract more tourists.

■ Guides lead tourists down into a mine shaft, approximately 26 feet (8 meters) deep, to explore a section of the underground labyrinth.

■ The area above ground is littered with tailings, which are left-over scraps of flint generated during the knapping process thousands of years ago.

In addition, since trees absorb carbon dioxide, the ongoing destruction of the world's rain forests adds huge additional amounts of CO_2 to the atmosphere. Although CO_2 is a natural component of the Earth's atmosphere, the freeing of this huge additional amount is altering the chemical composition of the air. Carbon dioxide, only one of the absorbing gases involved in the greenhouse effect, permits solar short-wave heat radiation to reach Earth's surface but acts to block or trap long-wave outgoing radiation, causing a thermal imbalance and global heating.

To begin to mitigate the effects of the environmental crises caused by global climate change, 38 industrialized countries signed the Kyoto Protocol in 2001. This document, originally discussed in Kyoto, Japan, in 1997 and renewed in Durban, South Africa, in 2011, binds these countries to reducing emissions of greenhouse gases so that their future levels of emissions will be less than their 1990 levels. The United States agreed to this protocol in 1997, but has since changed its position and is no longer a party to the agreement, claiming that it would cause economic setbacks and does not do enough to curtail the high emission rates of poor nations. It is true that the large emerging economies in the Global South discussed earlier are also adding quickly to the world's greenhouse gas emissions, and these nations must be central partners in any agreement to address global climate change. For this reason, Mexico, South Africa, Qatar, and South Korea have recently hosted or are scheduled to host global climate change summits.

THE ENVIRONMENTAL CONSEQUENCES OF POWERING INDUSTRIALIZATION

The extraction of natural resources from the Earth, and their transformation into the myriad consumer goods we now rely on, depend on the provision of energy to run factories and transport finished products. Consumers the world over have grown accustomed to heated and cooled dwellings, artificial lighting, and running water, among other amenities. The agricultural activities needed to feed the world's more than 7 billion inhabitants require plenty of power as well.

So-called **alternative energy** sources—that is, alternatives to fossil fuels—have long been pursued in an attempt to provide cleaner, cheaper, renewable power sources. These alternatives include hydroelectric, solar, wind, geothermal, nuclear, biofuel, and "clean" coal. However, the expense involved in building, running, and maintaining sites for harnessing and distributing these power sources; their intermittent nature in some cases, such as solar and wind; and the potential hazards associated with some of them have deterred their widespread adoption. As a result, the world still relies heavily on fossil fuels to power economic activities: 80 percent of the world's energy, in fact, still comes from fossil fuels. Not incidentally, powerful oil, gas, and electrical industry groups have spent millions of dollars to lobby politicians, especially in the United States, to reject policies favorable to alternative energy. The advent of hydraulic fracturing (*fracking*)—which uses pressurized liquid to fracture rock in order to extract natural gas and petroleum—is the latest example of how the United States has, as a nation, largely chosen to continue to extract fossil fuels rather than switch to other fuel sources.

As the emerging nations of the Global South become increasingly important producers of goods and services, how they choose to fuel their economic growth will be of paramount importance. As we have already seen, South–South cooperation in alternative environmental technologies is an important growth area for these emerging economies. It is entirely possible that the next wave of alternative energy ideas will emerge from technical innovations developed in the Global South. Lesotho exports water to South Africa, while Paraguay, Bhutan, and Mozambique are all net exporters of electricity to neighboring countries.

Some nations in wealthier Northern regions of the world are also noted for their use of renewable energy. Iceland, the world's largest volcanic island, has abundant geothermal and hydro power. Iceland's electricity is clean: it is produced entirely by renewable energy. Samsø, an island off the coast of Denmark, has become a wind power showcase for the world. Hot water is produced by wood chips and straw provided by local farmers, as well as by solar

> **alternative energy** Alternatives to traditional fossil fuels, including hydroelectric, solar, wind, geothermal, nuclear, biofuel, and "clean" coal.

panels. As a result, Samsø can boast its status as carbon-neutral and a net exporter of clean electricity (**Figure 9.31**. Nations the world over—North and South—are at the vanguard of developing alternative energy sources to fuel their activities.

China's Harmful Air Quality China provides an example of the negative environmental consequences of industrialization. Rather than pursuing clean, alternative, innovative sources of energy, China has decided to chart the same course, energy-wise, as the now wealthy nations of the North did to fuel their industrial development. In 2006 China's greenhouse gas emissions exceeded those of the United States for the first time, and the gap has continued to accelerate since then. China is now the world's biggest emitter of pollution.

The resulting air contamination has caused regular shutdowns of businesses, public places such as schools, and transportation venues from bus service to international airports in cities such as Shanghai and Beijing (**Figure 9.32**). Not only is human health severely affected by the thick soup of particulate

FIGURE 9.32 **China's deteriorated air quality.** This photo was taken at Beijing's Tiananmen Square during the winter, when air contamination from burning coal is at its worst. *(Lintao Zhang/Getty Images.)*

matter given off by cars, buses, household heating and cooking, and industry; visibility is also reduced so much that drivers cannot see across a roadway, or a pilot down a runway to safely land a plane. On a particularly bad day in October 2013, the level of fine particulate matter—those particles small enough to lodge deep inside human lungs—as measured in the northeastern city of Harbin was 40 times higher than the international safety standard set by the World Health Organization.

China's air (and water and ground) pollution is the result of its rapid, sustained rate of economic growth, which has averaged 10 percent annually since the late 1970s. This growth has lifted many Chinese out of poverty, and has transformed the country's cultural landscape from a primarily rural to a primarily urban one. Yet, the economic toll when business and transportation activities must be halted due to contamination is steep, as are the long-term costs to human health. The National Academy of Sciences reported that a person alive in the 1990s in northern China will live, on average, 5.5 years less than her counterpart in southern China. This is directly attributable to air contamination in northern China where industrial activities are centered. During the harsh northern winters, people routinely burn

FIGURE 9.31 **Europe's renewable energy island.** Samsø, located off the coast of Denmark, is a model of alternative energy in Europe. This photo depicts some of the island's wind-powered turbines, which supply 100 pecent of the 4,200 residents' electrical needs. Solar power and wood chips are used to provide heating. In all, Samsø removes more carbon dioxide from the atmosphere than it contributes. *(Slim ALLAGUI/AFP/Getty Images.)*

traditional (not "clean") coal—a highly contaminating fuel source—to keep warm. Financial experts attribute the recent slowdown in China's economic growth, to around 7 percent in 2012, to the economic costs of pollution resulting in disruptions to business, protests leading to factory closures, and increased health care expenses.

In 2012 China released a national plan to reduce its carbon emissions and to utilize more nonfossil fuels. In addition, the large northern industrial cities are participating in a **carbon cap and trade** program. These programs limit (cap) carbon emissions and lower the limit over time, as well as providing economic incentives for companies to find innovative ways to remain below their carbon cap. If a company finds it difficult to meet its cap, that company can buy allowances from other companies that find it easier to reduce their emissions. In this case, a sort of "carbon market" develops.

> ■ **carbon cap and trade**
> Programs that limit (cap) carbon emissions, and lower the limit over time, as well as providing economic incentives for companies to find innovative ways to remain below their carbon cap.

FIGURE 9.33 **Mining landscape.** This image shows a conveyor moving ore from an open pit mine to a processing plant. Such bleak landscapes are common in areas of primary industry, which involves extracting natural materials from the Earth. *(Thomas Moore/iStock/360/Getty Images.)*

⊞CULTURAL LANDSCAPE

9.5 LEARNING OBJECTIVE
Observe how development projects have shaped the landscapes around us.

It is difficult to look at any particular landscape and not see some evidence of economic development. Factories, shopping malls, housing developments, industrial agriculture, cities: all provide examples of landscapes that arise over the course of economic development. There is literally no place on Earth that has not been touched in some way by human activity.

Primary industries produce perhaps the most drastic impacts. The resulting landscapes contain slag heaps, clear-cut commercial forests, strip-mining and open-pit mining scars, and forestlike clusters of oil derricks (**Figure 9.33**). The destruction of nature is less apparent in other types of primary industrial landscapes. The fishing villages of Portugal and Newfoundland continue to attract tourists even though the fisheries are depleted (**Figure 9.34**). In still other cases, efforts are made to restore the preindustrial landscape or to repurpose it into something more attractive. Examples include the establishment of artificial new grasslands in old strip-mining areas of the American Midwest, and the creation of recreational ponds in old mine pits along interstate highways.

Among the most obvious features of the landscapes of secondary sector activities (manufacturing) are factory buildings. Early- to mid-nineteenth-century industrial landscapes are easy to identify, because the technologies that ran these factories were reliant on water power. Hence, the mill buildings that housed the machinery were designed in linear form to take full advantage of the turbines that were connected to waterwheels. Given this locational requirement, these mill complexes were most often built in rural areas. Entire communities, usually with housing for the workers, thus had to be constructed along with the mills. These mill towns dotted the landscape of England, Scotland, and New England, the site of the first industrial developments in the United States (**Figure 9.35** and Mona's Notebook, page 405). Later—as water power was supplanted by steam, coal, and then electric power—factories were located near urban areas, taking advantage of the housing supply already there and the proximity to a large consumer base. These industrial landscapes were often located at the edge of downtowns, lining the railroad routes into the city, and

Given the degree of deindustrialization in certain parts of the old core industrial regions, it is not surprising that many of these factory complexes, particularly those that date from the nineteenth and early twentieth centuries, are derelict or are being repurposed for housing or commercial uses. Others now house historical museums depicting past industrial technologies and ways of life. Lowell, Massachusetts, an early (mid-nineteenth-century) planned industrial town that produced textiles, is now the site of a national park. A good percentage of its factories, canals, and housing complexes have been preserved and can be toured by visitors. In Great Britain, many sites of industrial history are now preserved as museums (see World Heritage Site, pages 398–399).

Service industries, too, produce a cultural landscape. Its visual content includes elements as diverse as high-rise bank buildings, fast-food restaurants, gasoline stations, strip malls, and the concrete and steel webs of highways and railroads. Some highway interchanges can best be described as a modern art form, but perhaps the aesthetic high point of the industrial landscape may be found in bridges, which are often graceful and beautiful structures. The massive investment in these transportation systems has even changed the way we view the landscape. As geographer Yi-Fu Tuan

FIGURE 9.34 **A fishing village in Newfoundland.** Primary industrial landscapes can be pleasing to the eye. *(Courtesy of Terry G. Jordan-Bychkov.)*

were surrounded by working-class housing. In the second half of the twentieth century—with the development of the interstate highway system and the trucking industry, as well as the switch to high-tech industries such as electronics—industrial landscapes took on a different form. Factories began to move to industrial parks at interstate exchanges, and their architecture began to resemble other types of mass-produced architecture (**Figure 9.36**).

FIGURE 9.35 **Mill buildings in central Massachusetts.** Most textile mills in New England were abandoned by the mid-twentieth century, as textile production moved to southern states and then to countries outside the United States. *(Courtesy of Mona Domosh.)*

FIGURE 9.36 **Footwear factory in Picardie, France.** This factory is located right along the highway, providing easy access for its employees and for the trucks that transport its products. *(FORESTIER YVES/SYGMA/ CORBIS.)*

commented, "In the early decades of the twentieth century vehicles began to displace walking as the prevalent form of locomotion, and street scenes were perceived increasingly from the interior of automobiles moving staccato-fashion through regularly spaced traffic lights." Yet, the ubiquity of such landscapes throughout the world has led to a sense of loss of the distinctiveness from one place to another. What geographers call **placelessness**—a disorienting sense of sameness in the cultural landscape— is on the rise (see Chapter 2).

> **placelessness** A disorienting sense of sameness in the cultural landscape.

Producer services related to financial activities, such as legal services, trade, insurance, and banking, traditionally were located in high-rise buildings in urban centers, but with suburbanization they have taken on a nonurban form. Many are now located in five- or six-story buildings, along the interstates surrounding cities, in what we've called high-tech corridors. Other producer service industries choose to maintain their downtown location for symbolic reasons. Some consumer services, particularly retailing, have created distinctive and, within the American context at least, socially important landscapes. Shopping malls are now dominant features of the North American suburb and often serve as catalysts for suburban land development, in effect creating entirely new landscapes, all geared toward consumption. As more and more of this activity moves online, it is interesting to contemplate what sorts of landscapes will be created.

Indeed, the quaternary sector landscapes of intellectual and informational activities—including the development of technology that allows the other sectors to conduct business in an online environment—are notoriously difficult to see. Yet, these landscapes do exist. Named after a place that founder Jeff Bezos thought to be "exotic and different," Amazon.com began life as an online bookstore. Today, you may purchase just about anything you can think of from Amazon.com. Its headquarters are located in Seattle, Luxembourg, and Belgium. Nevertheless, its physical locations also include a vast global network of mammoth warehouses and fulfillment centers, and software development and customer service offices. Much of Amazon.com's business, however, is conducted in noncommercial locations such as households: with

FIGURE 9.37 **Delivery by drone.** This drone is a prototype used by the online retailer Amazon.com. Though the service has not yet been approved by the Federal Aviation Administration, Amazon envisions using drones to deliver packages to customers' doorsteps. (AMAZON/UPI/Newscom.)

customers ordering online in their pajamas, for instance. It also occurs in nonplaces. Think, for example, of Amazon's cloud computing services, or the books and video only available in digital format. The company is envisioning same-day delivery of orders via drone in some urban markets as early as 2015 (**Figure 9.37**). This reshaping of the quaternary sector cultural landscape echoes the vertical landscapes of privilege discussed previously.

CONCLUSION

Economic development has profoundly reshaped each of the five themes of contemporary human geography. Regions are defined and reconfigured on the basis of their relative location on the axis of development. The opportunities afforded by economic development facilitate human mobility, but also constrain it. In many ways, economic development is by definition a global phenomenon. Nature, its transformation, and the relationship of nature to culture have since the beginning constituted a central dimension of economic development. And finally, economic development activities have literally manufactured landscapes in their service, transforming the world around us.

Mona's Notebook

Imagining the New England Landscape

Mona Domosh. *(Courtesy of Mona Domosh.)*

When most Americans imagine the New England landscape, the picture that comes to mind often resembles the December page of an illustrated calendar—church steeples, white clapboard houses, village greens. Yet, we know that America's Industrial Revolution started in New England, first in areas close to Boston, like Lowell and Lawrence, but even as far north as Hanover, New Hampshire, where I teach at Dartmouth College. The water power needed to run mills could be supplied by the considerable number of rivers large and small, tempting industrialists to set up shop. When industries moved to urban areas (and, later, to different parts of the country), many of these mills went out of business, but still it's difficult to drive more than a couple of miles from where I live without seeing some relic of the region's industrial past: waterwheels to harness the power of the river; long, red-brick mill buildings; and rows of identical small houses for the mill workers.

Through my geographical reading and travels in the region, I've learned to "see" and appreciate this industrial landscape, and I've realized that it characterizes the New England landscape as much as white church steeples do. I've also learned that it is common for images of places to not always correspond to their geographical realities. For example, think of the place you live in and of its different landscapes. How do people who haven't visited your town or region describe it? Do you think their image corresponds to your own? Often outsiders' images of places and regions are based on commercial representations like those on TV, films, postcards, and guidebooks (and calendars!). Part of what is so exciting about seeing with a geographer's eye is that we can often see beyond these images to what is sometimes hidden to other people.

Some of the existing buildings that housed the Lebanon Woolen Mill located along the Mascoma River in Lebanon, New Hampshire. These buildings are now home to offices and some small shops. *(Courtesy of Mona Domosh.)*

GEOGRAPHY @ WORK

Matthew Toro. *(Miguel Asencio.)*

Matthew Toro

Maps and Imagery Services (MIUS) Coordinator and Research Associate, Florida International University

Education:
MA Geography—University of Miami

BA Geography and International Relations—Florida International University

Q. *Why did you major in geography and decide to pursue a career in the geography field?*

A. I wanted to understand the way the world works, and no other field seemed to offer the sort of comprehensive, multidimensional, multiscale answers afforded by the space-based perspective of my geography courses. Other disciplines I explored seemed to look at one or a few isolated components of social or "natural" reality. Geography was able to interweave seemingly unrelated human and geophysical systems, helping me see more completely the way the world system operates. I'll never succeed at ever comprehending the world in its entirety, but my geographic perspective gives me an insight that I'm confident no other field can match.

Q. *Please describe your job.*

A. As the MIUS Coordinator and Research Associate here at the Florida International University GIS Center, my primary responsibilities include teaching free and fee-based workshops on the applications of geospatial technologies (e.g., GIS, remote sensing, GPS, cartography geovisualization) to the FIU student and faculty community. I also perform a range of geospatial data acquisition, geoprocessing, and analysis tasks for various clients based within and outside of my university. Essentially, I'm here to serve the university community in optimizing their use of maps, imagery, and geospatial datasets. I'm also sure to pursue my own academic and community-based research projects.

Q. *How does your geographical background help you in your day-to-day work?*

A. My job is at the heart of geographical research, both my own, and that of others wishing to take advantage of digital data like remotely sensed raster imagery and vector files used for mapping and spatial analysis. The work is often technical in nature, and virtually every skill I possess came from my geography coursework and insights from my professors, including theoretical courses meant to sharpen one's critical thinking skills. It's important to match technical skills with fundamental information-processing skills.

Q. *In terms of employment, what advice do you have for students considering a career in the geography field?*

A. Exploit geography's multi- and transdisciplinary nature to the fullest. Gain exposure to multiple sorts of research methods, while specializing in a couple that will help you realize your long-term research goals. Be entrepreneurial and ambitious, and don't be afraid to market yourself. Appreciation for geographers—and how we think and what we can do—is growing tremendously across the globe.

DOING GEOGRAPHY

The Where and Why of What You Wear

The textile and garment industries are a large component of some countries' secondary sector economic activities. China, Vietnam, Bangladesh, and Sri Lanka, for instance, are several of the world's most significant garment producers. Textile and garment industries have historically been very mobile, able to move production facilities to locations that suit manufacturers, often because of the low cost of labor. In today's global economy, that tendency is even more pronounced, with manufacturers of clothing and other garment-related materials outsourcing many aspects of production and often subcontracting with other companies to complete different tasks in various parts of the world. It's possible, then, that the clothes and shoes you are wearing right now were designed in one place, woven into fabric in another place, and assembled in yet another.

This exercise is about tracing the "origins" of the clothes and shoes you are wearing right now and asking "Why?" The cost of labor, as we've learned, is important, but it certainly isn't the only factor in determining where garments and shoes are made. In manufacturing in general, other factors include the locations of markets as well as state and international policies, such as NAFTA. Sometimes it is the lack of policies—or lack of enforcement—that manufacturers find attractive. This can lead to unsafe conditions for laborers, as happened in Bangladesh in 2013 when a factory building collapsed, killing 1200 garment workers.

For clothing and shoes, another important factor is fashion. Styles change often, and there might be a need for manufacturers to be able to change their production quickly. This would lead companies to locate manufacturing facilities close to their main markets, which might mean in the United States or Canada.

Steps to Tracing the Origins of What You Wear

Step 1: Locate as many labels as you can from the shoes and all the items of clothing you are wearing right now.

Step 2: Read the labels carefully and create a list of the places mentioned in the "Made in" section of the labels.

Step 3: Locate those places on a map. This step alone should give you a good sense of the global nature of this industry.

Now that you have a better understanding of the global nature of the clothing industry, consider these questions:

- Why do you think certain items are manufactured in particular places?
- What are the similarities in these countries? What are the differences?

Consider each of the locational factors we've mentioned—labor, markets, state policies, consumer trends—and try to generalize the why of what you wear.

Workers in Bangladesh (*top*), produce clothing that more than likely will end up in the stores where this young woman (*bottom*) is shopping. *Think about all the other processes (packaging, shipping, advertising, etc.) and other forms of labor that enable clothing produced in Bangladesh to make its way to your mall (or mailbox).* (Top: MUNIR UZ SAMAN/AFP/Getty Images; Bottom: KidStock/Blend Images/Getty Images.)

SEEING GEOGRAPHY
Mapping the Life Course of a Smartphone
Have you ever wondered where your smartphone started its life, and where it will go when it dies?

Clockwise from top left: China's Mongolia mining region; Foxconn workers assembling smartphones; using the latest iPhone; children sifting through electronic waste in Manila, Philippines. *(Ren Junchuan/Xinhua/Sipa Press/Newscom; YM YIK/EPA/Newscom; Sergio Azenha/Alamy; REUTERS/Cheryl Ravelo/ Newscom.)*

If you are like most students, your smartphone never leaves your side. Perhaps you keep it nearby even when you are asleep! Given your current inseparability, it is perhaps disconcerting to think about where exactly your smartphone was before you got it, and where it will go when it has reached the end of its useful life with you.

The photo montage at the beginning of this chapter depicts some of these places. To begin with, no smartphone would be truly cool without so-called rare earth elements. Yttrium, lanthanum, terbium, cerium, and five other elements are used to manufacture a smartphone's screen, speakers, and vibration unit. The properties of rare earth elements are what make your phone "light, bright, and loud," according to Cecilia

Jamasmie. Ninety-five percent of rare earth minerals are mined in China's Inner Mongolia region, both because these elements are concentrated there but also because of the environmental damage caused by the techniques used to extract them, which are not allowed in other places.

The actual assembly of your smartphone occurs in a factory setting. Foxconn is the largest manufacturer, making smartphones for Apple and BlackBerry, as well as other electronic devices. Foxconn's largest factory, shown in the chapter opening, is located in Longua, a city in China's Shenzen province. No one knows exactly how many employees labor here, but it is estimated that nearly half a million people work at this factory. In fact, this place is commonly referred to as "Foxconn City" (or "iPod City"). Workers sleep in company dormitories, shop in stores, and engage in leisure activities, all located within the boundaries of the factory premises. Foxconn has been criticized for harsh labor practices resulting in employee suicides, unrest, and exposure to dangerous conditions. The stress of meeting production demands when, for instance, a new iPhone version is released can be overwhelming for employees. Literally millions of new phones can be sold in a single weekend.

What happens to old smartphones? According to a survey conducted by Google in 2013, 51 percent simply end their lives in a drawer or closet. Some are refurbished and resold, while others are simply too old or broken for that. These turn into **electronic waste.** Because of the hazardous components in smartphones, contamination of water and soil can result if they are simply thrown into the trash. Some phones or components are sent to places like Manila, in the Philippines, where children disassemble them to scavenge copper wire, and then burn the rest. Burning e-waste is an increasing health hazard in places such as the one shown here.

> **electronic waste**
> Used electronics, such as mobile phones, computers, office equipment, television sets, and refrigerators, which can no longer be used for their intended purpose and are discarded. Also known as e-waste.

<table>
<tr><td colspan="2">Chapter 9
LEARNING OBJECTIVES REEXAMINED</td></tr>
</table>

Chapter 9
LEARNING OBJECTIVES REEXAMINED

9.1 Discuss the history of development and how it has shaped, and reshaped, regions.
See pages 372–380. What is meant by the First, Second, Third, and Fourth Worlds? How have economic advances shaped how we view these "worlds"?

9.2 Specify the ways in which mobilities of various sorts—of individuals as well as things and ideas—are important for development.
See pages 380–387. What impact does the U.S. transportation infrastructure have on development today?

9.3 Identify how formerly poor places are today significant agents of development.
See pages 388–396. What is meant by the Global South and what have been its effects in the world?

9.4 Recognize the role that nature has played over time as concepts and practices of development have changed.
See pages 396–402. What is *global climate change* and how is it impacting the natural environment?

9.5 Observe how development projects have shaped the landscapes around us.
See pages 402–404. What are some significant changes to the landscape as a result of development?

KEY TERMS

Development Geography on the Internet

You can learn more about economic geography on the Internet at the following web sites:

Center for Global Development
http://www.cgdev.org/page/mdg-progress-index-gauging-country-level-achievements
On this website, the Center for Global Development provides information on progress toward achieving MDG goals. This link leads to an interactive map that can be sorted by goal.

Digital Death
http://buylife.org/browse-sacrifices.php
This is the official site for the "Is Dead" Campaign launched on World AIDS Day, 2010. Clicking on a celebrity's casket photo brings up a link to the "last tweet and testament" videos.

ForeignAssistance.gov
http://foreignassistance.gov/countryintro.aspx
Interactive map shows U.S. foreign assistance by country for the year you choose.

In Plain Sight: Poverty in America
http://www.inplainsight.nbcnews.com/
This NBC News blog chronicles poverty and inequality in the United States.

TED
http://www.ted.com/talks/hans_rosling_on_global_population_growth.html
The Swedish scholar Hans Rosling simply illustrates the connections between global population growth, health, and well-being using Ikea storage boxes as props.

Sources

Anderson, Kyle, and Gil Kaufman. 2010. "Looking Back at Live Aid, 25 Years Later: We Remember the Charity Concert That Broke the Mold on July 13, 1985." *MTV Online,* 13 July. http://www.mtv.com/news/articles/1643506/looking-back-at-live-aid-25-years-later.jhtml.

Austin, Gareth. 2010. "African Economic Development and Colonial Legacies." *International Development Policy* 1: 11–32.

Booz & Company. 2012. *Empowering the Third Billion: Women and the World of Work in 2012.* http://www.booz.com/media/file/BoozCo_Empowering-the-Third-Billion_Full-Report.pdf.

Bridgman, Benjamin, Andrew Dugan, Mikhael Lal, Matthew Osborne, and Shaunda Villones. 2012. "Accounting for Household Production in the National Accounts, 1965–2010." *Survey of Current Business* 92(5): 23–36.

Dolcourt, Jessica. 2012. "Your Smartphone's Secret Afterlife." *CNET,* 2 December. http://www.cnet.com/8301-17918_1-57556225-85/your-smartphones-secret-afterlife-smartphones-unlocked/.

Gose, Ben, Sarah Frostenson, and Marisa López-Rivera. 2013. "10 Companies That Gave the Most Cash in 2102." *The Chronicle of Philanthropy,* 14 July. http://philanthropy.com/article/10-Companies-That-Gave-the/140261/.

Greene, Jay. 2012. "Riots, Suicides, and Other Issues in Foxconn's iPhone Factories." *CNET,* 25 September. http://news.cnet.com/8301-13579_3-57515968-37/riots-suicides-and-other-issues-in-foxconns-iphone-factories/.

Grigoriadis, Vanessa. 2011. "Our Lady of Malawi." *New York Magazine,* 1 May. http://nymag.com/news/features/madonna-malawi-2011-5/.

Jamasmie, Cecilia. 2013. "Infographic: The Periodic Table of Smartphones." *Mining.com,* 5 February, http://www.mining.com/infographic-the-periodic-table-of-smartphones-56390/.

Lagarde, Christine. 2013. "Women and the World Economy." *Project Syndicate,* 24 September. http://www.project-syndicate.org/commentary/how-to-increase-women-s-participation-in-the-workforce-by-christine-lagarde.

Lim, Hank. 2008. "Infrastructure Development in Singapore." In N. Kumar (ed.), *International Infrastructure Development in East Asia: Towards Balanced Regional Integration and Development,* pp. 228–262. ERIA Research Project Report 2007-2. Chiba, Japan.

Mwenda, Andrew M. 2013. "Madonna and Africa's 'Celebrity Saviors.'" *CNN,* 17 April. http://www.cnn.com/2013/04/16/opinion/madonna-charity-africa-mwenda/.

Peter, Tom A. 2010. "Finding a Better Way to Bridge the Digital Divide." *Christian Science Monitor,* 2 June. http://www.csmonitor.com/Innovation/Tech/2010/0602/Finding-a-better-way-to-bridge-the-digital-divide.

Puffert, Douglas J. 2009. *Tracks across Continents, Paths through History: The Economic Dynamics of Standardization in Railway Gauge.* Chicago: University of Chicago Press.

Puri, Hardeep S. 2010. "Rise of the Global South and Its Impact on South-South Cooperation." *Development Outreach,* 18 October. Washington, D.C.: World Bank Institute. http://siteresources.worldbank.org/WBI/Resources/213798-1286217829056/puri.pdf.

Rodgers, Dennis. 2004. "'Disembedding' the City: Crime, Insecurity and Spatial Organization in Managua, Nicaragua." *Environment and Urbanization* 16: 113–123.

Rostow, W. W. 1962. *The Process of Economic Growth.* New York: W.W. Norton.

Sen, Amartya. 1999. *Development as Freedom.* New York: Alfred A. Knopf.

The Economist. 2011. "America's Transport Infrastructure: Life in the Slow Lane." 28 April. http://www.economist.com/node/18620944.

Torero, Maximo, and Shyamal Chowdhury. 2005. *Increasing Access to Infrastructure for Africa's Rural Poor.* Washington, D.C.: International Food Policy Research Institute.

Trachtenberg, Stephen Joel, Gerald B. Kauvar, and E. Grady Bogue. 2013. *Presidencies Derailed: Why University Leaders Fail and How to Prevent It.* Baltimore: Johns Hopkins University Press.

Trivett, Vincent. 2011. "25 US Mega Corporations: Where They Rank If They Were Countries." *Business Insider,* 27 June. http://www.businessinsider.com/25-corporations-bigger-than-countries-2011-6?op=1.

Tuan, Yi-Fu. 1989. "Cultural Pluralism and Technology." *Geographical Review* 79: 269–279.

Urbina, Ian. 2013. "U.S. Flouts Its Own Advice in Procuring Overseas Clothing." *The New York Times,* 23 December. http://www.nytimes.com/2013/12/23/world/americas/buying-overseas-clothing-us-flouts-its-own-advice.html.

Vince, Gaia. 2012. "How the World's Oceans Could Be Running Out of Fish." *BBC News,* 21 September. http://www.bbc.com/future/story/20120920-are-we-running-out-of-fish/all.

World Bank. 2012. *Information and Communications for Development: Maximizing Mobile.* Washington, D.C.

Zerbisias, Antonia. 2010. "Feminomics: Calculating the Value of 'Women's Work.'" *Toronto Star,* 30 October. http://www.thestar.com/news/insight/2010/10/30/feminomics_calculating_the_value_of_womens_work.html.

Ten Recommended Books on Development Geography

(For additional suggested readings, see the *Contemporary Human Geography* LaunchPad: http://www.macmillanhighered.com/launchpad/DomoshCHG1e.)

Alcock, Susan E., John Bodel, and Richard J. A. Talbert (eds.). 2012. *Highways, Byways, and Road Systems in the Pre-Modern World.* Hoboken, N.J.: John Wiley & Sons.

Collins, Daryl, Jonathan Morduch, Stuart Rutherford, and Orlanda Ruthven. 2009. *Portfolios of the Poor: How the World's Poor Live on $2 a Day.* Princeton, N.J.: Princeton University Press. Basing their findings on extensive interviews with poor people in Bangladesh, India, and South Africa, the authors argue that the poor utilize effective systems of microfinance to provide for their families and plan for the future.

Diamond, Jared. 2005. *Collapse: How Societies Choose to Fail or Succeed.* New York: Viking Press. Climate change, environmental problems, and failure to adapt to environmental changes are identified as three of the top five reasons why even the most powerful civilizations die out.

Kapoor, Ilan. 2012. *Celebrity Humanitarianism: The Ideology of Global Charity.* London: Routledge. Celebrities, according to Kapoor, engage in so-called humanitarian causes primarily to enhance their personal brands, without truly confronting the global injustices that they seek to redress.

Kristof, Nicholas D., and Sheryl WuDunn. 2010. *Half the Sky: Turning Oppression into Opportunity for Women Worldwide.* New York: Vintage Books. The authors approach women's rights as human rights in their exploration of injustices and innovations in violence against women, human trafficking, maternal health, and finance.

Mortenson, Greg, and David Oliver Relin. 2006. *Three Cups of Tea: One Man's Mission to Fight Terrorism and Build Nations . . . One School at a Time.* New York: Penguin. Set in rural Afghanistan and Pakistan, this book explores the connections between education, development, and the rejection of terrorism.

Sen, Amartya. 1999. *Development as Freedom.* New York: Alfred A. Knopf. Despite its liberating premise, globalization has not led to individual freedom for many of the world's inhabitants. Sen argues that freedoms both economic and political are the key to ethical societal advancement.

Stiglitz, Joseph. 2012. *The Price of Inequality: How Today's Divided Society Endangers Our Future.* New York: W.W. Norton. The fact that the wealthiest 1 percent of Americans controls 40 percent of the nation's wealth illustrates how, within the United States., there are forces of underdevelopment at work that if left unaddressed will undermine the nation's achievements.

Turkle, Sherry. 2011. *Alone Together: Why We Expect More From Technology and Less From Each Other.* New York: Basic Books. Presents the other side of the digital divide, by arguing that having too much access to technology can paradoxically impoverish us socially and emotionally.

Wimmer, Nancy. 2012. *Green Energy for a Billion Poor.* Germany: MCRE Verlag. Operating in rural Bangladesh, the Grameen Shakti Company helps local entrepreneurs and engineers establish sustainable social businesses bringing solar energy to poor customers.

Journals in Development Geography

Economic Geography. Published by Clark University. Volume 1 was published in 1925.

Geography, Planning and Development. Published by Elsevier Science. Volume 1 was published in 1985.

Journal of Transport Geography. Published by Elsevier Science. Volume 1 was published in 1993.

What are some of the major environmental and social impacts of an increasingly urbanized world?

A view of Rio de Janiero, Brazil, looking toward Sugarloaf Mountain. *(isitsharp/Vetta/Getty Images.)*

Go to "Seeing Geography" on page 453 to learn more about this image.

URBAN GEOGRAPHY
A World of Cities

Imagine the 2 million years that humankind has spent on Earth as a 24-hour day. In this framework, settlements of more than a hundred people came about only in the last half hour. Towns and cities emerged only a few minutes ago. Yet, it is during these "minutes" that we see the rise of civilization as we know it. *Civitas,* the Latin root word for *civilization,* was first applied to settled areas of the Roman Empire. Later it came to mean a specific town or city. *To civilize* meant literally "to citify."

Today, just over half—54 percent—of human beings live in cities. By the year 2050, the United Nations estimates that this figure will stand at 70 percent (**Figure 10.1**). The World Bank estimates that cities grow by a combined total of 3 million people each week. The cultural geography of the world will change dramatically as we become a predominantly urban people. (Figure 10.1)

In this chapter, we apply the five themes of human geography to consider overall patterns of urbanization, learn how urbanization began and developed, and discuss the differing forms of cities in the developing and developed worlds. We will see the many ways that migration has shaped the world's urban population, as well as how mobility within cities is facilitated or hindered. The spectacular rise of the so-called global city, along

LEARNING OBJECTIVES

10.1 Describe the historical and contemporary regional patterns of urbanization, and the models geographers use to understand cities.

10.2 Identify the ways that migration has shaped the growth of cities, how cities have grown and changed, and how humans get around within cities.

10.3 Discuss the notion of global cities and some of the problems associated with them.

10.4 Recognize the ways that human and natural worlds interact in cities, and the effects on the health and well-being of those living in cities.

10.5 Read an urban landscape and identify the trends shaping cities today and into the future.

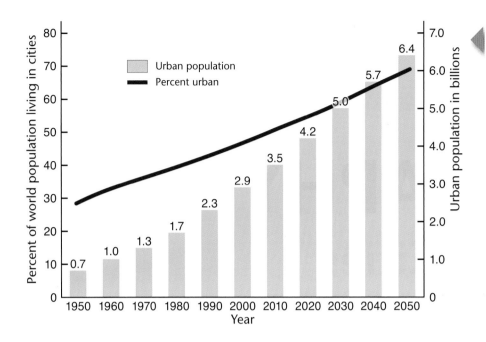

FIGURE 10.1 **People living in cities.** This graph depicts the changing percentage, and total number, of people living in urban areas across the century spanning 1950 to 2050. *(Source: United Nations, Department of Economic and Social Affairs, Population Division.)*

with its universally recognizable form but also the many social problems that go along with it, will be considered. The interaction of built environments and natural elements and processes that occur in cities is discussed. Finally, we turn to processes currently shaping cities, as well as how you might "read" urban landscapes from a cultural geography perspective.

REGION

10.1 **LEARNING OBJECTIVE**
Describe the historical and contemporary regional patterns of urbanization, and the models geographers use to understand cities.

We all know from our own travels, locally or internationally, that some regions contain lots of cities, while others contain relatively few, and that the size, population density, and look and feel of those cities can vary greatly. How then can we begin to understand the location, distribution, and size of

cities? We start with a consideration of global patterns of urbanization, examining the general distribution of urban populations around the world. A quick look at **Figure 10.2** reveals differing patterns of **urbanized population**—the percentage of a nation's population living in towns and cities—around the world. For example, the countries within Europe, North America, Latin America, and the Caribbean have relatively high levels of urbanization, with approximately 75 percent of each country's population living in urban areas. The nations of Africa and Asia, on the other hand, are less urbanized, with many countries in these regions having 40 percent or less of their population residing in urban areas. How do geographers explain these varying regional patterns of urbanization?

> **urbanized population** The proportion of a country's population living in cities.

PATTERNS AND PROCESSES OF URBANIZATION

According to United Nations estimates, almost all the worldwide population growth in the next several decades will be concentrated in urban areas, with the cities of the less developed regions accounting for most of that increase (**Figure 10.3**). The reasons for

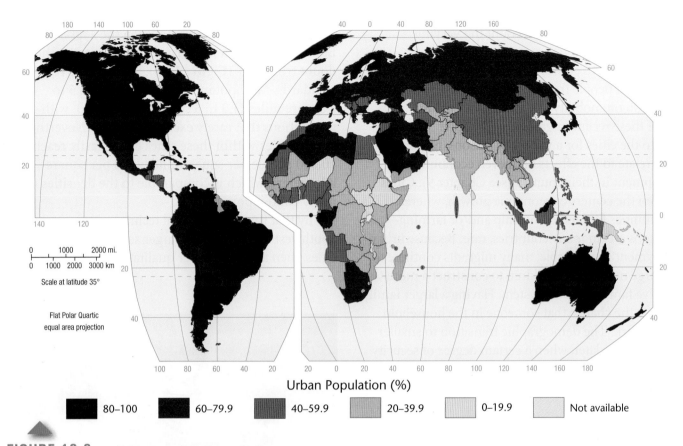

Urban Population (%)

■ 80–100	■ 60–79.9	▨ 40–59.9	▨ 20–39.9	▨ 0–19.9	□ Not available

FIGURE 10.2 **Urbanized population in the world.** Just over half—54 percent—of the world's inhabitants live in cities. *(Source: United Nations, Department of Economic and Social Affairs, Population Division.)*

this urban population growth and its uneven distribution around the world vary, as each country's unique history and society present a slightly different narrative of urban and economic development. Making matters more complex is the lack of a standard definition of what constitutes a city. Consequently, the criteria used to calculate a country's urbanized population differ from nation to nation. Using data based on these varying criteria would result in misleading conclusions. For example, the Indian government defines an urban center as an area having 5000 inhabitants, with an adult male population employed predominantly in nonagricultural work. In contrast, the U.S. Census Bureau defines a city as a densely populated area of 2500 people or more, and South Africa counts as a city any settlement of 500 or more people. It is important to remember, then, that an international

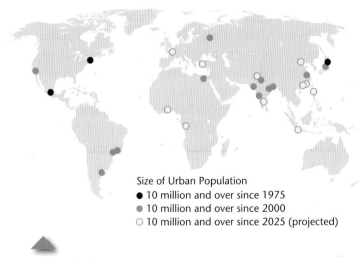

Size of Urban Population
● 10 million and over since 1975
● 10 million and over since 2000
○ 10 million and over since 2025 (projected)

FIGURE 10.3 **The world's largest cities in 1975, 2000, and 2025.** Large cities are located in almost every world region. However, Asia and Africa will be home to the large cities of the future. *(Source: Population Reference Bureau.)*

comparison of urbanized population data can be made only by taking into account the varying definitions of a city.

Urban growth comes from two sources: the migration of people from rural areas to cities (see also the section on mobility) and the higher natural birth rates of these recent migrants (see Chapter 3). People move to the cities for a variety of reasons, most of which relate to the effects of uneven economic development in their country (see Chapter 9). Cities are often the centers of economic growth, whereas opportunities for land ownership and/or farming-based jobs are, in many countries, rare. Because urban employment is unreliable, many migrants continue to have large numbers of children to construct a more extensive family support system. Having a larger family increases the chance that someone in the household will get work. The demographic transition to smaller families comes later, when a certain degree of security is ensured. Often, this transition occurs as women enter the workforce (see Chapter 3).

A BRIEF HISTORY OF URBANIZATION

Humans have not always lived in cities. Indeed, for most of our existence we humans have lived in small, nomadic groups of 25 to 50 hunter-gatherers. Such groups eventually became sedentary and formed into small villages, but these were nothing like what you would recognize as a real city today.

In the Middle East, where the first cities appeared, a network of permanent agricultural villages developed about 10,000 years ago. These farming villages were modest in size, rarely with more than 200 people, and were probably organized on a kinship basis.

Although small farming villages predate cities, it is wrong to assume that a simple quantitative change took place whereby villages slowly grew, first into towns and then into cities. Neolithic settlements that date from around 9000 B.C.E. (Jericho, in the modern-day Palestinian territories on the West Bank) and 7500 B.C.E. (Çatalhöyük, in modern-day Turkey) are considered by some archaeologists to be the earliest real cities (**Figure 10.4**). Others do not believe they qualified as true cities, based on their small populations and lack of apparent social stratification.

Urban settlements dating to around 4000 B.C.E. in Mesopotamia (the area between the Tigris and Euphrates rivers in modern-day Iraq) were without doubt true cities (**Figure 10.5**). Small by current standards, Mesopotamian cities covered 0.5 to 2 square miles (1.3 to 5 square kilometers), with populations that rarely exceeded 30,000. Nevertheless, the densities within these cities could easily reach 10,000 people per square mile (4000 per square kilometer), which is comparable to the densities in many contemporary cities.

By that time, agricultural technologies in Southwest Asia permitted larger settlements. True cities, then and now, differ qualitatively from

FIGURE 10.4 **Ancient urban dwelling.** A group of archaeologists uncovers the interior of a small home in the ancient Anatolian city of Çatalhöyük, revealing a living room and small bedrooms. Excavations have revealed that people kept their homes scrupulously clean as very little trash has been uncovered. Occupants entered and exited their homes through holes in the roof. Since there was no space between houses, people got from house to house by walking on the rooftops as if they were streets. *(Images & Stories/Alamy.)*

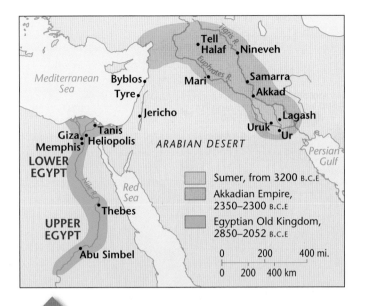

FIGURE 10.5 **Early Mesopotamian and Egyptian cities.** Note the location of these earliest cities: almost all are situated along rivers.

agricultural villages. All the inhabitants of agricultural villages were involved in some way in food procurement—tending the agricultural fields or harvesting and preparing the crops. Cities, however, were more removed, both physically and psychologically, from everyday agricultural activities. Food was supplied to the city, but not all city dwellers were involved in obtaining it.

Two elements were necessary for this dramatic social change: the creation of an **agricultural surplus** and the development of a stratified social system. Surplus food, which is a food supply larger than the everyday needs of the agricultural labor force, is a prerequisite for supporting nonfarmers—people who work at administrative, religious, military, or handicraft tasks. Social stratification, understood as the existence of distinct socioeconomic classes, facilitates the collection, storage, and distribution of resources through well-defined channels of authority that can exercise control over goods and people. A society with these two elements—surplus food and a means

> **agricultural surplus** The amount of food grown by a society that exceeds the demands of its population.

of storing and distributing it—was prepared for urbanization.

MODELS FOR THE RISE OF CITIES

One way to understand the transition from village to city life is to model the development of urban life assuming that a single factor is the trigger behind the transition. The question that scholars ask is: "What activity could be so important to an agricultural society that its people would be willing to give some of their surplus to support a social class that specializes in that activity?"

The **hydraulic civilization** model, developed by Karl Wittfogel, assumes that the development of large-scale irrigation systems was the primary driver of urbanization and that a class of technical specialists were the first urban dwellers. Irrigating agricultural crops yielded more food, and this surplus supported the development of a large nonfarming population. A strong, centralized government backed by an urban-based military could expand its power into the surrounding areas. Farmers who resisted the new authority were denied water. Continued reinforcement of the power elite derived from the need for organizational coordination to ensure continued operation of the irrigation system. Labor specialization developed. Some people farmed; others worked on the irrigation system. Still others became artisans, creating the implements needed to maintain the system, or administrative workers in the direct employ of the power elite's court.

> **hydraulic civilization** A civilization based on large-scale irrigation.

Although the hydraulic civilization model fits several areas where cities first arose—China, Egypt, and Mesopotamia—it cannot be applied to all urban hearths. In parts of Mesoamerica, for example, an urban civilization blossomed without widespread irrigated agriculture and therefore without a class of technical experts.

URBAN HEARTH AREAS

The first cities appeared in distinct regions, such as Mesopotamia, the Nile River valley, Pakistan's Indus River valley, the Yellow River valley of China,

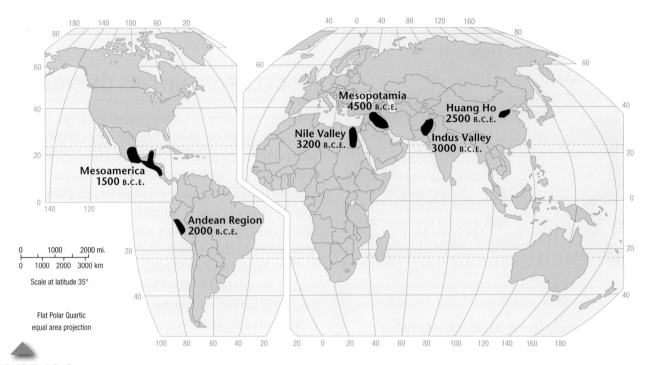

FIGURE 10.6 **The world's first cities arose in six urban hearth areas.** The dates shown are conservative figures for the rise of urban life in each area. For example, some scholars would suggest that urban life in Mesopotamia existed by 5000 B.C.E. Ongoing discoveries suggest that urban life appeared earlier in each of the hearth areas and that there are probably other hearth areas—in West Africa, for example (see also the World Heritage Site on The Great Zimbabwe National Monument in Chapter 1).

Mesoamerica, and the Andean highlands and coastal areas of Peru. These are called **urban hearth areas** (Figure 10.6).

Scholars have referred to these first cities in Mesopotamia and elsewhere as **cosmomagical cities**, cities that are arranged spatially according to religious principles. The spatial layouts of cosmomagical cities are similar in three important ways. First, great importance was given to the city's symbolic center, which was also believed to be the center of the known world. It was therefore the most sacred spot and was frequently identified by a vertical structure of monumental scale that represented the point on Earth closest to the heavens. This symbolic center, or **axis mundi**, took the form of the ziggurat (a massive structure having the form of a terraced step pyramid with receding levels) in

> **urban hearth areas**
> Regions in which the world's first cities evolved.
>
> **cosmomagical cities**
> Types of cities that are arranged spatially according to religious principles; characteristic of very early cities.
>
> **axis mundi** The symbolic center of cosmomagical cities, often demarcated by a large vertical structure.

Mesopotamia, the palace or temple in China, and the pyramid in Mesoamerica. Often this elevated structure, which usually served a religious purpose, was close to the palace or seat of political power and to the granary (a storehouse for grain). These three structures were often walled off from the rest of the city, forming a symbolic center that both reflected the significance of these societal functions and dominated the city physically and spiritually. The Forbidden City in Beijing remains one of the best examples of this guarded, fortress-like "city within the city" (Figure 10.7). The second spatial characteristic common to cosmomagical cities is that they were oriented toward the four cardinal directions. By aligning the city in the north–south and east–west directions, the geometric form of the city reflected the order of the universe. It was thought that this alignment would ensure harmony and order over the known world, which was bounded by the city walls.

In all these early cities, one sees evidence of a third spatial characteristic: an attempt to shape the form of the city according to that of the universe. The

FIGURE 10.7 **Beijing's Forbidden City.** This bronze lion guards the entrance to the Imperial Palace, shown in the background. *(Bernd Göttlicher/Alamy.)*

Cambodia that presents one of the best examples of this parallelism. An urban cluster that spread over 6 square miles (15.5 square kilometers), Angkor Thom was a representation in stone of a series of religious beliefs about the nature of the universe. Thus, the city was a microcosmos, a re-creation on Earth of an image of the larger universe.

IMPERIALISM AND URBANIZATION

Although urban life originated at several specific places in the world, cities are now found everywhere. How did city life come to these regions? There are two possible explanations. Cities could have evolved spontaneously in different places across the globe, as populations developed new agricultural technologies allowing for a surplus, and their societies differentiated. Or, contact with city dwellers, whether through trade, trans-oceanic voyages, or conquest, could have diffused the techniques and ideas associated with urban dwelling to other areas.

Cities and Conquest There is little doubt that diffusion has been responsible for the dispersal of the city in historical times (**Figure 10.8**), because the city

ordering of the space of the city was believed to be essential to maintaining harmony between the human and spiritual worlds. In this way, the world of humans would symbolically replicate the world of the gods. This characteristic may have taken a literal form—a city laid out, for example, in a pattern of a major star constellation. Far more common, however, were cities that symbolically approximated mythical conceptions of the universe. Angkor Thom was an early city in

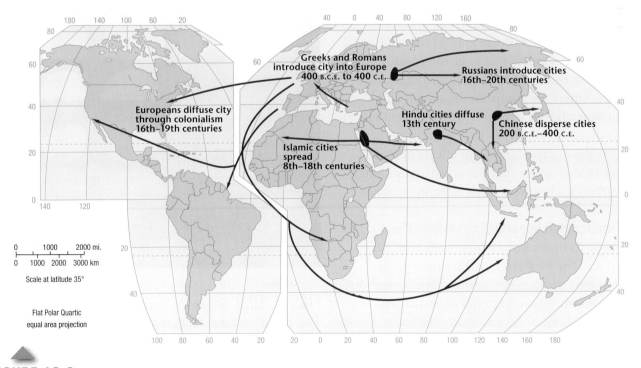

FIGURE 10.8 **The diffusion of urban life with the expansion of certain empires.** *What does this map tell us about the importance of urban life to military conquest?*

has commonly been used as a means for imperial expansion. Typically, urban life is carried outward in waves of conquest as the borders of an empire expand. Initially, the military controls newly won lands and sets up collection points for local resources, which are then shipped back to the heart of the empire. As the surrounding countryside is increasingly pacified, the new collection points lose some of their military atmosphere and begin to show the social diversity of a city. Artisans, merchants, and bureaucrats increase in number; families appear; the native people are slowly assimilated into the settlement as workers and may eventually control the city. Finally, the process repeats itself as the empire pushes farther outward: with first a military camp, next a collection point for resources, then a full-fledged city expressing a true division of labor and social diversity.

Such a process, however, did not always proceed without opposition. The imposition of a foreign civilization on native peoples was often met with resistance, both physical and symbolic. Expanding urban centers relied on the surrounding countryside for support. Their food was supplied by farmers living fairly close to the city walls, and tribute was demanded from the agricultural peoples living on the edges of the urban world. The increasing needs of the city required more and more land from which to draw resources. However, the peasants farming that land may not have wanted to change their way of life to accommodate the city. The fierce resistance of many Native American groups to the spread of Western urbanization is testimony to the potential power of indigenous society to defy urbanization, although the destructive long-term effects of this kind of resistance suggest that the organized military efforts of urban society were difficult to overcome.

Ancient Empires and Imperial Expansion   The ancient Greek civilization, to which both Western civilization and the Western city trace their roots, expanded in part through spreading cities throughout the Mediterranean. Greek imperial cities reached as far as the north shore of Africa, Spain, southern France, and Italy. These cities were of modest size, rarely containing more than 5000 inhabitants. Athens, however, may have reached a population of 300,000 in the fifth century B.C.E.

The Roman civilization supplanted ancient Greece, and was itself centered on a city: Rome. The Romans adopted many urban traits from the Greeks as well as the Etruscans, a civilization of central Italy that Rome had conquered. As the Roman Empire expanded, city life diffused farther into France, while also reaching Germany, England, interior Spain, the Alpine countries, and parts of eastern Europe—areas that had not previously experienced urbanization. Most of these cities were, initially, military and trading outposts of the Roman Empire.

Fundamental to Roman cities was the gridiron street pattern, composed of straight streets intersecting at right angles. Indeed, transportation lay at the heart of Roman cities. Water was transported into cities from remote regions using aqueducts, many of which still punctuate the landscapes of European cities today (**Figure 10.9**). The Roman Empire itself was held together by a complicated system of roads and highways linking towns and

FIGURE 10.9   **Ponte-du-Gard, near Remoulins, France.** This Roman-era aqueduct is a striking example of the infrastructure that extended throughout the Roman Empire. It was built in the first century C.E. and is one of the best-preserved aqueducts in existence. *(Lucyna Koch/Getty Images.)*

cities. In choosing a site for a new urban settlement, the Romans made access to transportation a major consideration. By 400 C.E., the Roman Empire had declined and so too urban life throughout much of Europe (see Chapter 3, Figure 3.34). The highway system that linked them fell into disrepair, so that cities could no longer exchange goods and ideas. Cities were either destroyed or left to decay on their own, becoming small villages where a few hundred people eked out a living. The urban areas, as with the rest of the territory formerly under Roman rule, fell into the Dark Ages.

Cities in other parts of Europe and in other world regions, however, thrived. Spanish cities, for example, flourished as centers of learning, commerce, and governance thanks to the North African Moorish imperial presence in Iberia from the eighth through fifteenth centuries. (See the World Heritage Site in Chapter 4 on Alhambra, Generalife, and Albayzín.) Chinese cities developed their own cosmomagical features, including sacred sites connecting residents to ancestors, pagodas that functioned as axis mundi, and fabulous palaces reflecting divine order on Earth (see Figure 10.7). In the Americas, several indigenous cultures established urban centers prior to their colonization by the Spanish in the fifteenth century. American cities were religious ceremonial centers, as well as control centers for the subjugation of surrounding conquered rural peoples. Some grew in size and complexity that outpaced anything that Europeans of the time had experienced.

Tenochtitlan—the Aztec capital now known as Mexico City—was founded in 1325 on an island in the middle of Lake Texcoco, with canals providing transportation routes within the city and causeways reportedly wide enough to accommodate 10 horses leading to and from the mainland (Figure 10.10). By the time of the Spanish conquest of Mexico in 1521, the city boasted over 200,000 inhabitants, far outstripping any Spanish city at that time. Bernal Díaz del Castillo, a foot soldier in Hernán Cortés's army, described coming upon it in awe-inspired tones: "And when we saw all those cities and villages built in the water, and other great towns built on dry land, and that straight and level causeway leading to Tenochtitlan, we were astounded. These great towns and *cues* (temples) and buildings rising from the water, all made of stone, seemed like an enchanted

FIGURE 10.10 **Pre-conquest Tenochtitlan.** This image, a detail from a mural by Mexican artist Diego Rivera, shows Tenochtitlan in the background. Notice how orderly, clean, and happy this painting makes this ancient city—today known as Mexico City—look.
(Gianni Dagli Orti/The Art Archive at Art Resource, NY © 2014 Banco de Mexico Diego Rivera Frida Kahlo Museums Trust, Mexico, D.F./ Artists Rights Society (ARS), New York.)

vision. . . . Indeed, some of our soldiers asked whether it was not all a dream."

CITIES OF THE INDUSTRIAL REVOLUTION

The agricultural innovations mentioned in the previous section—primarily, technological innovations leading to the domestication of crops and animals that allowed for the production of a food surplus—permitted the world's first true cities to develop because they facilitated support of a dense human population in one place, the rise of a nonagricultural class, and the accompanying social differentiation. The connection of agricultural innovations with the rise of cities is often referred to as the **first urban revolution**.

> **first urban revolution** The connection between agricultural innovations with the rise of the world's first true cities.

Cities obviously did not stop evolving, as evidenced by the many differences between early cities such as Tenochtitlan or ancient

Rome and the contemporary cities you are familiar with today. Why and in what ways have cities changed over the last six millennia?

The answer to this question can be traced, in large part, to what scholars term the **second urban revolution**. Rather than being tied to technological innovations in agriculture, the second urban revolution arose as societies, initially those located in western Europe and later North America, became industrialized. At its heart, the second urban revolution developed from innovations in mining and manufacturing. As industrialization progressed, many transformations occurred in society that fundamentally reshaped cities, both spatially and in terms of the social, cultural, political, and economic functions assumed by cities.

> **second urban revolution** The connection between industrial innovations with the rise of capitalist cities.

The transformations that culminated in capitalism had begun to reshape the cities of western Europe from the mid-sixteenth to the mid-eighteenth century. Agriculture became increasingly specialized and commercialized. Land formerly owned and worked collectively became enclosed and owned by individuals who purchased it and then paid others to work the land for them. The very landscape of the European countryside became reorganized (see Chapter 8, Seeing Geography).

Perhaps of greatest significance was how the capitalist mentality introduced a notion of urban land as a source of income. Proximity to the center of the city, and therefore to the most pedestrian traffic, added economic value to the land. Other specialized locations, such as areas close to the river or harbor, or along major thoroughfares into and out of the city, also increased land value.

In the emerging capitalist city, the ability to pay determined where one would live. The city's residential areas became segregated by economic class. The wealthy lived in the desirable neighborhoods; those without much money were forced to live in the more disagreeable parts of the city—for example, low-lying flood-prone zones, places contaminated by factory waste, or parts of the city located at higher altitudes where roadways and urban services were not easily provided (**Figure 10.11**).

In earlier urban configurations, there was no separation between one's home and one's workplace.

FIGURE 10.11 **Early industrial-era London.** This painting depicts mid-nineteenth-century London, awash in industrial air and water pollution. *(SSPL via Getty Images.)*

For most people, work was organized by family units, it centered on the home, and the workday was governed by the seasons and the rising and setting of the sun. In the capitalist city, the place of work became spatially distinct from the home, such that the laborer had to travel back and forth between them. This spatial separation of work from home, of public space from private space, both reflected and helped to shape the changing worlds of men and women. In general, men generated monetary income from work outside the home and therefore became associated with moving about and occupying the public spaces of the city. Women, who primarily did domestic work, were considered the keepers of the private world of the home and family. They were not monetarily compensated for their labors. This association of women with private domestic space and men with public work space deepened and became more complex over the next few hundred years. (See also Chapter 3, pages 106–108, and Chapter 9, pages 392–394.)

The center of the capitalist city—its axis mundi—was not the ziggurat, palace, or cathedral, but instead the buildings devoted to business enterprises. A

THE VIDEO CONNECTION

A Short History of the Highrise

Part I: Mud (4:30 min): www.tinyurl.com/kqquv4b

Part II: Concrete (7:08 min): www.tinyurl.com/m9tth6q

Part III: Glass (4:20 min): www.tinyurl.com/ktfygst

Part IV: Home (5:52 min): www.tinyurl.com/odlp85a

This series of four videos profiles the evolution of high-rise residential buildings over time and in different places. Innovations in building materials and technology were key to enabling humans to build higher. In addition, social attitudes, conditions, and policies formed a crucial component of whether and how high-rise housing has been adopted in different places and times.

Thinking Geographically:

1. Figure 10.27 depicts how high-rise urban zones, and individual buildings, tend to look similar throughout the world. These videos, however, note that changing social policies, conditions, and attitudes in regions such as North America, Europe, and Asia have led to very different ways of utilizing— or not —high-rise housing. Give some specific examples from what you have just watched.

2. People from all urban-based socioeconomic classes, from the very wealthy to the impoverished, have lived in high-rise housing at different points in history. Give some specific examples of the ways in which poor, middle-income, and wealthy occupants have been included or excluded from these dwellings. How has this varied from place to place, as well as historically?

3. High-rise housing has long had a reputation of being impersonal and ugly. But, some of the buildings you've just looked at are very beautiful. What's beautiful and what's ugly, in your opinion, about the buildings you've seen, with respect to their interiors as well as their exteriors? How do people make these places into their homes?

---

downtown defined by economic activity emerged, and as industrialization progressed, it expanded and evolved into specialized districts. With the downtown devoted to commerce and industry, the upper classes moved to the outskirts of the city and built large houses that displayed their wealth.

URBAN LOCATION

The geographic criteria for city placement changed as industrialization took hold. Early cities tended to be located in places that were easily defended—on islands, hilltops, sheltered harbors, or river bends—or along trade routes (**Figure 10.12**). Industrial-era cities sought locations near sources of raw materials, such as coal, wood, or metals, used in factories. Location along trade routes continued to be important but the parameters broadened, inasmuch as the waterways, railways, and roadways needed to accommodate increased traffic associated with raw

FIGURE 10.12   The classic defensive site of Mont St. Michel, France. A small town clustered around a medieval abbey, which was originally separated from the mainland during high tides, Mont St. Michel now has a causeway that connects the island to shore, allowing armies of tourists to penetrate the town's defenses easily. *(Mopy/Rapho Agence/Science Source.)*

FIGURE 10.13 **Pittsburgh's Golden Triangle.** The Allegheny River meets the Monongahela River at the site shown in the photograph. Both rivers were early trade routes. Today, Pittsburgh's commercial center is located at their juncture. *(Radius/Radius/Superstock.)*

materials entering and finished goods exiting the sites of industrial production (**Figure 10.13**).

Geographers are, as you are no doubt aware by now, fond of using models—including maps—to represent how space is ordered in the abstract, as well as to predict and plan for its optimal ordering. Urban geographers are no exception, and they have used a wide variety of models to understand how urban space is arranged and how to guide its arrangement in better ways. **Central place theory** is the term given to a set of models that attempts to understand why cities are located where they are, as well as to help planners position cities in space most efficiently.

> **central place theory** A set of models designed to explain the spatial distribution of urban centers.

Central place theory is associated with Walter Christaller, a German urban geographer who first published his work on this concept in 1933. Christaller understood cities primarily as economic centers concerned with distributing goods to people, who would travel certain distances to acquire them. According to Christaller, people would travel only

short distances to acquire everyday low-order goods such as food and other regularly purchased household items. However, in order to obtain specialized or costly high-order goods and services such as automobiles, the advice of a cardiologist, or to interview for a passport, he hypothesized that people would travel much longer distances. These travel distances dictated the placement of different orders of urban settlements, ranging from tiny hamlets and villages on one end, through towns and cities, and finally to regional capitals (**Figure 10.14**).

Because people would travel only short distances to acquire the low-order goods sold in smaller urban settlements, there were many villages and hamlets (what Christaller termed fourth-order places) and to a lesser extent towns (third-order places) located relatively close to one another. Each one could be supported by the routine traffic in low-order goods that nobody would travel very far to purchase. In Christaller's model, smaller urban places have a lower

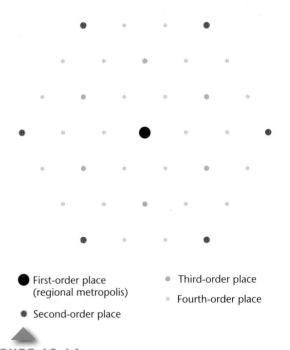

● First-order place (regional metropolis)

● Second-order place

• Third-order place

· Fourth-order place

FIGURE 10.14 **Christaller's hierarchy of central places shows the orderly arrangement of towns of different sizes.** This is an idealized presentation of places performing central functions. For each large central place, many smaller places are located within the larger place's hinterland.

threshold In central place theory, the size of the population required to make the provision of goods and services economically feasible.

range In central place theory, the average maximum distance people will travel to purchase a good or service.

threshold, the number of people required to make provision of a good feasible, and a smaller **range**, the distance people will travel to acquire a good.

Costlier and less frequent high-order purchases, however, could only be made in cities (second-order places) and regional capitals (first-order places). This is because people were willing to travel longer distances to obtain high-order goods and services (in other words, big cities have a larger range), while on the other hand, fewer urban areas could actually afford to carry these high-order goods and services because people didn't purchase them as frequently (in other words, the higher-order goods have a higher threshold).

The requirements of trade-based urban locations continue to shift to this day. For instance, San Francisco's importance as a West Coast port city was high before the advent of container ships. But once containerized cargo became the norm starting in the 1970s, San Francisco's piers were no longer needed and there was no room in the city to build the large holding lots needed for containers to go from port to rail. Oakland, San Francisco's rival city just across the bay, was quick to invest in the large cranes and other equipment needed to handle containers. Oakland was also able to accommodate containerized cargo because it filled in huge tracts of shallow bay lands, creating a massive area for the loading, unloading, and storage of cargo containers. As a result, San Francisco's importance as a port city declined, and Oakland is now the second-largest West Coast port city (**Figure 10.15**).

As capitalism shifts from a focus on manufacturing goods and exchanging them for money, and toward the provision of services and information, the question of what constitutes an ideal urban location changes yet again. It is no longer as critical to be located near physical transportation routes as it once was. Indeed, an argument may be made that other sorts of networks, particularly those associated with communications technology, are just as important (if not more so) criteria for city placement. It is worth considering whether the

FIGURE 10.15   **The seaport of Oakland.** Cranes are loading a large container ship. Note the ample area devoted to container movement and storage at this port. Oakland is an intermodal facility, which means that containers can be loaded to and from ships, trucks, and railways. *(David R. Frazier Photolibrary, Inc./Alamy.)*

historical legacy of where the world's major cities are located today will be outweighed by these new considerations, and whether the sites of the important cities of the future will shift accordingly.

MOBILITY

10.2 LEARNING OBJECTIVE
Identify the ways that migration has shaped the growth of cities, how cities have grown and changed, and how humans get around within cities.

As noted at the beginning of this chapter, we have entered an era when more people live in cities than rural areas: quite a historical feat. As we saw in the preceding section, when cities first sprung up, only a very small portion of the population lived in them. But over time, cities began to diffuse across the globe, and established cities grew larger. In this section, we look at the process of rural-to-urban migration that has, both in the past and today, fueled the growth of cities. Some cities have grown to a very large size, either in absolute population terms or relative to the other cities around them. We will take a look at where these very large cities are located, where they will appear in the future, and what role such cities play in a country's urban hierarchy. The role of transportation in facilitating the movement of people within cities has also shaped the size and layout of cities. In this chapter, we focus particularly on the advent of the automobile and its central role in shaping human mobility in and between cities.

RURAL-TO-URBAN MIGRATION

The increase in urban population went hand-in-hand with the growing number of cities across the globe. In today's world, however, increasing urbanization is caused by three related phenomena: natural population increase (see Chapter 3), rural-to-urban migration, and urban-to-urban migration. The last two phenomena can be domestic, as in a family moving from one city to another in the same country, or the result of international population movement from one country to another. Although the United Nations estimates that natural increase accounts for the majority of the recent growth in urban population, it was rural-to-urban migrations that historically brought millions of people into urban life.

As we have seen, wide-scale urbanization occurred hand-in-hand with industrialization. As capitalism took root and flourished, first in Europe and North America and subsequently shaping life everywhere in the world, cities allowed for the large workforces needed to run factories and to provide services of all sorts to factory owners and laborers. In the United States, rural-to-urban migration resulted in the growth of large cities such as Chicago in the nineteenth century, and large urban centers in Asia such as Tokyo and Mumbai in the twentieth century. Today, Europe and more recently North America as well as much of Latin America are highly urbanized societies, with about three-quarters of their populations living in cities. (See the World Heritage Site and Seeing Geography features in this chapter;

both examine one of the largest cities in Latin America, Rio de Janeiro.)

China presents a particularly interesting example of rural-to-urban migration. Until the late 1970s, the country was predominantly rural, and its communist government was explicitly anti-urban. Indeed, many urban Chinese were forced out of cities to work on farms during the decade of the Cultural Revolution from 1965 to 1975, under the logic that cities fostered class privilege. State policy reform after 1978 lifted restrictions on internal migration, which enabled its rural population to become more mobile. Many Chinese abandoned life in the countryside for futures in China's booming industrial cities, among them Guangzhou, Shanghai, and Beijing. Note the number of Chinese cities that appear on the map of world megalopolises in Figure 10.17. Though just over 50 percent of China's population is urban today, it is estimated that by 2025, 66 percent of China's population will be urban, and there will be 221 Chinese cities with a population over 1 million. China's "urban billion" is expected to materialize by 2030.

Today, particularly across Africa and Asia, large numbers of people still depart their rural villages and

FIGURE 10.16 **Beggar at an intersection in Islamabad, Pakistan.** Begging in traffic is common in major cities around the world. Beggars can be recent immigrants to the city or people who have difficulty securing employment. *(Farooq Naeem/AFP/Getty Images.)*

Population of Urban Agglomerations, 2010

⬤ Over 30 million

⬤ 20,000,000–29,999,999

⬤ 15,000,000–19,999,999

⬤ 10,000,000–14,999,999

• Under 10,000,000

FIGURE 10.17   **Map of the world's megalopoli.** This map shows the location of the world's truly large urban agglomerations. Notice how many of them are clustered in Asia. *(Source: United Nations, Department of Economic and Social Affairs, Population Division.)*

migrate to cities in search of better economic and social opportunities. The United Nations predicts that about four-fifths of the world's future urban growth will result from rural-to-urban migration in these two regions. Others will be forced from their lands due to rural conflict, natural disaster, climate change, or seizure of their land. Thus, cities can increase in size not necessarily because there is economic growth requiring urban laborers, but rather because conditions in the countryside are dismal. People leave their farms and rural villages to escape economic, political, and environmental hardships, in the hope that urban life will offer a slight improvement. Often it does. Given the new global economy, however, many of the employment opportunities available in such cities are low-skilled and low-paying manufacturing and service jobs with harsh working conditions. The result is that rural-to-urban migrants often find themselves either unemployed or with jobs that barely provide a decent living (**Figure 10.16**).

WHEN ARE CITIES TOO LARGE?

Although rural-to-urban migration affects nearly all cities in the developing world, the most visible cases are the extraordinarily large settlements we call **megacities**, those having populations of over 10 million. **Figure 10.17** shows the world's 30 largest cities, a majority of which are located in the developing world. This is a major change from 50 years ago, when the list was dominated by Western, industrialized cities.

> **megacities** Particularly large urban centers.

The urban population in the developing world is growing at impressive rates. Indeed, the World Bank estimates that most of the urban growth in the world today—over 90 percent of it, in fact—is occurring in the developing world. Seventy million people move into cities in the developing world each year. The world's two poorest regions, South Asia and sub-Saharan Africa, expect their urban populations to double by 2030.

World Heritage Site

This World Heritage Site encompasses a portion of the city of Rio de Janeiro, Brazil. (Take a look also at Seeing Geography at the end of this chapter.) Specifically, it includes the natural elements located within the urban boundaries that have fundamentally shaped the city of Rio: the peaks of the Tijuca National Park, the Botanical Gardens established in 1808, and the hills surrounding Guanabara Bay. These distinctive urban landscapes have been celebrated artistically in paintings, poetry, song, and literature

■ From 1763 to 1960, Rio de Janeiro was the capital of Brazil. In an effort to decentralize the national population away from the southeastern coast, the capital was relocated to Brasília, in the country's interior (see Figure 5.30).

■ Stretching from the mountains to the sea, this World Heritage Site is bounded by the best vantage points for appreciating the interaction of nature and culture.

■ As with many Latin American urban landscapes, Rio's is one of expensive high-rise housing located immediately adjacent to *favelas* (slums).

(Gardel Bertrand/Hermis/Corbis.)

Rio de Janeiro: Carioca Landscapes Between the Mountains and the Sea

CULTIVATING NATURE:

The word Carioca refers to a resident of Rio de Janeiro. It is derived from the indigenous Tupi word for "houses of the whites," referring to the Portuguese settlements built along the river in the sixteenth century C.E. Today, Carioca indicates both a dialect of Brazilian Portuguese, as well as the friendly, relaxed, and cosmopolitan personality associated with Rio de Janeiro. Carioca culture is defined by its outdoor living, framed by the beaches and mountains, but also the streets and parks. Urban planners have conscientiously highlighted the city's dramatic local natural features, as well as enhanced them through landscaping and constructed elements of the urban environment.

(Alex Robinson/JAI/Corbis.)

▪ This World Heritage Site is home to the largest urban forest on the planet: the Tijuca Forest. It is but a remnant of the Atlantic forest ecosystem, which once extended along the length of Brazil's coastline. From the founding of Rio in 1565 through the mid-1800s, trees were felled for timber and fuel for the expanding city. In 1860 Emperor Pedro II ordered the replanting of the forest with native species, acknowledging that the forest was essential for the city's environmental and cultural well-being.

▪ Although it is known for its natural features, this site also includes low-income urban neighborhoods. With their tendency to be built on the city's steep, forested hillsides, favelas such as Santa Marta put contemporary environmental pressures on the natural landscape.

TOURISM:

Brazil hosted nearly 6 million foreign visitors in 2012. Frequently referred to as Cidade Maravilhosa (The Marvelous City), Rio de Janeiro is the country's top destination for leisure travelers. It is widely described as one of the most beautiful cities in the world.

(Wilton Junior/dpa/Corbis/.)

▪ Rio has been a global tourist attraction since the early 1900s when it constructed one of the world's first aerial cable car lines to the top of a monolithic mountain called Sugar Loaf.

▪ Regardless of their religious affiliation, tourists from around the globe are keen to visit the world's largest statue of Jesus, located on Corcovado Mountain in the Tijuca Forest.

▪ Rio de Janeiro hosted the World Cup soccer games in 2014 and will host the 2016 Summer Olympics. The city's infrastructure is currently being upgraded to handle the expected influx of tourists.

▪ As with any large city, street crime is a concern. As of early 2014, assaults, theft, and muggings had increased dramatically in Rio in advance of the World Cup soccer games.

▪ Rio de Janeiro is known as one of the world's most beautiful cities in great part because of how planners have showcased its dramatic natural features.

▪ This city is home to the world's largest urban forest.

▪ The Carioca Landscapes Between the Mountains and the Sea were designated a World Heritage Site in 2012.

www.whc.unesco.org/en/list/1100

And with this incredible increase in the sheer numbers of urban dwellers in the less developed regions of the world comes a large list of problems. Unemployment rates in cities of the developing world are often over 50 percent for newcomers to the city; housing and infrastructure frequently cannot be built fast enough to keep pace with growth rates; water and sewage systems can rarely handle the influx of new people. Consequently, one of the world's ongoing challenges will be how to cope with the radical restructuring of population and culture as people in developing countries move into cities.

URBAN PRIMACY

The destination for much rural-to-urban migration in the developing world is the **primate city**. This is a city that, because it is so much larger than all others, dominates the economic, political, and cultural life of a country. Some scholars maintain that in order to qualify as a primate city, a city must be 10 times or more larger than the next largest city. With 13 million residents, Buenos Aires is an excellent example of a primate city because its population is exactly 10 times that of the next largest city, Rosario, which is the second largest city in Argentina with a population of 1.3 million. Although many developing countries are dominated by a primate city, often a former center of colonial power, urban primacy is not unique to these countries: think of the way London, Moscow, Athens, and Paris dominate their respective countries.

> **primate city** A city of large size and dominant power within a country.

These primate cities are often the ones that claim scholars' and policy makers' attention because of their dominance and large size. However, the United Nations finds that most of the world's population lives in mid-sized cities of 500,000 or less. Indeed, the majority of urban growth today is taking place in smaller, less dominant cities. Urban planners and policy makers are beginning to focus their attention on these mid-sized cities scattered throughout the world, particularly in developing countries, in order to examine the impacts of increased urbanization.

This is true of the United States, too, where so-called legacy cities—well-known, established, large urban centers such as San Francisco, New York,

FIGURE 10.18 **Houston's hipster gentrifiers.** These urbanites are partaking in Houston's happening cultural scene. This restaurant has more than 75 beers on tap. Houston has gone from cow-town to cool-town thanks to the influx of young gentrifiers. *(Richard Carson/Reuters/Corbis.)*

Chicago, and Los Angeles—are not growing nearly as fast as mid-sized cities, particularly those located in the Sunbelt, known as "aspirational cities"—among them, Raleigh, Houston, Phoenix, San Antonio, and Las Vegas. Legacy cities have simply priced all but the elites out of their housing markets. Lower-skilled people are attracted to aspirational cities because they offer employment, affordable housing, and a lower cost of living. In 2012 *Forbes* flagged Houston as the United States' "coolest city." Youth (the typical Houstonian's average age is 33), ethnic and racial diversity, and a growing slate of hip downtown amenities combined with employment and affordability have earned Houston this distinction (**Figure 10.18**).

URBAN TRANSPORTATION

Getting around inside the city, and transporting people and goods from city to city, has—for as long as cities have existed—been a major concern. You will recall that Rome was famous for its engineering marvels, including grid-patterned streets facilitating traffic flow within cities, and the remarkable network of roadways connecting cities across the Roman Empire.

Railroads, streetcars and trolleys, light rail systems, airplanes, busses, and subways have all presented innovations in urban transportation modes and infrastructure. These innovations have shaped and reshaped the layout and size of cities over time. For instance, the advent of streetcars—fixed rail systems within many U.S. cities that arose in 1880s—led to the development of so-called **streetcar suburbs**. Streetcar suburbs allowed urban residents to live farther away from their workplaces than they could in the past, when commuting was done by foot or horse.

streetcar suburbs The extension of urban residential areas along streetcar lines in the late nineteenth century.

No innovation in transportation has had a more profound impact on cities, however, than the automobile. In the United States, Henry Ford's mass-produced Model T initiated the norm of automobile ownership among the middle classes, and by 1920 over a million had been sold. Automobiles increased personal mobility exponentially, and it has been said that Henry Ford "freed common people from the limitations of geography." Automobiles brought with them a network of roadways and, by the 1950s, interstate highways. Housing architecture was reshaped around the automobile with the addition of the now-ubiquitous garage. Drive-through commerce and architectural forms associated with it became a hallmark of America's obsession with cars (**Figure 10.19**).

Because so much of the middle class owned a car, it became possible to locate housing further from the urban **Central Business District (CBD)**, as its placement was not limited to locations close to existing train lines. As long as roads could be built to connect these settlements to the CBD, the new automobile suburbs would be viable. Cities grew ever-outward in a process that came to be known as **sprawl** (**Figure 10.20**). Some U.S. cities, such as Miami, Atlanta, Las Vegas, and Phoenix, are known as **automobile cities** because

central business district (CBD) A dense cluster of offices and shops located at the city's most accessible point, usually its center.

sprawl The tendency of cities to grow outward in an unchecked manner, which is particularly notable in automobile cities.

automobile cities Cities whose spatial layout—both in terms of extent and form—is dictated by the near ubiquity of individual automobile ownership.

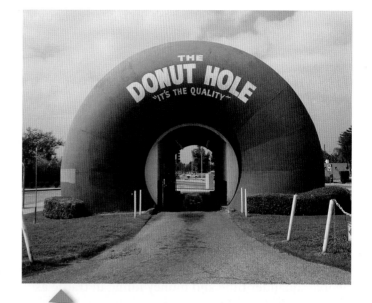

FIGURE 10.19 **Drive-through architecture.** This uniquely American architectural form emphasizes convenience and accessibility. As this photo shows, drive-through architecture can also be quirky. The Donut Hole drive-through is located in Los Angeles, California: ground zero for the drive-through. *(Barry Winiker/Getty Images.)*

FIGURE 10.20 **Aerial view of downtown Phoenix, Arizona.** Phoenix is a quintessential automobile city. You can discern this by the straight streets that intersect at right angles, and the fact that it sprawls out into the surrounding desert for miles. Only with access to an automobile or another form of transportation such as a bus, train, or subway can an individual truly move about comfortably in such a city. The distances are simply too far to bike or walk. *(iStockphoto/Getty Images.)*

they have been significantly built up recently enough that their layouts are primarily oriented around cars.

THE DOWNSIDES TO URBAN CAR CULTURE

Automobiles also brought to cities undesirable aspects. Cities became increasingly polluted with the fossil fuel residue released by the automobile's internal combustion engine, impacting the health of city residents. Safety, too, became a concern, and automobile fatalities—for both passengers and pedestrians—is the leading cause of accidental death. Traffic congestion is a major complaint, and many people squander time and money idling in traffic as they make their way to work, commuting from far-out suburban residences to workplaces (**Table 10.1**). In addition, the norm of individual commuters who spend a great portion of their day in a car has led to social isolation and inactivity. Some urban and suburban roads no longer have sidewalks, which has made pedestrian traffic—and the face-to-face social encounters seen by many sociologists to be at the heart of the urban experience—all but disappear.

FIGURE 10.21 **Bike sharing.** These men are paying for bike rental in New York City's Greenwich Village. When they are done riding, they can simply park the bikes at any kiosk. Bike rental and sharing services are helping to reduce the congestion, contamination, and dangers associated with automobile traffic. *(Waring Abbott/Getty Images.)*

| TABLE 10.1 Top Ten Worst Commutes in the United States* | |
City	Time (hours) Spent in in Traffic Annually
Austin	39.9
Boston	34.4
Bridgeport	40.9
Honolulu	51.5
Los Angeles	61.4
New York	52.6
San Francisco	51.8
San Jose	33.4
Seattle	34.7
Washington, D.C.	41

*Montreal and Vancouver would bump Boston and D.C. off the top ten.
Source: 12 months trailing June 2013, http://www.inrix.com/Scorecard/default.asp.

In fact, there seems to be a growing disenchantment with automobiles, particularly among young people in Europe and the United States. French teenagers, for example, are less likely to hold a driver's license: in 1983, 35 percent of 18- and 19-year-olds did, but by 2010 only 19 percent did. The numbers for U.S. teens show a similar decline, from 69 percent of 17-year-olds in 1983 to 46 percent in 2010. Researchers speculate that concerns over the costs of owning, operating, insuring, and maintaining a car have driven this decline, as has a disenchantment among today's youth with the negative environmental impacts of automobiles along with a growing use of online technology as the preferred way to connect with friends.

This phenomenon has led to a rise in alternate modes of urban transportation. So-called millennials—teens and people in their twenties—are opting to bike or take public transportation rather than to drive or even own cars. Car- and bike-sharing services have sprung up in cities across the globe (Figure 10.21). For a nominal fee, a user can avoid the cost and hassle of owning a bike or car, and simply rent it for the time needed, or subscribe for a monthly fee. In the United States, reservations and payment are often made through smartphones and the user simply parks the vehicle at a rental station when done. In Europe, the process is even more flexible, as drivers locate the nearest available car using their smartphone and simply pick up the vehicle wherever in the city the previous user parked it. As a result, fewer people own cars, and fewer families own more than one car, which reduces the need for parking spaces and curtails pollution.

:::: GLOBALIZATION

10.3 LEARNING OBJECTIVE
Discuss the notion of global cities and some of the problems associated with them.

Many of the globalizing forces we have already discussed in this book—the integration of international economies, the interaction of different peoples, and the reshaping of culture—are centered in urban life. In other words, cities are the places where one can see both the multitude of benefits that derive from globalization, as well as the numerous pitfalls that it can bring. In this way, all cities can be considered global: they are places being shaped by the new global forces.

GLOBAL CITIES: YESTERDAY, TODAY, AND TOMORROW

global cities Cities that are control centers of the global economy.

Global cities are defined as those cities that have become the control centers of the global economy—the sites of major decisions about the world's commercial networks and financial markets. Rather than being just big cities, global cities house a concentration of multinational and transnational corporate headquarters, international financial services, media offices, and related economic and cultural services. According to sociologist Saskia Sassen, who coined the term "global city" in 1991, there were at that time only three such cities operating at this level: New York City, London, and Tokyo. These cities had become, in many ways, the headquarters for a global economy and form the top level of a hierarchical global system of cities.

Yet, an argument could certainly be made that cities have been, for at least six thousand years, a globally relevant mode of human dwelling in the world. To be sure, there were not as many cities nor were they as large as today's megacities, but—as we have seen—cities in the ancient world were certainly powerful places. As industrialization progressed, cities became larger, more powerful centers of decision making, and through rural-to-urban migration affected the lives of greater numbers of people.

European colonialism from roughly the sixteenth through nineteenth centuries, and later manifestations of imperialism, can be understood as primarily an urban-based endeavor. In other words, the conquest of territories and economies was an enterprise driven largely through cities. Colonial cities were established throughout Latin America and the Caribbean, Asia, and Africa by colonizing European powers, particularly Spain, Portugal, England, France, Germany, Belgium, but also Japan. It was through these colonial urban centers that the subjugated territories and their inhabitants were governed, colonial economies and political systems were shaped, and the language and cultural patterns of the conquering power were imposed (Figure 10.22).

Contemporary definitions of a global city focus on the possession of certain characteristics. Among them are economic might, political power, information exchange, and research and development activity, as well as less tangible qualities such as

FIGURE 10.22 **Garrison area.** This heavily fortified structure, consisting of a thick wall and cannons, is part of the colonial-era architecture constructed by the British to defend their colonial possessions in the Caribbean. It is located in Bridgetown, Barbados.
(AA World Travel Library/Alamy.)

influence, engagement, the ability to attract talented individuals, diversity, cultural experience, accessibility, and environmental quality. Being labeled a global city is desirable, and therefore many urban officials lobby to make the list of top global cities and—unsurprisingly—the number of lists has proliferated. Yet, the same 20 or so cities appear on most of the lists, though their exact place on a certain list may vary (**Figure 10.23**).

Urban geographers have broadened the analysis of global cities in order to also understand the next tier of cities within the urban hierarchy—those that contain a large percentage of producer service firms (law,

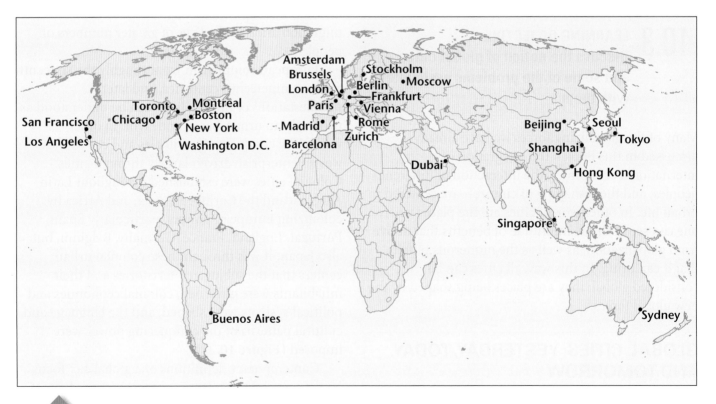

FIGURE 10.23 **Top 30 global cities.** This map shows the location of the world's top 30 global cities according to A. T. Kearney. That list used the following factors to determine global city status: business activity, human capital, information exchange, political engagement, and cultural experience. Note the world regions that have the most global cities, as well as those having none. *(Source: A.T. Kearney and The Chicago Council on Global Affairs–2012 Global Cities Index and Emerging Cities Outlook, pg. 3.)*

accounting, advertising, financial services, consulting) that are international. In this framework, cities are categorized by the number and type of transnational firms they house—not only headquarters but also regional and national offices. This analysis has identified globally and regionally dominant cities and those that are major participants in the new global economy. Geographers Yefang Huang, Yee Leung, and Jianfa Shen refer to these as international cities—places that are significant because they are centers of the new international economy. Their analysis allowed them to identify degrees of internationalization based on the number and locations of the international firms each city contained. The result is a very interesting list organized into classes of international cities, dominated by six in class A (London, New York, Hong Kong, Tokyo, Singapore, and Paris), followed by 10 in class B and 44 in class C. In this way, we can begin to understand globalization as a multidimensional process that has impacts on many cities previously left unexamined.

THE GLOBALIZATION OF URBAN WEALTH AND POVERTY

Whatever the precise parameters used to rank them, it is the interconnected and powerful nature of global cities that makes them stand out from all other urban centers. And with the rise of the global city comes the rise of the global cosmopolitan class, consisting of those individuals who, in the words of geographers Sam Schueth and John O'Loughlin, "belong to the world." Rather than being especially identified with one country, the global cosmopolitan moves with ease from place to place.

Ease of movement is usually only possible when one possesses wealth, and global cities are home to a great deal of wealthy individuals. Indeed, the presence of significant numbers of high net worth individuals is one criterion for inclusion as a global city, because this indicates a spatial concentration of economic power (**Figure 10.24**). And, though New York City

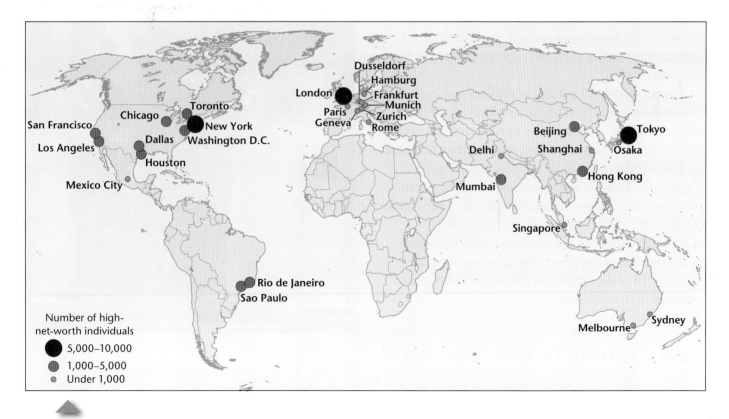

FIGURE 10.24 **Leading global cities of the super-rich.** This map of the world's top 30 global cities shows urban clusters of HNWIs (High Net Worth Individuals). HNWIs are those who possess more than $30 million in net assets. There are fewer than 200,000 HNWIs globally, which amounts to .003 percent of the world's population. One-third of them reside in the United States. (*Source: http://www.citylab.com/work/2013/03/global-cities-super-rich/4951/.*)

tops the list when it comes to concentration of high net worth individuals, the super-wealthy—those with over $30 million in net assets—may be found nearly worldwide.

The tastes and needs of the global cosmopolitan class have shaped the global city in particular ways. One of these is the prevalence of highly securitized residential enclaves known as **gated communities**. Cities throughout the world are home to residential enclaves that are more or less privately governed: they have their own services such as security and landscaping, their own oversight board that sets and enforces policies, and their own amenities such as clubhouses, golf courses, and swimming pools. Some communities have walls or fences around them, use surveillance technology to track entry and exit, and restrict access through gates to those who live there or have been approved for temporary entry, such as the guests of residents. Others do not have obvious barriers setting them apart from the surrounding urban areas. Regardless, according to geographer Renaud Le Goix and urban planner Chris Webster, "The issue is not really the gate as such, it is the fragmentation of the urban governance realm into micro-territories."

> **gated communities**
> Highly securitized residential enclaves that are more or less self-governing.

Approximately 15 percent of the housing stock in the United States is privately governed. And while it may be tempting to assume the gated community is a U.S. export, the researchers note it is an urban form that has arisen independently in cities for diverse reasons (**Figure 10.25**). Most of the reasons, however, can be traced to rising income inequality and fear. The wealthy in global cities have responded to this situation by attempting to shut out the have-nots, both from their residential spaces and from their urban governance processes. In some ways then, the urban wealthy have opted out of the global city.

GLOBALIZATION OF URBAN POVERTY

Scholars of urban poverty have traditionally focused on the specific ways that poverty is manifest in urban environments in developing regions. But, recent economic hardship in Europe and North America has affected cities in these regions in many ways.

Unemployment, deterioration of housing and urban infrastructure, homelessness, and other manifestations of poverty in the urban landscape are quite visible (**Figure 10.26**). You have encountered many examples thus far in this textbook of how urban poverty can be a problem in wealthy places as well as poor ones (see, for instance, Figures 3.35 and 3.36 in Chapter 3 and Figures 9.21 and 9.24 in Chapter 9). So, as with the rise of global cities and the global cosmopolitan class discussed above, it is important to bear in mind that urban poverty is not limited to the developing world but is rather a global phenomenon. Most global cities have pockets of both extreme wealth as well as its opposite, extreme poverty.

However, in cities of the developing world, the combination of large numbers of migrants and widespread unemployment leads to overwhelming pressure for low-rent housing in ways not frequently seen in cities in wealthier regions. Governments have rarely been able to meet these needs through housing

FIGURE 10.25 **South African gated community.** This "security village" in Johannesburg is fortified with high walls, guarded gates, and electrical fencing. South Africa's well-to-do, concerned about urban crime, are driving a huge and growing security industry in the country. *(Eden Breitz/Demotix/Corbis.)*

FIGURE 10.26 Poverty amidst plenty. This trash-strewn vacant lot is located in Los Angeles' Watts neighborhood. Though Los Angeles is home to the glitterati of Hollywood and Beverly Hills, it also has poor neighborhoods characterized by higher than average numbers of gangs and households headed by single parents. *(Julie Dermansky/ Getty Images.)*

projects, so one of the most characteristic landscape features of cities in the developing world has been the spontaneous construction of impoverished housing in areas known as **shantytowns**.

> **shantytowns** Precarious and often illegal housing settlements, usually made up of temporary shelters and located on the outskirts of a large city.

Shantytowns, also called *favelas, barriadas,* or squatter settlements, are often simply referred to as slums—areas of degraded, precarious, inadequate, and often illegal housing (see also Chapter 3, pages 126–130). The United Nations defines a slum household as a cohabiting group that lacks one or more of the following conditions: an adequate physical structure that protects people from extreme climatic conditions, sufficient living area such that no more than three people share a room, access to a sufficient amount of water, access to sanitation, and secure habitation or protection from forced eviction. Over 1 billion people in the world live in slums: that

is one out of every seven people on Earth. In greater Manila alone, for instance, scholars estimate that almost half of the city's 19 million inhabitants live in slum conditions within shantytowns.

In his book *Planet of Slums,* urban theorist Mike Davis speculates that the proportion of humanity living in slums will incline sharply in the next few decades, to 2 billion or more, constituting a potentially volatile political group. Yet historically, shantytowns simply constitute the margin of urban growth in developing world cities. Because governments do not or will not expend the resources to build housing for the poor, the poor are forced to undertake this task themselves. Shantytowns usually begin as collections of crude shacks constructed from scrap materials; gradually, they become increasingly elaborate and permanent. Paths and walkways link houses, vegetable gardens spring up, and often water and electricity are bootlegged into the area so that a common tap or outlet serves a number of houses. At later stages, municipal governments install paved roads, electricity, and sewage systems and typically initiate some process—usually timed with the municipal elections cycle—to grant residents legal occupancy to shantytown lands. Thus, the city proper eventually takes most shantytowns under its wing, and in many instances these supposedly temporary settlements become permanent parts of the city, functioning as regularized and recognized urban neighborhoods. Some former slums in Mexico City, for example, have become so established that they are now attracting the wealthy and undergoing a process of gentrification (page 446).

Given their similarities across the globe, it is fair to ask whether global cities are converging in terms of how they look and "feel" to their residents (**Figure 10.27**). Is there, in other words, an emerging global city model? How much does it really matter if you are part of the global cosmopolitan class in Melbourne or Mumbai? Or, poor in Shanghai or São Paulo? This question is far too complex and subjective to be addressed appropriately here, but it is certainly worth considering.

FIGURE 10.27   **Global high-rise landscapes.** These photos were shot in Toronto, Dubai, Taipei, and Kuala Lumpur (clockwise from top left). ***Given the similarities of these skylines, could you have guessed where these photos were taken?*** *(Clockwise from top left: naibank/Moment/Getty Images; Iain Masterton/Getty Images; SeanPavonePhoto/iStockphoto/Getty Images; Martin Puddy/Getty Images.)*

NATURE–CULTURE

10.4   LEARNING OBJECTIVE
Recognize the ways that human and natural worlds interact in cities, and the effects on the human health and well-being of those living in cities.

At first glance, cities seem totally divorced from the natural environment. What possible relationships could the shiny glass office buildings, paved streets, and high-rise apartment complexes that characterize most cities have with forests, fields, and rivers? In this section, we examine several different ways to think about the relationships between urban life and its natural setting. We start by exploring the climate of cities, particularly the vegetation, weather, and hydrology unique to urban environments. We then turn to an examination of the reciprocal relationship that exists between increasing urbanization and global environmental problems such as climate change and vulnerability to natural disasters, and how cities are becoming increasingly resilient in the face of these challenges.

URBAN WEATHER AND CLIMATE

Cities alter virtually all aspects of local weather and climate. Temperatures are higher in cities, rainfall increases, the incidence of fog and cloudiness is greater, and levels of atmospheric pollution are much higher.

The causes of these changes are no mystery. Because cities cover large areas of land with streets, buildings, parking lots, and rooftops, about 50 percent of the urban area is a hard surface. Rainfall is quickly carried into gutters and sewers, so that little standing water is available for evaporation. Because evaporation removes heat from the air, when moisture is reduced, evaporation is lessened and air temperatures are higher.

Moreover, cities generate enormous amounts of heat. This heat comes not just from the heating systems of buildings but also from automobiles, industry, and even human bodies. The hard surfaces so prevalent in urban areas not only repel water, they also retain heat

very effectively. This results in a large mass of warmer air sitting over the city, called the **urban heat island** (**Figure 10.28**). The urban heat island causes yearly temperature averages in cities to be 3.5°F (2°C) higher than in the countryside; during the winter, when there is more city-produced heat, the average difference can easily reach 7°F to 10°F (4°C to 5.6°C).

> ■ **urban heat island** A mass of warm air generated and retained by urban building materials and human activities; it sits over the city and causes urban temperatures to be greater than those of surrounding areas.

Urbanization also affects precipitation. Because of higher temperatures in the urban area, snowfall will be about 5 percent less than in the surrounding countryside. However, rainfall can be 5 to 10 percent higher in cities as well as areas downwind from them. The increased rainfall results from two factors: the large number of dust particles in urban air and the higher city temperatures. Dust particles are a necessary precondition for condensation, offering a nucleus around which moisture can adhere. An abundance of dust particles, then, facilitates condensation. That is why fog and clouds are usually more frequent around cities.

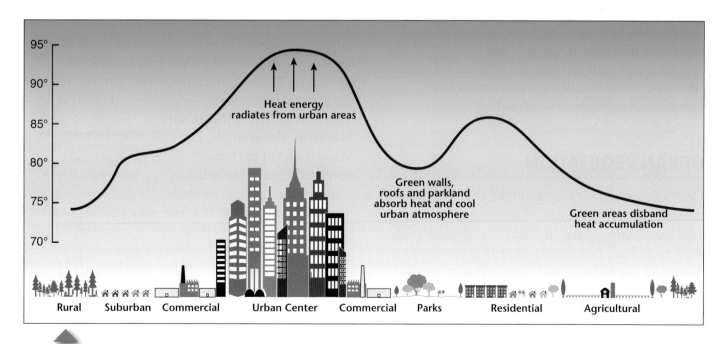

FIGURE 10.28 **Urban heat island.** Cities experience higher than normal air temperatures, thanks primarily to the heat retaining and deflecting properties of urban building materials. Cooling these overheated buildings stresses urban power grids, particularly in the summer. In addition, it is expensive and contaminates the atmosphere.

URBAN HYDROLOGY

Residential areas are usually the greatest consumers of water in urban areas. Water consumption can vary, but generally each person in the United States uses from 80 to 100 gallons (303 to 379 liters) per day in a residence. Residential demand is greater in drier climates as well as in middle- and high-income neighborhoods. Higher-income groups usually have a larger number of water-using appliances, such as washing machines, dishwashers, and swimming pools.

Not only is the city a great consumer of water, but it also alters water runoff patterns in a way that increases urban flooding. Urbanization can increase both the frequency and the magnitude of flooding because cities create large impervious areas where vegetation has been replaced with pavement, and water cannot soak into the earth. Instead, precipitation is converted into immediate runoff. It is forced into gutters, sewers, and stream channels that have been straightened and stripped of vegetation, which results in more frequent high water levels than may be found in a comparable area of rural land. Furthermore, the time between rainfall and peak runoff is reduced in cities; there is more lag in the countryside, where water runs across soil and vegetation into stream channels and then into rivers. So, because of hard surfaces and artificial collection channels, runoff in cities is concentrated and immediate, leading to urban flooding.

URBAN VEGETATION

Until a decade ago, it was commonly believed that cities were made up mostly of artificial materials: asphalt, concrete, glass, and steel. Studies, however, show that about two-thirds of a typical North American city is composed of trees and herbaceous plants (mostly weeds in vacant lots and cultivated grass in lawns). This urban vegetation, usually a mix of natural and introduced species, is a critical component of the urban ecosystem because it affects the city's topography, hydrology, and meteorology.

More specifically, urban vegetation influences the quantity and quality of surface water and groundwater; reduces wind velocity as well as turbulence and temperature extremes; affects the pattern of snow accumulation and melting; absorbs thousands of tons of airborne particulates and atmospheric gases; and offers a habitat for mammals, birds, reptiles, and insects, all of which play some useful role in the urban ecosystem. Furthermore, urban vegetation influences the propagation of sound waves by muffling much of the city's noise; affects the distribution of natural and artificial light; and, finally, is an extremely important component in the development of soil profiles that, in turn, control hillside stability.

Our urban settlements are still closely tied to the physical environment. Cities change these natural processes in profound ways, and we must understand these disturbances in order to make better decisions about adjustments and control.

CITIES AND ENVIRONMENTAL VULNERABILITY

As we examined above (see also Chapter 9), industrialization goes hand-in-hand with urbanization, so the environmental impacts of industrialization are often found in cities. But, urbanization generates its own set of environmental impacts: supplying enough energy, food, and water to large concentrations of people puts an array of stresses on the natural environment. Scholars refer to the extent of these varied impacts of urban areas on the environment as the **urban footprint**. For example, Las Vegas is one of the fastest-growing urban areas in the United States, yet it is located in a desert (**Figure 10.29**). The city's primary water source is the Colorado River, but in more ways than one the costs of delivering that water are extremely high. It is expensive to construct the dams and infrastructure necessary to move the water into the city, and the environmental damage to the region has been very costly. The dams alter the flows of water through the Colorado Valley, harming fish and disrupting aquatic life cycles, and the energy required to divert the water to the city leads to higher sulfur dioxide emissions, which have been shown to lead to

> **urban footprint** The spatial extent of the impacts of urban areas on the natural environment.

FIGURE 10.29 **Desert suburb.** This suburb of Las Vegas extends into the desert. Providing water to its residents is costly and causes environmental damage. *(Cameron Davidson/Getty Images.)*

global warming. Las Vegas's urban footprint, in other words, is large.

The environmental effects of increased urbanization—the urban footprint—can also spread beyond the immediate stresses caused by higher concentrations of population. Urban living often encourages rising levels of consumption, as new urbanites gain access to better jobs and acquire more disposable income. This growing demand for things like different foods can impact areas far from urban centers. For example, as the United Nations report *Unleashing the Potential of Urban Growth* suggests, tropical forests in Tabasco, an area 400 miles away from Mexico City, have been transformed into cattle-grazing areas in response to urbanites' demands for meat. In a second example, a major contributing factor to the deforestation of the Amazon is the increased demand for soybeans from the newly urbanizing regions of China as well as the urbanites of the United States, Japan, and Europe.

Urbanization, however, does not necessarily imply environmental degradation (see Subject to Debate on pages 444–445). For example, the higher densities of population in cities can be seen as a form of sustainable growth. Half the population of the world—the urban half—lives on approximately 3 percent of the landmass. The concentration of people in cities therefore opens up other areas that can be protected and left relatively free from human use. Many countries in the developing and developed world are working on local and regional planning projects that, on the one hand, maintain urban boundaries and prevent sprawl, and, on the other hand, protect natural environments outside urban areas from the sorts of environmental degradations we have just described. These sorts of urban environmental conservation projects will become increasingly important in future years.

NATURAL DISASTERS

If you follow the news, you are surely aware of the natural disasters that seem to strike urban areas with a certain vengeance: tornados, floods, hurricanes, landslides, and earthquakes are just some of them. For the most part, there is nothing particular to cities as such that makes them more vulnerable to natural disasters. Yet in urban areas, with their dense concentrations of people, disasters are more destructive because so many lives and livelihoods are at risk. In addition, since many cities are located near rivers or coasts, they often lie in the direct paths of disasters such as hurricanes and tsunamis.

Making matters even worse, scientists have recently noted an increase in the number of natural disasters, attributable in part, they believe, to global climate change (**Figure 10.30**). Sea level rise and storm surges have particularly impacted cities because, as we just noted, so many of them are located near coastlines. Rising temperatures have also lead to extended summer heat waves that affect city residents. This combination of increasing urbanization and a growing number of natural disasters means that more and more people's lives are being impacted adversely. It is most often poor people who are the most affected, since they tend to live in structures that are not well built and in sections of the

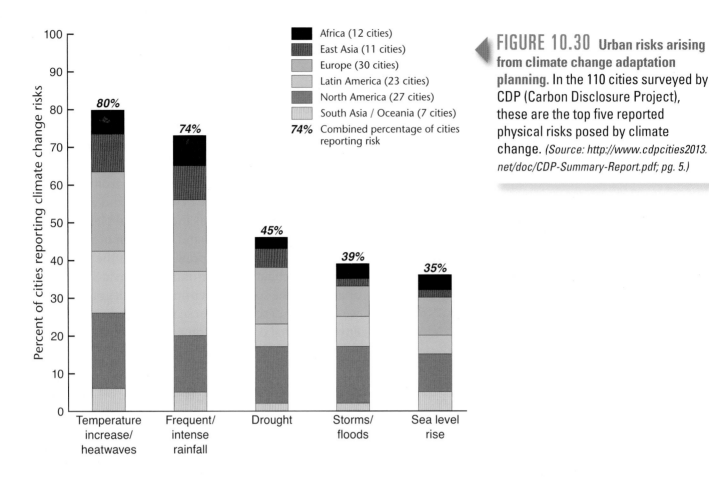

FIGURE 10.30 Urban risks arising from climate change adaptation planning. In the 110 cities surveyed by CDP (Carbon Disclosure Project), these are the top five reported physical risks posed by climate change. (*Source: http://www.cdpcities2013. net/doc/CDP-Summary-Report.pdf; pg. 5.*)

city that are more vulnerable. In cities in the developing world, shantytowns are frequently situated on steep hillsides or poorly drained areas, making these areas far more vulnerable to landslides, flooding, and other natural disasters.

The International Red Cross and Red Crescent Societies have termed this disparity in vulnerability to disaster "the urban risk divide" whereby "as the world's population becomes increasingly concentrated in large cities, we are seeing the urbanization of disasters and disaster risk." Some cities are better prepared for disasters when they strike—a component of resilience, discussed in the next section—while others, whether for lack of funding or concern on the part of government officials, are not. Compare, for instance, the human toll taken by two major earthquakes in 2010: in Chile, the death toll from a 8.8 magnitude quake was in the hundreds; while in Haiti, the death toll from a quake of slightly less magnitude was over 200,000 dead and more than 1 million homeless.

The wealthy, however, are not immune to natural disasters. Hurricane Sandy, which hit the coastline of New Jersey and New York in October 2012, disproportionately affected somewhat wealthier individuals. The storm caused $62 billion in damages and loss, with 375,000 homes destroyed, subway tunnels flooded, and business activities curtailed. (Also see Figure 3.27.) Without a doubt, global climate change and its urban impacts are a worldwide concern.

URBANIZATION, SUSTAINABILITY, AND RESILIENCE

How have cities—specifically, urban residents and governing bodies—responded to the negative environmental impacts associated with their presence and growth, and the challenges raised by natural disasters (particularly global climate change)? As with many geographic questions, the answer depends on where you look. Some cities have been leaders in creatively confronting environmental challenges and reducing risk, while others have lagged behind. You might be surprised by which areas of the world make the top of the list in the area of disaster readiness

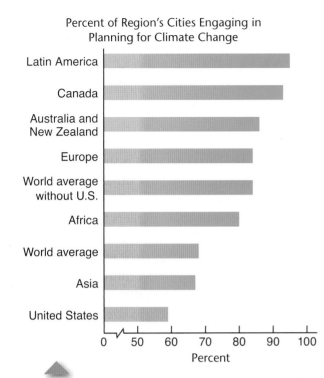

Percent of Region's Cities Engaging in
Planning for Climate Change

FIGURE 10.31 **Climate change adaptation planning.**
This chart shows the percentage of a region's cities that
are planning for climate change. The world average of
cities engaging in some planning, including U.S. cities, is
68 percent; without U.S. cities, the percentage rises to 84.
(Source: Carmin et al., 2012.)

mitigation measures can include restrictions on urban
industry regarding pollutant emissions, shifting to
low-carbon and renewable fuels, and reforestation to
reduce atmospheric carbon dioxide levels.

The very architecture of buildings can also be
reconfigured in ways that mitigate climate change.
For instance, the installation of so-called green roofs
and green walls provides a surface that does not
retain heat (thereby addressing the urban heat island
phenomenon), that absorbs water (addressing rapid
runoff) and carbon dioxide. In addition, these
surfaces calm urban noise levels and provide a
pleasant environment for urban humans, and refuge
for urban wildlife (**Figure 10.32**).

FIGURE 10.32 **Green walls.** The city of Sydney,
Australia, has developed a "Green Roofs and Walls"
initiative in an attempt to encourage more beautiful and
environmentally friendly urban surroundings. *(James D.
Morgan/Rex Features/AP Photos.)*

(**Figure 10.31**). As a region, Latin American cities
lead the pack, with 95 percent of the region's cities
engaging in some form of climate change adaptation
planning, while U.S. cities lag behind all other places
at 59 percent. Why is the United States so far behind
everyone else in this area? The reasons are varied and
can differ from city to city. Generally, however, the
conversation about global climate change is a
relatively recent one, coming at a time when many
U.S. cities are facing severe budgetary shortfalls and
even declaring bankruptcy. In addition, there are still
many local as well as national policy makers in the
United States who, for political or ideological reasons,
do not believe that global climate change is a reality.

Cities can take some measures that attempt to
mitigate, or lessen, the impact of climate change on
natural systems, humans, and the urban built
environment. Mitigation attempts to reduce the extent
of climate change by decreasing the emission of
greenhouse gasses into the environment by cities. Such

CAN URBANIZATION BE ENVIRONMENTALLY SUSTAINABLE?

The fact that more than half of the world's population now lives in cities rather than rural areas has refueled the debate over the environmental impacts of increasing urbanization. On the one hand, increasing urbanization can be seen as efficient. Cities concentrate human populations at one point in space, which leaves surrounding lands available for conservation, agriculture, or other low-density uses. Cities create economies of scale because more people can benefit from the urban infrastructure—transportation, power supply routes, water servicing, and so on—allowing these services to be delivered at a lower cost per person. On the other hand, these same concentrations of population lead to greater demands on resources that must be brought in from outside of the city, such as water, energy, and food. City residents, with their relative wealth, tend to demand more and better resources, and such consumer demands, in turn, impact large swaths of countryside outside the city that are recruited to meet these demands. Are cities, then, sustainable?

First, we have to think about what *sustainable* actually means, and this issue itself is subject to debate! For our purposes here, sustainable urbanization is the creation of a situation whereby a society can meet the needs of contemporary urban dwellers for water, food, and shelter, while not damaging the ability of future urban dwellers to meet their needs. It was only in the late nineteenth and the twentieth centuries, with industrialization (see Chapter 9), that cities in Europe and North America grew rapidly in size, adding stress to the environment. That stress has been further exacerbated with the recent and rapid growth of cities in the developing world. And, it has been this recent and rapid growth that has caused alarm among policy makers and scholars.

On one side of the debate are those who focus on the adverse environmental impacts of large concentrations of people. Not only do cities have a direct impact on the land, in terms of the amount of space they occupy and the pollution they create; they also have indirect impacts, since urban dwellers

typically have higher incomes and expectations in terms of material satisfaction than do rural dwellers, which places additional stress on the environment. More consumer demand for items such as meat, wood, and metals from city dwellers in, say, Vancouver, can have adverse environmental impacts throughout Canada and other parts of the Asia-Pacific Rim.

Those who are optimistic about creating a more sustainable future with increasing urbanization argue that higher-density human settlements are better for the environment in the long run than less dense settlements. Even with the current growth rate of cities in terms of size and number, recent estimates based on satellite data show that urban settlements occupy only 2.8 percent of the total land on Earth. Urban spatial expansion per se does not appear to be a major environmental stressor. Urban density actually helps maintain fragile ecosystems by keeping them free from human interference. Proponents of this side of the debate argue that policies geared toward dispersing the world's population away from big cities are misguided. The real problem, they argue, rests with unsustainable forms of production and consumption, and it is on managing these issues that the world's attention should be focused.

Continuing the Debate

The debate over the environmental sustainability of urbanization is certain to grow in intensity as the world's population continues its urban course. After all, many, many people are drawn to urban life. Given what you have read here, consider the following questions:

- Can an increasingly urban world become sustainable? How?

- Does your city have a sustainability plan? If so, how might you learn more about this plan and its feasibility?

- How vulnerable is your city to environmental or man-made threats? Is your city among those that have begun to engage in resilience planning (see the Nature–Culture section in this chapter)?

Aerial view of New York City. Because of the high density of its population and its mass transportation system, New York City is considered to be a relatively "green" city. Urban parks, like Central Park depicted here, often provide important sanctuaries for flora and fauna in the heart of the city. *(Panoramic Images/Getty Images.)*

Yet in many ways, the conversation on climate change is shifting from a focus on mitigation or even **sustainability**—attempting to preserve conditions as close as possible to what they have been in the past—to include the notion of resilience. **Resilience** involves the ability to better predict future trends rather than looking backward to how things used to be. Resilience involves an acceptance of climate change as the new normal, and an emphasis on how to most appropriately respond to threats—both human (e.g., terrorism) and natural—in order to minimize risk to human lives, livelihoods, and property.

> **sustainability** The ability to use resources in a way that does not deplete them over the long term.
>
> **resilience** The ability to recover quickly from adversity.

How cities respond to climate change will be a defining feature of life on Earth in the coming decades.

CULTURAL LANDSCAPE

10.5 LEARNING OBJECTIVE
Read an urban landscape and identify the trends shaping cities today and into the future.

As you probably already know from your local, regional, or international travels, the look and "feel" of cities varies from place to place. Some cities have bustling downtowns where most of the residents work and perhaps live, while others are characterized by nodes of activities clustered on the suburban fringe; some cities are built around a set of historical buildings with political or religious significance, while the centers of other cities are dominated by new skyscrapers housing financial offices. Finally, some cities' streets are laid out in a gridlike pattern, while in other cities streets are a crisscrossed jumble without any apparent order.

How can we make sense of this range of urban landscapes? In Doing Geography on page 452, you can try your hand at reading the urban landscape in the way that cultural geographers do. Here, we will discuss some of the major trends reshaping the look and feel of cities across the globe: gentrification, efforts to create more livable cities, and movements to

"take back the city" from developers who have increasingly privatized formerly public urban spaces.

GENTRIFICATION

Beginning in the 1970s, urban scholars began to observe what seemed to be a trend opposite to suburbanization. This trend, called **gentrification**, was the movement of upper-middle-class people into deteriorated areas of city centers (see also Chapter 2). Today, some older suburbs are becoming gentrified, as well as some neighborhoods that began their lives as shantytowns (as we saw earlier with respect to Mexico City; **Figure 10.33**). Gentrification often begins in an inner-city residential district, with gentrifiers moving into rundown housing or even spaces that didn't start out as housing—for example, warehouses. These places are more affordable than newer suburban housing and are often closer to workplaces or mass transit lines. The infusion of new capital into these neighborhoods usually results in higher property values, and this, in turn, often displaces existing residents who cannot afford the

> **gentrification** The displacement of lower-income residents and economic activities by higher-income residents and activities; frequently associated with a restoration of buildings in deteriorated areas of the city.

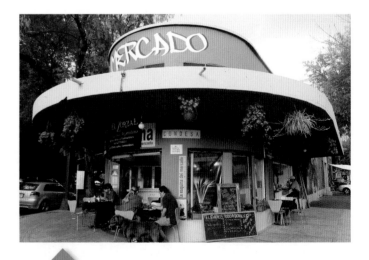

FIGURE 10.33 **Global gentrifiers.** Condesa is a neighborhood in Mexico City—probably not a place you associate with gentrification. Yet, several Mexico City neighborhoods, some that started their lives as lower-income settlements, have become targets of gentrification because of their desirable locations. *(Eye Ubiquitous/Alamy.)*

higher prices. The displaced residents are disproportionately poor, immigrant, and racialized populations. Their displacement opens up even more housing for gentrification, and the gentrified district continues its spatial expansion.

Commercial gentrification usually follows residential gentrification, as new patterns of consumption are introduced into the inner city by the middle-class newcomers. Pedestrian shopping districts with expensive boutiques and art galleries attract cultural consumers, and bars and restaurants catering to this new urban middle class provide entertainment and nightlife for the gentrifiers (see Figure 10.18).

The speed with which gentrification has proceeded in many of our cities, and the extent of changes—in the urban landscape, but also culturally, economically, and politically—that accompany gentrification are causing dramatic shifts in the urban fabric. What factors have led to gentrification?

Economic Factors Some urban scholars look to broad economic trends to explain gentrification. Throughout the post–World War II era in North America and Europe, most investments in metropolitan land were made in the suburbs; as a result, land in the inner city was devalued. By the 1970s, many home buyers and commercial investors found land in the city much more affordable, and a better economic investment, than in the higher-priced suburbs. This situation brought capital into areas that had been undervalued and accelerated the gentrifying process.

In addition, most Western countries have been experiencing **deindustrialization**, a process whereby the economy is shifting from one based on secondary industry to one based on the service sector. This shift has led to the abandonment of older industrial districts in the inner city, including waterfront areas. Many of these areas are prime targets of gentrifiers, who convert the waterfront from a noisy, commercial port area into an aesthetic asset. In Buenos Aires, for example, one of the new gentrified neighborhoods is Puerto Madero, an area that was once home to docking facilities and a wholesale market (**Figure 10.34**; see also Mona's

> **deindustrialization** The decline of industry, accompanied by a rise in service and information sectors of the economy.

FIGURE 10.34   Puerto Madero, Buenos Aires. These expansive dock facilities in Buenos Aires have been converted by the city and private entrepreneurs into a center of nightlife for the city, complete with clubs, restaurants, and shopping. New condominiums and hotels are now under construction. *(Courtesy of Mona Domosh.)*

Notebook, page 448). The shift to an economy based on the service sector also means that the new productive areas of the city will be dedicated to white-collar activities. These often take place in relatively clean and quiet office buildings, contributing to a view of the city as a more livable environment.

Political Factors Many metropolitan governments in the United States, faced with the abandonment of the central city by the middle class and therefore with the erosion of their tax base, have enacted policies to encourage commercial and residential development in downtown areas. Some policies provide tax breaks for companies willing to locate downtown; others furnish local and state funding to redevelop central-city residential and commercial buildings.

At a more comprehensive level, some larger metropolitan areas have devised long-term planning agendas that target certain neighborhoods for revitalization. Often this is accomplished by first condemning the targeted area, thereby transferring control of the land to an urban-development authority or other planning agency. Such areas are frequently older residential neighborhoods originally

Mona's Notebook

"Seeing" New Places

Mona Domosh.

As a cultural geographer, I've thought a bit about how people often see new places through the "eyes" of familiar ones. You, too, have most likely noticed this. For example, when you travel somewhere new, you often compare it to what you know, saying things like "This street reminds me of the one I grew up on" or "Don't we have almost an identical building in our town?" I shouldn't have been surprised then when on a trip to Argentina I noticed that parts of Buenos Aires looked like other cities I was familiar with, particularly the recently gentrified neighborhoods in the city (see Figure 10.34, page 447). The brick, four-story buildings that had once lodged warehouses and now were home to fancy apartments and upscale restaurants and bars looked to my North American eye to be almost identical to the gentrified waterfronts of Boston and New York City. But, they weren't. There are similarities, of course, since these waterfronts served similar purposes in the past, but each city and its history have put a particular stamp on its urban landscape. Boston's gentrified waterfront, for example, was built much earlier and in a style different from the Buenos Aires waterfront, and granite was used as a building material as much as brick.

I *was* surprised, however, when I visited other parts of Buenos Aires and realized that some neighborhoods were making explicit their similarity to and connection with North American cities through their place-names. Fairly recently, Palermo, a neighborhood in the process of gentrification in northeast Buenos Aires, has been subdivided into smaller districts, including Palermo Soho and Palermo Hollywood. The name Palermo itself is already a place-name taken from somewhere else (a relatively common feature of immigrant nations), but the twinning of that now well-established name with Soho and Hollywood was something new to me. As it turns out, I learned that other places in the world have taken the *Soho* name and used it to lend a certain trendy and hip connotation to their neighborhoods. Most likely, developers and entrepreneurs thought that using Soho, a term derived from *So*uth of *Ho*uston Street (a very chic neighborhood of lofts and trendy restaurants and bars in New York City), would boost the area's commercial potential. And, the global symbol of Hollywood, I realized, needed no explanation: it connoted a West Coast version of glamour. The local and the global often meet in very interesting ways!

The landscape of Boston's gentrified waterfront displays similarities to Puerto Madero, but it is also unique. *(Courtesy of Mona Domosh.)*

built to house people who worked in nearby factories, which have usually been torn down or transformed into lofts or office space. The redevelopment authority might locate a new civic or arts center in the neighborhood. Public-sector initiatives often lead to private investment, thereby increasing property values. These higher property values, in turn, lead to further investment and the eventual transformation of the neighborhood into a middle- to upper-class gentrified district.

Sexuality and Gentrification Gentrified residential districts are often correlated with the presence of a significant gay and lesbian population (Chapter 2, pages 56–59). It is fairly easy to understand this correlation. First, the typical suburban life tends not to appeal to people whose lifestyle is often regarded as different and whose community needs frequently diverge from those of people living in the child-centric suburbs. Second, gentrified inner-city neighborhoods provide access to the diversity of city life and amenities that often include gay cultural institutions. In fact, the association of urban neighborhoods with gays and lesbians has a long history. For example, urban historian George Chauncey has documented gay culture in New York City between 1890 and World War II, showing that a gay world occupied and shaped distinctive spaces in the city, such as neighborhood enclaves, gay commercial areas, and public parks and streets.

Yet, unlike this earlier period, when gay cultures were often forced to remain hidden, the gentrification of the postwar period has provided gay and lesbian populations with the opportunity to reshape entire neighborhoods actively and openly. Indeed, many geographers have explored how gays and lesbians played very influential roles in the gentrification of certain neighborhoods, which have become heavily identified with these populations. Other geographers have focused on black gentrifiers in New York's Harlem and Chicago's South Side neighborhoods. Both areas of work underscore that it is not always affluent whites who are the primary movers and shakers in gentrification. More recently, however, urban geographers have pointed out that while these contributions are important, it is vital to remember gentrification is mostly driven by and for affluent white urban populations.

The Costs of Gentrification Gentrification often results in the displacement of lower-income people who are forced to leave their homes because of rising property values. This displacement can have serious consequences for the city's social fabric. Because many of the displaced people come from disadvantaged groups, gentrification frequently contributes to racial and ethnic tensions. Displaced people are routinely forced into neighborhoods more peripheral to the city, a trend that only adds to their disadvantages. In addition, gentrified neighborhoods usually stand in stark contrast to surrounding areas where investment has not taken place, thus creating a very visible reminder of the uneven distribution of wealth within cities.

The success of a gentrification project is usually measured by its appeal to an upper-middle-class clientele. Indeed, the so-called hipster gentrifiers—artists, musicians, youth, and other creative types initially attracted to these lower-cost neighborhoods—are often displaced by later waves of wealthier incomers who find the "cool factor" of such up-and-coming neighborhoods irresistible. Richard Campanella refers to the final wave as "*bona fide* gentry, including lawyers, doctors, moneyed retirees, and alpha professionals from places like Manhattan or San Francisco." Thus, gentrification ultimately gravitates toward the suburban notion of residential homogeneity, eliminating what many people consider to be a great asset of urban life: its diversity and heterogeneity.

THE LIVABLE CITY

Probably for as long as humans have lived in cities, there have been some who have dedicated themselves to contemplating exactly what makes city life so attractive, pondering as well how we might go about making it better. Some five thousand years ago, urbanites in Mesopotamia, Egypt, and the Indus River Valley planned their cities to have straight streets that intersected at right angles and devised drainage systems. Since then, we have attempted to invent better—more beautiful and more efficient—ways to house, employ, and entertain city residents; to convey people and the things they need within cities and between them; and to fortify cities for defense from attack.

What makes a city livable? In many ways, the answer to that question depends on who you ask. Certainly, beautiful surroundings—natural and man-made—are important. Clean air and water, public transit, freedom from crime, historical importance, accessibility, affordability—all of these factors and more are also vital components of making a city not just livable, but truly great. Indeed, making the list of "the world's most livable cities" is something many cities would like to achieve; you might not be surprised to learn then that many such lists exist. And, given that there is no real way to measure and thereby quantify many of the positive aspects of cities listed above, a lot of subjectivity goes into compiling these lists. In 2012, for instance, Melbourne, Australia, was at the top of the Economic Intelligence Unit's "Most Livable Cities" list. Yet, other listings place Vienna or Copenhagen in the top spot. All of the lists, however, skew notably toward Australia, New Zealand, Canada, and continental Europe.

Refocusing cities on people seems to be at the heart of movements to make cities more livable. Moving away from individually owned and operated automobiles as the drivers, literally and figuratively, of our cities is part of the solution (**Figure 10.35**). As we discussed earlier in this chapter, more and more cities are adopting bike- and car-sharing programs and encouraging mass transit, carpooling, biking, and walking as inexpensive and less polluting ways to get around.

As our reliance on the automobile decreases, cities can be reshaped. No longer beholden to paved roadways, parking lots and garages, and long hours commuting alone in cars across sprawling urban jungles, people are freer to spend time in their neighborhoods. Space becomes available for meeting places such as plazas, parks, and pedestrian areas. With fewer automobiles, city spaces become safer for bicyclists and pedestrians. Danish architect Jan Gehl describes the multiple activities that city residents can engage in when they inhabit more livable city spaces: "purposeful walks from place to place, promenades, short stops, longer stays, window shopping, conversations and meetings, exercise, dancing, recreation, street trade, children's play, begging and street entertainment."

THE RIGHT TO THE CITY

Throughout this chapter, we have noted the processes that change cities—their layout, demographic profile, ecology, and what it means to live in them—affect people differently. Almost invariably, immigrants, racialized minorities, and the impoverished experience negative consequences stemming from urban transformations. These are the people who are displaced when neighborhoods gentrify. They are most at-risk from natural disasters and climate change. They have the least access to transportation, housing, and urban services.

We have also seen, for instance in the case of gated communities discussed earlier in this chapter, that some processes at work in cities are explicitly aimed at excluding disadvantaged people from access to certain parts of the city. To provide an additional example, think about the last time you visited a shopping mall. Did you see a sign like the one pictured in

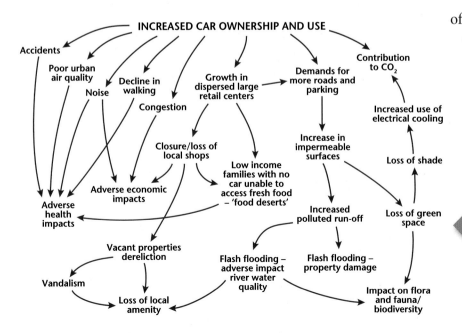

FIGURE 10.35 Cars and urban consequences. This flowchart, adapted from one produced by the government of the United Kingdom, shows the various negative environmental and social outcomes of increased car ownership and use.

FIGURE 10.36 **Code of conduct.** Though you may think of shopping malls as public places, they are not. Simon Property Group, a company that owns and operates many shopping malls in the United States, has posted a code of conduct for its visitors. Some malls have adopted parental escort policies or parental patrols, and ban groups of unaccompanied minors from occupying the mall during certain days or times. *(Courtesy of Patricia L. Price.)*

Figure 10.36? Though it may come as a shock to many young people, shopping malls are, in fact, not public spaces. While they may look open to the public—indeed, some open-air shopping malls are not enclosed by any physical barrier such as walls or fences—most shopping malls are owned by developers and function as private spaces. The mall developers look out for the business interests of their tenants: stores renting spaces in the mall. When groups of youths, or homeless individuals, use shopping malls for social gatherings, to take care of daily needs like sleeping or bathing, or as venues for begging, they are swiftly escorted from the premises. If they return, such individuals can be cited for trespassing or even disturbing the peace. Why? As one developer put it, his Albany, New York, mall "had become a babysitting service . . . with thousands of teen mall rats roaming around." Rambunctious crowds of teens at the food court and theater entrance were seen as frightening away older shoppers.

But, what about the truly public spaces of cities: the parks, pedestrian promenades, plazas, libraries, sidewalks, and beaches? Are certain people denied access to public urban spaces? The

right to the city The notion that all urban residents, not just the privileged, should be able to access city spaces and have a voice in how the city is shaped and used.

answer, unfortunately, is yes. What urban scholar Henri Lefebvre called the **right to the city**, specifically the right of all urban residents to be in and move through public urban places, is becoming increasingly curtailed. More of these spaces are becoming private property, and developers can then decide who stays and who goes. Some beaches, for instance, have been purchased by hotel and condominium developers who then restrict access to them to residents and guests only. Surveillance technologies such as drones and cameras may be used to watch people in public spaces and have them removed based on what is deemed threatening or disruptive behavior. The spaces themselves may be configured in ways that make them inaccessible to certain people or for certain purposes. This area of urban planning goes by the name of "crime prevention through environmental design." For instance, benches in waiting areas for public transit are often constructed in ways that deter people from sleeping on them (**Figure 10.37**).

Are our cities made more or less livable by such measures? What is lost when access to city spaces is curtailed? What is gained? Who benefits, and who

FIGURE 10.37 **Crime prevention through environmental design.** This bench, located at a Miami city bus stop, has bars placed across its surface to deter sleeping. These benches are referred to as "bum-proof." *(Courtesy of Patricia L. Price.)*

CONCLUSION

The first cities arose as new technologies—particularly the domestication of plants and animals—facilitated the concentration of people, wealth, and power in a few specific places. This transformation from village to city life was accompanied by increased social stratification. Although the first cities developed in specific hearths, urban life has now diffused worldwide, and the majority of the world's population has become urban.

The future of cities is an evolving question. Focused urban planning measures might alleviate many present-day ills, but long-range hope lies with endeavors to make cities more livable and resilient.

DOING GEOGRAPHY

Reading "Your" Urban Landscape

Even without extensive training in geography, most of us are avid "readers" of our landscapes. We drive through residential neighborhoods and make judgments about the types of people who live there; we walk downtown and, finding that the streets grow narrower, surmise that they were built a long time ago, before automobile traffic was a factor; we marvel at how high buildings rise into the sky, knowing that those with offices at the top make "top" dollar. But, reading landscapes can be tricky business. What we see in front of us may not always be as it appears, nor built when and for whom we think. This exercise, then, is about interpreting your own urban landscape and about understanding the limitations of that interpretation.

Your goal in this exercise is to create an interpretative walking tour of part of your city that could be used by tourists (or anyone, really) and that is both informative and engaging. In other words, the idea is to get people to look around them, think, and make connections—but also to understand the limitations of "just looking." To do this exercise, follow the steps below.

Steps to Reading Your City

Step 1: Pick your place. Decide on a route that takes people through a particular part of your city—it could be downtown, an ethnic landscape, a commercial area, or a local neighborhood. Almost anywhere will work. After reading this chapter, you should have a fairly good idea of the types of urban landscapes and those that are of interest to you.

Step 2: Walk the route yourself, marking places where you want people to stop, look, and think.

Step 3: Investigate the places you marked. You should be able to use the information in this chapter and elsewhere in the book to do a good bit of the investigative work, but the specifics of your city will require more information: archival research, perhaps informal interviews, or something else altogether. At each point, you should ask yourself: What can I learn from looking at this landscape and what would I never have guessed?

Step 4: Based on what you have learned, write your interpretative tour, providing information and questions for each particular stop on the route

Finally, try it out—give your route map and written descriptions and questions to a friend, and see what happens.

Observing the urban landscape. (Gerhilde Skoberne/Corbis)

KEY TERMS

agricultural surplus	p. 417	range	p. 425
automobile cities	p. 431	resilience	p. 446
axis mundi	p. 418	right to the city	p. 451
central business district (CBD)	p. 431	second urban revolution	p. 422
central place theory	p. 424	shantytowns	p. 437
cosmomagical cities	p. 418	sprawl	p. 431
deindustrialization	p. 447	streetcar suburbs	p. 431
first urban revolution	p. 421	sustainability	p. 446
gated communities	p. 436	threshold	p. 425
gentrification	p. 446	urban footprint	p. 440
global cities	p. 433	urban hearth areas	p. 418
hydraulic civilization	p. 417	urban heat island	p. 439
megacities	p. 427	urbanized population	p. 414
primate city	p. 430		

SEEING GEOGRAPHY Rio de Janeiro

What are some of the major environmental and social impacts of an increasingly urbanized world?

A view of Rio de Janiero, Brazil, looking toward Sugarloaf Mountain.
(isitsharp/Vetta/Getty Images.)

Few cities boast such a spectacular site as Rio de Janeiro, located between the mountains and Guanabara Bay along the Atlantic Ocean. This view highlights the dramatic siting: the bustling city, the beautiful beaches, the rugged mountain peaks against the sky. One can barely discern in this image the "other" side of Rio: the favelas, or slums, located on the mountainsides (see Figure 10.26 page 437, also the World Heritage Site feature for this chapter). Perched above the centers of economic activity and the middle- and upper-class residential areas located along the southern coastal areas of Rio, the favelas are an ever-present part of the urban landscape and home to approximately one-fifth of the city's population.

The information in this chapter should help us to "read" this image of Rio de Janeiro. As a city of the developing world, Rio experienced rapid population growth in the twentieth century, and its metropolitan area now exceeds 12 million people. More striking is the dramatic increase in the urban population of Brazil. The percentage of the population living in metropolitan areas rose from approximately 31 percent in 1940 to about 84.2 percent in 2010. In Rio, that population growth is evident in the intensity of land use within the city, which is marked by the presence of skyscrapers and high-rise apartment buildings, the metropolitan sprawl that extends well beyond the parameters of this image, and the presence of the favelas, home to many of the rural-to-urban migrants. Like other cities of the developing world, population growth has strained the city's infrastructure and its ability to provide services, which has led to traffic congestion, pollution, and crime. It has also exacerbated ecological problems. When vegetation covered the hillsides, the heavy summer rains Rio experiences were absorbed into the soil. Now the summer rains often flood the streets of the low-lying areas of the city and lead to landslides on the slopes that house the favelas. One such episode in 2011, for example, led to the death of 700 people. Yet Rio, like its larger neighbor 230 miles (370 kilometers) to the south, São Paulo, is now experiencing much slower population growth as a result of lower birthrates and less rural-to-urban migration.

Founded as a Portuguese colonial city in 1565, Rio grew quickly in the eighteenth century, when it became the primary port for exporting the gold and diamonds discovered in the interior of the country. Later, in the first decades of the twentieth century, Rio underwent industrialization. The southern and coastal portions of the city became home to the elite, while the factories and working classes moved north and west of the downtown.

Today, Rio is part colonial city and part global city. It is home to the regional headquarters of 10 multinational firms, hosted the World Cup soccer games in 2014, and will host the summer Olympic Games in 2016. Its beaches, particularly Ipanema and Copacabana, are icons for global jet-setters. Yet its favelas, some of which have now been recognized by the government as legal communities, continue to grow, with little infrastructure and few public services. Like other globalizing cities, it experiences both the bright lights and grimmer realities of the twenty-first-century economic order.

Chapter 10
LEARNING OBJECTIVES REEXAMINED

10.1 Describe the historical and contemporary regional patterns of urbanization, and the models geographers use to understand cities.
See pages 414–425. What is the hydraulic civilization model? Does this model address the development of all urban hearths? Why or why not?

10.2 Identify the ways that migration has shaped the growth of cities, how cities have grown and changed, and how humans get around within cities.
See pages 425–433. What is rural-to-urban migration and how has it affected the growth of cities?

10.3 Discuss the notion of global cities and some of the problems associated with them.
See pages 433–437. What defines a global city? Name three global cities.

10.4 Recognize the ways that human and natural worlds interact in cities, and the effects on the human health and well-being of those living in cities.
See pages 438–446. What is an urban footprint and what affect does it have on the natural environment?

10.5 Read an urban landscape and identify the trends shaping cities today and into the future.
See pages 446–451. What is gentrification and what positive and negative effects has it had on cities?

The City on the Internet
You can learn more about the city in time and space on the Internet at the following web sites:

A Shocking Satellite Tour of the World's Biggest Slums
http://www.businessinsider.com/worlds-biggest-slums-2011-2?op=1
Satellite images of the world's largest slums, showing detailed aerial views of their spatial layout, density, and relationship to the surrounding countryside.

Levittown: Documents of an Ideal American Suburb
http://tigger.uic.edu/~pbhales/Levittown/
Compiled by an art historian, this web site documents the cultural history of Levittown, New York—the first mass-produced automobile suburb in the United States.

McKinsey & Company: Urban World
http://www.mckinsey.com/insights/urbanization/urban_world
This site hosts an interactive map of the next generation of economically dynamic cities; in addition, there are interesting slideshows and other information showcasing the role of cities and urban dwellers as consumers.

United Nations Habitat for a Better Urban Future
http://www.unhabitat.org/
Provides a wealth of resources on the world's cities, focusing in particular on urban environmental challenges and climate change.

Urban Resilience
http://urbanresilience.net/
Maintained by Jon Coaffee, an urban geographer in the United Kingdom, this website hosts a blog and a variety of other resources on the topic of resilience in the face of natural as well as man-made threats.

Sources

Brennan, Morgan. 2012. "Houston Tops Our List of America's Coolest Cities." *Forbes,* 26 July. http://www.forbes.com/sites/morganbrennan/2012/07/26/houston-tops-our-list-of-americas-coolest-cities-to-live/.

Campanella, Richard. 2013. "Gentrification and Its Discontents: Notes from New Orleans." *NewGeography* blog post, 1 March. http://www.newgeography.com/content/003526-gentrification-and-its-discontents-notes-new-orleans.

Carmin, JoAnn, Nikhil Nadkarni, and Christopher Rhie. 2012. *Progress and Challenges in Urban Adaptation Planning: Results of a Global Survey.* Cambridge, Mass.: Massachusetts Institute of Technology. http://www.icleiusa.org/action-center/learn-from-others/progress-and-challenges-in-urban-climate-adaptation-planning-results-of-a-global-survey.

Chauncey, George. 1994. *Gay New York: Gender, Urban Culture, and the Making of the Gay Male World 1890–1940.* New York: Basic Books.

Christaller, Walter. 1966. *The Central Places of Southern Germany.* C. W. Baskin (trans.). Englewood Cliffs, N.J.: Prentice-Hall.

Davis, Mike. 2006. *Planet of Slums.* London: Verso.

Diem, William. 2013. "Decline of Car Culture Under Scrutiny in France." *WardsAuto,* 11 June. http://wardsauto.com/europe/decline-car-culture-under-scrutiny-france.

Florida, Richard. 2013. "Global Cities of the Super-Rich." *The Atlantic Cities: Place Matters,* 21 March. http://www.theatlanticcities.com/jobs-and-economy/2013/03/global-cities-super-rich/4951/.

Garthwaite, Josie. 2011. "Bike Share Schemes Shift into High Gear." *National Geographic Daily News.* http://news.nationalgeographic.com/news/energy/2011/06/110607-global-bike-share/.

Huang, Yefang, Yee Leung, and Jianfa Shen. 2007. "Cities and Globalization: An International Perspective." *Urban Geography* 28: 209–231.

International Federation of Red Cross and Red Crescent Societies. 2010. *World Disasters Report 2010: Focus on Urban Risk.* http://www.ifrc.org/Global/Publications/disasters/WDR/WDR2010-full.pdf.

Kearney, A. T., and the Chicago Council on Global Affairs. *2012 Global Cities Index and Emerging Cities Outlook.* http://www.atkearney.com/documents/10192/dfedfc4c-8a62-4162-90e5-2a3f14f0da3a.

Kotkin, Joel. 2013. "Houston Rising: Why the Next Great American Cities Aren't What You Think." *The Daily Beast,* 8 April. http://www.thedaily-beast.com/articles/2013/04/08/houston-rising-why-the-next-great-american-cities-aren-t-what-you-think.html.

Le Goix, Renaud, and Chris J. Webster. 2008. "Gated Communities." *Geography Compass* 2(4): 1189–1214.

Sassen, Saskia. 1991. *The Global City: New York, London, Tokyo.* Princeton, N.J.: Princeton University Press.

Schaefer, Kathleen. 2010. "New Policies Exterminating Teen Mall Rats." *ABC News,* 23 September. http://abcnews.go.com/Business/shopping-malls-increasingly-putting-restrictions-teens/story?id=11701470.

Schueth, Sam, and John O'Loughlin. 2008. "Belonging to the World: Cosmopolitanism in Geographic Contexts." *Geoforum* 39(2): 926–941.

Schwartz, John. 2013. "Young Americans Lead Trend to Less Driving." *The New York Times,* 13 May. http://www.nytimes.com/2013/05/14/us/report-finds-americans-are-driving-less-led-by-youth.html?pagewanted=all&_r=0.

United Nations, Department of Economic and Social Affairs, Population Division. 2009. *World Urbanization Prospects: The 2009 Revision.*

United Nations Human Settlements Programme (UN-HABITAT). 2010. *State of the World's Cities 2010/11: Cities for All: Bridging the Urban Divide.*

Wittfogel, Karl. 1957. *Oriental Despotism: A Comparative Study of Total Power.* New Haven, Conn.: Yale University Press.

World Bank. 2009. *Systems of Cities: Harnessing Urbanization for Growth and Poverty Alleviation.* Washington, D. C.

Ten Recommended Books on Urban Geography

(For additional suggested readings, see the *Contemporary Human Geography* LaunchPad: http://www.macmillanhighered.com/launchpad/DomoshCHG1e.)

Boo, Katherine. 2012. *Behind the Beautiful Forevers: Life, Death, and Hope in a Mumbai Undercity.* New York: Random House. Written by a journalist, this book details the lives of Mumbai slum dwellers and sheds light on the huge gap between rich and poor in India's cities.

Brown, Michael, and Richard Morrill (eds.). 2011. *Seattle Geographies.* Seattle: University of Washington Press. University of Washington geographers Brown and Morrill, alongside many U.W. geography faculty and graduate students, unpack the richly layered historical, cultural, and urban geographies of the Emerald City.

Cole, Teju. 2011. *Open City.* New York: Random House. This novel follows Jules, a Nigerian immigrant, in his walks around Manhattan and through cities of his past.

Davis, Mike. 2006. *Planet of Slums.* New York: Verso. A devastating overview of people's everyday struggles to maintain decent lives in the world's urban slums.

Fussey, Pete, Jon Coaffee, Gary Armstrong, and Dick Hobbs. 2011. *Sustaining and Securing the Olympic City.* Farnham, U.K.: Ashgate. Examines how cities that host the Olympic Games are profoundly impacted both spatially and socially.

Gehl, Jan. 2010. *Cities for People.* Washington, D.C.: Island Press. Architect Gehl draws on lessons from his involvement in urban planning in Melbourne and Moscow to discuss what it is that makes cities livable.

Peake, Linda, and Martina Rieker (eds.). 2013. *Rethinking Feminist Interventions into the Urban.* Abingdon, U.K.: Routledge. A collection of essays that utilizes feminist insights to challenge the conceptual and practical divisions between "developed" cities in the Global North and "developing" cities in the Global South.

Ross, Andrew. 2011. *Bird on Fire: Lessons from the World's Least Sustainable City.* New York: Oxford University Press. Provides a sobering look at Phoenix, Arizona––a sprawling automobile-oriented city of 4 million that has recorded the hottest summertime temperatures of any North American city and is poised for an even dryer, hotter, and less sustainable future as climate change progresses.

Rybczynski, Witold. 2010. *Makeshift Metropolis: Ideas about Cities.* New York: Scribner. Argues that market forces, rather than urban planning, are shaping today's cities.

Speck, Jeff. 2012. *Walkable City: How Downtown Can Save America, One Step at a Time.* New York: Farrar, Straus and Giroux. Acknowledges that we already know what is wrong with American cities, and makes practical suggestions for how cities can prioritize pedestrians in their attempts to become more livable and welcoming places.

Journals in Urban Geography

Environment and Urbanization. Published by SAGE, Thousand Oaks, Ca. Volume 1 appeared in 1989.

International Journal of Urban and Regional Research. Published by Edward Arnold, London. Volume 1 appeared in 1987.

Urban Geography. Published by Taylor and Francis Group, Lanham, Md. Volume 1 appeared in 1976.

Urban Studies. Published by Routledge, New York. Volume 1 appeared in 1964.

What do these images convey about empire and globalization, and their similarities and differences?

Two images of global reach, more than a century apart (Queen Victoria and world map; a J advertisement). *(Left: Getty Images/Enslow Publishers; Right: Raymond Boyd/Michael Ochs Archives/Getty Ima*

Go to "Seeing Geography" on page 486 to learn more about these images.

ONE WORLD OR MANY?

The Cultural Geography of the Future

This final chapter is about the future. Unless one is precognizant, writing about the future is "a fool's game." The best we can do is collect evidence about past and present conditions, examine ongoing trends, and then let our imaginations go to work. Even when we do this well, it is difficult to predict the outcome of anything with certainty. Who, for example, would have thought that we would still be combating slavery and the slave trade in the twenty-first century? It is even more difficult to predict wholly new phenomena. Who would have predicted the dramatic social changes that the revolution in information technology is bringing about, especially the ascendance of social media and their impacts on politics and culture around the world? Despite the lack of certainty, looking ahead is a necessary activity if we want to be at all prepared for the world in which we will build our careers, raise our families, and work toward the fulfillment of ourselves and others. The pervasive trends in globalization, whose effects we have featured throughout the pages of this text, will have a profound influence on the cultural geographies of the future. As noted in Chapter 1, the phenomenon of globalization has created an interlinked world of instantaneous communication where people, goods, and money move increasingly as if international borders did not exist. While there's no denying the general trends produced through globalization, we will see in this chapter that the future is far from predetermined by them.

LEARNING OBJECTIVES

11.1 Describe the possible futures of place and region, and explain their driving forces.

11.2 Explain how current trends will shape future mobilities.

11.3 Analyze the relationship of contemporary globalization to historical processes and its likely effects on the future.

11.4 Relate how the continuation of current globalizing trends will shape future nature–culture relations.

11.5 Identify likely future landscapes.

We are posing perhaps the most essential and difficult cultural geographical questions of all in this chapter. Will our future contain one world or many? Will we face a world where more people will have more opportunities and choices, or where deprivation and powerlessness will spread? Will we witness a global-scale process of cultural diversification or of homogenization? Are we headed toward cultural diversity or cultural monotony?

Check the latest news online almost any day and you can read about globalization, though the word may not always be used. New suburbs in India are built to resemble those in Southern California; the citizens of the United Kingdom lament membership in the European Union, worrying that the forces of globalization will erode their distinct English identity; commercial loggers deforest Rendova Island in the Solomon Islands and in the process destroy the culture and livelihoods of the native Haporai people; and so on—endlessly, it seems.

But, does globalization truly have the ability to render cultural differences irrelevant? Can't groups of people selectively choose from what globalization has to offer and still retain control of their cultural identity and attachment to place (**Figure 11.1**)? Does globalization act to homogenize the world or, instead, to widen the differences between the haves and have-nots? In fact, much of the resistance to globalization is based on the belief that it enriches and empowers the few at the expense of the many, heightening class differences.

Many—perhaps most—geographers believe that the future will continue to contain many worlds. They speak of profound and irreducible cultural contrasts and of the capacity of indigenous peoples to weave Western elements into their own cultures, creating new and distinct hybrids. Globalization, they argue, produces different results in different lands, while globalization itself is subject to transformation through interactions with diverse cultures. The global need not and cannot abolish the regional or the local. In fact, throughout the chapter, we will explore examples of groups using the technologies and networks of globalization in campaigns that defend local cultural geographies and reassert the relevance of the local. Often local cultures are not merely defending ancient tradition but rather are interacting

FIGURE 11.1 **A Mauritanian camel pastoralist pauses to use a mobile phone.** The blending of modern technology and traditional culture is increasingly common under globalization. *(Hubert Moal/AWL Images/Getty Images.)*

with outside influences to produce new and distinctive forms of cultural expression. So which view is correct or most likely? Will it be one world or many? Perhaps the forces of both diversification and homogenization will continue to operate with unpredictable results. The future is unknowable, but it can be enlightening to speculate about the possibilities, beginning with the future of regions.

REGION

11.1 **LEARNING OBJECTIVE**
Describe the possible futures of place and region, and explain their driving forces.

Is it possible to use maps to detect future trends related to culture regions? Some geographers look at maps and do indeed see regional homogenization. David Nemeth, for example, expresses the view that regions are "being crushed and recycled into a bland, ambiguous amalgam," producing "something akin to a vast parking lot of global scale." Without question, people in all parts of the world must confront the choice between their traditional attachment to place and local culture on the one hand and the attractions of globalization-driven cosmopolitanism on the other. Indeed, the fear of losing the distinctiveness of place—and with it, the associated cultural diversity—seems to be a recurring and longstanding theme in modern life. Geographer Wilbur Zelinsky observed that the lament over placelessness and vanishing diversity can be traced back in time for decades if not centuries.

Zelinsky argued that although there has been a long-expressed fear of homogenization, it is difficult to find the empirical evidence to demonstrate that it has yet come to pass. He sought to answer the question "Over the past few centuries what effects has the enormous process of modernization—and globalization in particular—had on the character, life, and durability of all places and regions?" Focusing on the United States, he first undermines the assumption that the country was historically a disparate collection of diverse regional cultures. In fact, standardizing and homogenizing practices have been widespread since the eighteenth century, addressing language, communications, transportation, and land registry to name just a few. During the same historical period, however, one can also find identifiable culture areas, such as New England and Dutch-Flemish New York.

Building on this historical insight, Zelinsky assesses the contemporary status of a variety of cultural-geographic phenomena, from architectural forms to food preferences. His findings show that the present parallels the past; while homogenizing forces have grown stronger and more pervasive, differentiation is evident everywhere. He concludes that place-making and regionalization in the United States are increasing and "promises to do so indefinitely" into the future. In other words, he found, paradoxically, that convergence and divergence are both occurring. If we extend Zelinsky's U.S. study to other parts of the world, it appears that the future homogenization of regional and cultural difference is far from predetermined, because the forces of globalization operate to simultaneously standardize and diversify cultural practices and meanings.

THE UNEVEN GEOGRAPHY OF DEVELOPMENT

Most cultural geographers expect that regional diversity will persist. In particular, the familiar core–periphery concept is just as relevant as ever, especially where development is concerned. Some regions—forming the core—are moving ahead and prospering, while many others—the periphery—fall further and further behind. Global interdependence has not evened out the differences between the haves and have-nots; in many ways, it increases them. What do we mean by falling behind or moving ahead? One clear measure is global wealth distribution. There is now a transnational class of the super-rich, the 0.25 percent of the world's population that owns as much wealth as the other 99.75 percent. Most of them live in a First World core consisting of Anglo America, the European Union (EU), the coastal zone of East Asia, and Australia–New Zealand. As the greatest beneficiaries of globalizing processes, these regions are moving ahead of the remaining Third World peripheral lands.

Not all observers, however, agree that the global income gap is widening. In *The New Geography of Global Inequality*, Glenn Firebaugh argues that world core–periphery differences are not actually increasing. He points out that if one takes into account only the fact that the incomes of the richest nations are increasing at a greater rate than those of the poorest, it does look as if the rich are becoming richer faster, while the poor remain poor. Yet, this overlooks another fact: the many poorer countries contain only 10 percent of the world's population, while those that are developing very rapidly (mostly Asian countries) contain more than 40 percent. Taking national population numbers into account, we see that global income gaps are, in fact, decreasing.

In the meantime, income gaps *within* nation-state boundaries are growing rapidly. Instead of core–periphery, another way to think about the unevenness of development is in terms of "fast" and "slow" worlds. In many ways, economic development speeds up daily life. The fast world is at the forefront of instantaneous global communication, where one can download the latest music and films from around the world or play the Tokyo stock market from one's home computer in Seattle. The fast world is connected to high-speed broadband Internet; based in the world's megacities; adaptable to rapid shifts in global trends of investment, trade, production, and consumption; and home to the transnational super-rich. The slow world consists of the hollowed-out rural landscapes, declining or abandoned manufacturing zones, and slums and shantytowns (**Figure 11.2**). Computers and Internet broadband service, even if available, are priced out of reach of most of the inhabitants of the slow world. Pieces of the periphery that lie adjacent to the core, such as Mexico's northern border, scramble to join the privileged part of the world, to be fast rather than slow. As Mexican border cities' recent experience with job loss to China demonstrates, for such regions, membership in the fast world can be fleeting.

British geographer Rob Shields, in *Places on the Margin*, concentrates on an array of places and regions of varying size that have been "left behind in the modern race for progress." He finds peripheries even within the core. Often these places and regions become sites of illicit or stigmatized activities, such as the international trades in sex and illegal drugs. Says Shields, such "margins become signifiers of everything centers deny or repress."

Tragically, places on the margin have also become key sites in the international trafficking of slave labor. Slavery, it turns out, is not a shameful practice consigned to the distant past, but an increasingly common phenomenon under globalization. Some believe that as many as 27 million people live in slavery worldwide, on every continent in the world save Antarctica. Traffickers prey on the most desperate and powerless of society's castoffs. In the United States, California, Texas, and Florida lead the way in the increase in slave labor cases. Some of Florida's famous orange juice, for example, was recently found to be harvested by illegal Mexican immigrants held against their wills in remote rural camps (**Figure 11.3**, page 461). The insatiable international demand for Brazil's timber and beef has been met through the labor of captive workers held deep in the Amazon. Labor recruiters in Brazilian cities lure jobless workers to cut remote forestlands through false promises of good wages and housing. Once there, workers' wages are withheld and they are prevented from leaving. Some are forced to work for years clearing forests to make way for cattle ranching. According to one Brazilian labor official, "Slave labor in Brazil is directly linked to deforestation." The example of slavery shows how illicit activities can thrive in peripheries, such as rural Florida and the Brazilian Amazon.

FIGURE 11.2 **Left behind in globalization's wake.** This abandoned industrial site and the communities that housed its workers belong to the slow world. *(Courtesy of Roderick Neumann.)*

ONE EUROPE OR MANY?

Looking at just one region of the world, geographer Ray Hudson echoes our question: "One Europe or many?" Local identities are asserting themselves within states at the same time that supranationalism is touted as the path to "one Europe." He opts decidedly for "many," saying that one main role of the European Union should be to promote "complex geographies of identities." Hudson also concludes that power and wealth will not be evenly distributed geographically in Europe, contributing to the maintenance of many cultural regions.

A similar outlook leads geographer Michael Keating to speak of a "new regionalism" in Europe, and David

FIGURE 11.3   **The mobile Florida Modern Slavery Museum in front of the United States Capitol building.** The museum has been driven to cities across the United States to bring the labor abuses of Florida's agricultural sector to public attention and promote reform. ***What reasons can you think of to explain the existence of modern slavery in agriculture?*** (Fritz Myer.)

Hooson went so far as to suggest that globalization actually *strengthens* people's bond between place and identity. This strengthening is suggested by the rise of ethnic separatism in countries as diverse as Spain, the United Kingdom, and Serbia and Montenegro. As people find that the forces of globalization are eroding cultural norms, customs, and practices, local political movements that seek to reinvigorate ethnic identities rooted in place may arise in reaction. Major political-geographic events such as the disintegration of the Soviet Union or the creation and expansion of the EU can have complex effects on cultural identities. The recent expansion of the EU is likely to bring a host of unintended and surprising outcomes for culture regions. In some newly independent former Soviet territories, such as Estonia, ethnic Russian immigrants have become a disadvantaged minority group. New post-Soviet states have used their membership in the EU to reassert national ethnic heritage and redefine themselves culturally as "Western."

Moreover, globalization means that local economies are linked more tightly with the economic fortunes of distant countries. Global economic downturns can thus produce powerful centrifugal forces. This was brought home to EU citizens everywhere in the aftermath of the Great Recession of 2008–2009. A 2013 Pew Research poll of EU citizens found that the economic crisis has become a political crisis for the future of the EU. Public support for EU integration plummeted since its highs in the first decade of the twenty-first century (Figure 11.4). In addition, attitudes toward the EU project are diverging among member states, with Germans showing strong support and a French majority expressing dissatisfaction. Most worrisome for future EU unity are the results that show support for the EU dropping among 18- to 29-year-olds in all countries. While few observers expect the EU to split up, these results make clear that national cultural differences remain strong and that the transition to a single European identity remains somewhere in the distant future.

GLOCALIZATION

The theme of region, then, seems to suggest that a new human mosaic is forming in the age of

FIGURE 11.4   **Centrifugal forces in the European Union.** Students in Athens, Greece burn the EU flag in protest of unpopular EU economic policy. In the context of continuing economic hardship, support for EU membership is dropping fastest among Europe's youth. (EPA/ORESTIS PANAGIOTOU/LANDOV.)

globalization. Rapid change is pervasive, but its direction differs from one location to the next. The future will not be like the past, but it will also not be monochromatic.

The interaction between global and local prompted geographer Erik Swyngedouw to promote the term **glocalization** to describe the consequences. In brief, he argues that the outcome of this interaction involves change both in the regional way of life *and* in the globalizing force. For example, transnational corporations often have to adapt their product lines to local norms and preferences or adjust their production practices to local labor and environmental laws. Put differently, glocalization is a process that ensures the survival of culture regions and places in the future. These considerations, prompted by the culture region theme, lead us to

glocalization The process by which global forces of change interact with local cultures, altering both in the process.

conclude that a planetary culture is almost certainly illusory and that potent forces are at work to prevent homogenization.

THE GEOGRAPHY OF THE INTERNET

Has the Internet reduced the importance of geography or the viability of regions? Journalist Thomas Friedman makes exactly that claim in *The World Is Flat,* his international best-selling book on globalization's effects. He singles out the Internet as a key technology that allows the global interconnectivity of individuals and institutions. Once place-bound human activities now occur on the Internet almost instantaneously on a global scale. The Internet can cross borders, both political and cultural; it breaks down barriers; it eliminates the effects of distance. According to Friedman, the Internet, among

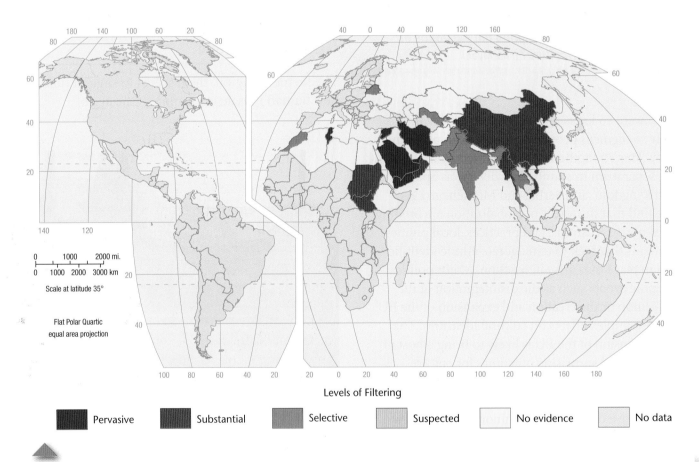

Levels of Filtering

Pervasive | Substantial | Selective | Suspected | No evidence | No data

FIGURE 11.5 **This map of government filtering practices shows that nation-state boundaries still matter in the global flow of information on the Internet.** *Based on your knowledge of world affairs, do you find the various government filtering activities depicted here predictable or surprising?* (Source: Open Net Initiative web site.)

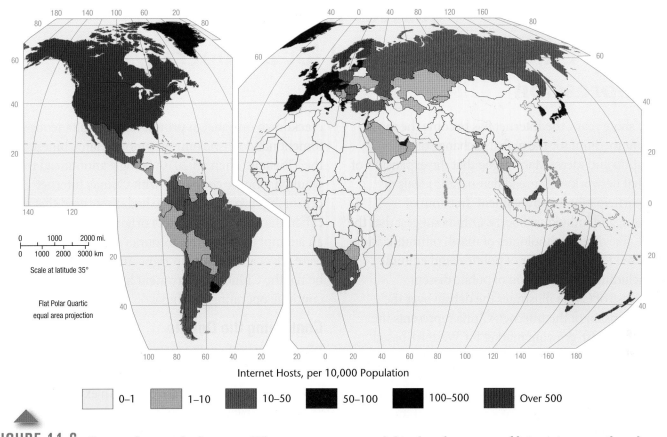

FIGURE 11.6 Connections to the Internet. *What consequences might a low frequency of Internet connections have on a country's economic, political, and cultural future?*

other globalizing phenomenon, is leveling the differences among nations and their citizens.

Many observers disagree with Friedman's assessment. The Nobel Prize–winning economist Joseph Stiglitz suggests that "in many ways the world is getting less flat" under globalization. The reality of the Internet's influence is very complex and ultimately leaves culture regions intact. We must remember that—like many technological innovations—the Internet is a tool with many, often unanticipated, uses. From al-Qaeda to French ultra-nationalists to First Nations in Canada, the Internet is used as a tool to emphasize and defend geographies of cultural difference. Rather than making national borders irrelevant, the Internet can sometimes highlight differences between different countries. For example, governments can monitor, censor, regulate, and block access to the Internet (**Figure 11.5**; see Subject to Debate, page 464). Ironically, Friedman's home paper, the *New York Times,* is periodically edited or blocked

when it publishes articles unfavorable to Chinese government figures.

Although the constraints of geography are increasingly recognized, there is no question that the Internet has been very effective in creating new forms of social interaction, including the creation of virtual communities. We might ask, as John Barlow does, "Is there a *there* in cyberspace?" Does the Internet contain a geography at all? Certainly, *places*—at least as understood by cultural geographers—cannot be created on the Internet. For starters, these "virtual places" lack a cultural landscape and a cultural ecology. In the broader context, on a worldwide scale, human diversity is poorly portrayed in cyberspace. "Old people, poor people, the illiterate, and the continent of Africa" seem not to be "there," as Barlow notes (**Figure 11.6**). Users usually end up "meeting" others much like themselves on the Internet. More important, the breath and spirit of place cannot exist in cyberspace. Barlow, resorting to a Hindu term,

SUBJECT TO *Debate*

THE INTERNET: Global Tool for Democracy or Repression

In a few short years, the Internet has become the most important medium for the global exchange of information and ideas. Many argue that the expansion of this global network will spread democratic ideals of citizen participation in government, freedom of expression, and individual liberty. Others suggest that the Internet, like other technologies of mass communication, can be used for any number of political and cultural ends, including the repression of public dissent.

The case of China illustrates the debate over the global expansion of the Internet. China represents the largest and fastest-growing market for Western-based Internet companies, and industry giants such as Google and Yahoo! have established subsidiaries there. These and other Western companies stress the power of the Internet to promote freedom of thought, creativity, and social equity through the widespread dissemination of information. The companies argue that their very presence and the access they provide to more information for increasing numbers of Chinese citizens are in and of themselves politically and culturally liberating.

There are signs, however, that the Internet in China can also function as an effective government tool to suppress dissent. In 2004 Yahoo! provided personal information to the Chinese government about a journalist's online pro-democracy writings. The government courts sentenced the journalist to 10 years in prison. Google was allowed to establish a subsidiary in China in 2006 only by complying with the government's censorship policies. The company developed a Chinese version of its search engine that automatically blocks results from searches on terms such as *Tibet* or *Tiananmen Square*. Cisco Systems designed the software behind the "Great Firewall of China," the feature of the Internet in China that filters all online information as it crosses the country's borders. The Great Firewall can block access to web sites as well as individual pages from a site. Terms such as *democracy* and *human rights* tend to trigger the Firewall.

In 2008, the Chinese government used its control of the Internet to suppress information on its military crackdown on Tibetan protests and simultaneously foment nationalistic outrage among Han Chinese against the Tibetans. Then, in 2010, Google announced that it would no longer cooperate with China's Internet censorship program. Google had discovered that a cyberattack on its computers originating from China was an attempt to access the accounts of Gmail users who were advocates of human rights in China. For the time being, the Chinese government has allowed Google to continue operating in China.

Continuing the Debate

What does the case of Google in China tell us about the power of the Internet? Consider these questions:

- Is China a unique case or just an extreme example of repressive uses of the Internet?

- Do private corporations have an ethical responsibility to citizens living under authoritarian governments? If so, what form should corporate action take?

- Should Internet companies agree to the Communist Party's censorship policies? Is such an agreement balanced by the benefits of greater access to the Internet by Chinese citizens?

- How can companies limit the harm done through the provision of Internet services?

Google China headquarters in Beijing's Tsinghua Science Park. *(Frederic J. Brown/AFP/Getty Images.)*

China's Web Junkies
http://www.tinyurl.com/pajv7ht

In Chapter 11, we note the central role that the Internet will play in our global future and raise questions about the local social and cultural effects. This video describes one way that China views those effects. In 2008, reports the video, China declared the prolonged recreational use of the Internet to be a clinical disorder similar to other behavioral addictions, such as gambling. The video examines a Chinese bootcamp-like center outside Beijing where youth are confined for treatment of "Internet addiction." Teenagers (predominantly male) are brought to the center by their parents; there, they are subjected to a combination of therapy and military-like discipline for up to four months.

Thinking Geographically:

1. One scene in the video shows hundreds of Chinese youth gaming in an expansive, warehouse-sized Internet café. A camp patient describes how on one occasion he "stayed there for three days" before he was forced into treatment. How would you describe the geography of recreational Internet use shown in China in comparison with what you are familiar with in the United States? Think in terms of domestic versus public spaces. How might the geography of Internet access in China versus the United States influence the tensions between children and parents over excessive Internet use? If in the future China's Internet access becomes more individualized (i.e., more domestic, less public), how might that alter Chinese perceptions of Internet addiction?

2. We have often emphasized the complexities of geographic variation under globalization. In the case of China's approach to prolonged Internet use, what differences can you identify in comparison to other countries and regions? For example, no global medical consensus exists on Internet addiction as there is for, say, opium addiction. Do you think similar teenage behavior (e.g., anger toward parents, prolonged absence from the house) in Europe or the United States will be blamed in the future on Internet addiction as has happened in China? How would you account for the regional differences that can be identified?

---

calls this missing essence *prana*. These are not real places, nor can they ever be.

Other critics point out that virtual communities do not have the defining qualities of geographic communities: communion among citizens, shared responsibilities, and civic duty. Communication across the Internet does not a community make. Nevertheless, people can use virtual communities to establish bonds that carry over into the real world. To keep things in historical perspective, we should remember that throughout the modern period, families and individuals have left their geographic communities and dispersed widely. However, these people have constructed social networks to maintain cultural and emotional ties with their geographic communities of origin. From this perspective, we might view the Internet as just another vehicle for maintaining those ties and preserving cultural identities (see Subject to Debate, page 464).

MOBILITY

11.2 LEARNING OBJECTIVE
Explain how current trends will shape future mobilities.

Movement on a global scale—of ideas, people, money, and commodities—in many ways will characterize the future. The computer, the Internet, satellite television, smartphones, and other globally available technologies greatly facilitate access to and the diffusion of ideas and information, while also accelerating their spread. At the same time, the spread of new ideas and innovations often produces disruptions that change the world in unexpected and unintended ways. At the very least, the theme of mobility cautions us against counting on predetermined outcomes.

MOBILITY IN THE DIGITAL AGE

The Internet allows ideas, music, artwork, money, and indeed any form of cultural expression that can be turned into digital data to be transmitted around the world in a matter of seconds. With the development of the Internet and rapid global transportation systems, transnational corporations are able to advertise, market, and deliver virtually anywhere. Almost all transnational companies manage globally accessible web sites that not only sell products, but also sell the company "brand" (its identity and business model) in an effort to build a global base of loyal customers. Any entrepreneur in any part of the world can open a web site, allowing him or her to market cultural products directly to consumers in every corner of the globe. What's more, nongovernmental organizations, art museums, indigenous peoples, individual artists and musicians, and just about all imaginable culture groups or producers of culture have their own web sites. Increasingly, social media such as Facebook, Twitter, and Instagram are linking individuals, communities, governments, and corporations in a complex global network of information sharing. In 2013 industry analysts estimated that Twitter was registering 135,000 new users every day!

The Internet clearly has profoundly altered the speed and character of cultural interaction, but to what effect? The spread of more democratic forms of governance, for example, has accompanied the more rapid movement of information over the Internet. Dictatorships thrive by controlling and manipulating information—an increasingly difficult task in the age of the Internet. And, the Internet is certainly not the only means by which information spreads rapidly. Smartphones that utilize microwave towers and Wi-Fi and global television networks that utilize satellite communications are integral to a worldwide network of digital transmission that is growing in complexity and reach. Social media and mobile devices have played visible roles in organizing new social movements, such as the so-called Arab Spring and Occupy Wall Street (**Figure 11.7**). Although the importance of social media in fomenting democratic revolutions in North Africa and the Middle East is a continuing debate, the Egyptian and Libyan governments felt threatened enough to shut down

FIGURE 11.7 **Social media in social movements.** Increasingly, all sorts of emerging social movements, such as Occupy Wall Street, are mobilized and directed through the use of new social media. *(Nano Calvo/Newscom.)*

Internet access nationwide during the height of street protests calling for reform.

The theme of mobility also applies to the spread of the Internet itself. If the future is to bring the universal diffusion of cultural elements, then surely this new homogenizing tendency ought to be revealed in the spread of this most essential element of globalization (**Figure 11.8**). The diffusion of the Internet has now spread across much of the world, following the models established in Chapter 1.

Although the Internet now reaches into almost every land, its use varies profoundly (see Figures 11.6 and 11.8). Barriers to access include inadequate infrastructure, poverty, and tyrannical governments. Nonetheless, current trends suggest a near universally connected world. For example, by one industry estimate, there are already 650 million mobile phone users in Africa, the world's poorest region. As the price of smartphones and other mobile computing devices declines, the Internet will become available to hundreds of millions more people in Africa. Google's CEO believes that the spread of mobile devices will make it possible for virtually the entire world to be connected through the Internet by 2020. Speculation on how these developments in digital mobility will shape the cultural geography of the future would fill another textbook.

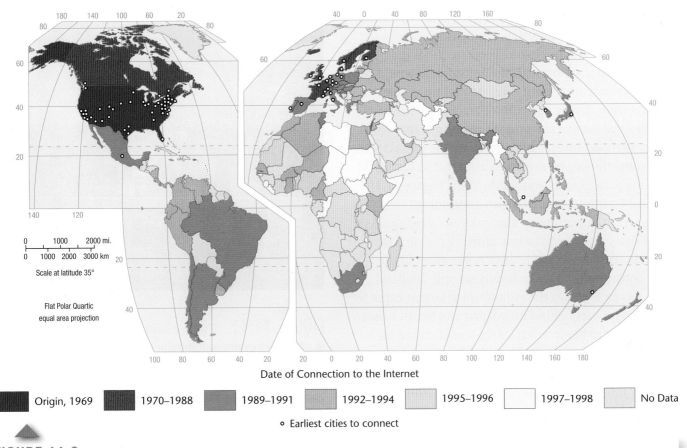

Date of Connection to the Internet

| ■ Origin, 1969 | ■ 1970–1988 | ■ 1989–1991 | ■ 1992–1994 | ■ 1995–1996 | □ 1997–1998 | □ No Data |

○ Earliest cities to connect

FIGURE 11.8 **Diffusion of the Internet.** This diffusion seems to have followed a rather typical pattern, with the earliest connections in core world regions and the latest in the periphery. *(Source: Crum, 2000; see also Press, 1997.)*

NEW (AUTO) MOBILITIES

Back in the mid-twentieth century, there was much popular debate about the future of transportation in the United States. People speculated that anything from a national network of high-speed railways to nuclear-powered flying machines would replace automobiles (**Figure 11.9**). Hardly anyone predicted

FIGURE 11.9 **A futuristic vision of urban transportation.** Such imaginative portrayals of the near future were common in the mid-twentieth century, but relatively little has changed. Our twenty-first-century transportation mode is more similar to Henry Ford's Model-T than to this flying machine. *(Glen Wexler/Masterfile.)*

that the twenty-first century would be more of the same old thing: the internal combustion engine–propelled family car. No one predicted that the American fascination with the automobile would diffuse to East and South Asia. At that time, these regions were largely agricultural, facing recurring famine, and among the poorest regions of the world.

In 2010, however, China overtook the United States to become the largest automobile market in the world in terms of both sales and production (Figure 11.10). By 2012 China accounted for an astounding 23.6 percent of worldwide auto sales. As a region, Asia now boasts the world's fastest-growing national auto markets, including such relative newcomers to the mass consumption race as Indonesia and Thailand. India has emerged as the region's third largest exporter of passenger cars behind only Japan and South Korea. It is forecast to become one of the world's three big auto manufacturers by 2030, with China number one and the United States number two.

The transformation in China's urban culture is as profound as it is sudden, with cars now clogging city streets built for pedestrians and bicycles (Figure 11.11). At the end of the twentieth century, only state

FIGURE 11.11   **Traffic congestion in Shanghai. China is now the largest automobile market in the world. Here, automobiles and bicycles compete for space in the increasingly congested streets of Shanghai, China.** *(Jeff Greenberg/Photolibrary/Getty Images.)*

bureaucrats bought cars and there were only six models from which to choose. Now Volkwagen alone has twenty models on the market and dozens more foreign and domestic automakers are on the scene offering hundreds of models. Businessmen, flush with profits from China's economic boom, are paying cash. Sales for premium (i.e., luxury) autos are growing at a faster rate than overall sales. China has already surpassed the United States as BMW's largest market, and it is likely that China will also surpass America in total luxury car sales by 2017.

Crowding and associated pollution will only intensify in the future as cities such as Beijing register 1000 new drivers *every day.* The environmental consequences of expanding automobile sales are numerous and of global importance. Beijing, Shenyang, and Shanghai are ranked among the world's 10 worst cities for air pollution. Among the biggest concerns is the increase in carbon dioxide emissions, caused by increased car exhausts, auto-manufacturing activities, and power plant growth to meet manufacturing demands for electricity. Carbon dioxide is a greenhouse gas that is the main contributor to the

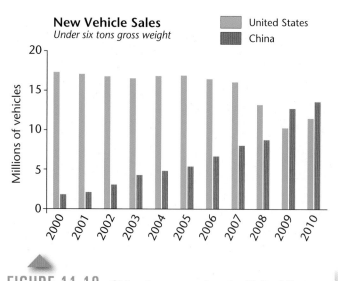

FIGURE 11.10   **China is surpassing the United States as the world's leader in car sales.** This is a phenomenal turn of events, considering China's level of economic development only a generation ago. *(Source: J.D. Power and Associates.)*

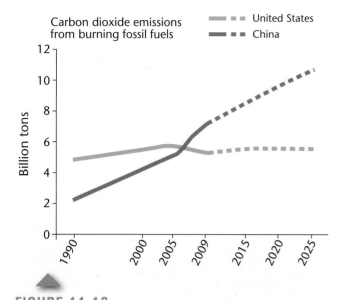

Carbon dioxide emissions
from burning fossil fuels

- - - United States
- - China

FIGURE 11.12 **China's greenhouse gas emissions are skyrocketing, surpassing the former world leader, the United States.** China's per capita output, however, remains far below that of the United States. This chart reflects projections that assume the success of a national efficiency campaign; experts in both China and the West, however, are increasingly concerned that emissions may rise even faster than projected here. *(Source: International Energy Association, www.iea.org.)*

problem of global warming. China has overtaken the United States as the world's number-one source of carbon dioxide and its 2000 emissions are expected to double by 2020 (**Figure 11.12**).

American auto manufacturers have scrambled to gain a foothold in emerging markets such as China. General Motors Corporation, for example, began establishing manufacturing plants in China in the late 1990s (**Figure 11.13**). It now is involved in nine joint ventures and two wholly owned foreign enterprises in China covering all aspects of the auto industry, including vehicle manufacturing, sales, financing, and parts distribution. Shanghai went from 5 GM dealerships in 2000 to 27 in 2010! Nationwide, GM's current network of about 3500 dealerships rivals its U.S. network of 4400. Carmakers are betting that the symbolic importance of the luxury automobile will diffuse to not only China but also other regions of East and South Asia. For example, Buick, GM's top model in China, is perceived as plush, elegant, and an indicator of high social standing. It appears as though a prime status symbol in American middle-class culture is poised to become the same for a new global middle class (**Figure 11.14**).

FIGURE 11.13 **GM plant in Shanghai.** Factory workers check a car body on the assembly line at the Shanghai General Motors factory in Pudong, China. Other American, Japanese, and European carmakers are building factories throughout China. *(David G. McIntyre/Newscom.)*

FIGURE 11.14 **General Motors Corporation launches its Cadillac line at the Beijing Auto Show.** *What brand name better symbolizes America's car culture than Cadillac?* *(Ng Han Gua/AP Photo.)*

MIGRATION FUTURES

What can we say about the range of possible futures for human migration? Looking at past experience is helpful, but imprecise. As long as there are environmental disasters, global-scale economic inequalities, and civil wars, we know diasporas will occur. Predicting where diaspora cultures will emerge and when, however, is much more difficult. Unpredictable events aside, tracking trends can provide insight on the future. We would, nevertheless, have to answer a range of questions about demographics, national economic trends, and basic geography, such as location and proximity.

Mexican immigration to the United States illustrates the complexities of trying to predict migration futures. By the start of the twenty-first century, Mexico was by far the single biggest source of the foreign-born population in the United States, exceeding all other Latin American and Caribbean countries combined. However, a close examination of current trends suggests that the main migration from Mexico to the United States has already peaked and is rapidly subsiding. For instance, demographic trends in Mexico show a rapidly declining fertility rate and leveling population growth rate. Mexico's basic socioeconomic indicators, such as education and income levels, are trending upward. The Mexican national economy continues to diversify, creating more domestic jobs for an expanding middle class, and the wage gap between Mexico and the United States has decreased. Moreover, political and economic changes throughout Latin America have resulted in new migration patterns, particularly an increase in movement within regions. In short, many of the push and pull factors that led to the explosion of Mexican migration to the United States from the 1970s through the 1990s have dissipated. We would need to do similar, and much more complex, comparative analyses of national demographics and socioeconomic indicators to make specific predictions about future international migrations.

Trends of greater temporal and spatial scale than the case of Mexico and the United States allow for more certainty in predicting migration futures. The rural-to-urban migration trend that began with the Industrial

FIGURE 11.15 **Lagos, Nigeria, the world's third largest city.** Africa is following the global trend of rapid urbanization driven by rural to urban migration. *(Wayne Parsons/Getty Images.)*

Revolution in Western Europe and became a global trend in the twentieth century will likely continue through the twenty-first. Countries in East Asia, such as China and Indonesia, are urbanizing faster than any prior country in history. Africa is following close on the heels of Asia. Populations are urbanizing across the continent and Lagos, Nigeria, became the world's third largest city in 2015 (Figure 11.15) On a global scale, humanity hit a benchmark in 2008 when for the first time in history over half of the world's population were living in cities. By 2030 the United Nations predicts that at least 60 percent of the world's population will live in cities, with further urbanization expected through the century. We can be relatively certain that rural-to-urban migration will continue to transform the human experience worldwide.

THE PLACE(S) OF THE GLOBAL TOURIST

Tourism will be one of the key types of mobility of the twenty-first century. By some estimates, tourism is or soon will be the largest industry in the world, second only to oil as the most important source of revenue for the Third World. Cultural geographers are increasingly interested in understanding this phenomenon. What sort of cultural interactions will take place under global tourism? Will the end result be the preservation, conversion, or elimination of distinct culture regions?

As with many global-scale patterns of mobility, there are distinct differences in power that structure the tourist experience. The most glaring inequality is that between the visiting and visited cultures. The vast majority of global tourists are from the First World. The cultures visited are often located in the marginalized sites of the global economy, such as on popular African safaris or spring-break excursions to Cancún, Mexico.

The global tourist industry presumes the existence of unique places that appear "exotic" to the Westerner—or at least serve to emphasize the difference between the experience of home and the experience of the travel destination. Tourism can thus be a double-edged sword. On the one hand, there is an imperative to preserve and nurture folk and indigenous cultures, and, on the other, an imperative to merely create the illusion of cultural difference. The former imperative can gain support from tourism, such as in the case of agricultural tourism, where traditional but uneconomical food production persists with the aid of profits gained from tourist visits. For example, the European Union provides funds to member states to support the production of a range of traditional agricultural products such as local cheeses and rare livestock breeds. The EU believes that these investments keep rural economies intact and spark the tourist trade because urbanites and foreign tourists are attracted to the countryside to sample unique and historically significant agricultural products. The latter imperative is often driven by commercial enterprises that make their profits from entertaining their customers with sights and sounds of the "exotic." In such cases, "traditional" dress and performance is for the benefit of tourists, not a part of everyday life (**Figure 11.16**). Some cultural geographers have concluded from these observations that the tourism business packages its offerings to conform to the tastes

FIGURE 11.16   Global tourism and local culture.
Tourists line up to photograph colorfully dressed "natives."
What do such encounters suggest about the effects of global tourism on local cultures? (Michael Coyne/Lonely Planet Images/Getty Images.)

of the affluent global tourist and so results in an "inauthentic" experience of place.

It would be wrong, however, to simply view local cultures as "made up" or "invented" by the global tourism industry or to suggest that only inauthentic cultural interactions are possible (see Rod's Notebook, page 472). Geographers Peggy Teo and Lim Hiong Li's study of global and local interactions at a tourist site in Singapore—a former private mansion and fantasy garden—provides an excellent case in point. They favor the view that local groups can use the forces of global tourism to strengthen their cultural identities and traditions. Their study focused on a government-sponsored effort to create from the mansion and garden an "Oriental Disneyland," a Western-style theme park loosely based on Disney's technologically enhanced tourist attractions. After the venture failed, local people successfully lobbied the government to redesign the site to more closely replicate its original condition. The motivation behind these efforts was to defend local cultural and

Rod's Notebook

Rethinking Global Tourism on the Quetzal Quest

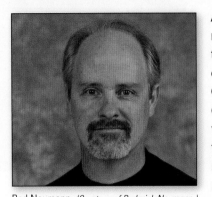

Rod Neumann. *(Courtesy of Roderick Neumann.)*

As dusk became night on the narrow two-lane highway climbing Costa Rica's Cerro de la Muerte (Mountain of Death), we leaned toward the windshield, scanning for some sign of our destination. We knew from the web site we had visited back home that we could expect a crude wooden sign signaling the turnoff. We had come for a glimpse of the resplendent quetzal (*Pharomachrus mocinno*), a bird revered by ancient Aztecs and contemporary nature tourists alike, though no doubt based on very different sets of cultural values. We found our way down a muddy track to a random collection of rough-hewn wooden buildings. The owner of the *finca* (farm), a solidly built farmer with a crooked smile, served us rice and beans and cold beer. It was a delightful and memorable meal.

Our experience in the quetzal's cloud forest home led me to ponder the meaning of the term *global tourism* and Western tourists' endless quest for the authentic encounter. Global tourism seems to imply huge institutions or corporations shuttling masses of camera-toting vacationers to every corner of the world. Clearly, there were many multinational corporations involved in this tourist encounter, beginning with the airline that had transported us from Miami to San José. On the other hand, the key global entity in this story was the worldwide Internet, which put us in direct contact with a peasant farmer high in the mountains 1995 kilometers away. And although we were "global tourists," we were also guests of the farm owner. Our host had been born and raised on this steeply tilted piece of forest that

happened to be key quetzal habitat. Our conversations at mealtimes and predawn bird-watching adventures had a feeling of intimacy absent in large, packaged tours. There was no sense of alienation or performance. We talked about his childhood in the mountains, farming, and forest conservation. I came away from our experience thinking that the degree of authenticity in the tourist encounter has a lot to do with who owns the means of producing the tourist experience.

The resplendent quetzal, which frequents the cloud forests of Central America. *(Danita Delimont/Alamy.)*

historical meanings associated with the site. They conclude that "the global does not annihilate the local." Rather, tourism can result in "unique outcomes in different locations."

GLOBALIZATION

11.3 **LEARNING OBJECTIVE**
Analyze the relationship of contemporary globalization to historical processes and its likely effects on the future.

How does globalization operate to shape the future? How does it relate to phenomena of the past, such as European colonialism? These are some of the questions geographers and other scholars are investigating in their research. We can start by looking at some of this groundbreaking work.

UNDERSTANDING THE FUTURE EFFECTS OF GLOBALIZATION THROUGH THE PAST

American Commodities in an Age of Empire, Mona Domosh's book on the United States' unique form of imperialism, illustrates the continuities between historical experience and globalization's future. Domosh shows how U.S. overseas imperialism was conducted not through military conquest and occupation, but through the international commercial activities of private corporations. By 1915 U.S. companies dominated all aspects of world trade. Many of them had established overseas manufacturing, developed international sales and financial networks—aided greatly by global telegraph communications—and functioned in a way that disregarded political borders. In other words, the early-twentieth-century international business looked a lot like what we call globalization.

Domosh's book provides historical evidence for thinking about the future of cultural difference under globalization. For example, in the early twentieth century, U.S. companies in search of international profits needed to market their commodities—soap, tractors, sewing machines, and much more—to a wide variety of cultures. Domosh shows how companies used existing categories of racial difference to pitch their products by offering the promise of transformation. That is, by consuming U.S. products, the Chinese, Indian, or African individual could become more modern and civilized. This is the paradox of global advertising. Cultural differences are highlighted with the promise of their erasure through the consumption of mass-produced commodities. See the World Heritage site on pages 474–475 for a historical look at the origins of globalization.

We can think of any number of advertising campaigns today that operate with a logic similar to that of early-twentieth-century international marketing. From soft drinks, to clothing, to automobiles, global advertising highlights cultural difference while promising unity through consumption. Of course, in today's ads the crude racial stereotypes and hierarchies of the past are nowhere to be seen, or are at least tucked well below the surface. Indeed, multiculturalism is celebrated in a sort of unity-in-diversity. The Italian multinational clothing firm United Colors of Benetton is renowned for this advertising strategy (**Figure 11.17**).

FIGURE 11.17 Multiculturalism in mass advertising. Ideas of racial and ethnic difference have historically shaped the messages of mass advertising. That's still the case, but now many multinational corporations celebrate diversity as a marketing strategy. *(Loop Images Ltd/Alamy.)*

World Heritage Site

Cidade Velha, Historic Centre of Ribeira Grande, preserves a key location of globalization's origin. Ribeira Grande was the first European colonial town in the tropics. If globalization is the force shaping our present and future, globalization itself has roots in the heritage preserved at Ribeira Grande. Inscribed as a World Heritage Site in 2009, it was selected as an early image of transcontinental geopolitical visions.

▥ Portuguese maritime explorers discovered then uninhabited Santiago Island in the Cape Verde archipelago in 1460. They established the port city of Ribeira Grande on its southern shore a few years later. Commercial activity shifted to the newer port of Praia in the 1700s. Ribeira Grande was salvaged for building materials and renamed Cidade Velha (old town).

▥ Ribeira Grande was a major crossroads for Portuguese maritime trade, which very quickly became globalized. By the mid-1500s, fleets from Africa, India, Brazil, Southeast Asia, and Europe traveled there to trade cargos and information.

▥ Because of its geostrategic location in relation to Europe, Africa, and the Americas, Ribeira Grande became a key connection in the African slave trade. Europeans brought African slaves to the port where they were traded in the slave market and then shipped to the Americas. It was also a key location of global plant exchange and agricultural experimentation that combined enslaved labor with the cultivation of imported tropical and subtropical crops.

▥ The blending of European and African cultures from the fifteenth century onward gave birth to the world's first Creole culture, now familiar throughout the Caribbean region. Languages, religions, foods, architectures, and music from the two continental traditions have mixed on Santiago to create a distinct Cape Verde Creole culture.

(Anthony Asael/DanitaDelimont.com.)

Cidade Velha, Historic Centre of Ribeira Grande

IN OLD TOWN:

When the Portuguese relocated commerce and government to Praia, they dismantled many of Ribeira Grande's historic structures for use in building the new port. However, some residents stayed behind and key elements of the old town's design and architecture have endured for centuries.

▦ The great wealth concentrated at this colonial outpost attracted the unwanted attention of mariners from competing European countries. Thus, the old town features the ruins of an elaborate system of walls and forts. After Sir Francis Drake sacked Ribeira Grande in 1585, the Portuguese fortified their defenses by building the royal fortress of São Felipe in 1593, known then as one of the world's strongest. It has been partially restored.

(Orlando Rodrigues/AFP/Getty Images.)

▦ The Portuguese centered the old town on Pillory Square, so named for the marble column raised in the early 1500s. The pillory is the oldest remaining monument of the town and was installed to symbolize Portuguese power and authority. Recalcitrant slaves were publically punished at the pillory. It is one reason Ribeira Grande became part of UNESCO's Slave Route project, an initiative to promote important memorials and places in the African slave trade.

▦ Ribeira Grande featured many religious buildings including churches and chapels, a cathedral, and a convent. Only the church of Nossa Senhora do Rosário, dating from the late fifteenth century, and the chapel of the Convent of São Francisco, built in the seventeenth century, exist in restored condition. The remainder of the site's religious buildings are in ruins or consist only of archaeological artifacts. Of the cathedral, only its walls, constructed in the mid-sixteenth century from stone imported from Portugal, stand today.

TOURISM:

Cidade Velha is a short distance from Praia, the capital city of the Republic of Cabo Verde, which is served by ferries and an international airport. Tourism is one of the republic's most important industries.

(AP-Foto/Alamy.)

▦ The historic ruins of the fifteenth-century church and the royal fortress, the old stone houses, and Pillory Square are main draws for tourist traffic.

▦ The local tourism industry in Cidade Velha is relatively underdeveloped and the Cabo Verde government has only recently attempted to encourage it. While the site has tourist appeal due to the antiquity of the town and stunning island location, it is also tainted by its association with slavery.

▦ Some might label a trip to Cidade Velha as a form of "dark tourism," because of the site's central role in the horrific history of the global trade in African slaves. However, UNESCO's Slave Route project terms it "memorial tourism," encouraging host countries to promote such sites for educational purposes and to memorialize the lives lost.

▦ Cidade Velha, in the Cape Verde Islands, was Europe's first tropical colony and a major crossroads for global maritime trade by the early 1500s.

▦ The site features many defensive fortifications, reminders of its former status as one of the world's wealthiest trading outposts.

▦ Cidade Velha is part of UNESCO's Slave Route project, an initiative to preserve sites of the African slave trade and promote "memorial tourism."

http://whc.unesco.org/en/list/1310

Geographers are interested in globalization because it bears so profoundly on the central geographical issue of human diversity. It produces a multitude of historically unique, contradictory, and paradoxical cultural conditions. The most evident paradox is that globalization—which would seem to erase cultural difference—has been accompanied by a reassertion of distinct religious, national, and ethnic identities. The start of the twenty-first century has been marked by heightened ethnic divisions and conflicts, the resurgence of nationalism, and numerous highly visible identity movements—gay, feminist, green, and religious fundamentalist. Contradictions and paradoxes such as these lead many geographers to conclude that the future effects of globalization will not be predominantly homogenizing. Moreover, people are self-aware actors who may view, for many different reasons, the effects of globalization as something that should be resisted or defended against.

GLOBALIZATION AND ITS DISCONTENTS

Some of the features of globalization—the consolidation of transnational corporate power, global environmental change, the spread of consumer culture to every corner—are viewed by people worldwide as threatening to established values, ways of life, and economic livelihoods. For many people, the future under globalization does not look bright. As geographer Matt Sparke observes, responses to globalization or to globalization's effects come in many forms and from disparate political corners. From the right comes ultra-nationalist reaction, which includes calls for protection of national industries against global competition and limits to immigration and migrant rights. France's National Front, led by Marine Le Pen, exemplifies the response from the right. In a 2013 speech, Le Pen compared globalism to a "totalitarian" ideology and called "for Europe to halt immigration." Religious conservatives also strongly react against globalization. Osama bin Laden viewed many aspects of the global economy as violating the sacred teachings of the Qur'an, as do present-day Islamic leaders associated with al-Qaeda.

As Sparke noted, there is an obvious contradiction between al-Qaeda's opposition to globalization and its use of global networks of communication and terrorist training.

On the political left are critics who see globalization as leading to a future of increased economic inequalities across all geographic scales and increased empowerment of transnational corporations at the expense of local access and control. The left calls for checks on unfettered trade liberalization to strengthen worker rights, for greater focus on and access to public services and spaces, and for the enhancement of local food security. The position on the left is oriented more toward altering the way globalization operates than toward simple opposition. The Occupy movement that started as a protest in 2011 in New York City's financial district, Wall Street, and has since grown internationally is a good example. The movement slogan, "We are the 99%," is meant to signify that globalization has led to the enrichment of a tiny minority while negatively impacting an overwhelming majority. Through the strategic use of social media such as Twitter and Facebook, the Occupy movement very quickly moved from Wall Street to a global network that organized a day of street protests in cities around the world.

The indigenous rights movement is difficult to classify on the left to right political spectrum, but it has nevertheless presented a constant challenge to economic globalization. Mexico's Zapatistas are a good example. On January 1, 1994—the day that the North American Free Trade Agreement (NAFTA) among Canada, the United States, and Mexico took effect—hundreds of indigenous Mayan peasants forcibly occupied government offices in the southern Mexican state of Chiapas. Calling themselves *Zapatistas*, they declared that they had no choice but to take up arms (**Figure 11.18**). The impoverished Mayan farmers saw globalization in the form of NAFTA as sounding the death knell for their culture and way of life. Globalization meant that cheap corn from the midwestern region of the United States would flood Mexico, making it impossible for Mayans to continue farming for a living and to maintain their communities. Interestingly, the initial armed rebellion soon gave way to a skillful campaign that used key instruments of globalization—satellite TV and the

FIGURE 11.18 **Local resistance to globalization.**
Zapatista commanders sit at the negotiating table in San Andres Larrainzar, Mexico, in 1996. The Mayan-based Zapatista leaders have been periodically involved in talks with the Mexican government over cultural and economic rights since NAFTA took effect in 1994. *(AP Photo/Scott Sady.)*

Internet—to gain worldwide support for their campaign. The Zapatistas have won some concessions from the Mexican government, but their struggle to protect their culture and homeland continues. Check their web site or follow them on Twitter for details on the latest developments.

Many more examples of efforts to resist globalization or at least mitigate its worst effects could be recounted. Some movements mobilize globally; others, locally. Some movements take up arms; others work through existing democratic institutions. In most cases, however, these movements seek to gain a voice in altering and controlling the speed and extent of globalization's transformative forces. As a consequence, the effects of globalization are not predetermined; cultural homogeneity is not the only possible outcome. Rather, the actions of local culture groups—be they indigenous peoples, urban workers, or rural farmers—also shape its effects. The local-global link, it seems, operates in two directions to shape an unfolding future.

BLENDING SOUNDS ON A GLOBAL SCALE

On the popular music scene, debates about the effects of globalization abound. These debates follow the general terms of the broader globalization debates. Is globalization a force for homogenization or new forms of diversification, hybridity, and synthesis? If we think about music historically, synthesis and hybridization have deep roots. It is actually fairly difficult to find a truly "authentic" or "pure" locally bound musical genre that has not been influenced by extra-local musical forms. Distinctive regional sounds, such as Memphis soul or Tibetan throat singing, can definitely be identified. In many cases, these musical genres are associated with culture regions. At the same time, musical genres rarely develop in geographic isolation. Prior to globalization, migration and movement produced all manner of blending of ideas, styles, and genres— probably since the time humans began imitating the rhythms and sounds of nature. Modern technological innovations, from the invention of the gramophone to the release of online file-sharing software, have accelerated, intensified, and added to the complex processes of musical hybridization.

Take the case of *soukous,* a musical genre centered in Africa's Congo River basin. Soukous originated in the folk music and dance traditions of various Congolese ethnic groups. Some of these performers began to adopt Western instruments and jazz arrangements during the period of Belgian colonization in the early twentieth century. As the population urbanized in the 1950s, record companies found a burgeoning market and began importing Cuban "rumba" 78-rpm vinyl recordings. These "Latin" rhythms were incorporated and helped launch soukous as the first pan-African sound. It is interesting to note that this transfer was part of a historical process of multidirectional diffusion, since the rhythms imported from Cuba originated among African slave laborers who had carried them from West Africa a century earlier. Congolese musicians were given record contracts and brought to Paris studios, where they continue to blend new sounds and techniques and pump out the pulsing soukous beat for the world music scene. One can get dizzy just trying to keep track of the many multidirectional pathways of cultural interaction across time and space.

Cultural interaction in music has been ongoing for centuries but has been greatly speeded up and geographically expanded under globalization. From one perspective, globalization has sparked a creative cultural interaction by mixing musical traditions from around the world to produce new hybrid forms, many of which are highly localized. As a result, new and innovative regionally based soundscapes emerge continuously, reinforcing old or helping to construct new culture regions. From another perspective, globalization has enriched First World transnational entertainment corporations without providing due compensation for the creative labor of local cultures. The debate over which perspective best reflects the actual effects of globalization on music is complex and will undoubtedly continue.

NATURE–CULTURE

11.4 LEARNING OBJECTIVE
Relate how the continuation of current globalizing trends will shape future nature–culture relations.

One might ask how the theme of nature–culture is related to the question of "one world or many." The central issues between the two include the impact of globalization processes on ecosystems and local communities, and the promise and peril of new technologies for the natural world. Globalization represents for some an ever-expanding world economy. Where will the natural resources come from to fuel this expansion? Global fossil fuel consumption is rising dramatically in response to the demands of China's and India's rapidly expanding economies. As a result of past and ongoing industrialization, the atmospheric carbon dioxide level is higher today than it has been for at least 2 million years. Physical geographers now think in terms of "no analog future" scenarios for the Earth's ecosystem. In other words, there is no past experience in human history to guide

our understanding of a future ecosystem transformed by the Industrial Revolution.

SUSTAINABLE FUTURES

Whether fearful or hopeful, most forecasters agree that recent trends indicate rising levels of consumption worldwide. Specifically, the cultures of mass consumption that developed first in the United States and Europe are spreading to every corner of the globe. China is a case in point. As the twenty-first century unfolds, China has become the world's largest consumer of coal and automobiles. The cultures of mass consumption at the heart of globalization require enormous amounts of natural resources and produce prodigious quantities of pollutants. Given the ecological problems associated with the mass consumption of commodities such as cars, refrigerators, and so forth, we are compelled to ask ourselves whether current trends in consumption are sustainable for much longer.

The question of sustainable development on a global scale has been around since the 1970s. Cambridge geographer William Adams has produced the most comprehensive and carefully researched history of sustainable development. He suggests that the first attempt to create a plan for globally sustainable development came in 1980 with the publication of the World Conservation Strategy. This idea was refined a few years later by the UN-sponsored World Commission on Environment and Development. The commission brought sustainable development into the mainstream with its 1987 book *Our Common Future*. It identified poverty as a fundamental cause of the world's ecological problems and concluded that to reduce global ecological problems, we need to reduce global poverty through the promotion of economic development. Today, just about everyone from the barefoot "tree hugger" to the well-heeled international bank executive advocates sustainable development. This is what Adams labeled the "mainstream" version of sustainable development, by which he meant that the concept had been redefined in a way that posed no serious challenge to the status quo of continual global expansion of consumption and economic growth.

But, let us take a closer look at the mainstream approach to sustainable development. In essence, it says that the familiar industrial model of continual economic expansion is the cure for both global poverty and ecological problems. However, although globalization has brought new prosperity and higher levels of consumption to some parts of the world, ecological problems only seem to be increasing. China's rising coal consumption has led to increased emissions of carbon dioxide, a leading greenhouse gas. It is now the world's largest producer of carbon dioxide and is projected to far exceed the United States, now in second place. Furthermore, as we noted previously, the rewards of globalization are distributed unevenly, with the majority of poor people remaining poor and only a few people becoming richer.

Some observers conclude that the term *sustainable development* is an oxymoron. Given the ecological record of modern industrialization, the faith in economic growth as a cure for environmental ills seems misplaced (**Figure 11.19**). Cultural and political ecologists have been strong critics of the mainstream approach to sustainable development. They claim that it does not take into account the larger-scale historical and structural causes of poverty, such as the lasting effects of European colonialism. Unless these are addressed, it is unlikely that the mainstream approach to sustainable development will substantially decrease poverty levels. A further criticism is that the focus on the link between poverty and ecological degradation downplays the environmental impact from high levels of consumption in affluent countries. For instance, Americans alone consume one-fourth of the world's petroleum output and generate one-fourth of the carbon dioxide pollution.

One suggested alternative is to formulate sustainable development "from below" rather than through a top-down global program. The idea is to

FIGURE 11.19 **Mountaintop mining in West Virginia.** Landscapes and ecosystems are permanently altered to meet the industrial demand for increasing amounts of fossil fuels. A popular technique in coal mining involves the removal of entire mountaintops to access the deposit. *(Melissa Farlow/National Geographic Creative/Getty Images.)*

assist local initiatives and employ local knowledge to craft different economic paths for developing countries and communities that will not degrade land and resources along the way to higher living standards. Such alternatives, because they are informed by the communities they most immediately concern, would also be designed to maintain cultural identities, landscapes, and regions. Across the globe there are now hundreds of such efforts, which go by a variety of titles, such as "community conservation," "joint forest management," and "indigenous peoples' reserves." Approaching sustainable development from below means that the future will be defined by cultural heterogeneity, not homogeneity.

FIGURE 11.20   **Solar panels in the Shanghai skyline.** China is one of the world's top three markets for renewable energy, including solar and wind power. *(Jeff_Hu/iStockphoto/Getty Images.)*

Another suggested alternative is to switch our main energy source from fossil fuels to renewable resources such as wind and solar. Again, China, which a 2012 Pew Charitable Trusts report labeled "the world's clean energy leader" in every measure of alternative energy growth, is an important signpost in forecasting the future. China is now the fastest growing market for solar power, third in the world in installations behind Germany and Italy (Figure 11.20). It is also the world's largest market for wind power. Moreover, the government launched a major incentive program for the development of "new energy" vehicles, including hybrid and all-electric cars. Our sustainable future will

likely be determined by how quickly the rest of the world adopts less environmentally harmful technologies and abandons the energy sources that have so far fueled the ongoing Industrial Revolution.

THINK GLOBALLY, ACT LOCALLY

Perhaps sustainable development from below can mitigate widespread poverty while minimizing the ecological degradation that accompanies many top-down development initiatives. But what about the environmental effects of modern affluence—how can those be addressed? Throughout the 1990s, a series of international conferences were held that sought to identify the world's most pressing environmental problems and propose global-scale initiatives to address them. The most prominent of these was the United Nations Conference on Environment and Development, held in Rio de Janeiro in 1992. Now known simply as the Rio Conference, it produced international conventions or agreements that sought to reduce global ecological problems such as species extinction and global warming. The most prominent of these was the Convention on Biological Diversity. This convention committed states that signed on to protecting wildlife habitats and pursuing economic development policies that minimize species loss.

The approach taken at the Rio Conference is best captured in the slogan "Think Globally, Act Locally." If we plan carefully, actions taken at the local scale in places around the world will collectively result in an improved global environment. For example, the establishment of local parks and reserves will provide a global network of protected habitats that will help to maintain the Earth's biodiversity.

The introduction of hybrid cars—vehicles that combine traditional fossil fuel engines with electric motors to greatly reduce gas consumption and pollution—is another case in point. When the state of California passed a law requiring 10 percent of all car sales to be hybrids, it forced carmakers to produce more fuel-efficient cars. Since California is the largest car market in the United States, its action has had ripple effects nationwide and, ultimately, worldwide. China is now promoting hybrid cars for its burgeoning market, and several of the new auto factories will produce hybrids. These local- and national-level initiatives should help considerably in reducing global levels of carbon dioxide and other pollutants.

CULTURAL LANDSCAPE

11.5 LEARNING OBJECTIVE
Identify likely future landscapes.

We can get some clue about the kind of cultural landscapes we might expect in the future and the future of existing cultural landscapes by observing the cultural landscape under globalization. Philip Kelly even speaks of "landscapes of globalization." He is referring mostly to the urban, corporate architecture that has sprung up on every continent, replicating itself in city after city. Such homogenization is often resisted by local communities that hope to preserve existing cultural landscapes.

FIGURE 11.21 **Modern office buildings in Buenos Aires, Argentina.** The forces of globalization have accelerated the construction of landscapes such as this one near the docklands area of Puerto Madero Harbor basin. *(wim wiskerke/Alamy.)*

GLOBALIZED LANDSCAPES

Geographer David Keeling sought visible evidence of globalization in the landscape of Buenos Aires, Argentina—the capital of a country that desperately wants to become enmeshed in the world economy but struggles, perhaps in vain, against its peripheral location. He found abundant evidence of "a homogenized landscape" of glass-and-steel corporate office towers, of luxury hotels and conference centers for the corporate power brokers of the world economy (**Figure 11.21**).

Clearly, urban landscapes—cityscapes—can serve as an index to the level and type of engagement with the globalization process. We have now begun witnessing the diffusion of California-style suburban housing tracts to affluent regions of India. Still, Keeling notes, these homogenizing processes are neither omnipresent nor omnipotent. Certain other Latin American capitals—such as Quito, Ecuador; La Paz, Bolivia; and Havana, Cuba—reveal minimal global influences in their cityscapes. Given the improbability of a global culture, visible differences among cities seem likely to persist. Moreover, people all over the world value their cultural landscapes,

whether as visual reminders of their heritage or as lucrative attractions in the global tourism industry, and will thus want to preserve them. Landscape and place still matter in a globalizing world.

STRIVING FOR THE UNIQUE

Urban landscapes in the age of globalization reveal another element: the enduring spirit of place. One city after another has preserved or erected some building or monument so unique as to be a symbol or icon of that particular city. When you see a photograph of this structure, you know at once where it is located (**Figure 11.22**). Television journalists often stand in front of such visible icons to prove to viewers that they are actually reporting on location. Examples would include the Petronas Towers in Kuala Lumpur, the arch linking east and west in St. Louis, the Space Needle in Seattle, and the Eiffel Tower in Paris. True, many or most of these icons predate the era of globalization, but that is not the issue. Rather, their retention and protection offer the relevant message. In the 1960s, the city of Copenhagen restored Nyhavn (New Harbor), then a derelict seventeenth-century canal, into one of the

FIGURE 11.22 **The Petronas Towers in Kuala Lumpur** have become a symbol for the city. Their unique design helps establish Kuala Lumpur's identity as a city different from others. Uniqueness of design, a feature of much modern architecture, stands in opposition to cultural homogenization. *(age fotostock/SuperStock.)*

most visually striking and instantly recognizable historical ports in the world (**Figure 11.23**). The Kremlin walls in Moscow may retain few if any of their original red bricks, as they fall victim to weathering and are replaced, but the structure is renewed and endures as a symbol of the city.

Neolocalism is the desire evident in many local communities to embrace the uniqueness and authenticity of place. Governments and electorates at all levels—from local to national—have a far greater say about globalization than one might imagine. A backlash against chain stores and conformity can find strength in local ordinances. A community can actually prevent McDonald's or Walmart from establishing outlets.

> **neolocalism** The desire to re-embrace the uniqueness and authenticity of place, in response to globalization.

Neolocalism, then, pits the cultural power of place against the economic power of globalization.

WAL-MARTIANS INVADE TREASURED LANDSCAPE!

Local communities and their governments attempt to preserve cultural landscapes in numerous ways. Zoning laws, architectural guidelines, minimum lot-size requirements, building codes, and conservation easements can all be used to maintain the distinct character of landscapes. At the global scale, we have World Heritage Sites, places that the United Nations Educational, Scientific and Cultural Organization (UNESCO) has deemed to be of such cultural significance that they should be given international protection. The temples at Angkor Wat in Cambodia (**Figure 11.24**) and the "Old Stone Town" on Zanzibar Island, Tanzania, are two examples of the hundreds of UNESCO World Heritage Sites worldwide.

FIGURE 11.23 **Copenhagen's Nyhavn district.** A key port from the 1600s to the early 1900s, Nyhavn was abandoned by the shipping industry as ship size increased after World War II. In the mid-1960s, the city slapped on fresh coats of bright paint and hauled in antique sailing ships to create one of the most recognizable historic ports in the world. *Can you think of other historic waterfronts that have been similarly restored to create iconic urban landscapes?* *(Courtesy of Roderick Neumann.)*

FIGURE 11.24   **Angkor Wat, UNESCO World Heritage Site.** Located 192 miles from Phnom Penh, Cambodia, and built between 1113 and 1150 C.E., this temple is regarded as the pinnacle of the Khmer Empire's architecture. UNESCO named it and surrounding structures a World Heritage Site in 1992. Such a designation helps efforts to safeguard and restore cultural landscapes of global significance. *(Courtesy of Ari Dorfsman.)*

Globalizing processes are putting new pressures on cultural landscapes. Often small community groups and town governments are pitted against powerful transnational corporations whose investment choices can profoundly transform a landscape. The phenomenal expansion of Walmart Stores, Inc., is an often cited example of how small-town landscapes are transformed by corporate investment. The biggest impact is the "hollowing out" of small-town main streets. Owner-run small businesses cannot compete with the giant retailer and soon have to lock their doors. Since "big box stores" such as Walmart are typically located outside of city centers, the old downtowns are turned into empty shells.

Some communities have welcomed Walmart and others have tried hard to keep out the chain. Perhaps the most novel opposition campaign was conducted in Vermont. In a confrontation that author Barbara Ehrenreich labeled "Earth People vs. Wal-Martians," the National Trust for Historic Preservation declared the entire state of Vermont to be an endangered landscape. Vermont is the only state ever to make the list of endangered historic places, which generally comprise individual buildings and historic urban districts. According to the National Trust, building more supersized Walmart stores in Vermont would degrade the state's "sense of place." The National Trust had employed a similar strategy in 1993, which forced Walmart to build stores more appropriate to Vermont's landscape (**Figure 11.25**).

PROTECTING EUROPE'S RURAL LANDSCAPE

Another aspect of globalization, the drive to eliminate territorial barriers to the free trade of commodities, threatens rural landscapes in Europe. The fear is that cheap food imports will put small farmers out of business, which in turn will lead to the demise of treasured rural landscapes (**Figure 11.26**). European farmers and their national and EU representatives have argued that agriculture is not solely about food production—that it performs multiple functions, such as maintaining cultural landscapes and providing environmental services. Geographer Gail Hollander has observed that farmers and their advocates are using what they call agriculture's "multifunctionality" to gain exemptions from the WTO's strict rules on free trade. The exemptions are warranted, they argue, in order to preserve the character of cultural landscapes that agriculture supports. Hollander concludes that multifunctionality could be used to preserve the landscapes of a few communities in Europe or, in stronger form, to challenge the very logic of the WTO's rules on the global trade of agricultural products. Given their symbolic value, the preservation of cultural landscapes may be a key tool used to slow or mitigate globalization's homogenizing effects.

NO ANALOG LANDSCAPE FUTURES

The trends in fossil fuel consumption all point to a global atmosphere with carbon dioxide concentrations far beyond what the human species has ever experienced. The consensus among physical geographers and other climate scientists is that higher carbon dioxide concentrations are closely correlated with rising global temperatures. How high and how fast of a temperature rise remains subject to debate, but the effects on cultural landscapes are already

▶ **FIGURE 11.25** **Residents and tourists believe the historic town squares of New England are threatened by the construction of Walmart stores.** Defenders of Vermont's cultural landscape argue that large retail ventures like Walmart are both visually disruptive and economically destructive of local businesses. ***What do you think might happen to small town centers when large retail stores open nearby?*** *(Left: Alan Copson/AWL Images/Getty Images; Right: © Chuck Franklin/Alamy.)*

▲ **FIGURE 11.26** **Rural town in Andalusia, Spain.** The ancient olive terraces, gardens, woodlots, and pastures of this town are typical of much of Mediterranean Europe. Globalization and other economic forces are making the traditional, extensive agricultural systems that produced and support such landscapes increasingly obsolete. *(Courtesy of Roderick Neumann.)*

observable. From these observations, we can speculate about landscape futures, albeit with no analogs in past landscapes to guide us.

Rural geographer Michael Woods surveyed a range of studies in an effort to forecast rural futures under conditions of global warming. Perhaps the most straightforward impact will be a latitudinal movement of agricultural landscapes northward in the Northern Hemisphere. As temperatures rise, growing seasons will lengthen, so we will see a movement of the various belts—Wheat Belt, Corn Belt, Cotton Belt—northward. More complex climate effects imply more complex outcomes. For example, climate models show local and regional rainfall patterns shifting under global warming. Increased drought occurrence may lead to the spread of irrigation, but only where local cultural and political institutions can manage the needed levels of organization and cooperation. Woods observes that in some cases climate change is exacerbating land abandonment and accelerating rural-to-urban urban migration, leading to an emptying out of rural landscapes. Efforts to expand wind power and reduce

FIGURE 11.27 **The Maeslant Barrier in Rotterdam, Netherlands.** This huge movable structure, the largest of its kind in the world, creates a barrier to high seas when closed. Such landscape features will become more numerous and prominent in the future as sea levels rise. *(frans lemmens/Alamy.)*

fossil fuel consumption have produced local conflicts over the placement of wind generators that disrupt the vision of idyllic rural landscapes. These examples merely hint at the complexity of forces shaping rural landscape futures.

Urban landscape futures are similarly confronted by the environmental changes wrought by global warming. Most critically for cities will be sea-level rise. Superstorm Sandy provided a glimpse of what the future holds for coastal cities when it hit lower Manhattan on October 28, 2012, bringing a record tidal surge of nearly 14 feet. The current trend indicates that sea level is rising faster than in past centuries and the rate of rise appears to be accelerating. In 2013 geographers Susan Cutter and William Soleki convened a panel of experts to report on the urban impacts of climate change for the U.S. National Climate Assessment and Development Advisory Committee. Sea-level rise, they observe, has the direct effect of coastal flooding, but the indirect effects, such as power outages and damaged sewer and water systems, extend well beyond flooded areas. A team of researchers recently projected the likely exposure to sea-level rise in 2070 of the world's largest port cities. Worldwide, they expect that 150 million people and over $3000 billion in property will be exposed to damaging storm surges. A wide range of responses to sea-level rise will drastically alter coastal urban landscape futures. Most dramatic will be population migration inland. In the wake of Sandy, for example, state governments are considering the abandonment of coastal zones for housing and the implementation of open-space green belts at the seashore to reduce the exposure of life and property. A wide variety of floating structures are now appearing on urban development plans. Infrastructure changes, including raised roadways, elevated power systems, and coastal protection structures such as breakwaters, dikes, and floodgates, are in the works (**Figure 11.27**). The future landscapes of coastal cities will no doubt include a wide variety of new visual elements.

CONCLUSION

As powerfully transformative as the processes of globalization are, they are unlikely to result in cultural homogenization any time soon. As we learned by examining globalization's interaction with each of cultural geography's five themes, we do not yet live in a placeless world. In fact, we have observed many trends suggesting that globalization will produce more geographic diversity and many unintended and unforeseen outcomes in our future.

In closing, it is our hope that we have excited your interest in the world's cultural diversity. To paraphrase the words of Aldous Huxley, we hope that our vicarious world travels have left you "poorer by exploded convictions" and "perished certainties," but richer by what you have seen. Perhaps we, like Huxley, set out on this journey with preconceptions of how people should best "live, be governed, and believe." When one travels—even if just through the pages of a geography book as we have—such convictions often get mislaid. The main message of *Contemporary Human Geography* is that we will best be prepared to thrive in this new millennium if we maintain a willingness to question even our most closely held convictions and remain open to the boundless capacity of human cultural expression to surprise and amaze.

SEEING GEOGRAPHY   Global Reach

What do these images convey about empire and globalization, and their similarities and differences?

Two images of global reach, more than a century apart (Queen Victoria with world map; a JVC advertisement). *(Left: Getty Images/Enslow Publishers; Right: Raymond Boyd/Michael Ochs Archives/Getty Images.)*

In his book *Apollo's Eye,* the late cultural geographer Denis Cosgrove attempted to interpret the power represented in images of the globe and to show how the practices of globalization are historically rooted in a Western cultural history of imagining, seeing, and representing the globe. We will try a little of this interpretive method on these two images.

The image on the left shows Queen Victoria circa 1850 in front of a world map oriented so that the majority of Britain's territorial empire is displayed. We know that during this period of European history, the queen was sovereign, meaning that she personified, even embodied, Britain and its colonial empire. The image is scaled such that the queen's arm span matches the span of the British Empire. She is positioned in front of the map, emphasizing her authority and power over it. With power and authority comes responsibility; the viewer is meant to read in the queen's pose and dress a moral role as protector and civilizing force. Foregrounded as she is, then, all lines of power, authority, and moral right and responsibility in empire run through her.

The image on the right is an advertisement for JVC, a transnational consumer electronics company that began as the Victor Talking Machine Company of Japan, Limited, in 1927. In the advertisement, photographed in 2004, the company's logo, "JVC," is scaled to continental size. The message conveyed is one of global dominance and global reach. It also conveys the message of one company bringing together the world, a goal that JVC's web site describes as "contributing to the global community through cultural activities" with corporate underwriting. Finally, the advertisement is meant to express through the image of the globe the fact that JVC now has a network of manufacturing sites throughout Asia, the Americas, and

Europe as well as sales subsidiaries in many more regions.

What do these images tell us about continuity and change from empire to globalization? We see a common theme in the claim of global reach. In the JVC ad, it is expressed as the company's ability to span the globe with its products and services. In the image with Queen Victoria, it is expressed in the cartographic representation of Britain's global empire. We can also identify significant differences between the emotional and affective meaning of British Empire and

globalization conveyed by these images. Under empire, the queen personified British imperial rule, and allegiance to the queen was required of all imperial subjects. Under globalization, on the other hand, allegiance is constructed between the consumer and transnational corporations. Corporations are faceless rather than personified, represented by abstract logos rather than by living, breathing sovereigns. In summary, we can see in these images both the roots of globalization in empire as well as the significant differences between the two kinds of global power.

DOING GEOGRAPHY

Interpreting the Imagery of Globalization

The mandate of large corporations today, nearly regardless of the type of service or good they produce, is to go global or go bankrupt. In addition to staying competitive, going global gives a company a certain cachet and consumer appeal, the way that being "modern" did in previous decades. Transnational corporations also stress other popular notions related to globalization, such as respect for the world's cultural diversity and concern for the global environment.

For this activity, you will use your interpretive skills to look at the way the processes of globalization are represented in corporate-produced visual imagery and text. Using the steps below, try to find representations of globalization in more than one medium, including product packaging, magazine advertisements, and corporate web sites.

Steps to Interpreting the Imagery of Globalization

Step 1: Concentrate on one type of industry, such as pharmaceuticals or automobiles, or several.

Step 2: Look for materials that include visual and textual representations of the globe, the Earth, or the world, and remember to make use of the five themes of cultural geography.

Step 3: Analyze and interpret each of the samples that you select by setting up a series of questions such as: What popular notions about globalization are emphasized? How much validity do these representations carry? This is not the same as asking about the truthfulness or falseness of an ad. Rather it is to ask, for example: What does going to the Hard Rock Café and buying a T-shirt have to do with "saving the planet"?

In the cases of visual imagery, look carefully at the way objects are arranged and scaled in relation to one another.

- How is power represented in the imagery?
- Does power lie with the individual consumer or the corporation?
- How are local-global linkages represented?
- Are the activities of global corporations given a moral authority and, if so, how?

Step 4: Provide a written summary of your findings and analysis. Include a copy of the image or images that you use and make sure to reference your sources.

As you do this exercise, bear in mind the power of transnational corporations in an era of globalization. Think about the importance of understanding how the images and texts they produce give meaning to our world, our places, and our landscapes.

Spanning the world. Corporations increasingly use such global imagery in advertising and public relations to signify global reach and engagement with the forces of globalization. *(Corbis/SuperStock.)*

Chapter 11
LEARNING OBJECTIVES REEXAMINED

11.1 **Describe the possible futures of place and region, and explain their driving forces.**
See pages 458–465. How has the Internet affected regional boundaries?

11.2 **Explain how current trends will shape future mobilities.**
See pages 465–473. What role has the automobile played in transforming China's urban culture?

11.3 **Analyze the relationship of contemporary globalization to historical processes and its likely effects on the future.**
See pages 473–478. What are the arguments from the left and right of the political spectrum in terms of globalization's effect on the future?

11.4 **Relate how the continuation of current globalizing trends will shape future nature–culture relations.**
See pages 478–481. What impact do recent globalization trends have on current and future sustainability?

11.5 **Identify likely future landscapes.**
See pages 481–485. How is globalization represented in urban landscapes?

KEY TERMS

glocalization   p. 462   neolocalism   p. 482

The Geography of the Future on the Internet

You can learn more about the geography of the future on the Internet at the following web sites:

An Atlas of Cyberspaces
http://www.cybergeography.org/atlas/atlas.html
Here, cyberspaces are made visible by Martin Dodge: graphic representations of the geography of the electronic territories of the Internet, the World Wide Web, and other new cyberspaces help you to visualize and comprehend the digital "landscapes" beyond your computer screen.

The Hawaii Research Center for Futures Studies
http://www.futures.hawaii.edu/
One of the best-known institutions for future studies. The Hawaii State Legislature created it in 1971 to train students in future thinking for work in government and business.

Millennium Project: World Federation of UN Associations
http://www.millennium-project.org/
A global, participatory, futures-research think tank of futurists, scholars, business planners, and policy makers. It produces numerous future scenarios and such documents as the annual "State of the Future" report.

World Future Society
http://www.wfs.org/
A scientific and educational association exploring how social and technological developments are shaping the future. The society serves as a clearinghouse for forecasts, recommendations, and alternative scenarios. It publishes, among other periodicals, the bimonthly journal *The Futurist*.

World Futures Studies Federation
http://www.wfsf.org/
Founded in 1967 to further research and education in future studies. It is made up of hundreds of individuals and institutions worldwide that together create a global network of researchers, teachers, policy analysts, and activists.

Sources

Adams, William. 2001. *Green Development: Environment and Sustainability in the Third World*. 2nd ed. London: Routledge.

Airriess, Christopher A. 2001. "Regional Production, Information-Communication Technology, and the Developmental State: The Rise of Singapore as a Global Container Hub." *Geoforum* 32: 235–254.

Barlow, John P. 1995. "Cyberhood Versus Neighborhood." Special issue of *Utne Reader* 68(3): 52–64.

Beaverstock, Jonathan, Phillip Hubbard, and John Short. 2004. "Getting Away with It? Exposing the Geographies of the Super-Rich." *Geoforum* 35(4): 401–407.

Bradsher, Keith. 2012. "China Blocks Web Access to Times After Article." *New York Times,* 25 October, p. A12.

Bruntland, H. 1987. *Our Common Future*. Oxford: Oxford University Press (for the World Commission on Environment and Development).

Bunge, William. 1973. "The Geography of Human Survival." *Annals of the Association of American Geographers* 63: 275–295.

Cosgrove, Denis. 2001. *Apollo's Eye: A Cartographic Genealogy of the Earth in the Western Imagination*. Baltimore: Johns Hopkins University Press.

Crouch, David (ed.). 1999. *Leisure/Tourism Geographies: Practices and Geographical Knowledge*. London: Routledge.

Crum, Shannon L. 2000. "The Spatial Diffusion of the Internet." PhD dissertation, University of Texas at Austin.

Cutter, Susan, and William Soleki. 2013. "Urban Systems, Infrastructure, and Vulnerability I." Federal Advisory Committee Draft Climate Assessment Report Released for Public Review. National Climate

Assessment and Development Advisory Committee, U.S. Global Change Research Program. http://ncadac.globalchange.gov/.

Domosh, Mona. 2006. *American Commodities in an Age of Empire.* New York: Routledge.

Drummond, Ian, and Terry Marsden. 1999. *The Condition of Stability: Global Environment Change.* New York: Routledge.

Durand, Jorge, and Douglas S. Massey. 2010. "New World Orders: Continuities and Changes in Latin American Migration." *Annals of the American Academy of Political and Social Science* 630: 20–52.

Ehrenreich, Barbara. 2004. "Earth People vs. Wal-Martians." *New York Times,* 25 July, p. 11.

Elwood, Wayne. 2001. *The No-Nonsense Guide to Globalization.* London: Verso.

Feld, Stephen. 2001. "A Sweet Lullaby for World Music." In Arjun Appadurai (ed.), *Globalization,* pp. 189–216. Durham, N.C.: Duke University Press.

Ferguson, Andrew. 2004. "Wal-Mart Opponents Launch Two-Front Attack." *Pittsburgh Post-Gazette,* 20 June, p. C2.

Firebaugh, Glenn. 2003. *The New Geography of Global Inequality.* Cambridge, Mass.: Harvard University Press.

Flack, Wes. 1997. "American Microbreweries and Neolocalism." *Journal of Cultural Geography* 16(2): 37–53.

Friedman, Thomas. 2005. *The World Is Flat: A Short History of the Twenty-First Century.* New York: Farrar, Straus and Giroux.

Fryer, Donald. 1974. "A Geographer's Inhumanity to Man." *Annals of the Association of American Geographers* 64: 479–482.

Gilbert, Anne, and Paul Villeneuve. 1999. "Social Space, Regional Development, and the Infobahn." *Canadian Geographer* 43: 114–117.

Hanson, Susan, Robert Nicholls, N. Ranger, S. Hallegatte, J. Corfee-Morlot, C. Herweijer, and J. Chateau. 2011. "A Global Ranking of Port Cities with High Exposure to Climate Extremes." *Climatic Change* 104: 89–111.

Hardt, Michael, and Antonio Negri. 2000. *Empire.* Cambridge, Mass.: Harvard University Press.

Hessler, Peter. 2007. "Wheels of Fortune: The People's Republic Learns to Drive." *The New Yorker,* 26 November, p. 104.

Hollander, Gail. 2004. "Agricultural Trade Liberalization, Multi-functionality, and Sugar in the South Florida Landscape." *Geoforum* 35: 299–312.

Hooson, David (ed.). 1994. *Geography and National Identity.* Oxford: Blackwell.

Hudson, Ray. 2000. "One Europe or Many? Reflections on Becoming European." *Transactions of the Institute of British Geographers* 25: 409–426.

Huxley, Aldous. 1926. *Jesting Pilate: An Intellectual Holiday.* New York: George H. Doran.

Iyer, Pico. 2000. *The Global Soul: Jet Lag, Shopping Malls, and the Search for Home.* New York: Knopf.

Keating, Michael. 1998. *The New Regionalism in Western Europe.* Northampton, Mass.: Edward Elgar.

Keeling, David J. 1999. "Neoliberal Reform and Landscape Change in Buenos Aires." *Yearbook, Conference of Latin Americanist Geographers* 25: 15–32.

Kelly, Philip F. 2000. *Landscapes of Globalization.* London: Routledge.

Kitchen, Rob, and Martin Dodge. 2002. "Emerging Geographies of Cyberspace." In R. J. Johnston, Peter Taylor, and Michael Watts (eds.), *Geographies of Global Change: Remapping the World,* 2nd ed., pp. 340–354. Oxford: Blackwell.

Knox, Paul. 2002. "World Cities and the Organization of Global Space." In R. J. Johnston, Peter Taylor, and Michael Watts (eds.), *Geographies of Global Change: Remapping the World,* 2nd ed., pp. 328–339. Oxford: Blackwell.

Kraus, Clifford. 2002. "Returning Tundra's Rhythm to the Inuit, in Film." *The New York Times,* 30 March, p. A4.

Kunstler, James H. 1993. *The Geography of Nowhere: The Rise and Decline of America's Man-Made Landscape.* New York: Simon & Schuster.

Leyshon, Andrew, David Matless, and George Revill (eds.). 1998. *The Place of Music.* New York: Guilford.

McDowell, Linda (ed.). 1997. *Undoing Place? A Geographical Reader.* London: Arnold.

Naughton, Keith. 2004. "China Hits the Road." *Newsweek,* 28 June, p. E22.

Nemeth, David J. 2000. "The End of the Re(li)gion?" *North American Geographer* 2: 1–8.

O'Loughlin, John, et al. 1998. "The Diffusion of Democracy, 1946–1994." *Annals of the Association of American Geographers* 88: 545–574.

Pew Charitable Trusts. 2013. *Who's Winning the Clean Energy Race? 2012 Edition.* http://www.pewtrusts.org/uploadedFiles/wwwpewtrustsorg/News/Press_Releases/Clean_Energy/clen-G20-report-2012-FINAL.pdf.

Pew Research Center. 2013. *The New Sick Man of Europe: The European Union.* http://www.pewglobal.org/files/2013/05/Pew-Research-Center-Global-Attitudes-Project-European-Union-Report-FINAL-FOR-PRINT-May-13-2013.pdf.

Poon, Jessie P. H., Edmund R. Thompson, and Philip F. Kelly. 2000. "Myth of the Triad? The Geography of Trade and Investment Blocs." *Transactions of the Institute of British Geographers* 25: 427–444.

Press, Larry. 1997. "Tracking the Global Diffusion of the Internet." *Communications of the Association of Computing Machinery* 40(11): 11–17.

Relph, Edward. 1976. *Place and Placelessness.* London: Pion.

Rohter, Larry. 2002. "Brazil's Prized Exports Rely on Slaves and Scorched Land." *The New York Times,* 25 March, pp. A1, A6.

Sessions, George (ed.). 1995. *Deep Ecology for the 21st Century.* Boulder, Colo.: Shambala.

Shields, Rob. 1991. *Places on the Margin: Alternative Geographies of Modernity.* London: Routledge.

Sparke, Matthew. 2013. *Introducing Globalization: Ties, Tensions, and Uneven Integration.* Oxford: John Wiley & Sons.

Stiglitz, Joseph. 2007. *Making Globalization Work.* New York: W.W. Norton.

Suvantola, Jaakko. 2002. *Tourist's Experience of Place.* Aldershot, U.K.: Ashgate.

Swerdlow, Joel L. (ed.). 1999. "Global Culture." Special issue of National Geographic 196(2): 2–132.

Swyngedouw, Erik. 1997. "Neither Global nor Local." In Kevin R. Cox (ed.), *Spaces of Globalization: Reasserting the Power of the Local,* pp. 137–166. New York: Guilford.

Teo, Peggy, and Lim Hiong Li. 2003. "Global and Local Interactions in Tourism." *Annals of Tourism Research* 30(2): 287–306.

"Used and Abused: Five Recent Cases with Slavery Convictions." 2003. *Palm Beach Post,* 7 December, p. 2.

Wood, William B. 2001. "Geographic Aspects of Genocide: A Comparison of Bosnia and Rwanda." *Transactions of the Institute of British Geographers* 26: 57–75.

Woods, Michael. 2012. "Rural Geography III: Rural Futures and the Future of Rural Geography." *Progress in Human Geography* 36(1): 125–134.

York, Christopher. 2013. "Marine Le Pen, France's Front National Leader, Claims Political Links With Nigel Farage's UKIP." *Huffington Post UK.* February2.http://www.huffingtonpost.co.uk/2013/02/20/marine-le-pen-ukip_n_2723386.html.

Zelinsky, Wilbur. 2011 *Not Yet a Placeless Land: Tracking an Evolving American Geography.* Amherst: University of Massachusetts Press.

Ten Recommended Books on the Geography of the Future

(For additional suggested readings, see the *Contemporary Human Geography* LaunchPad: http://www.macmillanhighered.com/launchpad/DomoshCHG1e.)

Appadurai, Arjun. 2013. *The Future as Cultural Fact: Essays on the Global Condition.* London: Verso. Using primarily examples from India, a renowned anthropologist presents essays on the past and future of the poor in a globalizating world. Features extended commentary on nationalism, violence, commodification, and terror.

Bagchi-Sen, Sharmistha, and Helen Smith (eds.). 2006. *Economic Geography: Past, Present and Future.* London: Routledge. Economic geographers point the way for the future development of the subfield in 20 wide-ranging chapters.

Firebaugh, Glenn. 2003. *The New Geography of Global Inequality.* Cambridge, Mass.: Harvard University Press. Makes the argument that income inequalities among and within world regions are misunderstood. There are a lot of economic data to wade through, but they make the case stronger. The author raises important questions about the effects of globalization.

Gabel, Medard, and Henry Bruner. 2003. *Global Inc.: An Atlas of the Multinational Corporation.* New York: New Press. A wonderful atlas mapping everything from the historical rise of multinational companies to the latest geographic expansions of Walmart. It's full of facts on every important global industry, including food, cars, and pharmaceuticals. It also maps the impacts of multinational corporations, including cultural and environmental ones.

Gornitz, Vivien. 2013. *Rising Seas: Past, Present, Future.* New York: Columbia University Press. A sobering survey of the evidence for and likely future impacts of accelerating sea-level rise in the twenty-first century. The book provides an indispensable foundation for understanding the future of coastal landscapes and nature–culture interactions.

Hastrup, Kirsten, and Karen Fog Olwig (eds.). 2012. Climate change and human mobility: global challenges to the social sciences. Cambridge, U.K.: Cambridge University Press. Geographers and anthropologists bring a range of perspectives in an attempt to understand the effects of climate change on future human migrations. The collection includes cases from the Arctic, Pacifica, and Africa.

Johnston, R. J., Peter Taylor, and Michael Watts (eds.). 2002. *Geographies of Global Change: Remapping the World.* 2nd ed. Oxford: Blackwell. Considers such issues as post–cold war geopolitics, global environmental governance, and cultural changes related to mass consumption, the Internet, and ethnic identity.

Jussila, Heikki, Roser Majoral, and Fernanda Delgado-Cravidao. 2001. *Globalization and Marginality in Geographical Space.* Aldershot, U.K.: Ashgate. Case studies from Europe, the Americas, Africa, and Australia illustrate how geographical research aids our understanding of the way in which the policies and politics of globalization affect the more marginalized areas of the world.

Miles, Malcolm, and Tim Hall (eds.). 2003. *Urban Futures: Critical Commentaries on Shaping Cities.* London: Routledge. This volume brings together experts from a range of disciplines to debate the cultures and forms of tomorrow's cities.

Schmidt, Eric, and Jared Cohen. 2013. *The New Digital Age—Reshaping the Future of People, Nations, and Business.* New York: Knopf. Eye-opening look at the new digital technologies just around the corner and what their introduction means for human affairs. Written by industry insiders, it reflects a modernist ideology that technology will deliver a better future for all.

The late, renowned cultural geographer Barney Nietschmann used to tell his students at the University of California at Berkeley, "The map is the starting point not the ending point for geographers." What he meant was that the practice of human geography is not to produce maps. (Map making is a technical field called cartography.) Rather, human geographers use the spatial arrangement of social and natural phenomena to formulate and answer questions about relationships, including causal relations. Maps are the common language for pursuing these questions.

For our purposes in *Contemporary Human Geography,* maps are two-dimensional representations of three-dimensional spatial phenomena. Cartographers have agreed on certain graphic conventions that are shared by all maps. These conventions allow cartographers to represent a wide range of spatial phenomena including movement, qualitative and quantitative variation across space, boundaries, and location, to name just a few. We will start with some of the basic conventions and then move on to discuss how different mapping choices can provide different answers.

scale The distance on a map in relation to the distance in actual space.

Most maps will have a **scale.** Map (or cartographic) scale is the distance on the map in relation to distance in actual space. For example, 1 centimeter on the map equals 1 kilometer of distance in actual space. Scale on a map can be represented by a bar, as we see in Figure 1.6, or it can be represented as a numerical ratio, as in 1/100,000 or 1:100,000.

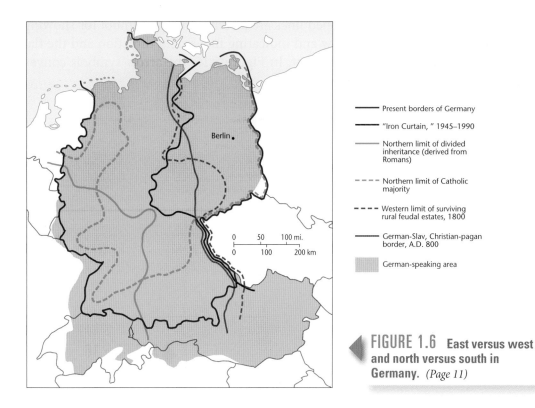

Berlin.

—— Present borders of Germany

—— "Iron Curtain, " 1945–1990

—— Northern limit of divided inheritance (derived from Romans)

-- -- Northern limit of Catholic majority

‒ ‒ ‒ Western limit of surviving rural feudal estates, 1800

—— German-Slav, Christian-pagan border, A.D. 800

German-speaking area

| 0 | 50 | 100 mi. |
| 0 | 100 | 200 km |

◀ FIGURE 1.6 **East versus west and north versus south in Germany.** *(Page 11)*

A map will also have a **legend.** Legends are essential for reading maps, for they tell what **map symbols** mean. Map symbols are graphic figures that represent any number of real-world phenomena. The legend for Figure 1.6 consists mostly of line symbols representing different types of boundaries (e.g., religious, ethnic) by using different types and colors of lines. Dots are used to represent specific location (in this example, Berlin) and are so intuitively understood that they often are not included in map legends. Finally, all maps have an **orientation,** which refers to the compass direction of the top of the map. Many map legends include an arrow indicating north so that the reader knows the map's orientation. The north arrow is essential in maps used for way finding, but less so for maps used for other purposes. Figure 1.6 does not have an arrow, because orientation is not particularly important to the map's purpose. Established convention dictates that maps are oriented to the north and so we can safely assume that is the case with Figure 1.6.

Maps are broadly categorized as either **baseline maps** (also called general maps) or **thematic maps.** Baseline maps provide basic information of coastlines, country boundaries, and so forth. Thematic maps emphasize a specific phenomenon or process, such as transportation, migration, and agricultural production. Most maps in *Contemporary Human Geography* are thematic.

The theme of Figure 5.16, for example, is the geographic displacement of people resulting from armed conflict. The theme requires that a fixed, two-dimensional map symbol somehow represent the mass movement of people in space. Arrow-tipped lines are a common map symbol for showing movement, with the arrow end indicating movement direction and the flat end indicating the origin point. In Figure 5.16, the arrow symbols convey

■ **legend** Explains what the symbols on a map mean.

map symbols Graphic figures that represent any number of real-world phenomena

orientation Refers to the compass direction of the top of a map.

■ **baseline maps** Also called general maps, baseline maps provide basic information such as coastlines, country, and boundaries.

thematic maps Maps that emphasize a specific phenomenon or process, such as transportation, migration, and agricultural production.

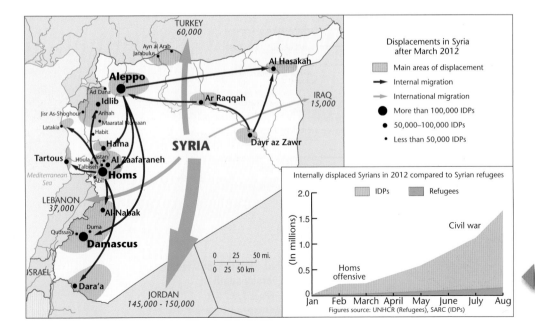

FIGURE 5.16
Syrians displaced by conflict. *(Page 197)*

other information in addition to origin and direction. The width of the line indicates quantity, in this case, the number of people displaced. The color of the line indicates either internal or external migration.

Note in the case of Figure 5.16 that the quantitative measure (number of people) represented by the two colors remains the same, while the qualitative information (locations of origin and destination) changes. Internal migrants are shown as leaving from and arriving in specific cities and regions, while external migrants are shown as leaving from and arriving in countries. This reflects the way displaced people's identities shift when they cross international borders. They become simply "Syrian refugees" and their particular home city or region becomes largely irrelevant to their new classification.

As we see in these examples, simple symbols can carry a wealth of information, and the more closely we look, the more we are able to discern patterns and relationships. We are also increasingly able to read the limitations of the symbols and to think about what maps hide as well as reveal. We see, for instance, how the same data can be rescaled, as in Figure 5.16's arrows, to tell a different story.

Let's look at Figures 6.10 and 6.12 to further illustrate this last point. These are common thematic maps known as **choropleth maps**. A choropleth map uses data aggregated over a predefined geographic area, typically a political designation such as a county, province, or state. Often colors are used to represent differences in values in different areas. In Figure 6.10, red represents a majority vote for the Republican Party and blue represents a majority vote for the Democratic Party.

choropleth maps Maps that use data aggregated over a predefined geographic area, typically a political designation such as a county, province, or state.

As noted in Chapter 6, the story of political party divisions within the United States changes depending on which geographic unit we decide to use. If we aggregate data at the state level, as in Figure 6.10, we tell one sort of story. If we aggregate the data at the county level, as in Figure 6.12, the story of political difference changes by becoming more nuanced and less starkly

FIGURE 6.10 **2012 presidential election results.** *(Page 237)*

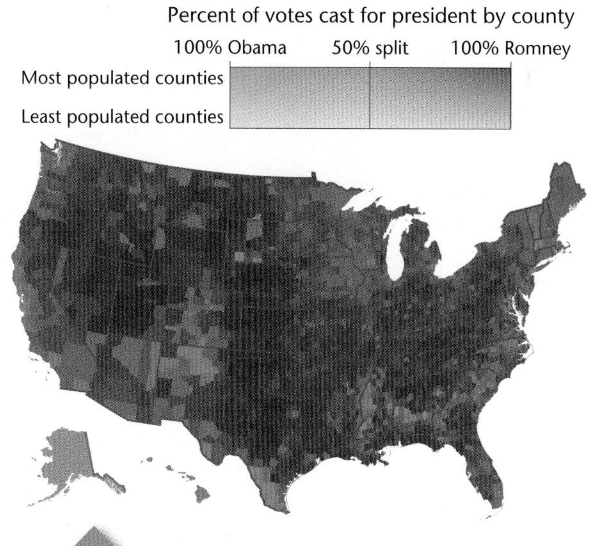

Republican
Democrat

Percent of votes cast for president by county

100% Obama   50% split   100% Romney

Most populated counties

Least populated counties

FIGURE 6.12 **Purple America.** *(Page 238)*

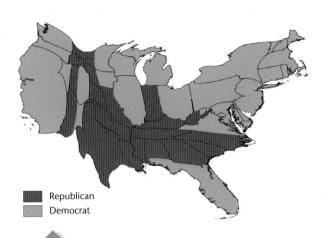

FIGURE 6.11 **Cartogram of the 2012 presidential election results.** *(Page 237)*

Legend:
- Republican
- Democrat

defined. In addition, the coloring technique changed from a color binary in Figure 6.10 to a color progression in 6.12. Progressive color shading is better able to represent the subtle geographic variation in voting patterns.

Another type of thematic map used in *Contemporary Human Geography* is the **cartogram.** A cartogram is a map that results from using some thematic variable—the number of people voting Republican or Democrat in the case of Figure 6.11—rather than geographic area or distance. (*Note:* Map scale is no longer relevant for such maps.) Once again, we see that the story of electoral geography in the United States changes when we represent the data differently through different mapping choice.

> **cartogram** A map that results from using a thematic variable rather than geographic area or distance.

Equipped with these basic map-reading skills, we are ready to use maps as tools for formulating and answering questions about spatial relations. Take a look at Figure 5.24, which accompanies our discussion of environmental racism in Chapter 5. The problem of environmental racism is fundamentally a spatial problem, the scope of which only becomes evident through mapping. Early researchers of the problem noticed that many toxic facilities were located near minority communities in the United States. Based on these observations, they asked what would happen if we created a map that included the location of toxic facilities and census data on residence by race. Maps such as 5.24 were the result, clearly demonstrating a close spatial relationship between minority neighborhoods and pollution sources.

Thus, we see how maps can be powerful analytical tools, not just handy aids for finding the nearest Starbucks. The environmental racism map, for example, both answered the question of scope and raised new questions about how structural racism operates geographically. Armed with a basic understanding of how information can be represented spatially in the form of maps, we are better able to interpret the message of maps, sometimes even finding undisclosed ideologies or political agendas. Just as importantly, we can begin to ask our own questions—sometimes based on everyday observations, sometimes in an effort to solve difficult real-world problems—and to discover important spatial relations that only maps can reveal.

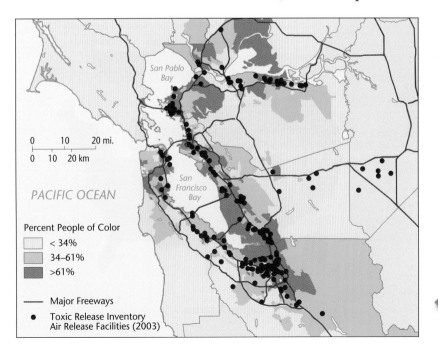

Legend:
PACIFIC OCEAN

Percent People of Color
- < 34%
- 34–61%
- >61%

— Major Freeways
• Toxic Release Inventory Air Release Facilities (2003)

FIGURE 5.24 **Industrial air pollution and minority neighborhoods.** *(Page 208)*

GLOSSARY

absorbing barrier   A barrier that completely halts diffusion of innovations and blocks the spread of cultural elements. (page 14)

acculturation   The adoption by an ethnic group of enough of the ways of the host society to be able to function economically and socially. (page 184)

adaptive strategy   The unique way in which each culture uses its particular physical environment; those aspects of culture that serve to provide the necessities of life—food, clothing, shelter, and defense. (pages 119, 295)

agribusiness   Highly mechanized, large-scale farming, usually under corporate ownership. (page 341)

agricultural landscape   The cultural landscape of agricultural areas. (page 356)

agricultural region   A geographic region defined by a distinctive combination of physical and environmental conditions; crop type; settlement patterns; and labor, cultivation, and harvesting practices. (page 318)

agricultural surplus   The amount of food grown by a society that exceeds the demands of its population. (page 417)

agriculture   The cultivation of domesticated crops and the raising of domesticated animals. (page 317)

agroforestry   A cultivation system that features the interplanting of trees with field crops. (page 75)

alternative energy   Alternatives to traditional fossil fuels, including hydroelectric, solar, wind, geothermal, nuclear, biofuel, and "clean" coal. (page 400)

amenity landscapes   Landscapes that are prized for their natural and cultural aesthetic qualities by the tourism and real estate industries and their customers. (page 83)

Anatolian hypothesis   A theory of language diffusion holding that the movement of Indo-European languages from the area in contemporary Turkey known as Anatolia followed the spread of plant domestication technologies. (page 145)

animists   Adherents of animism, the idea that souls or spirits exist not only in humans but also in animals, plants, rocks, natural phenomena such as thunder, geographic features such as mountains or rivers, or other entities of the natural environment. (page 285)

aquaculture   The cultivation, under controlled conditions, of aquatic organisms, primarily for food but also for scientific and aquarium uses. (page 328)

assimilation   The complete blending of an ethnic group into the host society, resulting in the loss of all distinctive ethnic traits. (page 184)

automobile cities   Cities whose spatial layout—both in terms of extent and form—is dictated by the near ubiquity of individual automobile ownership. (page 431)

axis mundi   The symbolic center of cosmomagical cities, often demarcated by a large vertical structure. (page 418)

baseline maps   Also called general maps, baseline maps provide basic information such as coastlines, country, and boundaries. (page A-2)

bilingualism   The ability to speak two languages fluently. (page 141)

biodiversity   Biological diversity of the entire living world, as measured at various scales including diversity among individuals, populations, species, communities, and ecosystems. (page 71)

biofuel   Broadly, this term refers to any form of energy derived from biological matter, increasingly used in reference to replacements for fossil fuels in internal combustion engines, industrial processes, and the heating and cooling of buildings. (page 355)

border zones   The areas where different regions meet and sometimes overlap. (page 9)

Brandt Line   Division of the countries of the world into two regions: wealthy and poor. The Brandt Line is named after former West German Chancellor Willy Brandt, and was a concept commonly applied in the 1970s through the 1980s. (page 376)

buffer state   An independent but small and weak country lying between two powerful countries. (page 229)

cadastral pattern   The shapes formed by property borders; the pattern of land ownership. (page 356)

carbon cap and trade   Programs that limit (cap) carbon emissions, and lower the limit over time, as well as providing economic incentives for companies to find innovative ways to remain below their carbon cap. (page 402)

carrying capacity   The maximum number of people that can be supported in a given area. (page 96)

cartogram   A map that results from using a thematic variable rather than geographic area or distance. (page A-4)

central business district (CBD)   The central portion of a city, characterized by high-density land uses. (page 431)

central-place theory   A set of models designed to explain the spatial distribution of urban service centers. (page 424)

centrifugal force   Any factor that disrupts the internal unity of a country. (page 231)

centripetal force   Any factor that supports the internal unity of a country. (page 231)

chain migration   The tendency of people to migrate along channels, over a period of time, from specific source areas to specific destinations. (page 195)

channelization   A process whereby a specific source region becomes linked to a particular destination, so neighbors in the old place became neighbors in the new place as well. (page 195)

choropleth maps   Maps that use data aggregated over a predefined geographic area, typically a political designation such as a country, province, or state. (page A-3)

circulation   A term that implies an ongoing set of movements of people, ideas, or things that have no particular center or periphery. (page 16)

cleavage model   A political-geographic model suggesting that persistent regional patterns in voting behavior, sometimes leading to separatism, can usually be explained in terms of tensions pitting urban against rural, core against periphery, capitalists against workers, and power group against minority culture. (page 249)

colonialism   The forceful appropriation of a territory by a distant state, often involving the displacement of indigenous populations to make way for colonial settlers (page 71); the building and maintaining of colonies in one territory by people based elsewhere. (page 225)

consumer nationalism   A situation in which local consumers favor nationally produced goods over imported goods as part of a nationalist political agenda. (page 68)

contact conversion   Spread of religious beliefs by personal contact. (page 286)

contagious diffusion   A type of expansion diffusion in which cultural innovation spreads by person-to-person contact, moving wavelike through an area and population without regard to social status. (pages 14, 244)

conventional agriculture   The widely adopted commercial, industrialized form of farming that uses a range of synthetic fertilizers, insecticides, and herbicides to control pests and maximize productivity; a term that emerged following the creation of alternative forms, such as organic farming. (page 354)

convergence hypothesis   A hypothesis holding that cultural differences among places are being reduced by improved transportation and communications systems, leading to a homogenization of popular culture. (page 67)

cool chain   The refrigeration and transport technologies that allow for the distribution of perishables. (page 339)

core area   The territorial nucleus from which a country grows in area and over time, often containing the national capital and the main center of commerce, culture, and industry. (page 239)

core–periphery   A concept based on the tendency of both formal and functional culture regions to consist of a core or node, in which defining traits are purest or functions are headquartered, and a periphery that is tributary and displays fewer of the defining traits. (pages 9, 338)

cornucopians   Those who believe that science and technology can solve resource shortages. In this view, human beings are our greatest resource rather than a burden to be limited. (page 117)

cosmomagical cities   Types of cities that are laid out in accordance with religious principles, characteristic of very early cities. (page 418)

creole   A language derived from a pidgin language that has acquired a fuller vocabulary and become the native language of its speakers. (page 140)

cultural diffusion   The spread of elements of culture from the point of origin over an area. (page 330)

cultural ecology   Broadly defined, the study of the relationships between the physical environment and culture; narrowly (and more commonly) defined, the study of culture as an adaptive system that facilitates human adaptation to nature and environmental change. (page 23)

cultural landscape   The visible human imprint on the land. (page 28)

cultural maladaptation   Poor or inadequate adaptation that occurs when a group pursues an adaptive strategy that, in the short run, fails to provide the necessities of life or, in the long run, destroys the environment that nourishes it. (page 205)

cultural practices   The social activities and interactions—ranging from religious rituals to food preferences to clothing—that collectively distinguish group identity. (page 3)

cultural preadaptation   A complex of adaptive traits and skills possessed in advance of migration by a group, giving it survival ability and competitive advantage in occupying the new environment. (page 205)

cultural simplification   The process by which immigrant ethnic groups lose certain aspects of their traditional culture in the process of settling overseas, creating a new culture that is less complex than the old. (page 197)

culturally preadapted   (See cultural preadaptation)

culture A total way of life held in common by a group of people, including such learned features as speech, ideology, behavior, livelihood, technology, and government; or the local, customary way of doing things—a way of life; a never-changing process in which a group is actively engaged; a dynamic mix of symbols, beliefs, speech, and practices. (page 2)

culture hearth A focused geographic area where important innovations are born and from which they spread. (page 286)

death rate The number of deaths per year per 1000 people. (page 98)

deindustrialization The decline of primary and secondary industry, accompanied by a rise in the service sectors of the industrial economy. (page 447)

demographic transition A term used to describe the movement from high birth and death rates to low birth and death rates. (page 100)

depopulation A decrease in population that sometimes occurs as the result of sudden catastrophic events, such as natural disasters, disease epidemics, and warfare. (page 125)

desertification A process whereby human actions unintentionally turn productive lands into deserts through agricultural and pastoral misuse, destroying vegetation and soil to the point where they cannot regenerate. (page 349)

dialect A distinctive local or regional variant of a language that remains mutually intelligible to speakers of other dialects of that language; a subtype of a language. (page 138)

diaspora culture Ethnic, racial, and national population concentrations of people displaced and geographically scattered from their homelands. Such displaced groups often maintain strong social and economic ties to their homelands. (page 49)

diffusion The movement of people, ideas, or things from one location outward toward other locations. (page 12)

digital divide A pattern of unequal access to advanced information technologies produced by socioeconomic inequalities and measured at scales ranging from the individual to countries and world regions. (pages 66, 391)

dispersed A type of settlement form in which people live relatively distant from each other. (page 34)

domesticated animal An animal kept for some utilitarian purpose whose breeding is controlled by humans and whose survival is dependent on humans; domesticated animals differ genetically and behaviorally from wild animals. (page 332)

domesticated plant A plant deliberately planted and tended by humans that is genetically distinct from its wild ancestors as a result of selective breeding. (page 330)

double-cropping Harvesting twice a year from the same parcel of land. (page 320)

ecofeminism A doctrine proposing that women are inherently better environmental preservationists than men because the traditional roles of women involved creating and nurturing life, whereas the traditional roles of men too often necessitated death and destruction. (pages 28, 75)

ecotheology The study of the influence of religious belief on habitat modification. (page 298)

effective sovereignty The idea that states' power to effectively enforce or ignore sovereignty claims irrespective of territorial boundaries varies in time and from country to country. (page 247)

electronic waste Used electronics, such as mobile phones, computers, office equipment, television sets, and refrigerators, which can no longer be used for their intended purpose and are discarded. Also known as e-waste. (page 408)

electoral geography The study of the interactions among space, place, and region and the conduct and results of elections. (page 235)

enclave A piece of territory surrounded by, but not part of, a country. (page 237)

environmental determinism The belief that cultures are directly or indirectly shaped by the physical environment. (page 23)

environmental perception The belief that culture depends more on what people perceive the environment to be than on the actual character of the environment; perception, in turn, is colored by the teachings of culture. (page 24)

environmental racism The targeting of areas where ethnic or racial minorities live with respect to environmental contamination or failure to enforce environmental regulations. (page 207)

environmental refugees People who are displaced from their homes due to severe environmental disruption. (page 121)

escalation diffusion The idea that civil wars may escalate through the diffusion of violence across ever-greater areas over time. (page 245)

ethnic cleansing The removal of unwanted ethnic minority populations from a nation state through mass killing, deportation, or imprisonment. (page 196)

ethnic flag A readily visible marker of ethnicity on the landscape. (page 209)

ethnic group A group of people who share a common ancestry and cultural tradition, often living as a minority group in a larger society. (page 180)

ethnic homelands Sizable areas inhabited by an ethnic minority that exhibits a strong sense of attachment to the region and often exercises some measure of political and social control over it. (page 184)

ethnic islands   Small ethnic areas in the rural countryside; sometimes called folk islands. (page 185)

ethnic neighborhood   A voluntary community where people of like origin reside by choice. (page 187)

ethnic religion   A religion identified with a particular ethnic or tribal group; does not seek converts. (page 273)

ethnic substrate   Regional cultural distinctiveness that remains following the assimilation of an ethnic homeland. (page 187)

ethnoburbs   A suburban ethnic neighborhood, sometimes home to relatively affluent immigration populations. (page 189)

ethnographic boundary   A political boundary that follows some cultural border, such as a linguistic or religious border. (page 229)

ethnolect   A dialect spoken by a particular ethnic group. (page 150)

Eurocentric   Using the historical experience of Europe as the benchmark for all cases. (page 102)

European Union   Established by European countries through a set of political, cultural, and economic treaties and supranational institutions. (page 63)

exclave   A piece of national territory separated from the main body of a country by the territory of another country. (page 227)

expansion diffusion   The spread of innovations within an area in a snowballing process, so that the total number of knowers or users becomes greater and the area of occurrence grows. (pages 13, 245)

extensive agriculture   The practices of farming and livestock raising using low levels of labor and capital relative to the areal extent of land under production, relying chiefly on natural soil fertility and prevailing climate. (page 318)

farm villages   Clustered rural settlements of moderate size, inhabited by people who are engaged in farming. (page 124)

farmstead   The center of farm operations, containing the house, barn, sheds, and livestock pens. (page 124)

federal state   An independent country that gives considerable powers and even autonomy to its constituent parts. (page 229)

feedlot   A factorylike farm devoted to either livestock fattening or dairying; all feed is imported and no crops are grown on the farm. (page 323)

first urban revolution   The connection between agricultural innovations with the rise of the world's first true cities. (page 421)

First World   A term used to describe the group of wealthy, democratic, capitalist nations across the world. (page 375)

folk architecture   Structures built by members of a folk society or culture in a traditional manner and style, without the assistance of professional architects or blueprints, using locally available raw materials. (page 76)

folk culture   A small, cohesive, stable, isolated, nearly self-sufficient group that is homogeneous in custom and race; characterized by a strong family or clan structure, order maintained through sanctions based in the religion or family, little division of labor other than that between the sexes, frequent and strong interpersonal relationships, and a material culture consisting mainly of handmade goods. (pages 45, 320)

folk fortress   A stronghold area with natural defensive qualities, useful in the defense of a country against invaders. (page 252)

foodborne disease outbreaks   When two or more people acquire the same illness from the same contaminated food or drink. (page 344)

foodways   Customary behaviors associated with food preparation and consumption. (page 213)

formal region   A cultural region inhabited by people who have one or more cultural traits in common. (page 8)

Fourth World   A term used to describe the world's poorest nations. (page 376)

functional region   A cultural area that functions as a unit politically, socially, or economically. (page 9)

fundamentalism   A movement to return to the founding principles of a religion, which can include literal interpretation of sacred texts, or the attempt to follow the ways of a religious founder as closely as possible. (page 274)

Gaia hypothesis   The theory that there is one interacting planetary ecosystem, Gaia, that includes all living things and the land, waters, and atmosphere in which they live; further, that Gaia functions almost as a living organism, acting to control deviations in climate and to correct chemical imbalances, so as to preserve Earth as a living planet. (page 300)

gated communities   Highly securitized residential enclaves that are more or less self-governing. (page 436)

GDP per capita   The mathematical result of dividing a country's GDP by its national population. (The Latin *per capita* means "per head.") (page 368)

gender roles   What it means to be a man or a woman in different cultural and historical contexts. (page 107)

generic toponym   The descriptive part of many place-names, often repeated throughout a culture area. (page 164)

genetically modified (GM) crops   Plants whose genetic characteristics have been altered through recombinant DNA technology. (page 343)

genocide The systematic killing of a racial, ethnic, religious, or linguistic group. (page 202)

gentrification The displacement of lower-income residents by higher-income residents as buildings in deteriorated areas of city centers are restored. (page 446)

geography The study of spatial patterns and of differences and similarities from one place to another in environment and culture. (page 2)

geometric boundary A political border drawn in a regular, geometric manner, often a straight line, without regard for environmental or cultural patterns. (page 229)

geopolitics The influence of the habitat on political entities. (page 252)

gerrymandering The drawing of electoral district boundaries in an awkward pattern to enhance the voting impact of one constituency at the expense of another. (page 236)

ghetto Traditionally, an area within a city where an ethnic group lives, either by choice or by force. Today in the United States, the term typically indicates an impoverished African American neighborhood. (page 187)

global cities Cities that are control centers of the global economy. (page 433)

global climate change Also known as global warming, global climate change involves alterations in climate caused by human activity and, most particularly, by the greatly increased amount of carbon dioxide (CO_2) produced by burning fossil fuels. (page 397)

Global South A term that has largely replaced "Third World" when referring to regions of Latin America, Africa, and most of Asia, in recognition that much of the world's dynamism, growth, and power reside in these places. (page 390)

globalization The binding together of all the lands and peoples of the world into an integrated system driven by capitalistic free markets, in which cultural diffusion is rapid, independent states are weakened, and cultural homogenization is encouraged. (pages 18, 341)

glocalization Forces of change interact with local cultures, altering both in the process. (page 462)

Great Migration The twentieth-century movement of 6 million African Americans from the rural southern states to the cities of the midwestern and northeastern states. (page 17)

green revolution The recent introduction of high-yield hybrid crops and chemical fertilizers and pesticides into traditional Asian agricultural systems, most notably paddy rice farming, with attendant increases in production and ecological damage. (page 334)

greenhouse effect A process in which the increased release of carbon dioxide and other gases into the atmosphere, caused by industrial activity and deforestation, permits solar short-wave heat radiation to reach Earth's surface but blocks long-wave outgoing radiation, causing a thermal imbalance and global heating. (page 397)

Gross Domestic Product (GDP) The total dollar value of all goods and services produced in a country over a specific period, usually a year. (page 367)

hamlet A small rural settlement, smaller than a village. (page 356)

heartland The interior of a sizable landmass, removed from maritime connections; in particular, the interior of the Eurasian continent. (page 253)

heartland theory A 1904 proposal by Halford Mackinder that the key to world conquest lay in control of the interior of Eurasia. (page 252)

hierarchical diffusion A type of expansion diffusion in which innovations spread from one important person to another or from one urban center to another, temporarily bypassing other persons or rural areas. (page 14)

Human Development Index (HDI) A numerical value devised by the United Nations Development Programme that is used to measure how well basic needs are being met. It is a composite index, derived from three areas: life expectancy, education, and income. (page 369)

human geography The study of the relationships between people and the places and spaces in which they live. (page 2)

hunting and gathering The killing of wild game and the harvesting of wild plants to provide food in traditional cultures. (page 329)

hydraulic civilization A civilization based on large-scale irrigation. (page 417)

imperialism A relationship whereby wealthy nations dominate poor ones by controlling their economic, political, and cultural systems. (page 390)

independent invention A cultural innovation that is developed in two or more locations by individuals or groups working independently. (page 13)

indigenous culture A culture group that constitutes the original inhabitants of a territory, distinct from the dominant national culture, which is often derived from colonial occupation. (page 46)

indigenous technical knowledge (ITK) Highly localized knowledge about environmental conditions and sustainable land-use practices. (pages 72, 333)

Industrial Revolution Beginning in England in the early 1700s, the Industrial Revolution saw the rapid transformation of the economy through the introduction of machines, new power sources, and novel chemical processes. These replaced human hands in the making of products, initially in the cotton textile industry. (page 380)

infant mortality rate   The number of infants per 1000 live births who die before reaching one year of age. (page 107)

intensive agriculture   The expenditure of much labor and capital on a piece of land to increase its productivity. In contrast, *extensive* agriculture involves less labor and capital. (pages 318, 348)

intercropping   The practice of growing two or more different types of crops in the same field at the same time. (page 318)

interdependence   Relations between regions or countries of mutual, but not necessarily equal, dependence. (page 19)

internal migration   Human migration that occurs within the borders of a country. (page 17)

internally displaced persons (IDPs)   Persons or groups that have been forced to flee their homes due to conflict, natural disaster, or persecution but who remain within the borders of their home country. (page 196)

international migration   Human migration across country borders. (page 16)

involuntary migration   Also called forced migration, refers to the forced displacement of a population, whether by government policy (such as a resettlement program), warfare or other violence, ethnic cleansing, disease, natural disaster, or enslavement. (page 195)

isogloss   The border of usage of an individual word or pronunciation. (page 147)

Kurgan hypothesis   A theory of language diffusion holding that the spread of Indo-European languages originated with animal domestication in the central Asian steppes and grew more aggressively and swiftly than proponents of the Anatolian hypothesis maintain. (page 145)

land-division patterns   A term that refers to the spatial patterns of different land uses. (page 34)

language   A mutually agreed-on system of symbolic communication that has a spoken and usually a written expression. (page 137)

language family   A group of related languages derived from a common ancestor. (page 142)

language hotspots   Those places on Earth that are home to the most unique, misunderstood, or endangered languages. (page 157)

legend   Explains the meaning of symbols on a map. (page A-2)

leisure landscapes   Landscapes that are planned and designed primarily for entertainment purposes, such as ski and beach resorts. (page 73)

lingua franca   An existing, well-established language of communication and commerce used widely where it is not a mother tongue. (page 141)

linguistic refuge area   An area protected by isolation or inhospitable environmental conditions in which a language or dialect has survived. (page 160)

livestock fattening   A commercial type of agriculture that produces fattened cattle and hogs for meat. (page 323)

local consumption cultures   Distinct consumption practices and preferences in food, clothing, music, and so forth formed in specific places and historical moments. (page 67)

Malthusian   Those who hold the views of Thomas Malthus, who believed that overpopulation is the root cause of poverty, illness, and warfare. (page 116)

map symbols   Graphic figures that represent any number of real-world phenomena. (page A-2)

marchland   A strip of territory, traditionally one day's march for infantry, that served as a boundary zone for independent countries in premodern times. (page 229)

mariculture   A branch of aquaculture specific to the cultivation of marine organisms, often involving the transformation of coastal environments and the production of distinctive new landscapes. (page 328)

market gardening   Farming devoted to specialized fruit, vegetable, or vine crops for sale rather than consumption. (page 322)

material culture   All physical, tangible objects made and used by members of a cultural group, such as clothing, buildings, tools and utensils, instruments, furniture, and artwork; the visible aspect of culture. (page 44)

mechanistic view of nature   The view that humans are separate from nature and hold dominion over it and that the habitat is an integrated mechanism governed by external forces that the human mind can understand and manipulate. (page 27)

megachurch   Large Protestant church structures, usually located in suburban areas of the United States, that have large congregations (2000–10,000 members) and utilize business models to tailor their spaces and services to their congregations' needs. (page 301)

megacities   A term that refers to particularly large urban centers. (page 427)

migrant workers   Most broadly, this term refers to people working outside of their home country. Migrant workers are particularly critical to large-scale commercial agriculture. (page 334)

migrations   The large-scale movements of people between different regions of the world. (page 16)

mobility   The relative ability of people, ideas, or things to move freely through space. (page 12)

model   An abstraction, an imaginary situation, proposed by geographers to simulate laboratory conditions so that they can isolate certain causal forces for detailed study. (page 5)

monoculture   The raising of only one crop on a huge tract of land in agribusiness. (page 342)

monotheistic religion   The worship of only one god. (page 273)

national culture   The controversial idea that citizens possess a set of recognizable values, behaviors, and beliefs—often including the same ethnic and linguistic traits—that express the core culture of each modern nation. (page 46)

nationalism   The sense of belonging to and self-identifying with a national culture. (page 226)

nationalization   Government takeover of the transportation infrastructure and fuel sources from private foreign ownership, occurring in many previously colonized nations in the twentieth century. (page 381)

nation-state   An independent country dominated by a relatively homogeneous culture group. (page 226)

natural boundary   A political border that follows some feature of the natural environment, such as a river or mountain ridge. (page 229)

natural hazard   An inherent danger present in a given habitat, such as floods, hurricanes, volcanic eruptions, or earthquakes; often perceived differently by different peoples. (page 24)

nature–culture   A term that refers to the complex relationships between people and the physical environment, including how culture, politics, and economies affect people's ecological situation and resource use. (page 21)

neolocalism   The desire to reembrace the uniqueness and authenticity of place, in response to globalization. (page 482)

neo-Malthusians   Modern-day followers of Thomas Malthus. (page 117)

node   A central point in a functional culture region where functions are coordinated and directed. (page 9)

nomadic livestock herder   A member of a group that continually moves with its livestock in search of forage for its animals. (page 326)

nonmaterial culture   The wide range of tales, songs, lore, beliefs, values, and customs that pass from generation to generation as part of an oral or written tradition. (page 44)

nucleation   A relatively dense settlement form. (page 34)

organic agriculture   A form of farming that relies on manuring, mulching, and biological pest control and rejects the use of synthetic fertilizers, insecticides, herbicides, and genetically modified crops. (page 354)

organic view of nature   The view that humans are part of, not separate from, nature and that the habitat possesses a soul and is filled with nature-spirits. (page 27)

orientation   Refers to the compass direction of the top of a map. (page A-2)

orthodox religions   Strands within most major religions that emphasize purity of faith and are not open to blending with other religions. (page 273)

paddy rice farming   The cultivation of rice on a paddy, or small flooded field enclosed by mud dikes, practiced in the humid areas of the Far East. (page 319)

peasant   A farmer belonging to a folk culture and practicing a traditional system of agriculture. (page 320)

permeable barrier   A barrier that permits some aspects of an innovation to diffuse through it but weakens and retards continued spread; an innovation can be modified in passing through a permeable barrier. (page 15)

physical environment   All aspects of the natural physical surroundings, such as climate, terrain, soils, vegetation, and wildlife. (page 4)

pidgin   A composite language consisting of a small vocabulary borrowed from the linguistic groups involved in commerce. (page 139)

pilgrimages   Journeys to places of religious importance. (page 289)

place   A term used to connote the subjective, humanistic, and culturally oriented notion of a specific location. (page 6)

placelessness   A spatial standardization that diminishes regional variety; may result from the spread of popular culture, which can diminish or destroy the uniqueness of place through cultural standardization on a national or even worldwide scale. (pages 48, 404)

plantation   A large landholding devoted to specialized production of a tropical cash crop. (page 321)

plantation agriculture   A system of monoculture for producing export crops requiring relatively large amounts of land and capital; originally dependent on slave labor. (page 321)

political geography   The geographic study of politics and political matters. (page 223)

polyglot   A mixture of different languages. (page 143)

polytheistic religion   The worship of many gods. (page 273)

popular culture   A dynamic culture based in large, heterogeneous societies permitting considerable individualism, innovation, and change; having a money-based economy, division of labor into professions, secular institutions of control, and weak interpersonal ties; producing and consuming machine-made goods. (page 47)

population density   A measurement of population per unit area (e.g., per square mile). (page 94)

population explosion   The rapid, accelerating increase in world population since about 1650 and especially since about 1900. (page 115)

population geography   The study of the spatial and ecological aspects of population, including distribution, density per unit of land area, fertility, gender, health, age, mortality, and migration. (page 93)

population pyramid   A graph used to show the age and sex composition of a population. (page 106)

possibilism   A school of thought based on the belief that humans, rather than the physical environment, are the primary active force; that any environment offers a number of different possible ways for a culture to develop; and that the choices among these possibilities are guided by cultural heritage. (page 24)

postdevelopment   A theoretical approach that is critical of standard development practices, asserting that traditional development efforts do not work because they are ultimately about controlling, not empowering, poor nations and people. (page 389)

primary sector   The set of economic activities that involves extracting natural resources from the Earth. (page 376)

primate city   A city of large size and dominant power within a country. (page 430)

proselytic religions   Religions that actively seek new members and aim to convert all humankind. (page 273)

protracted refugee situation (PRS)   Results when people are born and mature into adulthood in refugee camps with no promise of becoming citizens in a new country and little hope of returning to their home country. (page 243)

proxemics   The study of the size and shape of people's envelopes of personal space. (page 133)

push-and-pull factors   Unfavorable, repelling conditions (push factors) and favorable, attractive conditions (pull factors) that interact to affect migration and other elements of diffusion. (page 112)

quaternary sector   Intellectual and informational economic activities. (page 377)

race   A classification system that is sometimes understood as arising from genetically significant differences among human populations, or from visible differences in human physiognomy, or as a social construction that varies across time and space. (page 178)

racism   The belief that human capabilities are determined by racial classification and that some races are superior to others. (page 180)

ranching   The commercial raising of herd livestock on a large landholding. (page 327)

range   In central-place theory, the average maximum distance people will travel to purchase a good or service. (page 425)

refugees   Those fleeing from persecution in their country of nationality. The persecution can be religious, political, racial, or ethnic. (pages 112, 196)

region   A grouping of like places or the functional union of places to form a spatial unit. (page 1)

regional trading blocs   Agreements made among geographically proximate countries that reduce trade barriers in order to better compete with other regional markets. (page 235)

relic boundary   A former political border that no longer functions as a boundary. (page 229)

religion   A social system involving a set of beliefs and practices through which people seek harmony with the universe and attempt to influence the forces of nature, life, and death. (page 271)

relocation diffusion   The spread of an innovation or other element of culture that occurs with the bodily relocation (migration) of the individual or group responsible for the innovation. (pages 13, 245, 333)

renewable resources   Those primary sector products that can be replenished naturally at a rate sufficient to balance their depletion by human use. (page 396)

resilience   The ability to recover quickly from adversity. (page 446)

return migration   A type of ethnic diffusion that involves the voluntary movement of a group of migrants back to their ancestral or native country or homeland. (pages 17, 196)

right to the city   The notion that all urban residents, not just the privileged, should be able to access city spaces and have a voice in how the city is shaped and used. (page 451)

rimland   The maritime fringe of a country or continent; in particular, the western, southern, and eastern edges of the Eurasian continent. (page 253)

sacred spaces   Areas recognized by a religious group as worthy of devotion, loyalty, esteem, or fear to the extent that they become sought out, avoided, inaccessible to nonbelievers, and/or removed from economic use. (page 305)

satellite state   A small, weak country dominated by one powerful neighbor to the extent that some or much of its independence is lost. (page 229)

scale   The distance on a map in relation to the distance in actual space. (page A-1)

seasonal migration   Usually associated with crop harvest periods, migrants move according to seasonal changes in weather. (page 17)

second urban revolution   The connection between industrial innovations with the rise of capitalist cities. (page 422)

Second World   A term used during the Cold War era to describe the group of countries that were communist or socialist in their governmental philosophy, and had centrally planned rather than free-market economic systems. (page 375)

secondary sector   The set of economic activities that involves processing the raw materials extracted by primary sector activities into usable goods. Commonly referred to as manufacturing. (page 376)

sedentary cultivation   Farming in fixed and permanent fields. (page 326)

settlement forms   The spatial arrangement of buildings, roads, towns, and other features that people construct while inhabiting an area. (page 34)

sex ratio   The numerical ratio of males to females in a population. (page 106)

shantytowns   Precarious and often illegal housing settlements, usually made up of temporary shelters and located on the outskirts of a large city. (page 437)

slang   Words and phrases that are not part of a standard, recognized vocabulary for a given language but that are nonetheless used and understood by some of its speakers. (page 149)

South–South cooperation   Trade, technological innovation, and other forms of exchange and assistance that occur between the larger and wealthier nations of the Global South and poorer nations. (page 390)

sovereignty   The right of individual states to control political and economic affairs within their territorial boundaries without external interference. (page 224)

space   A term used to connote the objective, quantitative, theoretical, model-based, economics-oriented type of geography that seeks to understand spatial systems and networks through application of the principles of social science. (page 5)

sprawl   The tendency of cities to grow outward in an unchecked manner, which is particularly notable in automobile cities. (page 431)

state   A centralized authority that enforces a single political, economic, and legal system within its territorial boundaries. Often used synonymously with "country." (page 223)

step migration   A process by which a group proceeds to its final destination via a series of intermediate migrations. (page 195)

stepwise migration   (See step migration)

stimulus diffusion   A type of expansion diffusion in which a specific trait fails to spread but the underlying idea or concept is accepted. (page 14)

streetcar suburbs   The extension of urban residential areas along streetcar lines in the late nineteenth century. (page 431)

subcultures   Groups of people with norms, values, and material practices that differentiate them from the dominant culture to which they belong. (page 43)

subsistence agriculture   Farming to supply the minimum food and materials necessary to survive. (page 319)

subsistence economies   Economies in which people seek to consume only what they produce and to produce only for local consumption rather than for exchange or export. (page 73)

suitcase farm   In American commercial grain agriculture, a farm on which no one lives; planting and harvesting are done by hired migratory crews. (page 325)

supranational organization   A group of independent countries joined together for purposes of mutual interest. (page 233)

supranationalism   A situation that occurs when states willingly relinquish some degree of sovereignty in order to gain the benefits of belonging to a larger political-economic entity. (page 233)

survey pattern   A pattern of original land survey in an area. (page 356)

sustainability   The survival of a land-use system for centuries or millennia without destruction of the environmental base, allowing generation after generation to continue to live there. (pages 347, 446)

swidden cultivation   A type of agriculture characterized by land rotation in which temporary clearings are used for several years and then abandoned to be replaced by new clearings; also known as slash-and-burn agriculture. (page 318)

symbolic landscapes   Landscapes that express the values, beliefs, and meanings of a particular culture. (page 33)

syncretic religions   Religions, or strands within religions, that combine elements of two or more belief systems. (page 273)

territorial fission　Occurs when an ethnic or linguistic minority population seeks to secede from the state or to alter state territorial boundaries to promote cultural homogeneity and political autonomy. (page 231)

territoriality　A learned cultural response, rooted in European history, that produced the external bounding and internal territorial organization characteristic of modern states. (page 225)

tertiary sector　The set of economic activities that refers to all the different types of work necessary to move goods and resources around and deliver them to people; in other words, service activities. (page 376)

thematic maps　Maps that emphasize a specific phenomenon or process, such as transportation, migration, and agricultural production. (page A-2)

Third World　A term used to describe the world's poor nations as a group. During the Cold War era, these were also the nations that were not communist or socialist; in other words, they were not aligned with either the First or Second World. (page 375)

threshold　In central-place theory, the size of the population required to make provision of goods and services economically feasible. (page 425)

time-distance decay　The decrease in acceptance of a cultural innovation with increasing time and distance from its origin. (page 14)

toponym　A place-name, usually consisting of two parts, the generic and the specific. (page 163)

total fertility rate (TFR)　The number of children the average woman will bear during her reproductive lifetime (15–44 or 15–49 years of age). A TFR of less than 2.1, if maintained, will cause a natural decline of population. (page 97)

transculturation　The notion that people adopt elements of other cultures as well as contributing elements of their own culture, thereby transforming both cultures. (page 184)

transnational migrations　The movements of groups of people who maintain ties to their homelands after they have migrated. (page 17)

uneven development　The tendency for industry to develop in a core–periphery pattern, enriching the industrialized countries of the core and impoverishing the less industrialized periphery. This term is also used to describe urban patterns in which suburban areas are enriched while the inner city is impoverished. (page 19)

unitary state　An independent state that concentrates power in the central government and grants little authority to the provinces. (page 229)

universalizing religions　Also called proselytic religions, they expand through active conversion of new members and aim to encompass all of humankind. (page 273)

urban agriculture　The raising of food, including fruit, vegetables, meat, and milk, inside cities, especially common in the Third World. (page 327)

urban footprint　The spatial extent of the impacts of urban areas on the natural environment. (page 440)

urban hearth areas　Regions in which the world's first cities evolved. (page 418)

urban heat island　A mass of warm air generated and retained by urban building materials and human activities; it sits over the city and causes urban temperatures to be higher than those of surrounding areas. (page 439)

urbanized population　The proportion of a country's population living in cities. (page 414)

vernacular (culture) region　A (culture) region perceived to exist by its inhabitants, based in the collective spatial perception of the population at large, and bearing a generally accepted name or nickname (such as "Dixie"). (pages 10, 60)

World Heritage Sites　Places (e.g., buildings, cities, forests, lakes, deserts, archeological ruins) that the UN's International Heritage Programme judges to possess outstanding cultural or natural importance to the common heritage of humanity. (page 6)

zero population growth　A stabilized population created when an average of only two children per couple survive to adulthood, so that, eventually, the number of deaths equals the number of births. (page 98)

INDEX

Note: Page numbers followed by f indicate figures; those followed by t indicate tables.